SOIL, PLANT AND WATER ANALYSIS

SOIL, PLANT AND WATER ANALYSIS

DR. P.C. JAISWAL
M.Sc., Ag., Ph.D.,
Professor Emeritus
Department of Soil Science
School of Forestry and Environment
Sam Higginboltom Institute of Agriculture,
Technology & Science Deemed-to-be-University
(Formerly Allahabad Agricultural Institute)
Allahabad-211007, India

KALYANI PUBLISHERS
LUDHIANA - NEW DELHI - NOIDA (U.P.) - HYDERABAD
CHENNAI - KOLKATA - CUTTACK - GUWAHATI - KOCHI - BANGALORE

KALYANI PUBLISHERS

Head Office :
B-I/1292, Rajinder Nagar, **Ludhiana**-141 008
e-mail : kalyanibooks@yahoo.co.in
Administrative Office :
4779, Ansari Road,
23, Daryaganj, **New Delhi**-110 002
Works :
B-16, Sector 8, **Noida** (U.P.)
Branches :
1, Mahalakshmi Street,
T. Nagar, **Chennai**-600 017

110/111, Bharatia Towers Badambadi
Cuttack-753 009 (Orissa)

3-5-1108, Narayanaguda, **Hyderabad**-500 029 (A.P.)

10/2B, Ramanath Mazumdar Street,
Kolkata-700 009

1st Floor, Saraswati Road, Pan Bazar,
Guwahati-781 001
Convent Road, **Kochi**-682 035

No. 24 & 25, Ist Floor, Hameed Shah Complex,
Banpet Main Road, **Bangalore**-560 002

KPPJ 13507 05
O-27981 05
U-41770 05

First Edition 2003
Second Revised Edition 2006
Third Revised Edition 2011
Reprinted 2013 ***Reprinted, 2024***

Typesetting & Diagrams by :
(SO-PL-WA)
Datalink Computer
213-Sangam Complex, Milap Chowk,

ISBN 978-93-272-1017-0

PRINTED IN INDIA
At Process World India C-90 Sec-63 Noida

PREFACE

In recent years, soil, plants and water analysis has become a scientific approach for understanding of the problems concerning soil, plant and water and their management for augmenting agricultural production on a sustained basis. The problems encountered may be of varied nature requiring thorough investigations for proper diagnosis and remedial measures. Apart from problems of nutrient deficiencies and nutrient imbalances, there exist problems related to salinity, sodicity, acidity, poor soil physical conditions and proper use of water of poor quality or scarce supply.

Many students of soil science, soil-plant relationship and agronomy having deep interest in soil & plant analysis expressed the need for a comprehensive book on the above subject to the author. The author being devoted to teaching and research as well as to soil and plant analysis gradually took the work of preparing the manuscript.

The text includes four sections. The first section is devoted for soil physical analysis, and the remaining three sections deal with the chemical analysis of soil, plant and water. Each determination has been written under the headings: Principle, Reagents/Equipments, Procedure, calculation, comments and references. The procedure has been presented under logical sub-headings. The comments includes neccesary precautions and suggestions, functions of some reagents, interpretation of results based on rating criteria, critical levels, conditions under which deficiency or toxicity is encountered, and striking diagnostic criteria under field conditions.

The author is very much indebted to Prof. S.G. Mishra, former Director, Shiela Dhar Institute of Soil Science, Allahabad University, Allahabad for going through the entire manuscript. Apart from constructive suggestions, Prof. (Dr.) Mishra expressed his satisfaction and appreciation which encouraged the author in the endevour. The author gratefully acknowledges a great deal of assistance rendered by his daughter, Jyotsana.

It is felt that the students, researchers as well as teachers would find the book useful in the field of their interest in India as well as else where. The author will be grateful to readers giving suggestions for further improvement of the subject.

Allahabad Agricultural Institute
(Deemed University).
Allahabad– 211007, India

Dr. P.C. Jaiswal
Former Professor & Head
Department of Agronomy

CONTENTS

Physical Analysis of Soil

1.1 SOIL SAMPLING FOR PHYSICAL ANALYSIS

A portion of the entire soil mass taken from its natural occurrence for analysis in order to describe the properties of the soil is referred as soil sample. The purpose of the sampling is to estimate the properties of the entire soil mass in question with a certain degree of accuracy at the lowest possible cost. This can be achieved only when the sample is so selected that it represents the entire soil mass.

The sampling procedure described in section on soil chemical analysis is also satisfactory for soil physical analysis such as soil colour, soil texture by feel method, particle size analysis, hydraulic conductivity, moisture equivalent and soil moisture tension relationship by pressure plate and pressure membrane apparatus.

The sample for moisture determination need to be taken with a narrow stainless steel soil auger inside a close fitting but free stainless steel cylinder described in section 1.13.1. The sample need to be quickly taken and immediately transferred to a tared aluminium can with tight fitting lid to prevent loss of moisture by evaporation during its transport from the field to the laboratory for weighment and subsequently drying in an oven.

As moisture content vary with the soil depth, the soil sample need to be taken by driving the soil auger to the different depth as desired at the same boring. As moisture content may vary considerably within a very short distance, 3 to 4 borings would be desireable at each location for the collection of soil samples for moisture determination in order to obtain a reliable moisture content.

Undisturbed soil sample be taken with a core sampler or with a copper cylinder for the desired depth for moisture determination on volume basis, and in terms of height of water column as well as for the determination of bulk density, % total pore space, capillary pore space and for air capacity of soil.

The determinations such as infiltration rate, soil compaction, measurement on field capacity, soil moisture and tension relationship with tensiometer, water content of soil by resistance block or by neutron thermalization method are done at the selected site. However, soil samples may be collected when required by 3-4 boring with soil auger around the selected site. In case of determination of water content by neutron thermalization method, boring made is quite deep as well as large and aluminium hollow pipe slightly larger than the diameter of bore is inserted inside the boring for lowering the neutron probe with detector to the desired depth for moisture determination at the desired interval.

For aggregate analysis of soil by dry or wet sieving, usually surface soil after tillage in the tillage experiments are gathered from four different places by moving a rectangular tray across the ploughed layer. Each sample is air dried separately for aggregate analysis.

The sampling technique in physical analysis of soil may differ with the parameter to be determined, but normally 4 samples to serve as replicates is very much desireable for reliable results.

REFERENCES

Black, C.A. 1965. Methods of Soil Analysis Vol. 1. Amer. Soc. Agron. Madison, Wisconsin, USA.

Jackson, M.L. 1958. Soil Chemical Analysis, Prentice Hall of India Private Limited, New Delhi (1973).

Muhr, G.R., N.P. Datta, H. Sankarasubramoney, V.K. Leley and R.L. Donahur 1965. Soil Testing in India, US., Agency for Intern Development Mission to India, New Delhi.

Richards, L.A. (ed.). 1954. Diagnosis and Improvement of Saline and Alkali Soils. U.S.D.A. Hand Book No., 60. Oxford & IBH Publishing Co., New Delh..

1.2 SOIL COLOUR-ANNONYMOUS (1971)

1.2.1 Principle

The colour of soil samples are measured by matching with a colour chart such as Munsell Soil Color Chart.

The book "Munsell Soil Color Chart" contains seven set of color charts for measuring soil colour. Additional charts are also made available by the publisher for determining the color of particularly tropical, sub-tropical and submerged soils.

The colour chips on each chart is arranged by three simple variables known as the Hue, Value and Chroma. The Hue notation of a colour indicates its relation to red, yellow, green, blue and purple. The Value notation indicates its lightness and the Chroma notation indicates its strength (saturation) or the degree of departure of a particular hue from a neutral grey of the same value. The Hue is indicated by a symbol in the upper right hand corner of the card. The value notation of each colour chip is indicated by vertical scale in the far left column of the chart. The chroma notation is indicented by the horizontal scale across the bottom of the chart.

The colours become successively lighter from the bottom of the card to the top by visually equal steps. As the chroma increase to the right, the colours become greyish to the left. The soil colour names and their limits are found in the diagrams which appear opposite each chart. One such diagram is shown in Fig. 1.2.1.

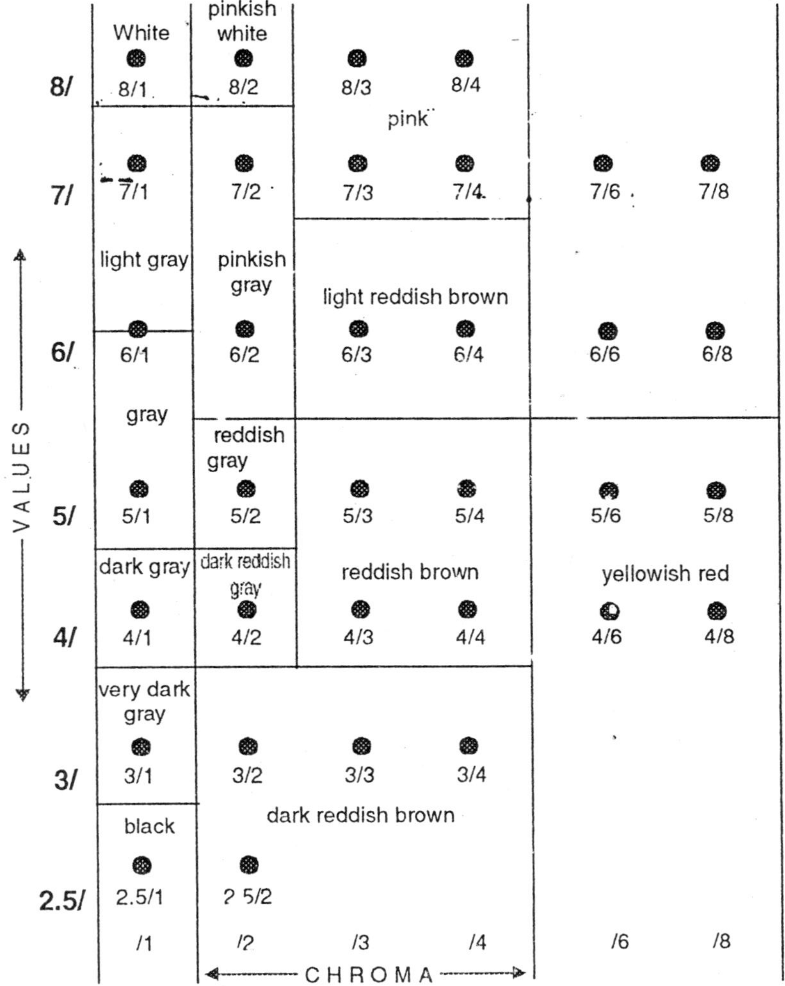

Fig. 1.2.1 Diagram opposite the color chart of HUE 5 YR from Munsell Soil Color chart, 1971.

1.2.2 Requirements

Munsell Soil Colour Chart

1.2.3 Procedure

Select an appropriate Hue Card and hold the dry or moist soil sample directly behind the apertures separating the closest matching colour chips. Cover the colour chips by the transparent polythene sheet and place the appropriate mask (black mask for dark samples and grey mask for intermediate and light samples) over the polythene sheet before matching the soil colour with that of the color chips. The aperture in the mask allows exposure of four adjacent colour chips. The remaining colours in the field are covered to eliminate the possibility of disturbing contrasts.

On locating the color chip with which the soil colour matches closely, note the color name, and its Hue, Value and Chroma from the diagram opposite the color chart. The Munsell notations are written in parenthesis after the colour name. In writing the Munsell notation, the order is Hue, Value, Chroma with a space between the Hue letter and the succeeding value number, and a virugula between the two numbers for value and chroma. Thus a dry soil of Hue 5 YR, value 5, chroma 6, and colour yellowish red should be described as yellowish Red (5YR 5/6 dry). The moist soil colour be described in the similar manner.

1.2.4 Calculation : NIL

1.2.5 Comments

The colour of soil samples may rarely give a perfect match by any colour in the chart and so one should look for a closest match. As moist soils have a darker colour than a dry soil, it is necessary to specify whether the colour description is that of a moist or dry soil.

The soil colour varies with the mineralogical composition, content and kind of humus and the degree of oxidation and hydration of the soil constituents. The colour of red and yellow soils found in tropical regions are due to the presence of haematite and limonite respectively. Higher oxides of iron and titanium impart a black colour to the soil as noticed in black cotton soils of India which are also exceptionally high in expanding type of clay minerals. Iron under reduced conditions impart greenish cast or mottling which may serve to identify the problem of poor drainage and aeration. The presence of soluble salts in significant amount impart a greyish white colour to the soil as noticed in saline soils. Some of the sodic soils develop bands of greenish-black colour on wetting.

The soil colour often varies with the texture. The coarser textured soils are of lighter colour than the fine textured soils in the Indo-gangetic basin. The fine textured soils apart from being richer in clay fractions are also higher in humus content. The humus composed of fulvic acid is yellow to yellow brown whereas humus composed of humic acid is dark brown to grey-black. The development of pigments in humus of tropical region is very slight. Prairie soils of temperate regions have a dark brown colour and chernozems have a black A_1 horizon due to accumulation of humus. Under warmer conditions, the soils tend to develop a reddish tinge. Soils of brick Kiln also have a reddish brown shade.

The soil colour also varies with the depth as a result of differences in content of organic matter and due to eluviation and illuviation of soil constituents. Surface soils have a darker shade than the sub-soils because of higher content of organic matter. A_2 horizon of Podsols are ashy-grey in colour due to eluviation of iron to lower layers. Accumulation of calcium carbonate and gypsum in the soil profile imparts a greyish white colour to the horizon. Dark coloured soils are often more productive than light coloured soils.

REFERENCES

Anonymous. 1971. Munsell Soil Colour Chart. Munsell Color Company Inc. 2441 N, Calvert Street, Baltimore, Maryland 21212, USA.

1.3 SOIL TEXTURE BY FEEL METHOD

1.3.1 Principle

The physical properties of soils such as cohesion, adhesion, plasticity and feel to the fingers largely depend on the relative proportion of different sizes of mineral fractions i.e. soil texture. Instead of estimating the mineral fraction of different size groups in the soil, the above mentioned physical properties are assessed from ball formation, ribbon formation, staining or stickiness, and the relative coarseness, grittiness and the smoothness of the soil.

1.3.2 Reagents : Nil

1.3.3 Procedure

Dig the soil and collect handful of field moist soil and rub it between the finger. Add water with a wash bottle to make it wet. Again rub the wet soil mass in between the fingers observing whether the feel is rough, gritty, coarse or smooth, very smooth, buttery or else the feel is neither coarse nor smooth.

Then collect a fresh soil mass in your hand, and if it is dry make it moist but not wet. Then try to make a ball by squeezing the soil mass. Observe whether the soil does not make a ball or else it forms a very weak ball which crumbles readily on applying pressure. Is the ball neither weak nor firm, or else the ball is very firm, and on drying the ball will be as hard as stone.

Collect another sample of fresh soil and make it wet. Touch the soil mass with your fingers and pass your fingers over the wet soil mass. Alternatively place a wet soil mass on your palm. Observe whether your fingers or palm is not stained, slightly stained or definitely stained.

Add more water to the wet soil after judging its stickiness. Take the wet soil mass in your palm. Cover the soil with your fingers and then press it so hard with the fingers so that the wet soil mass squeezes out into a ribbon.

Alternatively, wet the soil mass and press it between the forefinger and the thumb moving the thumb from down upward to form ribbon.

Observe the tendency of the wet soil to form ribbon. If there is no ribbon formation or there is slight tendency to ribbon with flaky surface. Does the wet soil ribbon out on squeezing but ribbon breaks easily, ribbon surface remains flaky or ribbon formed is long (2.5 - 7.5 cm) and smooth with no flaky surface.

Table 1.3.1 Physical properties of soils of different textural classes by feel method

Textural class	Physical properties of soils			
	Feel of fingers	**Ball Formation**	**Stickness**	**Ribbon Formation**
Sand	Very gritty	Does not form ball	Does not stain fingers	No
Loamy sand	Very gritty	Forms very easily broken ball	Very little. Stains fingers slightly	No
Sandy loam	Moderately gritty	Forms fairly firm ball but is easily broken	Definitely stains fingers	No
Loam	Neither very gritty nor very smooth	Forms firm ball	-do-	No
Silt loam	Smooth or slick buttery feel	Forms firm ball	-do-	Slight tendency to ribbon with flaky surface.
Clay loam	slightly gritty feel	Moderately hard ball when dry	-do-	Ribbons out on squeezing but ribbon breaks easily
Silly Clay Loam	Very smooth	-do-	-do-	Shows some flaking on ribbon surface similar to silt loam
Clay	-do-	Forms hard ball which when dry cannot be crushed by fingers.	-do-	Squeezes out at right moisture into long (2.5 - 7.5 cm) ribbons

1.3.4 Calculation : NIL

1.3.5 Comments

The method is simple, inexpensive as well as reliable and suitable for assessment in the field.

REFERENCES

Perur, N.G.,C.K. Subramaniam, G.R. Muhr, and H.R. Ray. 1973. Soil Fertility Evaluation to Serve The Indian Farmers. Mysore University of Agricultural Sciences, U.S. Agency for Intern Development, Banglore, India.

1.4 PARTICLE SIZE ANALYSIS OF SOIL

1.4.1 Soil separates Bouyoucos hydrometer method -Bouyoucos (1927)

1.4.1.1 Principle

The soil sample made free from organic matter is dispersed in the medium of distilled water after treatment with a dispersing agent. The homogenized dispersed soil water suspension is placed in a tall soil testing cylinder with a mark at 1000 ml. After mixing the contents thoroughly, the cylinder is placed on a firm table and the time is noted in hours, minutes and seconds because mineral particles begin to settle at different velocity which brings a change in the suspension density with time. The settling velocity is defined by the Stokes' law :

$$V = \frac{2}{9} \frac{(D_P - D)}{\eta} r^2 g$$

Where V is settling velocity in cms/sec. Dp is particle density, D is density of liquid, η is viscosity of the liquid in poisse, r is the radius of particles in cm and g is acceleration due to gravity 980.7 cm sec^{-2}.

The buoyant force on a hydrometer vary with the suspension density which in turn depend on the concentration of suspended mineral particles. A hydrometer placed in the soil water suspension give reading inversely related to the depth of immersion of the hydrometer bulb which is equal to the length from the level of soil-water suspension up to the centre of the hydrometer bulb. By use of a soil hydrometer such as Bouyoucos hydrometer, the stem of which is calibrated to give concentration of mineral particles in grams per litre, it is possible to estimate fraction of total mass of soil remaining in suspension at the desired time.

1.4.1.2 Equipments and Reagents

Reagents

1. 5% Sodium hexametaphosphate

   Dissolve 25g of sodium hexametaphosphate in distilled water and make the volume to 500 mL with distilled water.

2. 30% H_2O_2 (hydrogenperoxide) - reagent grade.
3. Approx. 2N HCl Dilute 16.7 ml of AR conc. hydrochloric acid to 100 ml with distilled water.
4. 1N NH_4OH.

Equipments & Glasswares

1. Bouyoucos soil hydrometer
2. Soil testing cylinder with mark at 1000 ml length 30 ± 2 cm from inside bottom with rubber stopper.
3. Fahrenheit scale thermometer.
4. Electric stirrer or homogenizer

1.4.1.3 Procedure

1. Oxidation of organic carbon in soil sample by treatment with 30% H_2O_2. Weigh 40g of soil sample (already passed through 2 mm sieve) in duplicate and transfer it to separate tall beakers of large capacity (600 to 1000 ml). Add 40 to 80 ml of distilled water and mix the contents using a glass rod. If the soil is alkaline, add approx. 2.0 M Hydrochloric acid to make the suspension acidic to litmus paper.

Then add 30% H_2O_2 in portions of 5 ml to oxidize organic carbon present in the soil sample. Observe the visual effects of H_2O_2 additions on the soil suspension as indicated by frothing, evolution of gas and change in the soil colour. Continue adding 30% H_2O_2in portions of 5 ml until subsequent additions fail to show any visual effect on the soil water suspension. Stir the suspension by glass rod after each addition of H_2O_2. Usually 15 ml of 30% H_2O_2 is sufficient for oxidizing organic carbon in normal soils containing less than 0.5% organic carbon.

2. Removal of excess H_2O_2 added and estimation of oven dry mass of organic C free soil. After adding the last portion of 30% hydrogen peroxide, and when the evolution of gas has subsided, add ammonium hydroxide to make the solution alkaline to the litmus paper. Then place the beakers on a water bath or on a hot plate at low heat for 3 to 4 hours. Alternatively place the beakers after adding sufficient water in an oven at 70°C. On completion of the digestion period, heat the contents to boil gently for 10 minutes to expel any remaining H_2O_2. Keep the beakers covered with a watch glass during gentle boiling of the contents. Do not allow the contents to become dry. Let the contents cool after boiling. Transfer one of the beaker to an oven set at 105°C and let it remain there until it dries to a constant weight. Weigh the contents after cooling preferably in a desiccator, and determine the oven dry mass of organic carbon free soil.

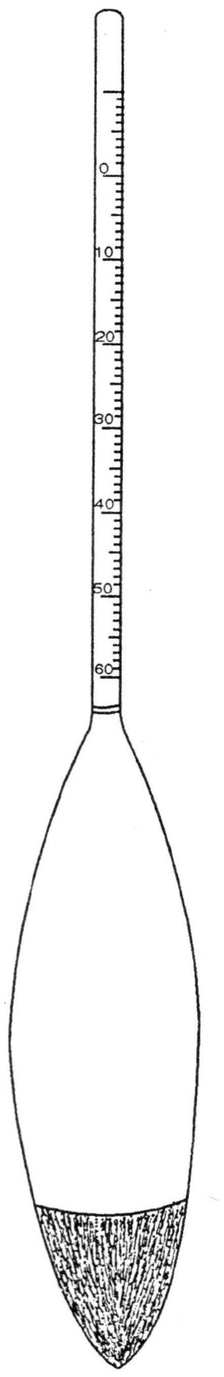

Fig. 1.4.1 Bouyoucos Hydrometer

3. Treatment of organic carbon free soil with dispersing agent and homogenization of sodium hexametaphosphate adsorbed soil. Add 100 ml of 5% sodium hexametaphosphate solution to the second beaker containing soil previously treated with 30% hydrogen peroxide. After allowing several hours (4 hours or more preferably overnight) for the adsorption of sodium by the soil colloids, transfer the soil-water suspension to the dispersion cup of a homogenizer or electric stirrer or mixer. Transfer the contents quantitatively using distilled water. Continue adding distilled water until the level of the soil-water suspension reaches 3.75 cm below the rim of the dispersion cup. The volume of the suspension may reach a level of about 500 to 600 ml. Homogenize the soil-water suspension at moderate speed for 5 to 30 minutes depending upon the soil texture (5 minutes for sandy, 10 minutes for loamy and 20-30 minutes for clayey soils).

4. Transfer of homogenized soil-water suspension into the soil testing cylinder and its further preparation prior to taking of hydrometer reading. Transfer the homogenized and dispersed soil-water suspension from the dispersion cup into a soil testing cylinder quantitatively. Flush the soil sediments from the dispersion cup into the soil testing cylinder by the jet of a wash bottle filled with distilled water. Then fill the cylinder with distilled water making the level of the suspension upto the 1000 ml mark.

In another cylinder of the same size and make, add 100 ml of 5% sodium hexametaphosphate solution and dilute it to the 1000 ml mark with distilled water. Take the temperature of the salt solution as well as the soil-water suspension in degrees Fahrenheit. Place a rubber stopper tightly into the mouth of the cylinder containing the soil-water suspension and mix the contents by turning the cylinder upside down six times. In absence of a rubber stopper of the required size, one may place large size palm of a wet hand over the mouth of the cylinder tightly before turning the cylinder upside down. Make sure that the suspension does not spill out of the cylinder. Alternatively place a plunger inside the cylinder and move it up and down holding the cylinder so as to mix the contents thoroughly without spilling the contents. Use strong upward strokes of the plunger to lift particles lodged at the bottom. Note down the time in hours, minutes and seconds, the moment cylinder is placed on a firm table.

5. Immersion of hydrometer in soil-water suspension and measurement of the hydrometer reading. Lower the Bouyoucos hydrometer gently into the soil-water suspension 30 seconds before the time of taking hydrometer reading. After taking the hydrometer reading of the soil-water suspension (R_{surp}) immerse the hydrometer gently in the salt solution i.e. in the second cylinder containing 1000 ml of 0.5% dispersing agent solution and note down the hydrometer reading placing eye slightly above the level of the salt solution.

For determining merely the content of sand (2.00 to 0.02 mm), silt (0.02 to 0.002 mm), and clay (< .002 mm) in the soil sample by the simple procedure, take the hydrometer reading after allowing a settling time of 4 minutes and 120 minutes without remixing the soil-water suspension between the measurements. Record the temperature also corresponding to each hydrometer reading. The settling time is measured from the moment the soil testing cylinder is placed on a firm table after mixing of its contents (dispersed and homogenized soil-water suspension) by turning it upside down. The content of soil + clay is estimated from the first hydrometer reading and the content of clay is estimated from the second hydrometer reading.

1.4.1.4 Calculation

$$\% \text{ Silt + Clay} = \frac{\left\{\left(R^{1}_{susp} - R^{1*}_{salt}\right) - (68 - T_{susp}\,°F) \times 0.2\right\} \times 100}{\text{Mass of oven dry organic C free soil in g}}$$

$$\% \text{ Clay} = \frac{\left\{\left(R^{2}_{susp} - R^{2*}_{salt}\right) - (68 - T_{susp}\,°F) \times 0.2\right\} \times 100}{\text{Mass of organic C free soil in g}}$$

% silt = % silt + clay – % clay

% sand = 100 – % silt + clay

where

$R^1{}_{susp}$ and $R^2{}_{susp}$ denote hydrometer readings of the dispersed and homogenized soil-water suspension after allowing settling time of 4 and 120 minutes respectively.

$R^{1*}{}_{salt}$ and $R^{2*}{}_{salt}$ denote hydrometer readings of 1000 ml of 0.5% salt solution in 1000 ml cylinder after settling time of 4 and 120 minutes duly adjusted to the temperature of the soil-water suspension. The hydrometer reading of the salt solution be adjusted employing the following equation.

$$R_{salt}{}^{*} = R_{salt} - (T_{susp} - T_{salt}) \times 0.2$$

Where

T_{susp} and T_{salt} denote temperature of the soil-water suspension and 0.5% salt solution in degress Fahrenheit.

1.4.1.5 Comment

The determination of percent sand, silt and clay in a soil sample by the Bouyoucos hydrometer from the change in the composition of particles in suspension which alter the suspension density on allowing particles to settle in a stream line at different velocity depending on the size of particles in the medium of water is simple and reliable.

However, once the results are obtained, the textural class of the soil be determined by reference to the textural triangle and the result be compared with the soil texture as determined by the feel method.

Besides, the reproducibility of the results be checked by estimating the coefficient of variation under similar experimental conditions. Care be taken to take the hydrometer reading at the predetermined time causing minimum disturbance to the suspension. The hydrometer reading be carefully taken at the level of suspension not from above or from below.

REFERENCES

Black, C.A. (ed). 1965. Methods of Soil Analysis. Vol. 1. Am. Soc. Agron., Madison, Wisconsin, U.S.A.

Bouyoucos, G.J. 1927. The hydrometer as a new method for the mechanical analysis of soils. Soil Sci. 23 : 343-353.

Dakshinamurti, C. and R.P. Gupta 1968. Practicals in Soil Physics. Indian Agric. Res. Inst. New Delhi -12.

1.4.2 Soil separates analysis by pipette method-Black (1965)

1.4.2.1 Principle

Soil freed from organic matter and bases by treatment with H_2O_2 and by leaching with dilute acid is homogenized in the presence of a dispersing agent to achieve complete dispersion of the soil particles. The dispersed soil-water suspension is transferred to a 36 cm tall soil testing cylinder (dia 7.5 cm) and diluted to 1000 ml with distilled water. After thorough mixing of the suspension, the cylinder is placed on a firm table to allow sedimentation of the mineral particles. Mineral particles of different diameter settle at different velocities which is defined by the Stokes' law. It is possible to estimate in how much time particles of a diameter would settle to a pre-determined depth in cm (h) from the following equation :

$$T = \frac{0.301\eta h}{(pS - pL)d^2}$$

where

T = time in minutes taken by particles of d diameter in mm to settle to a depth of h cms.

η = Viscosity of water in poise at the suspension temperature.

h = settling depth in cms of the particles of d diameter in time T minutes.

pS = particle density of soil minerals in g/ml.

pL = density of the liquid in g/ml at the suspension temperature.

d = diameter of the mineral particles in mm.

Thus, sedimentation eliminates from the depth h in a time T all particles having settling velocities greater than $\frac{h}{T}$, while retaining at that depth the original concentration of particles having settling velocities less than $\frac{h}{T}$.

The taking of a small volume element by a pipette at a depth h at time T furnishes a sample from which all particles coarser than d diameter (as determined by Stokes' equation) has been eliminated and in which all particles finer than that size are present in the same amount as initially.

In the pipette method, time required for soil particles of the diameter 0.02 mm and 0.002 mm to settle to a depth of 5 or 10 cm from the surface of suspension in the soil testing cylinder at the desired suspension temperature is estimated from the above given equation or else noted from Table 1.4.2. As the time approaches to the predetermined settling time, a 10 ml pipette is lowered to h cm depth by an adjustable pipette stand fitted into a mechanical assembly. An aliquot of the sample is drawn by the pipette and transferred to preweighed large crucibles. The contents are then dried in an oven at 105°C to a constant weight. The first aliquot contains salt + mineral particles less than 0.02 mm, and the second aliquot contains salt + mineral particles less than 0.002 mm. The salt carried into the soil particles is determined and substracted from the mass of salt + mineral particles. The mass of mineral particles obtained is expressed as a percentage of organic carbon free and hydrogen saturated soil.

1.4.2.2 Equipments and Reagents

Equipments. Buchner funnel fitted on a suction flask and connected to a suction pump, Mechanical stirrer or homogenizer, soil testing cyclinder of 500 ml or 1000 ml capacity which could be held by the clamp provided in the mechanical assembly with adjustable pipette stand, steam bath or hot plate, stop watch and oven.

Reagents. 30% H_2O_2, 0.1 N HCl, glacial acetic acid, 0.5% phenolphthalein indicator, 0.1N NaOH and 5% sodium hexametaphosphate solution.

1.4.2.3 Procedure

1. Preparation of soil sample. Dry the soil sample collected by spreading in thin layer under shade for a couple of days. Then take 50g soil sample and pass through 2 mm sieve. Portion of soil retained on 2 mm sieve be recorded as gravel and the soil passing through 2 mm sieve be preserved for the particle size distribution analysis.

2. Treatment of a portion of 2 mm soil sample by 30% H_2O_2. Weigh 10.0 g of air dry 2 mm soil samples in duplicate and transfer the weighed samples into separate 600 ml pyrex beakers. Add 20 ml distilled in both the beakers, and make the contents acidic to the litmus paper by adding 0.1N hydrochloric acid. Stir with a glass rod. Then add 5ml portions of 30% H_2O_2. Stir the contents with a glass rod until the reaction has subsided. Add another 5 ml portions of 30% H_2O_2 and stir with a glass rod. Observe the contents for the evolution of CO_2 and change in the soil colour. If CO_2 evolution is negligible, add 30 ml distilled water in both beakers. Cover with watch glass and place them in an oven set at 70°C for digesting the contents overnight. On the following day, add 20 ml of 30% H_2O_2 and stir the contents for 10 minutes. Make the solution alkaline to litmus paper by adding 0.1 N NaOH. Place the beakers on a hot plate and heat to boil the contents gently for 20-30 minutes to expel the hydrogen peroxide added in excess. Add sufficient water before heating so that the contents do not dry on boiling. Cool the beaker after boiling the contents.

3. Saturation of the hydrogen peroxide treated soil with hydrogen ions. Add 25 ml of 0.1 N HCl into beakers containing organic carbon free soil and stir the contents with a glass rod. Cut a fine grained filter paper (Whatman No. 42) to the

size of the Buchner funnel and set it properly in the Buchner funnel applying some distilled water. Then transfer the acidified organic carbon free soil-water suspension into the Buchner funnel fitted on a suction flask connected to a suction pump. Switch on the suction pump to accelerate the filtration. When leaching of soil with first 25 ml of 0.1 N hydrochloric acid is over, leach the soil with three more 25 ml portions of 0.1 N hydrochloric acid. Thereafter, leach the soil with 125 ml of distilled water in portions of 25 ml. Transfer H^+ saturated soil of one Buchner funnel into a preweighed aluminium can and place it in an oven to dry at 105°C for 24 hours or more to a constant weight. Cool the aluminium can in a desiccator before weighment. Record the mass of organic carbon free and H^+– saturated oven dry soil in four figures after decimal.

4. Saturation of H-saturated soil with sodium by treatment with dispersing agent and homogenization of the suspension to effect complete dispersion of the soil particles.

Transfer other sample of H-saturated soil into 600 ml beaker. Use glass rod provided with rubber policeman. Wash the filter paper with distilled water and collect the washings in the same beaker. Add enough 0.1 N sodium hydroxide to make the suspension just alkaline to the phenolphthalein end point. Then add 10 ml of 5% sodium hexametaphosphate solution into the beaker and stir the contents.

Transfer the suspension from the beaker into the dispersion cup of the motor stirrer or homogenizer. Add distilled water to fill the dispersion cup upto 3.75 cm below the rim of dispersion cup. Switch on the motor stirrer or homogenizer at moderate speed for 10 minutes to effect complete dispersion of the soil particles.

5. Sedimentation of dispersed soil water suspension and removal of 10 ml aliquot from the suspension by pipette at pre-determined time from predetermined depth of suspension.

Note down the temperature of temperature control room or constant temperature bath and find out the time required for 20 and 2 micron mineral particle to settle to a depth of 5 or 10 cm from the initial level of suspension in the soil testing cylinder.

Table 1.4.2.1 Time of sampling at 5 and 10 cm depth for different temperatures and fractions

Temp.	5 cm depth				10 cm depth			
	20 micron		2 micron		20 micron		2 micron	
0°C	minutes	seconds	hours	minutes	minutes	seconds	hours	minutes
20	2	19	3	53	4	38	7	45
22	2	14	3	44	4	27	7	27
24	2	8	3	34	4	15	7	07
26	2	2	3	24	4	03	6	48
28	1	56	3	14	3	52	6	28
30	1	50	3	9	3	40	6	18

Thereafter, transfer the contents of the dispersion cup quantitatively into the 1000 ml soil testing cylinder. Wash the dispersion cup with distilled water and transfer the washings also into the soil testing cylinder. Make the volume to 1000 ml mark with distilled water and let the suspension come in equilibrium with the constant room temperature or constant temperature water bath.

Mix the contents of the cylinder thoroughly using a plunger or turn the cylinder upside down for 60 seconds after closing the mouth with a rubber stopper. Remove the stopper or plunger and immediately place the soil testing cylinder on a firm table or in the constant temperature water bath. Switch on the stop watch or note down the time in hours, minutes and seconds the moment the cylinder is kept on a firm table to initiate the sedimentation of mineral particles.

Twenty seconds before the predetermined time, lower the 10 ml pipette into the suspension upto the desired depth by lowering the adjustable pipette stand provided in the mechanical analysis pipette assembly. Withdraw 10 ml aliquot just at the predetermined time for collecting particles less than 20 micron in diameter. Transfer the aliquot in a pre-weighed large crucible. Wash the pipette with a jet of distilled water and collect the washing in the same large crucible. Similarly draw another 10 ml aliquot out of the suspension from the same depth for collecting mineral particles < 2 micron in diameter. The time of sampling may be noted from the Table 1.4.2.1 provided or by direct estimation from the equation. Transfer the aliquot into another preweighed large crucible. Wash pipette with a jet of distilled water and collect the washings in the same crucible in which the aliquot was transferred.

Transfer both the crucibles in an oven at 105°C and evaporate the contents to a constant weight. Cool the crucibles in a desiccator before weighment. Weigh on a chemical balance accurate upto four figures after decimal. Transfer 10 ml aliquot of 0.05% sodium hexametaphosphate solution into a pre weighed evaporating dish. Place the evaporating dish in an oven at 105°C. Dry the contents to a constant weight and determine the mass of dry salt. Tabulate the observations.

S.No.	Particulars	< 20 micron sample	< 2 micron sample
1.	Mass of crucible		
2.	Mass of crucible + mineral fraction + salt in 10 ml aliquot		
3.	Mass of mineral fraction + salt in 10 ml aliquot		
4.	Mass of salt in 10 ml aliquot		
5.	Mass of mineral fraction.		

1.4.2.4 Calculation

$$\% \text{ silt + clay} = \frac{\text{Mass of mineral fraction} < 20 \text{ micron diameter} \times 1000 \times 100}{\text{Mass of organic carbon free \& H-saturated soil in g} \times 10}$$

$$\% \text{ Clay} = \frac{\text{Mass of mineral fraction} < 2 \text{ micron diameter} \times 1000 \times 100}{\text{Mass of organic carbon free and H-saturated soil in g} \times 10}$$

% silt = % silt + clay – % clay

% sand = 100 – % silt + clay

1.4.2.5 Comments

In pipette method the accuracy desired in weighment is four figures after decimal. The samples be dried to a constant weight and the weightment be made after cooling in a desiccator. The method is considered to be reliable and satisfactory. For other comments see the section on particle size distribution analysis by hydrometer method.

REFERENCES

Black, C.A. 1965. Methods of Soil Analysis. Vol. I. Am. Soc. Agron., Madison, Wisconsin, USA.

Dakshinamurti, C. and R.P. Gupta 1968. Practicals in Soil Physics. Indian Agric. Res. Inst. New Delhi.

1.4.3 Particle size distribution analysis by Bouyoucos hydrometer–Black (1965)

1.4.3.1 Principle

Organic carbon free soil on being treated with dispersing agent is homogenized and transferred to a 36 cm tall soil testing cylinder of 1000 ml capacity. The soil water suspension is thoroughly mixed and then mineral particles are allowed to settle. Mineral particles settle at different velocity which brings about a change in the suspension density with time.

The buoyoant force acting on a hydrometer placed in the soil-water suspension vary with the suspension density which is directly proportional to the concentration of the mineral particles in suspension. The depth of immersion of the hydrometer bulb is inversely related to the buoyoant force. The buoyoant force decreases with the settling of mineral particles which result in an increase in the depth of immersion of hydrometer bulb and decrease in the hydrometer reading. The stem of Bouyoucos hydrometer is calibrated to give concentration of mineral particles in g/l at 68°F and as such it enables estimation of fraction of total mass of soil remaining in suspension at the desired time. The corresponding particle size in diameter in microns may be estimated by dividing θ a sedimentation parameter by the square root of the sedimentation time in minutes. The sedimentation parameter θ may be estimated by the following equation

$$\theta = 1000\left[\frac{30\,\eta h}{(pS - pL)g}\right]^{1/2}$$

Where η = viscosity of water in poise, h = settling distance in cm i.e. distance from the centre of hydrometer bulb to the hydrometer reading in cm, pS = particle density in g/ml pL = density of the liquid and g – acceleration due to gravity.

The above equation has been derived from the Stokes' law defining the velocity of fall of spherical particles in the medium of a liquid. The θ value for different settling depth h may be estimated at a particular temperature such as 68°F for Bouyoucos hydrometer is also calibrated at the same temperature. Besides, hydrometer reading taken at temperature other than 68°F is also adjusted to the temperature of the calibration of the hydrometer reading by use of temperature correction factor which is 0.2 for each degree deviation from 68°F.

1.4.3.2 Reagents and Equipments

Same as particle size analysis of silt + clay and clay.

1.4.3.3 Procedure

Steps 1-4. As described in section 1.4.1.3.

5. Several hydrometer readings are required for the complete particle size distribution analysis and so, the hydrometer reading be taken after 1/2, 1, 2, 4, 10, 30, 60, 90, 120, 240, 360 and 480 minutes completing the analysis in 8 hours period. Thereafter the hydrometer reading may change quite insignificantly with time. The temperature of the suspension be also recorded along with the hydrometer reading. As buoyoant force acting on the hydrometer is the result of concentration of the mineral particles and salt solution, the hydrometer reading of the salt solution be also taken simultaneously along with the temperature of measurement. The hydrometer be lowered just 30 seconds before taking the hydrometer reading. Care should be taken to cause minimum disturbance to the soil-water suspension in lowering and removal of the hydrometer. The hydrometer be removed from the soil-water suspension after taking the first-four observations at short interval. Thereafter, the hydrometer should be immediately removed after noting the hydrometer reading of the soil water suspension. Record the hydrometer reading and the temperature of measurement of soil-water suspension as well as salt solution corresponding to each settling time.

6. Estimation of % of different size of coarse mineral fraction. After taking the last hydrometer reading after 480 minutes of the start of sedimentation, pour the suspension and sediments directly from the cylinder into a .05 or .053 mm sieve. Discard the effluent. Agitate the residue on the sieve by running ordinary tap water into it direct from the tap or from a rubber tubing leading from the tap.

After washing the residue remaining on the sieve sufficiently, transfer the sieve into an oven and let it dry for 24 hours at 105°C to a constant weight. Arrange a set of sieves 2.0, 1.0, 0.5, 0.212, 0.100 and 0.053 mm in descending order on a mechanical shaker.

Transfer the oven dry soil residue composed of mineral matter on the 2.0 mm sieve. Place the cover lid and fasten the sieve properly on the mechanical shaker. Run the mechanical shaker at moderate speed for 10 minutes. Collect the soil fraction remaining on each sieve separately and weigh. Record the mass of soil residue corresponding the each size fraction and express the mass recorded as percentage of the organic carbon free soil.

Settling Time	Soil-water Suspension		0.5% solution of dispersing agent	
minutes	Hydrometer Reading	Suspension Temp. °F	Hydrometer Reading	Solution Temp. °F

	Diameter of mineral fraction in mm					
	2.0	1.0	0.5	0.212	0.100	0.053
Mass of minerals in g						

1.4.3.4 Calculation

$R^*_{susp} = (R_{susp} - R^*_{salt}) - (68 - \text{Susp. Temp. °F}) \times 0.2$

Where

R_{susp} is hydrometer reading of the soil-water suspension

R^*_{salt} is the hydrometer reading of salt solution at the suspn temp.

$= R_{salt} - (\text{Susp. temp °F} - \text{Salt solu temp °F}) \times 0.2$

R^*_{susp} is adjusted hydrometer reading of soil-water suspension at 68°F.

Settling depth i.e. depth of immersion of the hydrometer bulb, h.

$$h = H - \left(R^*_{susp} \times \frac{L}{60}\right) = 17.9 - \left(R^*_{susp} \times \frac{10.2}{60}\right)$$

Where

H = distance from the centre of hydrometer bulb to zero reading of hydrometer stem in cm.

R^*_{susp} = Adjusted hydrometer reading of the suspension at 68°F.

L = distance from zero to sixty reading of the hydrometer stem in cm.

Note. In Bouyoucos hydrometer (ASTM) of Taylor Industries, Rochester, N.Y., the value of H and L was found to be 17.9 cm and 10.2 cm respectively.

Sedimentation parameter θ

$$\underset{\text{at } 68°\text{ F}}{\theta} = 1000\left[\frac{30\,\eta h}{(pS - pL)\,g}\right]^{1/2}$$

Where

η = Viscosity of water in poise at 68°F

h = depth of immersion of the hydrometer bulb as estimated from the R^*_{susp}.

pS = particle density of the soil.

pL = density of 0.5% sodium hexameta phosphate solution at 68°F commonly used as dispersing agent.

g = acceleration due to gravity 980.7 cm sec^2.

Diameter of mineral particles in micrometer

$$x = \frac{\theta}{(t)^{1/2}}$$

Where θ is the sedimentation parameter

t is the settling time in minute.

$$\text{Diameter of mineral particles in millimeter} = \frac{x}{1000}$$

$$\text{\% of mineral fraction in soil sample} = \frac{R^{*}_{susp}}{\text{mass of organic C free soil}} \times 100$$

Where R^{*}_{susp} is the adjusted hydrometer reading at 68°F corresponding to each settling time in minutes.

% of the different size of coarse fraction

$$= \frac{\text{Oven dry mass of coarse fraction in different size sieves}}{\text{mass of organic carbon in g}} \times 100$$

Graphical Presentation of particle size distribution

1. Tabulate the results of particle size distribution analysis giving diameter of mineral particles in mm in ascending order and the corresponding percentage of mineral particles in the soil sample.
2. Draw the graph on a semi-log graph paper, plotting diameter of mineral particles on the x-axis in log scale and the corresponding percentage of mineral particles on the y-axis in the linear scale.

1.4.3.5 Comments

It is felt that pretreatment of the soil samples be done to the minimum. However the removal of free iron oxides in latosols and laterites, soluble salts in saline soils, calcium carbonate in Calcareous soils and organic matter in most soils will be advantagous.

The removal of free iron oxides require use of reducing agent eg. sodium dithionate as well as a chelating agent such as sodium citrate.

The soluble salts flocculate soil, and the presence of alkaline salts decompose hydrogen peroxide readily and therefore the suspension be made slightly acidic before adding hydrogen peroxide. The suspension be made again alkaline before gentle boiling to expel hydrogen peroxide added in excess. The dissolution of soluble salts in water is very effective in removal of soluble salts.

In the presence of calcium carbonate, dispersion of salt and clay particles become difficult. The carbonates can be removed by treatment with acid or acid sodium acetate.

The organic matter in soil has aggregating effect and so its removal is necessary for the separation of mineral particles. The organic matter is oxidized by treatment of the soil with 30% hydrogen peroxide. To overcome the effects of manganese oxide, sodium acetate (pH 5.0) be added prior to oxidation of organic matter by hydrogen peroxide. At high temperature, hydrogen peroxide is decomposed and so the digestion be done at temperature not exceeding 70°C. Hydrogen peroxide is a strong oxidizing agent and contact with the skin be avoided. If contact is made, affected parts should be washed with copious quantities of water.

The soil-water suspension treated with sodium hexametaphosphate should not be homogenized too long as soft minerals and rock fragments may desintegrate during mixing. At the same time the blades of electric homogenizer may deteriorate rapidly by abrasion and so the blade be replaced as they begin to show signs of wear.

REFERENCES

Black, C. A. 1965. Methods of Soil Analysis Vol. I. Am. Soc. Agron. Madison, Wisconsin, U.S.A.

Dakshinamurti, C. and R.P. Gupta 1968. Practicals in Soil Physics. Indian Agric. Res. Inst. New Delhi.

1.5 PARTICLE DENSITY OF SOIL - BLACK (1965)

1.5.1 Estimation of particle density by use of relative density bottle or volumetric flask

1.5.1.1 Principle

Particle density of soil is estimated by dividing the mass of oven dry soil by the volume of particles without pore space or else by the volume of water displaced by the soil. The product of specific gravity of soil and the density of water is equal to the particle density as the specific gravity is estimated by dividing mass of the soil by the mass of water displaced by the soil.

1.5.1.2 Apparatus & Equipment

25 to 50 ml R.D. bottle or 50-100 ml volumetric flask and weighing balance.

1.5.1.3 Procedure

Weigh a clean dry R.D. bottle and fill it with oven dry soil sample up to half of its volume. If a 100 ml vol. flask is used fill it with soil up to half of its volume. Clean the outside and neck of the R.D. Bottle to remove any soil adhering. Insert the stopper and weigh.

Then add boiled and cooled distilled water into R.D. bottle (partly filled with soil) until the level of water reaches 3/4 of the volume of R.D. bottle. Heat the R.D. bottle to boil the contents gently to remove the entrapped air. Frequently agitate the contents gently to prevent loss of soil by foaming. Cool the contents and fill the remaining space in R.D. bottle with distilled water. Insert the stopper. Dry outside of R.D. bottle by wiping it dry with a cloth and weigh.

Thereafter, remove the soil water suspension from the R.D. bottle, wash with tap water and rinse it with distilled water. Then fill the R.D. bottle completely with distilled water. Insert the stopper spilling some water through the mouth. Dry outside of R.D. bottle by wiping it dry with a cloth and weigh.

Transfer the contents to a 50 ml beaker and dip a thermometer in the soil water suspension and measure its temperature. Note the density of water at the observed temperature.

1.5.1.4 Calculation

$$\text{Particle density in } g/cm^3 = \frac{(W_2 - W_1)}{(W_4 - W_1) - (W_3 - W_2)} \times dW$$

Where W_1 is mass of empty R.D. bottle

W_2 is mass of R.D. bottle + soil up to half of its volume

W_3 is mass of R.D. bottle + $\frac{1}{2}$ soil and $\frac{1}{2}$ water by volume.

W_4 is mass of R.D. bottle completely filled with water.

dw is the density of water at the temperature of the soil-water suspension.

1.5.1.5 Comments

Use of water gives higher particle density for finely divided active powder than do weakly adsorbed non-polar liquids such as toluene, xylene or carbon tetrachloride. The estimation may be done in duplicate to check the reproducibility of the result.

1.5.2 Estimation of particle density of soil by use of 100 ml graduated cylinder

1.5.2.1 Principle as in section 1.5.1.1

1.5.2.2 Apparatus

100 ml graduated cylinder.

1.5.2.3 Procedure

Take a clean 100 ml capacity graduated cylinder. Add distilled water to fill the graduated cylinder upto 50 ml mark. Then weigh 40g of oven dry soil and transfer it into the 100 ml graduated cylinder filled with water up to 50 ml mark. Take a glass rod some what longer than the height of the graduated cylinder. Stir the contents thoroughly for about 10 minutes and allow the particles to settle. Note down the increase in the level of water on adding 40g of oven dry soil to measure the volume of water displaced by the soil in cm^3.

1.5.2.4 Calculation

$$\text{Particle density } (g/cm^3) = \frac{\text{Mass of oven dry soil in g}}{\text{Volume of water displaced by the soil in } cm^3}$$

1.5.2.5 Comments

The particle density estimated by the above method in samples from mineral soils is usually very close to the average particle density of the mineral soils.

REFERENCES

Black, C.A. (ed). 1965. Methods of Soil Analysis Vol.1. Am. Soc Agron. Madison, Wisconsin, U.S.A.

Brady, N.C. 1980. The Nature and Properties of Soils. Eurasia Publishing House Private Limited, Ram Nagar, New Delhi-110055

Dakshinamurti, C. and R.P. Gupta, 1968. Practicals in Soil Physics. Indian Agric. Res. Inst. New Delhi.

1.6 BULK DENSITY OF SOIL - BLACK (1965)

1.6.1 Estimation of the bulk density of soil by the core method

1.6.1.1 Principle

An undisturbed soil core is removed by driving a metal cylinder about 7.5 cm in dia and 7.5 cm in height on a level and neat soil surface. The metal cylinder filled with soil is removed and dried in an oven at 105°C to a constant weight. The volume of the soil core is estimated from the internal radius and height of the metal cylinder. The mass of oven dry soil core is divided by its volume in cm^3 to obtain bulk density of soil in g/cm^3.

1.6.1.2 Equipment and Reagents

Metal cylinder of desired dimensions, square metal plate with a rod at the centre, sliding weights, khurpi (shovel), vernier calipre, physical balance and oven.

1.6.1.3 Procedure

1. Collection of soil core samples. When the soil is neither dry nor wet, select a site which is level and free from weeds and crop residues. Place the metal cylinder at the selected site. On the metal cylinder, place square metal plate with a rod at the centre. Slide weights through the iron rod and let it fall on the square plate to drive the metal cylinder uniformly into the soil from the impacts of the weights on the metal plate.

When the metal cylinder is uniformly filled with the soil up to its top, remove the soil from outside the metal cylinder deeper than the cylinder with the help of a shovel (khurpi). Lift the cylinder and make a clean cut at the cylinder base removing extra soil adhering. Keep the metal cylinder filled with soil core inside a tray over wire gauze lined with a filter paper. Take more than one samples.

2. Drying and weighment of the soil core sample. Cover the tray with a waxed paper or else place the soil core with metal cylinder inside a waxed paper carton with lid and transport it to a laboratory. Transfer the metal cylinder with soil core after weighment into an oven at 105°C. Keep the wire gauze lined with filter paper below the metal cylinder. Let the soil core dry to a constant weight in 2-3 days time.

Remove the metal cylinder with soil core from the oven and let it cool to room temperature in about 3-4 hours time. Take the mass of metal cylinder + oven dry soil core. Remove the soil core from the metal cylinder and weigh the metal cylinder.

3. Oven dry mass and volume of the soil core sample. Substract the mass of metal cylinder from the mass of oven dry soil core + metal cylinder to obtain the mass of oven dry soil core.

As the volume of the soil core is equal to the internal volume of the metal cylinder, measure the radius (r) and height (h) of the metal cylinder in cm by a Vernier calipre and substitute the values in the equation $\pi r^2 h$ to obtain the volume of soil core in cm^3.

1.6.1.4 Calculation

$$\text{Bulk Density} = \frac{\text{Mass of oven dry soil core in g}}{\text{Volume of soil core in cm}^3} = g/cm^3$$

1.6.1.5 Comments

Soil core samples should not be taken from dry or wet soil. Soil level inside and outside the sample must remain the same if the sample is to be considered satisfactory.

A cylinderical waxed paper carton with lids of the desired size to cover the metal cylinder containing moist soil would be satisfactory to prevent loss of water from the soil during transportation to the laboratory for determining the mass of moist soil. Some times the samples for bulk density is also used to determine the moisture content of the soil on volume basis.

Bulk density of soils is influenced by green manuring, incorporation of organic manures, growing of perennial grasses, use of soil conditioners, tillage, settling of soil after rain or irrigation and compaction by wheels of heavy machinery. Bulk density of soil is inversely related to % pore space in soil.

1.6.2 Estimation of bulk density of soil by the excavation method

1.6.2.1 Principle

A moist soil is excavated by a table spoon and the soil excavated is collected in a metal container for drying in the oven at 105°C to a constant weight. The volume of soil excavated is determined by taking a known volume of fine, free flowing sand in a graduated cylinder and pouring it into the excavated portion to fill the soil uniformly.

1.6.2.2 Equipments and Reagents

Clean, dry, free flowing fine sand, a metal table spoon for excavating moist soil, metal container for collecting excavated soil sample, physical balance and oven.

1.6.2.3 Procedure

1. Collection of soil sample for estimation of bulk density. Select soils for collecting soil samples for the estimation of bulk density by the excavation method. Carry the metal container, graduated cylinder filled with clean, dry, free flowing fine sand and a table spoon.

Excavate the moist soil with the table spoon and transfer it into the metal container. Excavate the site sufficiently deep with clean edges. Collect all the soil excavated in the metal container. Fill the metal container with excavated moist soil upto few cms depth and place the lid on the metal container.

2. Estimation of the volume of soil excavated. Then pour fine free flowing sand from the graduated cylinder into the excavated portion of the soil to fill it uniformly up to the soil surface. Find out the volume of sand used in filling the excavated portion uniformly from the initial and final volume of sand in the graduated cylinder.

3. Drying and weighment of the excavated soil sample. Transfer the metal container after removing the lid and placing it below the bottom of container inside an oven at 105°C to dry the excavated soil. Dry the soil to a constant weight. The soil sample be dried in the oven at 105°C atleast for 24 hours. If the depth of soil in the container is more, time required for drying at 105°C will be greater than 24 hours.

1.6.2.4 Calculation

$$\text{Bulk density of soil } (g/cm^3) = \frac{\text{Mass of excavated soil on oven drying to a constant weight in g}}{\text{Volume of fine sand used in filling the excavated site in } cm^3}$$

1.6.2.5 Comments

The method is simple and satisfactory in absence of a soil core sampler. The precision will increase with the increase in the volume of excavated soil and with the increase in the number of samples. However, the practical feasibility should also be taken into consideration.

REFERENCES

Black, C.A. (ed). 1965. Methods of Soil Analysis, Vol. 1. Am. Soc. Agron., Madison, Wisconsin, U.S.A.

Brady, N.C. 1980. The Nature and Properties of Soils. Eurasia Publishing House (P) Limited, Ram Nagar, New Delhi - 110055.

Dakshinamurti, C. and R.P. Gupta 1968. Practicals in Soil Physics, Indian Agric. Res. Inst. New Dehli - 12

1.7 PERCENT TOTAL PORE SPACE–BLACK (1965)

1.7.1 Principle

The volume of void spaces found in an oven dry undisturbed soil expressed as a percentage of volume of the soil is termed % total pore space. It is estimated by substracting % solid space from 100. The ratio of bulk density to particle density is a measure of solid space in the soil.

1.7.2 Equipments and reagents

Same as given in section 1.5.2 and 1.6.2.

1.7.3 Procedure

Follow the procedure described in section 1.5.2.3. and 1.6.1.3.

Estimate % solid space and % total pore space by the following equation.

1.7.4 Calculation

$$\% \text{ solid space} = \frac{\text{Bulk density}}{\text{Particle density}} \times 100$$

$$\% \text{ total pore space} = 100 - \% \text{ solid space}$$

1.7.5 Comments

% total pore space is the sum of micro and macro pores in the soil. The micropores retain water against the gravitational pull and are referred as capillary pores. The macro pores do not retain water against the gravitational pull and so it remains filled with air most of the time. However, it permits exchange of gases between gases phase of soil and the atmospheric air in response to differences in their partial pressure of gases which is termed as soil aeration.

REFERENCES

Black, C.A. (ed). 1965. Methods of Soil Analysis. Vol. 1. Am. Soc. Agron., Madison, Wisconsin, U.S.A.

Brady, N.C. 1980. The Nature and Properties of Soils. Eurasia Publishing House (P) Limited, Ram Nagar, New Delhi-10055.

Dakshinamurti, C. and R.P. Gupta. 1968. Practicals in Soil Physics. Indian Agric. Res. Insti., New Delhi-12.

1.8 AIR CAPACITY OF SOIL BY FIELD METHOD-KOPECKY (1914, 1927)

1.8.1 Principle

Air filled pore space in soils is of significance because interchange of gases in between soil and atmosphere occur primarily through air filled pore space. Kopecky (1927) defined air capacity as the quantity of air in the soil which remains there after the soil has been saturated with water to the point of absolute water capacity. Following steps are involved in estimation of air capacity of soil.

(*i*) % Capillary water = % moisture in soil at Field Capacity – % Moisture at hygroscopic coefficient.

(*ii*) $\text{Mass of capillary water} = \frac{\% \text{ Capillary water} \times \text{Mass of oven dry soil}}{100}$

(*iii*) $\text{Volume of capillary water} = \frac{\text{Mass of Capillary Water}}{\text{Density of Water}}$

(*iv*) $\% \text{ Capillary pore space} = \frac{\text{Volume of capillary water} \times 100}{\text{Volume of soi.}}$

(*v*) $\% \text{ total pore space} = 100 - \frac{\text{bulk density}}{\text{Particle density}} \times 100$

(*vi*) Volume of total pore space $= \dfrac{\text{\% total pore space} \times \text{Volume of soil}}{100}$

(*vii*) Volume of air filled pore space at field capacity = Volume of total pore space – Vol. of capillary pore space.

(*viii*) Air capacity of soil. $= \dfrac{\begin{array}{c}\text{Volume of air filled pore space}\\ \text{at field capacity}\end{array} \times 100}{\text{Volume of soil}}$

1.8.2 Equipments and Reagents

Metal cylinder approx. 3.75 cm radius (r) 7.56 cm height (h) with accessories for core sampling, Polythene sheet to cover bunded area after irrigation, physical balance and oven.

1.8.3 Procedure

1. Collection of undisturbed soil sample at field capacity. The small bunded plots be irrigated to saturate the soil atleast up to 20 cm depth. Spread the polythene sheet to cover the irrigated plot. Place soil on polythene spread outside the bund to minimize evaporation loss. When the soil reaches field capacity, collect undisturbed soil sample in triplicate using metal cylinder. Cover soil sample with waxed paper carton. Also measure the soil temperature. Place metal cylinder containing undisturbed soil sample with paper carton inside a tray on a wire gauze lined with filter paper. Carry the tray to the lab. Weigh the metal cylinder plus moist soil and then transfer the cylinder + moist soil along with the wire gauze inside an oven at 105°C and let it dry for 4-5 days to a constant weight. Cool the cylinder + oven dry soil to room temperature. After weighment, remove the oven dry soil and crush it to pass through 2 mm sieve. Preserve 2 mm soil for subsequent analysis. Wash and dry the empty metal cylinder. Take its mass and measure its internal diameter and height in cms.

2. Estimate the following observations

(*i*) % moisture in undisturbed soil at field capacity : Substitute the observations taken on soil sample collected in the equation given below :

$$= \left[\frac{\text{Mass of moist soil collected at field capacity in g}}{\text{Mass of the soil after drying in the oven in g}} - 1{,}000\right] \times 100$$

(*ii*) % moisture in soil at hygroscopic coefficient

Transfer a portion of 2mm oven dry soil (about 20 g) into a saturated atmosphere and let it remain there for few days. Weigh and transfer the sample inside an even at 105°C for 24 hours. Cool the oven dry sample in a desiccator and weigh accurately and estimate the moisture content.

% moisture at hygroscopic coefficient

$$= \frac{\text{Loss of moisture on oven drying in g} \times 100}{\text{Oven dry mass of the soil sample}}$$

(*iii*) Bulk density of soil : Divide the mass of oven dry soil core by the volume of cylinder as estimated by the equation $\pi r^2 h$, where r is the internal radius in cm, and h is the height of the cylinder in cm.

(*iv*) Particle density of the soil : Follow laboratory technique described to estimate particle density of 2 mm oven dry soil sample.

(*v*) Density of water at the temperature of taking undisturbed soil sample. See physical table giving density of water at the observed temperature.

1.8.4 Calculations

Follow each step in the sequence given for the estimation of air capacity of soil listed in section 1.8.1.

1.8.5 Comments

During the course of imparting practical training, the procedure was employed to have an estimate of air capacity of soil.

In crops other than paddy, air capacity of soil should be greater than 10%. The requirement of optimum air capacity of soil by crops vary with the stage of plant development.

REFERENCES

Kopecky, J. 1914. Die Physikalischen Eigenschaften der Bodens. Intern. Mill. Bodenk. 4 : 138-198.

Kopecky, J. 1927. Investigations of the relations of water to soil. Proc. Ist Intern. Congr. Soil Sci. 1 : 495-503.

1.9 SOIL COMPACTION MEASUREMENT WITH PENETROMETER–BLACK (1965) AND SINGH (1972)

1.9.1 Principle

Pentrometer measures resistance offered by the soil to penetration of a metal cone or a thin metal circular rod designated as needle. The resistance offered by the soil is in proportion to soil strength. The soil strength is positively related to bulk density and is inversely related to moisture content of the soil. The bulk density is a direct measure of soil compaction. The bulk density of the entire soil profile can be mapped by cobalt 65 radiation across the soil column.

When the adverse effect of soil compaction is the result of increase in soil strength which reduce crop emergence and restrict development of roots and tuber crops, it would be desirable to measure soil strength in terms of resistance offered by the soil to penetration of a metal cone or narrow steel rod called needle by use of a penetrometer. The results are expressed in load in Kg/cm^2 of cross sectional area of cone base or steel rod i.e. needle.

1.9.2 Equipments and Reagents

Cone Penetrometer

Needle Penetrometer

1.9.3 Procedure

1. Use of Cone Penetrometer. It consists of a circular steel nickel plated spring to which a circular gauze meter is attached. The top portion of the nickle plated spring is fitted to a flat handle, and bottom of the nickel plated spring is connected to an extension iron rod 2.5 cm in dia. The lower end of the extension rod is provided with grooves for fitting a 30° stainless steel cone with a base 3.0 cm in dia.

The cone end point is placed on the soil surface with extension rod at right angle to the soil surface. Then the penetrometer handle is gently pushed with both hand to drive the metal cone into the ground up to the base of the cone or else to a greater depth (7.5 cm) as prescribed by the manufacturer. Then, the pressure on the handle is released. The reading at which gauze needle rest is recorded, and the gauze needle release button is pressed down so that the gauze needle returns to the initial position.

At least four observation on penetrometer reading be taken in the vicinity of the location. The soil samples be taken atleast in duplicate from each point of observation for the determination of moisture content of the soil. The penetrometer is lifted out of the soil. The cone and the metal extension rod may be detached from the circular nickel plated spring attached to the handle for storage in the lab. care be taken to prevent cone point becoming blunt.

2. Calibration of the cone penetrometer reading. Place on the handle of cone penetrometer load weighing 1.5, 3, 5, 10, 15, 20, 25, 35 and 40 kg one by one and record the corresponding circular nickel plated spring gauze reading. The value recorded for one of the cone penetrometer is presented below

Table 1.9.1 Load in kg placed on cone penetrometer handle and the corresponding circular nickel plated spring gauze reading

Load in Kg	Spring gauze reading	Load in kg	Spring gauze reading
1.5	0.12	20.0	1.38
5.0	0.32	25.0	1.73
10.0	0.66	35.0	2.48
15.0	1.05	40.0	2.80

Source. Singh (1972)

Work out the numerical relationship between load in kg (x) and the corresponding circular nickel plated spring gauze reading (y) by fitting a linear regression equation. In the data presented in the above table, the regression equation worked out to

$$y = 0.0674\,x \text{ or}$$

or $$x = 14.3368\,y$$

where x is the load in kg and y is the corresponding nickel plated spring gauze reading.

3. Use of needle penetrometer. About 8.0 cm long steel rod 0.635 cm (0.25 inches) in dia called needle is fitted into a rigid spring inside a 6.5 cm long hollow handle 2.50 cm in dia. The handle tapers off towards the steel rod so closely that it can slide on the steel rod on being pushed forward. The handle top i.e. the other side of the handle is closed. The lower end of the steel rod has a circular marking engraved 1 cm above the tip. At the juncture of steel rod and the handle a light fitting plastic ring is placed. The steel rod from the handle towards the tip is graduated in pounds of load per square inch of the cross sectional area of the tip.

The tip of the needle penetrometer is placed on the soil surface making a right angle with the soil surface and then the handle is slowly pushed to penetrate the needle up to 1 cm mark or deeper up to the desired depth as specified by the manufacturer. As pressure on handle is released, the handle reverts smoothly to its initial position leaving behind the plastic ring on the needle (i.e. steel rod) which is calibrated. The load in lbs/.inch^2 is noted from the position of the plastic ring. After recording the load, the plastic ring is pushed back to the juncture between the handle and the steel rod. Similarly additional three observations be taken from the vicinity and record the mean value. The soil samples be also taken atleast in duplicate to determine the moisture content of the soil.

1.9.4 Calculation

$$\text{Cone index} = \frac{\text{Load in Kg}}{\text{Cross sectional area of cone base}}$$

$$= \frac{14.8368y}{\pi r^2} = \frac{y}{.0674 \times \pi r^2}$$

$$= 2.100y = \frac{y}{.47618}$$

where y is nickel plated spring gauze reading and r is the radius of cone base *i.e.* 1.5 cm.

Needle index = load in kg/cm^2 = loads in lbs/inch^2× .0705

1.9.5 Comments

A calibration curve is required for estimating bulk density of soil from the penetrometer readings.

This can be done by taking 4 kg tins atleast 28 in number. Measure the volume of each tin by filling it with measured volume of water. Empty the tins and number the tins from 1 to 28. Allocate 4 tins for each bulk density. Estimate the amount of air dry soil to be filled in each 4 kg tin for obtaining bulk density of 1.1, 1.2, 1.3, 1.4, 1.5, 1.6 and 1.7 g/ml.

For uniform filling, divide the length of tin into four sections and estimate how much soil should be filled in each section for obtaining the desired bulk density of soil. Fill each section of the tin with the same amount of soil. When filling of soil in 4 kg tins for seven bulk densities replicated four times has been completed, take 2 or 3 measurements with penetrometer in each tin filled with soil, and record the mean value. Also collect at least two samples from the same soil depth for determining the moisture content of the soil. Take the mean of two determinations on moisture content of the soil sample. Tabulatate the mean penetrometer reading and percent moisture content of the four replications for each bulk density and then take the average of the four replications. The resulting data on penetrometer reading be tabulated for the seven bulk densities under the column air dry soil giving in bracket over all mean moisture content of air dry soil sample.

Thereafter, prepare a calibration curve on a linear scale graph paper. Set penetrometer reading scale on the ordinate and bulk density scale on the abscissa. Then plot the mean penetrometer reading on the ordinate corresponding to the bulk density on the abscissa. Connect the points plotted to obtain a calibration curve at the moisture content of air dry soil.

The experiment as described above need to be repeated several times by increasing moisture content of the soil progressively by 2.0% using the same soil until the moisture content of the soil approaches field capacity.

Tabulate the data on mean penetrometer reading for the seven bulk densities of the soil with the progressive increase in moisture content of the soil by 2.0%. Prepare a calibration curve on a graph paper for each set of observation on mean penetrometer reading corresponding to the 7 bulk densities. Plot mean penetrometer reading on the ordinate and the corresponding bulk density on the abscissa separately with progressive increase in moisture content by 2%.

When the above calibration curves are made, it would be possible to determine the bulk density of soil from the penetrometer reading provided percent moisture in soil sample has been also determined. This would, however, be valid for the same texture of soil for which the calibration curve has been prepared.

REFERENCES

Black, C.A. (ed). 1965. Methods of Soil Analysis Vol. 1. Am. Soc. Agron., Madison, Wisconsin., U.S.A.

Dakshinamurti, C. and R.P. Gupta 1968. Practicals in Soil Physics. Indian Agri. Res. Inst. New Delhi 12

Singh, Mahendra. 1972. Tractor efficiency of a small tractor tyre on deformable surface M.Sc. Agric Engr. Thesis, Alld. Univ.

1.10 AGGREGATE SIZE DISTRIBUTION ANALYSIS

1.10.1 Aggregate size distribution in soil by dry sieving Dakshinamurti and Gupta (1968).

1.10.1.1 Principle

Mineral particles in close contact with each other are held together by cohesion and by the action of organic ligands, iron oxides and calcium carbonate. Such a group of particles in soil is termed as aggregates.

The aggregate size distribution play a significant role in modifying a physical environment in terms of air and water movement, air capacity, water retention capacity, susceptibility to wind and water erosion and over all productivity of soil. There is effect of tillage, cropping, manurial and cultural practices on aggregation in soils.

The aggregate size distribution in soils is commonly determined by sieving with a nests of sieves arranged in descending order of theirs holes diameter from 2 mm to 0.053 mm. The aggregates larger than 5 mm are determined separately. In dry sieving sieves oscillate usually on a horizontal plane by manual or motor driven mechanical device whereas wet sieving is done in a tub filled with water up to the level of the top sieve in the upward position. The sieves oscillate on a vertical plane i.e. in upward and downward direction by an electric motor.

In wet sieving water from tub is drained. Sieves are removed and dried in an oven. Oven dry aggregates from each sieve is collected and weighed. The mass of aggregates corresponding to each class size is tabulated against mean diameter. Then, the mass of aggregates is expressed as percentage of the mass of sample taken for wet sieving. The data pertaining to aggregate size distribution by dry sieving is also tabulated in the same manner. Both the set of data may be further processed for the estimation of following parameters.

$$\text{MWD} = \sum_{i=1}^{n} \overline{X}_i W_i$$

The mean weight diameter (MWD) is equal to the sum of products of mean diameter ($\overline{X}_i$) of each aggregate size fraction, and the mass of soil aggregates W_i occurring in the corresponding size fraction. Σ denote the summation of the sum of products from i^{th} to n^{th}.

$$\text{GMD} = \exp\left[\frac{\sum_{i=1}^{n} W_i \log \overline{X}_i}{\sum_{i=1}^{n} W_i}\right]$$

The geometric mean diameter (GMD) is calculated approximately by the above equation where W_i is the mass of aggregates in a size class with a mean diameter X_i and $\sum_{i=1}^{n} W_i$ is the total mass of the sample.

Log GMD and log standard deviation

When mean diameter of each aggregate size class is plotted on a log scale and the corresponding cumulative frequency of % aggregates i.e., percent oversize is plotted on probability scale. The diameter of aggregates at 50% over size is log GMD and the log standard deviation i.e. σ_e is given by ratio :

$$\sigma_e = \frac{\text{Size at 84.15\%}}{\text{Size at 50\%}} = \frac{\text{Size at 50\%}}{\text{Size at 15.87\%}}$$

% aggregates greater than 0.212 mm

The above parameter is estimated from the results of dry sieving of aggregates. The soils having lower value are more susceptible to wind erosion.

The soil with a higher proportion of large size aggregates are considered to be more productive than soils having larger proportion of small size aggregates.

However, there is also need to give due importance to the stability of soil aggregate in water. In soils with higher proportion of water stable aggregates, the deterioration of the structural condition on wetting by rain or irrigation water will be much less as compared to soils with lower proportion of water stable aggregates.

1.10.1.2 Equipments and reagents

Set of sieves 20 cm dia, 5 cm height with 8, 5, 2, 1, 0.5, 0.212, 0.100, 0.053 mm openings with lid cover and bottom pan. Rotary sieve shaking machine. Flat rectangular four cornered spade or rectangular large size tray, aluminium can, brush and physical balance.

1.10.1.3 Procedure

1. Collection and preparation of soil sample. Collect the soil sample when the soil is reasonably dry to avoid break down or change in structure. However, as tillage is usually done at friable consistency and if the effect of tillage on structural condition of the soil is to be evaluated, the soil sample have to be collected immediately after the tillage.

For taking the sample from the soil, four cornered flat blade spade be pushed under the sample, and it be then lifted and placed into a large size tray from the desired depth. The samples can be also gathered by a rectangular large size tray from the ploughed depth.

Several samples be taken from the ploughed strip keeping the depth of sampling uniform.

Samples collected be spread in thin layer for drying in the air. A portion of the air dried sample is taken for estimation of the aggregate size distribution.

2. Sieving of air dried samples. Air dried, soil sample be first passed through 8 mm/sieve. The aggregates remaining on the sieve be broken lightly into units to be retained by 5 mm sieve, but which passes through 8 mm sieve.

Then weigh 100g soil sample in duplicate. Sieve one of the samples by a 5 mm sieve and weigh the fraction retained on the 5 mm sieve. Arrange the remaining sieves in descending order of their holes diameter eg. 2.00, 1.00, 0.50, 0.212, 0.100 and 0.053 mm. Attach a pan to the bottom sieve and place the nest of sieves on a rotary sieving machine. Transfer the fraction of soil sample not retained on 5mm sieve on the 2 mm sieve and place the cover lid. Fasten the sieves tightly into the sieve shaking assembly. Run the sieve shaking machine for 10 minutes. Remove the soil fraction on each sieve and in the pan attached to the bottom sieve with the help of brush and weigh each fraction. Similarly treat the other sample, weigh the fraction collected on each sieve. Take the mean of aggregates collected on sieving in duplicate and tabulate the results in Table 1.10.1.1.

1.10.1.4 Calculation

From the data obtained from dry sieving, one may estimate mean weight diameter (MWD), geometric mean diameter (GMD), log GMD, log standard deviation (GG), and % aggregates greater than 0.212 mm or any other desired diameter. See section 1.10.1.1 for the definition of each parameter.

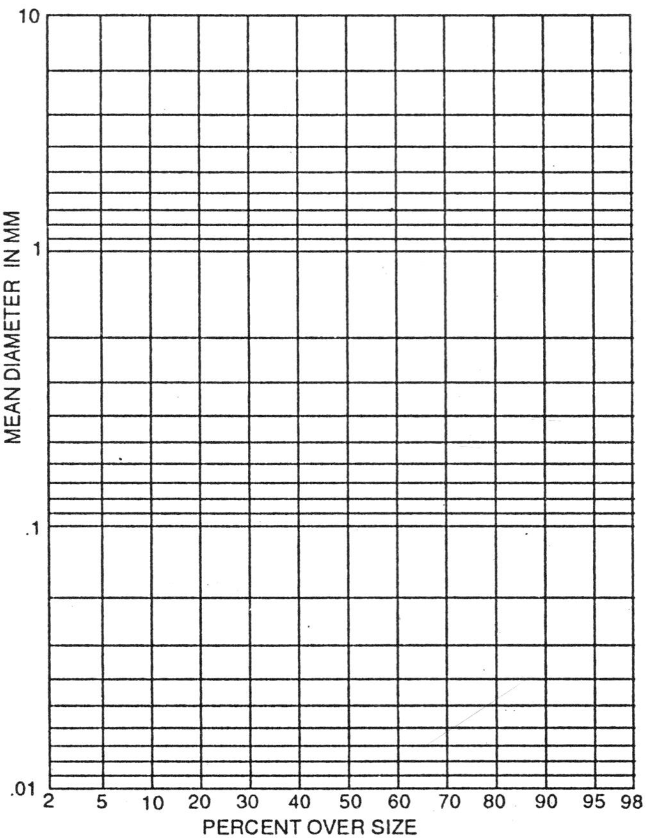

Figure 1.10.1.1 Log probability graph for plotting cumulative percent aggregate size distribution by weight (percent over size) corresponding to mean aggregate diameter. On using log scale for mean diameter and probability scale for percent over size.

Table 1.10.1.1.

Sieve Size Class mm	X Mean diameter mm	W Mass of aggregates g/100 g sample	P Cumulative percentage %	$\bar{X}_i \times W_i$	$W_i \times \log \bar{X}_i$
8.00 - 5.00	5.00				
5.00 - 2.00	2.00				
2.00 - 1.00	1.00				
1.00 - 0.50	0.50				
0.50 - 0.212	0.212				
0.212 - 0.100	0.100				
0.100 - 0.053	0.053				
0.053 - 0.000	0.000				
	Total	100.0			

1.10.1.5 Comments

There is close relationship between size distribution of aggregates by dry sieving and the susceptibility of the soil to erosion by wind.

The mass of sample size for dry sieving depend on the discretion of the investigator. However, it should not be less than 100 g.

Manual sieving introduces a nonconsistent factor which can cause variation in the result. The use of rotary sieve shaking machine is desirable for following reasons.

1. It is a most consistent dry sieving method.
2. It is not subject to a personal factor.
3. It gives fairly consistent results irrespective of the size of soil samples used.
4. It causes less break down of clods than by mechanical sieves.
5. It eliminates clogging of sieves.

1.10.2 Aggregate size distribution by wet sieving - Black (1965) & Russel (1949)

1.10.2.1 Principle : as described in section 1.10.1.1.

1.10.2.2 Equipments and reagents

Nests of sieves 20 cm dia & 5 cm height with screen opening of 5.0, 2.0, 1.0, 0.5, 0.212 and 0.053 mm. In addition one sieve with screen opening of 8 mm.

Sieving machine which raises and lowers the nests of sieves through water approximately 30 times per minute.

Buchner funnel, suction flask, suction pump and oven for drying soil sample and aggregates.

1.10.2.3 Procedure

1. Collection and preparation of soil sample. Collect the sample when soil is at friable consistency. Break the clods larger than 8 mm by hand so that it passes through the 8 mm sieve. Thereafter, pass the sample through 5 mm sieve breaking larger clods just to pass through 5 mm sieve. Sieving of sample through sieve with 5 mm screen is required only when diameter of sieve with 5 mm screen is different than the diameter of nests of sieves. Then spread the soil passing through 8 mm or 5 mm sieve in thin layers under a shade for drying in the air. Weigh 50 g of air dried sample in triplicate and transfer one of the sample to an oven for drying to a constant weight at 105°C. Keep the other two samples for wet sieving in duplicate.

2. Transfer of soil sample on screen of the top sieve and its treatment prior to sieving in water. Arrange the nests of sieves in descending order of their screen holes diameter eg. 5.0, 2.0, 1.0, 0.5, 0.212, 0.100 and 0.053 mm and place the receiving pan at the bottom. Then spread one of the sample weighing 50g on the screen of top sieve. Spray about 5 ml of water on the soil surface with a handy mist sprayer. Leave the sample for 15-30 minutes to come in equilibrium with water applied as a spray. After the lapse of 15-30 minutes, spray again about 5ml of water

on the soil surface and let the soil come in equilibrium with the water applied as a spray for 15-30 minutes.

Now remove the receiving pan and lower the nests of sieves in the tub of the sieve shaker and fasten it properly. Then fill the tub with salt free water ($EC_{25} < 10^{-5}$ mho/cm) at 25°C until the level of water in the tub reach just below the screen of top sieve when it is at the highest position. Then, lower the sieve so that the soil sample in top sieve becomes wet with the water in the tub. Allow 10 minutes time for the water to wet the soil sample. Bring the nests of sieves to the initial position. Then adjust the level of water in the tub so that screen in the top sieve becomes just covered with water in its highest position.

3. Wet sieving of soil sample. Now switch on the sieving machine to raise and lower the nests of sieves through water approximately 30-35 times per minutes. Run the sieving machine for 10 minutes. Remove the nests of sieves from the tub and drain off the water from the tub.

4. Drying the aggregates in an oven. Separate the sieves from each other. Then invert each sieve on to a paper towel. Fold the ends of paper up and clip the ends to the bottom rim of the sieve with a paper clip. Place the sieve in the oven for 24 hours to make the soil and paper towel dry. Then remove the sieve along with the paper towel. Remove the paper clip and tap on the screen lightly so that the soil from the screen falls on the paper towel. Brush the screen to remove any soil adhering to the screen and collect it on the same paper towel. Transfer the soil collected on paper towel into a preweighed aluminium can. Transfer the aluminium can into an oven at 105°C and dry the sample to a constant weight. Record the mass of oven dry aggregates against the corresponding screen diameter in a tabular form.

Alternatively, after 10 minutes of wet sieving, drain off the water from the tub. Then, remove the nest of sieves, and allow it to drain for several minutes. Thereafter, transfer the soil resting on each screen with a stream of distilled water into a Buchner funnel lined with a preweighed filter paper. Place the Buchner funnel on a suction flask connected to a suction pump. Switch on the suction pump for the partial removal of water from the soil. Then remove the filter paper along with the soil from the Buchner funnel and transfer it into an aluminium can. Place the aluminium can with the soil into an oven set at 105°C for 24 hours. Weigh the sample after drying and record the mass of soil aggregates corresponding to the screen diameter of sieve in a tabular form.

After tabulation of the mass of soil aggregates, estimate MWD, GMD, and plot the mean diameter on a log scale and the corresponding cumulative percentage of soil aggregates on a probability scale, and estimate the value of log GMD and the log standard deviation of the aggregates.

5. Dispersion of water stable aggregates retained on the different sieves dur ing wet sieving. If % aggregate stability is to be estimated, transfer the oven dry aggregates from all the sieves into a dispersion cup. Add 10 ml of 5% sodium hexametaphosphate solution in Ca saturated soil or 10 ml of 4% sodium hydroxide

solution in acidic soil. Then add distilled water to fill the dispersion cup upto 3.75 cm of the dispersion cup rim. Then, run the mechanical stirrer or homogenizer for 10 minutes. Pour the dispersed soil suspension on to a 0.25 or 0.212 mm sieve screen. Wash the residue on sieve by a stream of tap water. Place the sieve in an oven at 105°C for 24 hours. Collect the oven dry sand and weigh to an accuracy of two places after decimal. Estimate the % aggregate stability by use of equation given below :

$$\% \text{ Aggregate Stability} = \frac{\{(\text{Mass of aggregates} + \text{sand}) - (\text{mass of sand})\}100}{\text{Mass of soil sample} - \text{mass of sand}}$$

1.10.2.4 Calculation

Estimate the parameters as per definitions given in section 1.10.1.1.

1.10.2.5 Comments

Soils with high water stable aggregates will have a high proportion of large pores persisting in the soil when rain or irrigation increase the water content of the soil.

Large pores in the soil generally favour high infiltration rate, good tilth and adequate aeration for plant growth. Immediately after cultivation most soils contain an abundance of these large pores. Their continued existence in the soil depend on the stability of clods and aggregates.

Great care is required in the sampling procedure in order to minimize variation from sample to sample. When lumps of soil 73 mm are broken, it should pass 8 mm sieve but major portion be retained by 5 mm sieve. If major portion of the clod passes through 5 mm sieve, it will decrease the value of mean weight diameter.

Sodium hexametaphosphate is very effective in replacing Ca^{2+} by Na^{+} in initially Ca^{2+} saturated soil. Sodium hydroxide is a good dispersing agent in soils containing large amount of hydrogen ions.

In wet aggregate analysis, direct immersion of dry soil in water at atmospheric pressure causes great disruption of large aggregates into smaller aggregates. A much smaller degree of disruption is caused when soils are wetted slowly under tension by spraying with a fine mist.

The effect of tillage, cultural and manurial practices on soil aggregation may be evaluated by determining MWD, GMD, log GMD and % aggregates greater than a certain mean diameter. The above mentioned parameters are influenced to a varying degree by the proportion of different size aggregate. The variation caused by sampling should be minimized. The coefficient of variation should be preferably less than 10%.

REFERENCES

Black, C.A. (ed.) 1965. Methods of Soil Analysis. Vol. 1. Am. Soc. Agron, Madison, Wisconsin. U.S.A.

Dakshinamurti, C. and R.P. Gupta. 1968. Practicals in Soil Physics, Indian Agric. Res. Inst. New Delhi-12.

Richards, L.A. (ed.) 1954. Diagnosis and Improvement of Saline and Alkali Soils. U.S.D.A. Hand Book 60. Oxford & IBH Publishing Co., New Delhi.

Russel, M.B. 1949. Methods of measuring soil structure and aeration, Soil Sci. 68 : 25-35.

1.11 HYDRAULIC CONDUCTIVITY OF A DISTURBED SOIL SAMPLE BY CONSTANT HEAD METHOD-BLACK (1965)

1.11.1 Principle

Water flow through a column of soil takes place in accordance with the equation.

$$\frac{Q}{t} = KA \frac{\Delta H}{\Delta L}$$

where Q is the volume of water passing through the soil column in time (t), A is the cross sectional area of the soil column, K is the average hydraulic conductivity in the soil interval (ΔL) over which there is a hydraulic head difference (ΔH). Solving for hydraulic conductivity K, we have

$$K = \frac{Q}{t} \times \frac{\Delta L}{\Delta H \times A}$$

ΔH must be measured from the surface of water standing on the soil to the elevation at which water will stand during the flow test in a riser or manometer connected at the bottom of the soil column. The length of the soil column ΔL should be measured during or after water flow and not when the soil is dry.

1.11.2 Equipments

Glass buchner funnel with stand, burette stand, receiving beaker, graduated cylinder, vernier scale, stop watch.

1.11.3 Procedure

Place a glass Buchner funnel (8-10 cm in dia) on a tall tripod stand or else on a wooden stand provided with a hole less than the diameter of Buchner funnel. Place a burette stand just behind the Buchner funnel to hold an inverted 1l capacity volumetric flask with its mouth positioned inside the Buchner funnel.

Take a pair of Whatman No. 44 paper 12 cm in diameter and cut it evenly of the size of inner diameter of Buchner funnel at the base. Place one of the filter paper at the bottom of Buchner funnel. Place a plastic scale across the Buchner funnel and resting a vernier calipre at the plastic surface, measure the depth in cm from the plastic upper surface up to the surface of filter paper and designate the depth as A.

Transfer 150-200g of air dry < 2mm soil sample into the Buchner funnel spreading the soil uniformly. Tap the sides of the Buchner funnel for uniform spread of the soil sample. Then, place the second filter paper, cut of the size of internal diameter of the Buchner funnel, on the uniformly spread soil sample. Again place a plastic scale across the Buchner funnel. Then, place a vernier calipre on scale surface and measure the depth in cm from scale surface up to the surface of filter paper and designate it as B.

Now, fill 1l volumetric flask with tube well water of the same area or with distilled water. Place a strong paper on its mouth giving support with the fingers and set the volumetric flask into the burette stand clips in such a way that the volumetric flask mouth is positioned inside the Buchner funnel. Adjust the volumetric flask mouth close to the filter paper placed on the soil surface inside the buchner funnel. Remove the paper from the mouth of vol. flask to allow water to fill the Buchner funnel up to a certain height.

Place a beaker of convenient size below the Buchner funnel drain outlet to collect water flowing across the soil column.

Lift the mouth of volumetric flask if it is too near the filter paper placed on the soil surface. Let the water level in the Buchner funnel just 1.5 to 2.0 cm below the rim. Place the scale again across the Buchner funnel, and measure the depth in cm from the surface of scale to the surface of water standing on the soil column by a vernier calipre and designate it as C.

When the water begins to flow through the soil column more or less at uniform rate, collect the water discharged at 15-30 minutes interval depending upon the rate of discharge. In soil samples from alkali soils, the rate of discharge of water through the soil column may be so slow, that it may be desirable to measure the discharge at an interval of one hour or more.

Take three readings on water discharge through the soil column and then find out the mean discharge in time 't' and then convert the readings into mean discharge in cc/hour.

Now, remove the volumetric flask with its mouth closed carefully from the burette stand. Siphon away the water remaining in the Buchner funnel above the soil column on filter paper. Place a scale across the Buchner funnel. Set vernier calipre on the scale and measure the depth from scale surface to the surface of the filter paper and designate it B'.

1.11.4 Calculations

$$K = \frac{Q}{t} \times \frac{\Delta L}{\Delta H \times A}$$

where

K is the hydraulic conductivity of soil in cm/hour.

Q is the discharge in cc in time 't' in hours.

ΔL is the soil depth in cm *i.e.* A–B'

ΔH is the hydraulic head in cm *i.e.* A –C.

A is the cross sectional area of the soil column which is equal to πr^2 where r is the radius of soil column and π is equal to 22/7.

1.11.5 Comments

The hydraulic conductivity is affected by the temperature of the water. The results may be corrected to a standard temperature by use of the solution.

$$KST = KT\left(\frac{NT}{NST}\right)$$

Where K and N are the hydraulic conductivity and viscosity respectively and subscript T and ST denote the temperature of measurement and the standard temperature respectively. The viscosity of water is a function of temperature is usually taken from standard table.

When using the constant head method it is not necessary to standardize the time or the volume of water that is passed through the soil sample. It is suggested that about 100 cc of water be passed through the sample provided that this does not take longer than 15-30 minutes time. This would, however, not be applicable to alkali soils in which the hydraulic conductivity may be as low as .01 cm/ hour.

Intrinsic permeability to water is related to hydraulic conductivity. Hydraulic conductivity may be converted to intrinsic permeability by multiplying by the ratio η/dwg, where η is the viscosity of water at the recorded temperature dyne sec cm^{-2} (poise), dw is the density of water g/cm^3, g is acceleration of gravity 980.7 cm sec^{-2}.

The limits of hydraulic conductivity of different classes and permeability of the soil to water are presented in the table 1.11.1.

Smith and Browning (1946)* described six classes of permeability based on hydraulic conductivity which is presented in the Table 1.11.2.

Table 1.11.1 Hydraulic conductivity and Permeability classes

Classes	Hydraulic Conductivity cm/hour	Permeability cm^2
very slow	< .125	$< 3\times10^{-10}$
slow	.125–0.5	$3\times10^{-10} - 15\times10^{-10}$
moderately slow	.5–2.0	$15\times10^{-10} - 60\times10^{-10}$
moderate	2.0–6.25	$60\times10^{-10} - 170\times10^{-10}$
moderately rapid	6.25–12.50	$170\times10^{-10} - 350\times10^{-10}$
rapid	12.50–25.00	$350\times10^{-10} - 700\times10^{-10}$
Very rapid	> 25.00	$> 700\times10^{-10}$.

Table 1.11.2 Permeability classes based on hydraulic conductivity of soils

Permeability class	Hydraulic conductivity cm hr^{-1}
Extremely slow	< 0.0025
Very slow	0.0025 – 0.025
Slow	0.025 – 0.25
Moderate	0.25 – 2.5
Rapid	2.5 – 25.0
Very Rapid	>25.0

REFERENCES

Black, C.A. 1965. Methods of Soil Analysis Vol. 1 American Society of Agronomy, Madison. Wisconsin USA.

Dakshinamurti, C. and R.D. Gupta. 1968. Practicals in Soil Physics. Indian Agric. Res. Inst. New Delhi-12.

Fireman, M. 1944. Permeability measurements on disturbed soil sample. Soil Sci. 58 : 337–353.

Reeve, R.C. 1957. The measurement of permeability in the laboratory Agronomy Journal. 7 : 414-419.

Richards, L.A. (ed) ; 1954. Diagnosis and Improvement of Saline and Alkali Soils. USDA Hand Book 60. Oxford -IBH Publishing Co. New Delhi.

Smith, R.M. and D.R. Browning. 1946. Some suggested laboratory standards of sub-soil permeability. Proc. Soil Sci. Soc. Am. 11 : 21-26.

1.12 INFILTRATION RATE OF SOIL IN THE FIELD-BLACK (1965)

1.12.1 Principle

Infiltration rate is the rate of water entry into the soil when water covers the soil surface at a shallow depth and downward flow into and through the soil is non-divergent.

For practical purpose, the subsidence rate of its free water surface in a large basin is taken as a measure of infiltration rate. It has the dimension of velocity (LT^{-1}) and is normally expressed in cm/hour.

The effect of divergent flow increases as the ponded area decrease. If small basin or cylinder are used, it is difficult to determine the true infiltration rate unless divergent flow is minimized by ponding water in a gaurded ring or border area around the plot or cylinder.

The cylinder method of Musgrave (1940) is more commonly employed with a gaurd ring to minimize the effect of divergent flow in the determination of infiltration rate.

A metal cylinder of 30 cm dia is driven into the soil to a depth of 10 cm. Then, another metal cylinder of 50 cm dia is driven to the same depth around the first cylinder. The soil surface inside the two cylinders are covered with cloth or heavy paper to protect soil surface from being puddled at the time of shallow flooding with irrigation water.

The water is first added to outer ring to a depth of 5 cm and then the smaller cylinder is filled with the irrigation water upto 5 cm or more as desired. Thereafter, the rate of subsidence of water level inside the smaller cylinder is measured at certain time interval untill the subsidence of water level becomes constant. Then, the infiltration rate as well as the cumulative infiltration rate is plotted corresponding to the time interval.

1.12.2 Equipments

Metal cylinders - 30 & 50 cm in dia, 25 cm in length, bottom edge sharp, fabricated from 16 gauze iron or galvanized sheet with hook gauze.

Metal sheet plate - 60 cm in dia, 1.25 cm thick with tugs for keeping steel plate on the cylinder.

Sliding weight hammer - 20 to 30 kg.

Stop clock and thermometer.

1.12.3 Procedure

1. Selection of site. Select a representative area for the purpose of measurement. Also take soil samples from the soil profile to describe texture, structure, water content and bulk density with depth, pH and adsorbed Na from 0 to 30 cm depth. Record the crop and its stage of growth, characteristics of soil surface which may have effect on infiltration rate. Also select the representative sites for replicated tests.

2. Installation of cylinder infiltrometer. Set a 30 cm dia cylinder in place and press it firmly into the soil. Then, place the driving metal plate on the metal cylinder and drive the cylinder into the soil by tamping the metal plate with the driving hammer. Drive the cylinder vertically and uniformly. Use carpenter's level to check uniform driving. There should be also good bond between cylinder wall and the soil. Drive the cylinder upto a depth of 10 cm. If the cylinder enters the soil at an angle, remove the cylinder and resett in another location. Tamp soil into the space between the soil column and the cylinder. If the space is greater than 0.3 cm, the cylinder should be resett.

3. Installation of buffer cylinder. Place 50 cm dia buffer cylinder around the 30 cm measuring cylinder so that the buffer cylinder wall remains 10 cm away from the measuring cylinder. Place the buffer cylinder firmly into the soil. Then, place the 60 cm dia. metal plate on the buffer cylinder and drive the cylinder into the soil up to a depth of 10 cm by tamping the plate around its circumference with the driving hammer. The driving of the buffer cylinder into the soil may not be as perfect as the driving of the measuring cylinder.

As a substitute of buffer cylinder, one may construct a buffer pond by construction of a 15 cm high dike around the measuring cylinder. The inside toe of the dike be atleast 15 cm away from the cylinder. Avoid disturbing the soil inside the dike.

When cylinder infiltrometer is not available, circular dike 15 to 20 cm high enclosing 30 cm diameter soil surface for the study of rate of subsidence level on ponding water at certain time interval in the presence of buffer pond around the enclosed area may also prove to be satisfactory.

4. Application of water into the buffer pond and into the cylinder infiltrometer. The soil surface enclosed by the buffer area and the cylinder infiltrometer should be protected from being disturbed or puddled at the time of water application for ponding. It is desirable to spread cloth or thick paper on soil surface before filling it with water. The protective material be removed after the ponding of water.

First fill the buffer pond outside the cylinder infiltrometer to a depth of 5 cm, and maintain the depth throughout the period of observation.

Immediately after adding water to the buffer pond, fill the cylinder infiltrometer with water to a depth of 5 to 7.5 cm and then remove the soil surface protective material and make measurements on rate of water subsidence level by a hook gauze or staff gauze. The cylinder edge may be taken as the reference level in absence of any other reference level provided in the cylinder infiltrometer. Mark the cylinder so that the subsequent measurement can be made from the same point.

5. Measurement of water level. The observations on subsidence of water level may be taken at the end of 1, 3, 5, 10, 20, 30, 45, 60, 90 and 120 minutes and thereafter at hourly interval. As a general rule, the intake between the measurements should not exceed 2.5 cm. Continue measurements until the rate of intake is almost constant.

6. Refilling the cylinder infiltrometer and the buffer pond. When the water level in cylinder infiltrometer falls by 2.5 to 5.0 cm, refilling of the cylinder infiltrometer be quickly done without any loss of time to its initial level. Record the water level and time just before filling, and the water level after filling. Immediately, thereafter, the buffer pond be also filled approximately to the initial level.

If a constant water level in the cylinder infiltrometer or basin is maintained by a float value, measure the rate of subsidence of water level in the supply tank by a hook gauze or manometer. When water subsidence is recorded in cc. Then the volume recorded in cc be divided by cross sectional area of the cylinder infiltrometer i.e. 706.5 cm^2 will give water subsidence in cm.

7. Replications. 6 to 10 separate measurements may be taken for working out a satisfactory infiltration rate.

8. Graphical Presentation. Prepare a plot of cumulative water intake rate versus time on a graph paper. It should give a smooth curve, the slope of which will give infiltration rate at different time. Prepare another plot of infiltrate rate versus time. It will give a curve showing the change in tendency of the soil to absorb water with time. The area under the curve represents the capacity of the soil to absorb water up to the point in question. This infiltration capacity curve is normally the final result.

1.12.4 Calculation

Estimate infiltration rate in cm/hour at different time and not the initial infiltration and the final intake rate.

1.12.5 Comments

The infiltration studies are confined to small plots or else to cylinder infiltration relatively small in diameter. Under such condition divergence flow of water has to be minimized by the construction of a buffer pond in the f· .n of a gaurd ring or a border area. Still the reliability of the result will depend on the permeability of the sub-soil. If permeability increases with depth, the divergence flow will be minimized and the result will be satisfactory. However, when permeability decreases with depth, divergence flow will still be considerable and the result may be misleading.

The infiltration studies related to leaching operation be preferably undertaken in test plots measuring 1.8m×3.6 m with 90 cm wide buffer pond around the entire test plot.

The water movement in soil is also influenced by the level and the composition of salt. It is, therefore, necessary that water likely to be used either for irrigation or leaching be used for studies on infiltration rates.

The depth of filling of water in cylinder infiltrometer be equal to the depth of each irrigation of crops grown, and the depth of ponding for infiltration studies in relation to leaching operation should be equal to the full depth of leaching water for each leaching period in the test plot.

When the depth of initial water level is 5 cm, refilling be done without any loss of time when water level falls by 2.5 cm. However, when depth of flooding is 7 to 10 cm, refilling be undertaken on 5 cm subsidence of water level.

Normally subsidence of water level by evaporation loss is negligible. When evaporation loss is likely to be appreciable, water surface may be covered with a thin film of oil.

The density and viscosity of water changes with the temperature, and so the temperature of water in bulk be measured at the time of adding water in cylinder infiltrometer as temperature is likely to influence water movement in the soil.

Both high and low infiltration rate result in soil, crop and water management problem. Both irrigation and water conveyance efficiency is low when infiltration rate is high. Low infiltration rate result in greater run off, more erosion, poor aeration and shallow irrigation. The infiltration rate is low, in alkalis soils, in fine textured and compact soil as well as in soils with impermeable layer or hard pan. Permeability classes have been recognized but the limits of infiltration rates has not been worked out.

REFERENCES

Black, C.A. 1965. Methods of Soil Analysis. Vol. I. Am. Soc. of Agron. Madison, Wisconsin, USA.

Dakshinamurti, C. and R.P. Gupta. 1968. Practicals in Soil Physics. Indian Agric. Inst. New Delhi-12.

Musgrave, C.W., 1935. The infiltration capacity of soils in relation to the control of surface run off and erosion. Amer. Soc. Agron. J. 271 336-345.

Richards, L.A. (ed). 1954. Diagnosis and Improvement of Saline and Alkali Soils. USDA Hand Book No. 60, Oxford & IBH Publishing Co., New Delhi.

1.13 WATER CONTENT OF SOIL

1.13.1 Water content of soil by gravimetric method

1.13.1.1 Principle

In gravimetric method, the determination of water content in moist soil sample is based on the loss in weight of moist soil on drying. For drying, soil sample is commonly placed in an oven at 105°C for 24 hours to a constant weight. Heat supplied at the above temperature is effective in evaporating water held in soil pores and on soil surface of small soil samples spread in thin layers inside an open container. The organic matter is, however, not lost from the soil. The water content of soil is expressed in percent which is the loss in weight of moist soil on oven drying expressed as percentage of oven dry mass of soil. It is usually designated as P_w.

However, water content of soil on volume basis i.e. P_v, or else depth of water per unit soil depth i.e. P_d are considered more useful for irrigation purposes, but for estimating P_v and P_d, the volume of moist soil escavated be known in addition to depth of soil important for irrigation. For the determination of P_v, undisturbed soil sample be collected by use of hollow metal cylinder with 7.5 cm i.d. and 7.0 cm in length or else the volume of moist soil excavated by spoon be determined by filling the excavated portion of the field with measured volume of free flowing fine sand.

1.13.1.2 Equipment and Reagent

Equipment

(*i*) Soil auger. Preferably a stainless steel auger with closely fitting hollow stainless steel cylinder with marking at 5, 10, 15, 20 and 25 cm from lower end upward. The cylinder is closed at the upper end leaving the centre open slightly greater than the diameter of extension rod with a slit downward. During drilling of soil auger, the cylinder is held in position by a metal point at the upper end above the auger cutting groove on the extension rod. The metal point gets inside the slit.

(*ii*) Aluminium can or Metal container with tight fitting lid 6. cm. i.d. and 4 cm in height.

(*iii*) Metal hollow cylinder. 7.5. cm i.d. and 7.0 cm in height for taking undisturbed soil sample.

(*iv*) Metal plate 1.25 cm thick, with metal rod fixed at the centre to slide circular weight with hole at the centre.

(*v*) Single pan balance weighing up to 1 kg with precision of 0.1 g.

(*vi*) Oven with proper temperature control to maintain 105°C inside.

1.13.1.3 Procedure

1. Collection of undisturbed soil sample, and its drying in an oven. Select a level surface free from any extraneous material, and place a tared hollow metal cylinder firmly on the level surface. Then, place the thick metal plate on it. Now, put sliding weight on the iron rod and let it slide down to drive the hollow metal cylinder uniformly into the soil. Similarly drop again, the sliding weight until the metal cylinder is driven to a depth of 7.0 cm.

Remove the soil outside the periphery of the cylinder up to a depth of 8 cm. Then, place the blade of khurpi or shovel below the bottom of the metal cylinder, and lift the metal cylinder filled with soil up to the cylinder surface. Cut a slice across the cylinder bottom and place on filter paper at the cylinder bottom to hold the soil. Keep the metal cylinder with soil on a wire gauze kept inside a tray. Cover the tray with a polythene sheet to minimize evaporation and carry the sample to the laboratory.

Weigh the hollow metal cylinder filled with moist soil along with the filter paper and record the mass as well as the cylinder number. After weighment place the cylinder on wire gauze and then transfer the metal cylinder with wire gauze into an oven at 105°C. Let the undisturbed soil sample dry continuously for 4 days or longer to a constant weight. Thereafter, allow it to cool to room temperature. Thereafter, record the mass of cylinder plus oven dry soil.

With the help of a small wooden mallet, strike the soil core inside the cylinder to push out the soil core. Wash the hollow metal cylinder with tap water. Then, take the mass of empty hollow cylinder and measure its internal diameter and height in cm with a Vernier calipre.

2. Collection of disturbed soil sample and its drying in an oven. Select a level surface free from extraneous material at the location site. Drive the soil auger to the desired depth and then take out the soil auger.

Push back the close fitting stainless steel hollow cylinder exposing the auger cutting groove filled with moist soil. Remove the soil with the fingers and transfer about 25-30 g soil sample immediately into a preweighed aluminium can/steel container. Place the tight fitting cover lid. Then record the number of aluminium can/steel container, soil depth, date and place from where the soil has been collected.

Carry the container filled with soil sample for moisture determination without loss of time to the laboratory. Weigh the container filled with soil sample.

Remove the cover lid and place it below the container bottom and then transfer the container into an oven at 105°C to dry for 24 hours or longer to a constant weight. Weigh the oven dry soil sample after cooling in a desiccator. Record the oven dry mass of soil sample.

Similarly one may take more than one sample from the same boring with the successive increment is soil depth.

1.13.1.4 Calculation

$$P_w = \frac{(\text{mass of moist soil in g} - \text{mass of oven dry soil in g})}{\text{Mass of oven dry soil in g}} \times 100$$

$$\therefore P_w = \left\{\frac{\text{mass of moist soil in g}}{\text{mass of oven dry soil in g}} - 1.00\right\} \times 100$$

$$P_v = \frac{\text{Volume of moisture lost on drying in cm}^3}{\text{Volume of oven dry soil in cm}^3} \times 100 \text{ (in undisturbed soil)}$$

$$P_v = P_W \times \text{Apparent specific gravity of soil i.e. AS}$$

$$P_d = \frac{P_v}{100} \times \text{soil depth in cm} \quad (\text{cm } H_2O/x \text{ cm soil depth})$$

where

P_w and P_v is percent moisture in soil on oven dry mass and on volume basis.

P_d is water content in height of water column in cm per x cm soil depth.

AS is the ratio of mass of oven dry soil to the mass of water equal to the volume of soil core i.e. mass of water displaced by the soil.

When the density of water is 1.0, the numerical value of bulk density is equal to the numerical value of AS.

1.13.1.5 Comments

The heat supplied in an oven at 105°C for 24 hours is sufficient to evaporate moisture present in soil pores and on soil surface of 25-30 g moist soil samples provided over loading of oven is avoided.

The drying at 105°C for 24 hours removes capillary and hygroscopic water only. At 105°C there is no loss of organic matter.

Even when the soil contains only hygroscopic water, the sample be dried for 24 hours at 105°C. The oven be not loaded when drying is in progress half way or more. With the decrease in the content of moisture in soil sample, water is held with greater tenacity i.e. with greater suction or negative pressure, and more energy is required for its removal.

When the oven is fully loaded with moist samples, the total energy required for drying the soil samples completely may exceed the heat energy supplied at 105°C for 24 hours. Under such a situation, the samples have to be dried for a longer period until constant weight is attained.

The temperature inside the oven should not be raised above 105°C, because it may result in loss of organic matter making the result on moisture content unreliable.

An oven dry soil is capable of adsorbing water molecules from a humid atmosphere. Therefore, oven dry soil be cooled in a desiccator before weighment.

The moisture content in soil may show large variation even in the close proximity. It is, therefore, preferable to take 4 or 5 samples from each location.

For the determination of volumetric water content of soil, accurate measurement of volume of soil sample is necessary. Therefore, undisturbed soil sample with core sampler would be preferable.

REFERENCES

Black, C.A. 1965. Methods of Soil Analysis. Vol. 1. Amer. Soc. Agron. Madison, Wisconsin, U.S.A.

Dakshinamurti, C and R.D. Gupta 1968. Practicals in Soil Physics. Indian Agric. Res. Inst. New Delhi-12.

1.13.2 Water content of soil by burning alcohol–Gardner (1965) and Singh & Rajput (1970)

1.13.2.1 Principle

Burning alcohol after wetting moist soil with alcohol raises soil temperature to about 130°C and higher in some cases. Heat produced on burning alcohol is dissipated in quickly evaporating soil moisture. The soil gets quickly dried, and thus, it enables determination of water content of the soil in about 20-30 minutes time. The method is particularly suitable for mineral soils low in organic matter as there is some risk of oxidation of organic fraction.

1.13.2.2 Equipments/Reagents

Ethyl alcohol technical grade

1.13.2.3 Procedure

Weigh a clean dry aluminium can of the size of about 6.5 cm in diameter and 4.5 cm in height with a cover lid. Transfer about 10g of moist soil into the aluminium can. Put the cover lid and take the mass of aluminium can plus moist soil in g.

Now, place the aluminium can on a wire gauze placed on a tripod stand. Remove the cover lid. Pipette 10 ml ethyl alcohol into the aluminium can wetting the soil sample uniformly. Ignite the alcohol dropped in the aluminium can. When all the alcohol has burnt completely, let the aluminium can cool. Place the cover lid for some time to estinguish any flame remaining.

Remove the cover lid again and pipette 10 ml alcohol drop by drop wetting the soil mass again uniformly. Place the aluminium can on the wire gauze and ignite the alcohol when alcohol added has burnt completely. Cool for a while on wire gauze and then transfer the aluminium can into a desiccator to cool. Thereafter, take the mass of aluminium can + dry soil.

1.13.2.4 Calculations

$$\text{\% Moisture in soil} = \frac{W_2 - W_3}{W_3 - W_1} \times 100$$

$$= \frac{\text{Loss of moisture on drying in g}}{\text{mass of dry soil in g}} \times 100$$

where,

W_1 is the mass of aluminium can in g

W_2 is the mass of aluminium can plus moist soil in g.

W_3 is the mass of aluminium can plus soil dried by burning alcohol.

1.13.2.5 Comments

In this method moisture is removed from soil sample at temperature in the range of about 130°C but in oven dry method soil sample is dried at 105°C.

There is some risk of burning of organic fraction when alcohol added to the soil is burnt. Consequently, it may give slightly higher moisture content than obtained by oven dry method. Singh *et. al.* (1970) worked out a regression equation given below to estimate Y, moisture content on oven dry basis from x, the moisture content by spirit burning on a loam soil.

$$Y = 0.989x - 0.80 \qquad r = -0.999^{xx}$$

The moisture content by burning alcohol takes about 30 minutes. However, the method is satisfactory for mineral soils low in organic matter.

The flame in aluminium can must be completely estinguished before adding alcohol for the second time. Do not add alcohol directly from the bottle. Eliminate risk of alcohol catching fire while adding into the aluminium can.

REFERENCES

Singh, M., S.B. Hukkeri & R.K. Rajput. 1970. Spirit Burning and Papadakis method for the determination of soil moisture as substitute for oven dry method. Indian J. Agric. Sci. 40 (1) : 71-78.

Black, C.A. 1965. Methods of Soil Analysis Vol. 1. Amer. Soc. Agron. Madison Wisconsin, U.S.A.

Dakshinamurti, C. & R.P. Gupta. 1968. Practicals in Soil Physics. Indian Agric. Res. Inst., New Delhi-12.

1.13.3 Water Content in soil by Papadakis method–Papadakis (1941) and Singh & Rajput (1970)

1.13.3.1 Principle

A 100 ml volumetric flask filled with 25g moist soil sample and distilled water up to 100 ml mark is weighed and designated as W_1 followed by weighment of the same flask with distilled water alone up to 100 ml mark designated as W_2.

The $W_1 - W_2$ is found less than 25 g and it is numerically equal to $\frac{W_3}{1} - \frac{W_3}{D_s}$

where W_3 is mass of soil present in 25g moist soil and D_s is the particle density of soil in g cm^{-3}. The loss in weight is thus equal to the volume of water displaced by the soil.

With the increase in the content of soil in 25 g moist soil sample, there will be increase in the value of $(W_1 - W_2)$ which will be accompanied with the decrease in water content in moist soil as well as in percent moisture in soil sample.

The inverse relationship between $W_1 - W_2$ and percent moisture in soil sample could be established by plotting the pairs of values on a graph paper to draw a calibration curve and by fitting a regression equation to work out numerical relationship between the two parameters. This will enable determination of percent moisture in moist soil sample from the value of $(W_1 - W_2)$ designated as x.

1.13.3.2 Equipments and Reagents

Equipments. 100 ml capacity volumetric flask and single pan balance with precision of 0.1 g.

1.13.3.3 Procedure

Weigh 25.0 g moist soil sample in a tared aluminium can and transfer the moist soil quantitatively into a 100 ml volumetric flask. The soil adhering to the aluminium can be also transferred to the same 100 ml volumetric flask with 5-10 ml portions of distilled water. Make the volume to about 50 ml and shake the contents to expel any air entrapped. Thereafter, add distilled water to make the volume to 100 ml mark. Allow 10 minutes time for the soil particles to settle. Check the level of water meniscus. Then weigh on a pan balance with a precision of 0.1g. Record the mass as W_1.

Now empty the volumetric flask by pouring out the contents. Then, wash it 2-3 times with sufficient tap water to remove any soil remaining in the flask. Rinse the volumetric flask with distilled water. Thereafter, fill it with distilled water up to the 100 ml mark. Place the glass stopper. Wipe off the outer surface of flask with a clean and dry cloth. Then, weigh the water filled 100 ml volumetric flask on a pan balance and record it as W_2.

Transfer about 30g of fresh moist sample from the same lot into a tared aluminium can. Weigh aluminium can plus moist soil, and then transfer the aluminium can with soil into an oven at 105°C for 24 hours to dry to a constant weight. Then weigh aluminium can plus oven dry soil after cooling in a desiccator. Determine percent moisture in soil sample on oven dry mass basis.

Calibration curve. Following the procedure described above, take observations on $W_1 - W_2$ and percent moisture in soil sample on oven dry mass basis by collecting soil samples from the same soil varying in moisture content. Tabulate the values of $W_1 - W_2$ *i.e.* "x" coresponding to percent moisture in soil i.e. "y".

Plot the pairs of values on the two parameters on a linear scale graph paper with "x" scale on the abscissa and "y" scale on the ordinate. Draw a calibration curve after fitting a linear regression equation. Also work out the regression coefficient and the test of significance. This would enable estimation of percent moisture in soil sample from the value of x i.e. $(W_1 - W_2)$.

1.13.3.4 Calculation

Illustration

(*i*) Value of x *i.e.* $W_1 - W_2 = 10.0$ g

(*ii*) Regression equation

$$Y = 82.8 - 5.10\,x \qquad r = -0.892^{xx}$$

Where Y is percent moisture in soil sample on oven dry mass basis.

Substituting the value of x in above equation, we have

$$y = 82.8 - 5.10 \times 10 = 82.8 - 51.0 = 31.8$$

1.13.3.5 Comments

The regression equation define the relationship between the difference between $W_1 - W_2$ i.e. "x" and percent moisture in soil sample on oven dry basis *i.e.* "y". The regression equation is most likely to vary with the mineralogical composition and the content of organic matter.

One such regression equation as worked out by Singh *et. al.* (1970) has been cited in the illustration on calculation. The method is simple, reliable, satisfactory and rapid after the preparation of the calibration curve and the regression equation.

REFERENCES

Singh, M., S.B. Hukkeri, & R.K. Rajput, 1970. Spirit burning and Papadakis method for the determination of soil moisture as substitute for oven dry method. Indian J. Agric. Sci. 40 : 71-78.

Papadakis J.A. 1941. A rapid method for determining soil moisture. Soil Sci. 51 : 279-281.

1.13.4 Water content of soil–Prihar & Sandhu (1967)

1.13.4.1 Principle

Water displaced by 20 g moist soil in especially designed PAU moisture gauze filled with 100 ml distilled water is measured by increase in the volume of water. The content of water present in the moist soil is estimated from the following relationship.

$$W_w = D_w (D_s V_i - W_{sm}) / (D_s - D_w)$$

where,

W_w is the mass of water in g present in soil sample

W_{sm} is the mass of moist soil sample taken in g.

V_i is the increase in volume of the water on adding 20 g moist soil

D_w & D_s being the density of water and soil respectively in g cm^{-3}.

The density of water may be taken as 1.0 g cm^{-3} and the density of mineral soil may be taken as 2.65 g cm^{-3}. The density of soil may be experimentally determined by recording water displaced by 40g oven dry < 2mm soil sample in a 100 ml graduated cylinder containing 50 ml distilled water. If water temperature is taken, D_w can be recorded from physical table giving density of water at different temperature.

When V_i, D_w and D_s is known, W_w can be estimated from the relationship given above. The substraction W_w from W_{sm} will give W_s. Then the percent moisture in soil sample can be estimated by $\frac{W_w}{W_s} \times 100$.

1.13.4.2 Equipments

PAU moisture gauze or 100 ml graduated cylinder, Pipette 25 ml capacity, 50 ml burette, and Pan balance with precision of 0.1 g.

1.13.4.3 Procedure

In absence of PAU moisture gauze, take 100 ml clean and dry graduated cylinder. Weigh 20 g moist soil in a aluminium can and transfer it into 100 ml graduated cylinder. The soil adhering to aluminium can be transferred quantitatively with three 5ml portions of distilled water taken from a burette. Transfer each washing to the same 100 ml graduated cylinder. Add 10 ml distilled water to the same cylinder and stir the contents with a long glass rod.

Then add 25 ml more distilled water through the burette into the 100 ml graduated cylinder, and allow 10 minutes time for the soil particles to settle. Read the increase in the volume of water above 50.0 ml and designate it as V_i.

Now weigh 40 g oven dry < 2 mm soil sample collected from the same site, and transfer it to 100 ml graduated cylinder containing 50 ml distilled water. Stir the suspension with a long glass rod for 5 minutes and then allow 10 minutes time for the soil particles to settle. Read the volume of water displaced by the soil *i.e.* V_s and determined D_s by dividing 40 by V_s.

Take the temperature of water by a thermometer, and find out its density from physical table giving the value of density of water at different temperature.

1.13.4.4 Calculation

$$W_w = D_w (D_s V_i - W_{sm}) \;/\; (D_s - D_w)$$

where,

W_w is g water present in 20 g moist soil.

W_{sm} is the mass of moist soil sample *i.e.* 20 g.

V_i is the increase in volume on displacement of water by 20 g moist soil sample.

D_w and D_s being the density of water and soil respectively.

$$\text{\% moisture in soil sample} = \frac{W_W}{W_{sm} - W_W} \times 100 = \frac{W_W}{W_S} \times 100$$

Where W_s is the mass of soil in g in 20 g of moist soil.

1.13.4.5 Comments

In PAU moisture gauze 20g moist soil sample is dispersed in 100 ml distilled water and when the soil particles settle, the increase in the volume of water above 100 ml is recorded as a measure of water displaced by the moist soil.

The volume of water displaced by 20 g moist soil sample can be also determined by dispersing the moist soil sample in 50 ml of distilled water inside a 100 ml graduated cylinder. The moist soil sample can be dispersed by stirring with a long glass rod for 5 minutes.

The method given for the determination of particle density of the soil sample has been found by the author to be simple, reliable, rapid and reproducible.

REFERENCES

Prihar, S.S. and B.S. Sandhu, 1967. Studies on immersion method of soil moisture determination. J. Res. 4 : 157 – 161.

Prihar, S.S. and B.S. Sandhu, 1968. A rapid method of soil moisture determination. Soil Sci. 105 : 142 – 146.

Prihar, S.S. and B.S. Sandhu. 1987. Irrigation of Field Crops. I.C.A.R., New Delhi.

1.13.5 Water content of soils by resistance blocks-Bouyoucos and Mick (1940)

1.13.5.1 Principle

The resistant blocks made from gypsum, nylon or glass fibre by embedding a pair of electrodes made of stainless wire screen soldered to the upper wire lead is installed into the soil at desired depth with the electrical lead above the ground.

The electrical resistance between the electrodes varies with the moisture content in porous blocks *i.e.* resistance blocks which is in equilibrium with the soil moisture. The resistance is measured by a resistance meter such as Bouyoucos moisture meter.

The resistance block need to be calibrated in the laboratory with soil sample obtained from the place of its future installation in the field. The calibration curve give relationship between the resistance of resistance block correspending to the moisture content of the soil or its soil moisture tension. Thus, it enables determination of moisture content of the soil in which the resistance block is placed by taking measurement on the resistance of resistance block installed in the soil.

1.13.5.2 Equipments

Soil auger, resistant blocks (gypsum, nylon or glass fibre), and resistance meter such as Bouyoucos soil moisture meter.

Wooden mould with inside dimension of 5.50×3.75×2.00 cm in two equal halves held together by rubber bands are laid on the working table. Two stainless wire screen electrodes of the size 3.75×1.25 cm soldered to copper wire lead are placed inside the wooden mould 2.00 cm apart. The plaster of paris slurry is poured to fill the wooden mould. As the plaster of paris slurry dries, a rectangular block is formed in which two electrodes remain embedded with its copper lead wire of sufficient length extending out of the block.

1.13.5.3 Procedure

1. Initial testing of resistance blocks. Place the resistance blocks in distilled water. Remove the resistance blocks from distilled water after 10 minutes, and then keep the porous blocks on a table for 10 minutes. Thereafter, place the resistance blocks again in distilled water, so that it gets saturated without entrapment of air in the porous block. Now measure the resistance of each porous block after numbering them in serial order by putting tag in its lead wire. Select resistance blocks not deviating by more than 100 ohms.

2. Calibration of the resistance blocks in the laboratory. Collect enough soil sample from the desired depth of site where resistance is to be installed. Spread the soil in thin layer on a clean news paper and allow it to dry in air for a week. When the soil has become air dry remove any extraneous material seen, and crush the larger aggregates lightly with a wooden mallet. Sieve the crushed soil through a 2 mm sieve having round holes and store the soil passing through 2 mm sieve after weighment.

Construct a rectangular container from a wire screen with one side open and then line it with fine cloth of sufficient length to cover all the four sides. Place the cloth along the three side leaving one side open for filling the soil.

Take out the air dry < 2mm weighed soil and fill the rectangular container to a depth of 7.5 cm with the air dry soil. Level the soil surface and make it some what compact. Then lay the resistance blocks after weighment keeping their lead wire out of the container. Now, cover the resistance blocks to a depth of 7.5 cm with the air dry soil uniformly and make it slightly compact. Record the mass of air dry soil, fine cloth, resistance blocks and wire screen rectangular container separately. Transfer a sample of air dry soil into a tared aluminium can. Weigh the aluminium can with soil sample and transfer it to an oven at 105°C to dry for 24 hours to a constant weight to find out the oven dry mass of the soil filled into the rectangular container.

Fill a sufficiently large trough with water up to a depth of 5 cm and lower the wire screen rectangular container with soil wrapped in fine cloth inside the trough and leave it overnight to get the entire soil mass saturated with soil.

On the following day of the entire mass of soil is found saturated with water, remove the rectangular wire screen container with saturated soil out of the trough and keep the container outside exposing all the sides for evaporation.

Periodically take the measurement on electrical resistance of each resistance block embedded in the soil and loss in weight of saturated soil mass. Estimate % moisture in soil from the loss in weight on air drying corresponding to each measuremeant on resistance block by a resistance meter.

Continue taking measurement on resistance of resistance blocks and the corresponding % moisture in soil as estimated by the loss in weight until no further change in moisture content or resistance of resistance block is noticed.

3. Preparation of calibration curve. Take a semilog graph paper. Set the resistance range on the log scale i.e. on the ordinate and the range of moisture content on the abscissa in a linear scale. Thereafter, plot the resistance in ohms on the ordinate corresponding to the moisture content on the abscissa. Connect the points plotted to obtain a smooth calibration curve.

4. Installation of the resistance block in the field. With a soil auger bore a hole in the field to the desired depth after planting a crop. The hole should be such as to facilitate the placement of the resistance block.

After lowering the resistance block to the desired depth, drop the soil mass taken out to fill the hole in portion compacting it also with each filling to prevent any entrapment of air. Fill the hole uniformly upto the soil surface.

5. Determination of soil moisture content in situ without sampling. Allow 10-12 hours time to resistance block after its installation into the soil to come in equilibrium with the soil moisture. Then take measurement on resistance between the two electrodes embedded in the resistance block installed by a resistance meter in ohms. Determine the moisture content of the soil in situ without taking out any soil sample by reference to the calibration curve prepared in the laboratory with the soil collected from the same site and with the same resistance block prior to its installation in the field.

1.13.5.4 Calculation

$$\% \text{ moisture in soil} = \frac{\text{Loss in weight of moist soil in g}}{\text{Oven dry mass of soil in g}} \times 100$$

1.13.5.5 Comments

For calibration curve, the resistance block can be installed at desired depth directly in the field. As soil dries, the measurement on resistance of the porous block and moisture content could be taken.

The resistance block can be calibrated also by placing it in soil studies on permanent willing percentage and on soil moisture and tension relationship.

Bouyoucos and Mick (1940) reported the use of gypsum block for determining moisture content of the soil by recording the resistance of a pair of electrodes embedded in the gypsum block.

Ewart and Baver (1950) reported that the resistance reading of gypsum block remained unaffected by increase in salt up to 0.2% in soil, and by use of irrigation water containing up to 2000 ppm salt as NaCl.

The fibre glass unit was found to be more sensitive in the wet range, but it was found to be extremely sensitive to the presence of small amount of electrolytes.

Bouyoucos (1949) found nylon block units to be highly sensitive to moisture changes in the wet range. The resistance recorded by nylon block at saturation, field capacity and wilting point was found to be 150 ohms, 2500 ohms, and 300,000 ohms respectively. However, nylon block was found as sensitive to the presence of electrolyte as glass fibre block.

The durability of the resistance block embeded in the field is important. The gypsum block shall be subject to some degree of solubilization by water in the soil, and so its durability will be relatively lower than resistance blocks made from nylon and glass fibre.

In Hawai, gypsum blocks are installed in the field in large sugarcane plantations for scheduling irrigation in sugarcane at resistance of 5000 ohms equivalent to 4 atmosphere soil-moisture tension.

Humbert and Cornelison (1952) reported that where the resistance recorded by gypsum block was 5000 ohms at 12 inches soil depth, the depletion of available water at 6, 12, 18 and 24 inches was found to be 87, 75, 67 and 55% respectively.

There is thus a scope for use of resistance blocks for scheduling irrigation in irrigated cash crops such as sugarcane, gram on fine textured soil which release a small fraction of available water at soil moisture tension less than 1 bar.

However, on coarse textured soil which release a large fraction of available water at lower moisture tension, tensiometer would be perferable.

REFERENCES

Black, C.A. (ed). 1965. Methods of Soil Analysis. Am. Soc. Agron., Madison, Wisconsin, U.S.A.

Bouyoucos, G.J. and A.H. Mick. 1940. An electrical resistance method for the continuous measurement of soil moisture under field conditions. Michigan Agr. Expt. Sta. Tech. Bull. 172.

Bouyoucos G. J. 1949. Nylon electrical resistance unit for continuous measurement of soil moisture in the field. Soil Sci., 67 : 319-330.

Dakshinamurti, C. and R.P. Gupta. 1968. Practicals in Soil Physics. Indian Agric. Res. Inst. New Delhi-12.

Ewart, G.Y. and L.D. Baver. 1950. Salinity effects on soil moisture-electrical resistance relationships. Soil Sci. Soc. Amer. Proc. 15 : 56-57.

Humbert, R.P. and A.H. Cornelison, 1952. Hawaiian Sugar Planters Assn. Exp. Sta. Com. Annual Reports, P. 19.

WTC. 1977. Water Requirement and Irrigation Management of Crops in India. I.A.R.I. Monograph No. 4 (New Series), Water Technology Centre, Indian Agric. Res. Inst. New Delhi.

1.13.6 Water content of soil by neutron thermalization method–Gardner (1965)

1.13.6.1 Principles

High energy neutrons (0.1 to 10 MeV) emitted from a radio active susbstance such as radium-beryllium or americium-beryllium are slowed down and its direction changed by elastic colloision with atomic nuclei such as hydrogen. This process is called thermalization in which neutron energy is reduced to about the thermal energy of atom in a substance at room temperature. Hydrogen nuclei has significant property of scattering and slowing neutrons. This property is exploited in neutron method of determining water content in soil.

With soil auger hole slightly smaller than the access tube-20 gauze stainless steel or aluminium pipe 1.9 in i.d. is drilled into the soil to the desired depth. The access tube 6 to 10 feet in length with lower end stoppered is forced into the hole with 20 cm of access tube remaining above soil surface. The neutron probe with shield is placed on the access tube so that access tube fits into the hole of probe shield.

Neutron probe lead wire from slow neutron detector tube is connected to scalar or rate meter. The scalar is switched on and the standard count is taken. Thereafter, the neutron probe is lowered into access tube to the desired depth.

The fast neutron emitted by radium in neutron probe penetrate into the soil mass across the access tube. On striking hydrogen nuclei of water molecule, fast neutrons are slowed down and the density of thermalized neutrons around slow neutron detector tube increases in proportion to the increase in water content of the soil. The scalar measures the density of thermalized neutrons expressed in counts/ 3 minutes which is reduced to counts/minute.

The soil sample is collected from 15 cm radius around access tube at the probe depth for determining the moisture content of the soil on volume basis. The pair of values of relative count *i.e.* count/min divided by the standard count and the volumetric water content at varying depth are plotted on a graph paper to obtain a calibration curve.

Thereafter, the moisture content of soil can be determined any time without taking soil sample just by lowering the neutron probe into the access tube, and reading the density of thermalized neutrons in the scalar expressed in unit of counts/min. The volumetric water content is determined by reference to the calibration curve.

1.13.6.2 Equipments

(*i*) Neutron probe with shield :

The neutron probe consist of (*a*) Fast neutron source. 2.5 mC mixture of radium-beryllium or 30 mC mixture of americium beryllium.

(*b*) Slow neutron detector tube. The tube is filled with BF_3 gas, and a high positive voltage is imposed across a wire running axially through the tube centre. The thermalized neutron is absorbed by B^{10} nuclei and then it emits alpha particle. The alpha particles move to charged wire creating an electrical impulse which is amplified before being picked by the scalar or rate meter. The slow neutron detector tube is placed just above the neutron source in the probe. The probe is provided with a long lead for connecting to the scalar.

(*c*) Probe shield. The probe is kept inside a probe shield 30 cm in diameter with vertical hole in the centre slightly greater than the diameter of probe. The shield is filled with lead up to a radius of 13 cm and paraffin is filled in the remaining space.

(*ii*) Access tube. 20 gauze stainless steel or aluminium hollow pipe 6′ to 10′ in length having slightly greater diameter than the probe so that the probe could be just lowered in the access tube.

(*iii*) Soil Auger. Soil auger is needed for drilling vertical hole up to the desired soil depth slightly smaller than the diameter of the access tube.

(*iv*) Portable scalar or rate meter. Battery operated.

1.13.6.3 Procedure

1. Installation of access tube at the selected site. Drill soil auger to make a hole slightly smaller in diameter than the diameter of access tube vertically downward several inches deeper than the desired depth. Then force into the hole made a stainless or aluminium pipe having outer diameter of the size of hole. If water table is high, stopper the lower end of pipe before forcing it into the hole.

Thereafter, take out the stainless or aluminium pipe inserted into the hole and then force into the hole access tube 20 gauge stainless or aluminium pipe having outer diameter slightly bigger than the diameter of the hole dug. Stopper the lower end of the access tube before forcing it into the hole. The access tube should be forced vertically downward straight without any bend or dent. When about 20 cm of the access tube remain above the soil surface stop forcing the access tube inside the hole and cover the pipe with an inverted can.

2. Measurement of the standard count i.e. count with probe inside the shield. Switch on the scalar or rate meter to allow it to warm. The warming up period may be omitted in transistorized unit. Connect the neutron probe lead wire to the scalar. Let the neutron probe remain inside the shield and place the neutron probe with shield on to the access tube. Take the measurement on density of slow neutrons expressed in counts/3 minutes and then reduce it to counts/minute.

3. Measurement of slow neutron density on lowering neutron probe from shield bottom into access tube at regular increments in depth. Lower the neutron probe inside the access tube 15 cm below the soil surface and measure the density of slow neutrons as measured by the scalar in counts/min. Thereafter, further lower the neutron probe inside the access tube each time by 15 cm after recording the density of slow neutrons expressed in counts/min.

When a set of observations on density of slow neutrons up to the desired depth has been taken, lift the neutron probe inside the probe shield and again take the standard count by way of checking. Then switch off the scalar. Keep the neutron probe inside the probe shield when not in use.

4. Collection of soil sample from a radius of 15 cm around access tube in the determination of moisture content. From a radius of 15 cm from the access tube, collect soil samples from four borings with successive increment of 7.5 cm in soil depth. Collect the entire soil removed from each 7.5 cm depth in a large aluminium can with cover lid; noting down the diameter of soil removed by soil auger. Weigh the aluminium can plus soil, and then transfer the soil sample into an oven at 105°C for 24 hours or longer to a constant weight. Remove aluminium can + oven dry soil from oven. Allow it to cool in a desiccator and weigh after cooling the soil sample. Record the mass of moist and oven dry soil corresponding to each soil depth together with the volume of soil excavated. Estimate the moisture content of soil from each depth on volume basis. Also estimate the mean of four samples and tabulate against the corresponding density of slow neutrons expressed in counts/minute.

5. Calibration curve. Divide the density of slow neutrons expressed in counts/min. by the standard count/min to obtain count ratio *i.e.* relative count. Tabulate the relative count against mean moisture content on volume basis for each soil depth.

Now set scale for relative count on the ordinate and set scale for mean moisture content on volume basis on the abscissa of a graph paper. Then plot, the values of relative count corresponding to mean moisture content on volume basis irrespective of the soil depth. Connect the points plotted to obtain calibration curve which may be compared with the calibration curve supplied by the manufacturer.

6. Moisture content of the soil by reference to the calibration curve. Record the relative count corresponding to each soil depth as per measurement taken by neutron moisture meter.

Locate the relative count value on the ordinate of a calibration curve drawn on a graph paper, and draw an imaginary line from the point parallel to the abscissa intersecting the calibration curve. From the point of intersection, draw an imaginary perpendicular to the abscissa, and note down the moisture content of soil on volume basis given on the abscissa at the point of intersection.

1.13.6.4 Calculation

$$\begin{array}{c}\text{\% moisture in soil}\\ \text{on oven dry basis}\end{array} = \frac{\begin{array}{c}\text{Loss in mass of soil sample}\\ \text{on drying in oven in g}\end{array}}{\text{Oven dry mass of soil in g}} \times 100$$

$$\begin{array}{c}\text{\% moisture in soil}\\ \text{on volume basis}\end{array} = \begin{array}{c}\text{\% moisture in sample}\\ \text{on oven dry basis } (P_w)\end{array} \times \begin{array}{c}\text{bulk density}\\ \text{of soil}\end{array}$$

$$= P_w \times \frac{\text{Oven dry mass of soil in g}}{\text{Volume of soil excavated in cm}^3}$$

1.13.6.5 Comments

The neutron probe should fill the access tube as closely as possible.

The procedure for use of a surface probe is primarily the same as well type probe.

The inaccuracies associated with the determination of water content on volume basis be strictly avoided to enhance the reliability of the calibration curve. There could be error in the determination of both moisture content as well as bulk density of soil.

There can be random variation in count rate due to random variation in fast neutron emission from the source. The timer in country instruments need to be correct. It is preferable to record counts/3 minutes and then reduce it to count/minute.

If the data on counting rate is taken in triplicate, the standard deviation may be estimated and the reliability may be tested on the basis of coefficient of variation.

The radium as well as americium being radio active material, reasonable attention be given to safety rules to prevent health hazards.

The neutron probe must be kept inside the shield all the time except when it is lowered in the access tube for the measurement of the density of slow neutrons.

The open end of the neutron probe should be kept away from the person to avoid exposure to small radiation.

The equipment may be taken to the field by loading on a cart, truck, luggage carrier on car roof. For a short distance, it may be kept on the rear seat of a car or on back of rear seat in Jeep.

The leakage should be periodically tested. The maintenance be performed by person trained in servicing radio active equipment.

The instrument be stored in the lab. under the lock with prominent label "Radio Active Material'. The operator should wear a film badge at waste level.

REFERENCES

Black,. C.A. (ed). 1965. Methods of Soil Analysis Vol. 1. Amer. Soc. Agron Madison, Wisconsin, U.S.A.

Dakshinamurti, C & R.P. Gupta. 1968. Practicals in Soil Physics. Indian Agric. Res. Inst. New Delhi.

Prihar, S.S. and B.S. Sandhu. 1987. Irrigation of Field Crops. Principles & Practices. I. C.A.R., Krishi Anusandhan Bhawan, New Delhi-12.

WTC, 1977. Water Requirement and Irrigation Management of Crops in India. W.T.C. Indian Agric. Res. Inst., New Delhi-12.

1.14 FIELD CAPACITY OF SOIL BY FIELD METHOD– VEIHMEYER AND HENDRICKSEN (1931, 1949)

1.14.1 Principle

When soil is saturated with water by rain or else by irrigation, both micro and macro pores get filled with water. Under the gravitational pull, water in the macro pores continue to percolate downward under unrestricted drainage and minimum evaporation. The moisture content of the soil decreases with time first at a higher rate and later at a decreasing rate. Eventually the moisture content becomes constant when the macropores become empty. At this stage water is present only in the micropores *i.e.* in capillary pores, and the soil is said to be at field capacity. In light to medium textured soil, it takes about 48 hours for a saturated soil to reach field capacity under unrestricted drainage.

1.14.2 Requirement

A small plot 3 m×3m surrounded by bunds in a field, polythene sheet larger than the size of the plot, source of irrigation, preferably screw type soil auger for collection of soil sample, aluminium can, physical balance, and oven.

1.14.3 Procedure

1. Collection of soil samples from a saturated soil. Irrigate the bunded plot slowly to make the soil saturated with water upto a depth of 30 cms. Spread polythene sheet on the plot and bury edges with the soil at the periphery of the plot. At the time of taking sample, remove the polythene sheet partly and drive screw type soil auger upto the depth of wetting by successive increments of 15 cm *i.e.* 0-15 cm first and then to 15-30 cm depth etc. Lift the soil auger and transfer the collected soil sample depth wise into tared aluminium cans. Collect soil sample from morning to evening at convenient interval *i.e.* from 2 to 4 hours for a period of 48 hours, transfer, the samples into aluminium cans and immediately place the lid and note down the number of aluminium cans, and time of sampling. Make at least five boringes if gravimatric procedure is used for the determination of moisture content.

2. Weightment and drying of the soil sample. Weigh the aluminium can + moist soil in laboratory. Then remove the lid and place it at the bottom of aluminium can. Transfer the aluminium can with mouth open into an oven at 105°C and let it dry to a constant weight. Remove the aluminium can from the oven and cool it in a desiccator. Thereafter weigh the aluminium can + oven dry soil. Tabulate the mass of moist and oven dry soil depth wise against time of sampling in hours after saturation of the soil.

1.14.4 Calculation

$$\% \text{ moisture in soil} = \frac{\text{Loss of moisture on drying in g}}{\text{mass of oven dry sample in g}} \times 100$$

$$= \left[\frac{\text{Mass of moist soil}}{\text{Mass of oven dry soil}} - 1.00\right] \times 100$$

Plot the % moisture in soil sample on y-axis of a linear graph paper against time in hours on the x-axis. Ascertain the time when decrease in moisture content of the soil ceases. Note the % moisture in soil sample against the time when decrease in soil moisture ceases, and record it as % moisture at field capacity of the soil.

1.14.5 Comments

Field capacity is an important soil moisture constant and care should be taken in its determination. It signifies the upper level of plant available water in soil.

For studiers on "Field Capacity" the soil upto the root zone be saturated by irrigation. The water determined 48 hours after irrigation under minimum evaporation and unrestricted drainage is ordinarily taken as field capacity.

Veihmeyer and Hendricksen (1931) have defined field capacity as "the amount of water held in the soil after gravitational water has drained away and after the downward movement of water has materially decreased.

This is the water held by the soil against the gravitational force. At this point capillary conductivity approaches zero. The 1/3 bar percentage has been found closely correlated to field capacity for many soils.

The moisture equivalent (ME) is a close approximation of field (FC). The numerical relationship between ME and FC is ME = (FC – 2.62) / 0.865.

REFERENCES

Dakshinamurti, C., & R.P. Gupta. 1968. Praticals in Soil Physics. I.A.R.I. New Delhi-12.

Veihmeyer, R.J. and A.H. Hendricksen (1949). Methods of measuring field capacity and permanent wilting percentage of soils. Soil Sci. 68 : 75-91

Veihmeyer, F.J. and A.H. Hendricksen 1931. The moisture as a measure of the field capacity of soils. Soil Sci. 32 : 181-194.

1.15 MOISTURE EQUIVALENT OF SOIL BY CENTRIFUGE METHOD-VEIHMEYER AND HENDRICKSEN (1931)

1.15.1 Principle

Moisture retained by a saturated soil on being centrifuged at a speed equivalent to a force of 1000 times that of gravity is termed moisture equivalent. The moisture content is expressed in percentage on oven dry basis.

1.15.2 Equipments

Soil Sieve, air dry soil passed through 2 mm sieve, boxes with wire gauze bottom and lids, centrifuge with a speed of 2440 rpm ; physical balance and oven.

1.15.3 Procedure

1. Saturation of the air dry soil. Cut filter paper (Whatman No. 42) of the size of wire gauze screen at the bottom, and place it on the screen. Fill the box with 2 mm dry soil to give a layer of 2 cm thick soil on the wire gauze lined with filter paper. Place the boxes into a tray filled with 1 cm of distilled water. When the soil inside the wire gauze box become saturated as judged by the wetness of the soil surface in the wire gauze box, transfer the boxes in a saturated atmosphere for 24 hours to drain off excess water.

2. Centrifugation of saturated soil. Remove the wire gauze bottom box containing saturated soil from the saturated atmosphere, and place it on the centrifuge drum and screw down the lid. Place the centrifuge cover and switch on the centrifuge. Regulate the speed knob to slowly attain a speed of 2440 revolutions per minute. Let the centrifuge run at the above mentioned speed for 30 minutes. Then reduce the speed of the centrifuge drum gradually until it comes to a rest.

3. Drying in oven after centrifugation and weighment of the oven dry sample. After centrifugation for 30 minutes at 2440 rpm, open the centrifuge cover and remove the wire gauze bottom boxes. Transfer the moist soil from the gauze bottom box into preweighed aluminium can and weigh the aluminium can + moist soil and record it together with the number of aluminium can. Open the lid of aluminium can and place it at the bottom of can. Transfer the can containing moist soil inside an oven at 105°C and let it dry to constant weight. Cool the can containing oven dry soil inside a desiccator and then weigh to find the mass of oven dry soil.

1.15.4 Calculation

Moisture at Moisture Equivalent (ME)

$$= \left[\frac{\text{Mass of moist soil after centrifugation}}{\text{Mass of oven dry soil in g}} - 1.00\right] \times 100$$

1.15.5 Comments

Moisture Equivalent is an important soil moisture constant, and its relationship with other soil moisture constants such as Field Capacity (FC), Wilting Coefficient (WC) and Hygroscopic Coefficient (HC) is given below :

$$FC = 2.62 + MC \times 0.865$$

$$WC = \frac{M.E.}{1.84}$$

$$HC = WC \times 0.684$$

REFERENCES

Meyer F.J. and A.H. Hendricksen, 1931. The moisture equivalent as a measure of the field capacity of soils. Soil Sci. 32 : 181 – 194.

1.16 PERMANENT WITLING POINT OF SOIL–VEIHMEYER & HENDRICKSEN (1949)

1.16.1 Principle

The permanent wilting point is defined as water percentage of a soil when plants growing in that soil are first reduced to a wilted condition from which they can not recover in an approximately saturated atmosphere. The soil moisture suction at this stage is approximately equal to 15 atmosphere.

For its direct determination, seeds of dwarf sunflower (*Helianthus annus*) is sown in soil filled container provided with tight fitting lid having one hole in the centre. After adding fertilizer solution, water is added to wet the soil to nearly field capacity and the container is weighed.

The best seedling is led through the lid hole above the lid. The remaining seedlings are cut off. The plant is allowed to grow in green house with regular watering maintaining the initial weight by adding water till three pairs of true leaves develop fully. Water is now added for the last time maintaining the initial mass of container +soil + seed and water.

The space between the plant stem and hole is plugged with cotton and the container with plant is transferred to green house or under cloth shade and is allowed to grow.

As the first pair of lower leaves wilt, the plant is placed in saturated atmosphere to check whether it will regain its turgidity. This checking is continued until all the three pairs of fully developed leaves wilt and do not regain their turgidity on being placed in saturated atmosphere over night or else for a period of 14 to 16 hours. When all the three pairs of leaves have wilted permanently, moisture content of the soil is determined by drying a sample in oven to a constant weight to estimate permanent wilting percentage. When the first pair of lower leaves wilt permanently, the plant reaches incipient wilting.

1.16.2 Equipment

Metal containers of the size of 7.5 cm dia and 15 cm in height provided with tight fitting lid having one hole in the centre to serve as pots.

1.16.3 Procedure

1. Setting of the experiment and after care. Take four metal pots, and transfer in each 600 g of air dry < 2 mm soil sample. Add 100 ml of fertilizer solution prepared by dissolving 0.4g of ammonium sulphate, 0.4g of single super phosphate and 0.2g of muriate of potash in 100 ml distilled water to each pot.

On the following day sow several seeds of dwarf sunflower (*Helianthus annus*) at a depth of 5 cm around the centre of each pot. Add 50 ml or more of distilled water as required to wet the soil to near field capacity and weigh each pot on a pan balance and designate it as initial mass of the pot with soil and seed.

Transfer the pots into green house or under cloth shade, and let it remain there for 10-12 days time for the seeds to germinate and emerge above the soil surface. During this period water may be applied once to compensate for the loss in mass of the pots due to loss of moisture by evaporation and transpiration.

Lead the best seedling in each pot through the lid hole above the lid and cut off the remaining plants. Transfer the pots again to the green house or under cloth shade till the development of three pairs of fully developed leaves. Water the plants every 2nd or 3rd day to compensate the loss in weight of pots. Do not add water in excess to cause water logging.

After the development of the third pair of leaves fully developed, water the plants again approximately to field capacity by placing the pot on a pan balance and then add water equal to the loss in mass of pot with soil and seed. Also plug the space between plant stem and hole with cotton to minimize loss of water by evaporation. Transfer the pots to green house or under cloth shade.

2. Checking of the incipient wilting of the plant. When the lowest one pair of true leaves wilt, transfer the pot with plant to a dark saturated atmosphere and let it remain there overnight or for a period of 14-16 hours. If the pair of true leaves regain their turgidity, transfer the pot with plant to the growth chamber. When the lowest pair of leaves wilt again, transfer the pot with plant to a saturated atmosphere and let it remain there overnight or longer. If the lower pair of leaves fail to regain their turgidity, the plant has reached incipient stage of wilting.

3. Checking of the permanent wilting of the plant. Transfer the pot with the plant to the growth chamber i.e. green house or under cloth shade. Let the plant remain there until the 2nd and 3rd pair of fully developed leaves wilt. Remove the pot with plant to a saturated atmosphere for overnight or longer to check whether the leaves regain their turgidity. If the leaves fail to recover, the plant has wilted permanently.

4. Determination of moisture content in soil at permanent wilting point. When the plant has wilted permanently, cut off the plant from the soil surface. Remove soil sample and make it free from roots without loss of time. Transfer about 30 g soil sample into a tared and numbered aluminium can. Cover it with the lid and weigh on a precision balance.

Transfer the aluminium can with soil placing the lid below the aluminium can into an oven at 105°C to dry for 24 hours or longer to a constant weight. Cool the aluminium can with oven dry soil inside a desiccator and weigh accurately with an accuracy of 0.1 g. Similarly take out the soil samples from the remaining three pots on permanent wilting of the plant. Dry the soil samples in an oven at 105°C for 24 hours to a constant weight. Record the mass of aluminium can (W_1), mass of aluminium can plus moist soil sample (W_2), and the mass of aluminium can plus oven dry soil (W_3). The weighments need to be accurate up to 0.1g. Determine % moisture in soil sample at the permanent wilting point. Take the mean of 2 or 3 closest values.

1.16.4 Calculation

Permanent Wilting percentage

$$= \frac{W_2 - W_3}{W_3 - W_1} \times 100 = \left[\frac{W_2 - W_1}{W_3 - W_1} - 1.00\right] \times 100$$

$$= \frac{\text{Loss in mass on drying}}{\text{Oven dry mass of soil}} \times 100 = \left[\frac{\text{Mass of moist soil}}{\text{Mass of oven dry soil}} - 1.00\right] \times 100$$

1.16.5 Comments

A cotton plug reduces evaporation loss from the soil without preventing adequate aeration.

A variety of sunflower with a larger top/root ratio is likely to be more sensitive to moisture stress and so it would be preferable for study related to wilting.

The wilting of plant is also influenced by the atmospheric evaporative demand relative to the water supply and the capillary conductivity of the soil.

The incipient wilting point as defined as the water content in the soil at which the lowest pair of true leaves of a particular kind of plant at a particular stage of growth wilts and fail to recover in a saturated atmosphere. The elongation of the plant ceases at about the same time that incipient wilting occurs. The concept of incipient wilting was first elucidated by Furr and Reeve (1945).

The water content in soil at the inception of permanent wilting and the water content in the soil at permanent wilting is termed wilting range.

The permanent wilting percentage vary with the soil texture, and the mineralogical composition of the soil.

The difference between water content at field capacity and permanent wilting point give percent available water in soil.

REFERENCES

Black, C.A. (ed). 1965. Methods of Soil Analysis. Vol. 1. Am. Soc. Agron. Madison, Wisconsin, U.S.A.

Dakshinamurti, C. and R.P. Gupta 1968. Practicals in Soil Physics I. A.R. I, New Delhi-12.

Furr, J.R. and J.D. Reeve, 1945. Range of soil moisture percentage through which plants undergo permanent wilting in some soils from semi-arid irrigated areas. J. Agric Res. 71 : 149-170.

Veihmayer, F.J. and A.H. Hendricksen. 1949. Methods of measuring field capacity and permanent wilting percentage of soils. Soil Sci. 68 : 75 – 95.

1.17 SOIL MOISTURE AND TENSION RELATIONSHIP

1.17.1 Soil moisture and tension relationship by use of tensiometer-Gardner (1965)

1.17.1.1 Principle

If a porous cup connected to a hollow plastic tube provided with a vacuum gauze near the upper end fitted with a screw cap is filled completely with water free from dissolved air and placed in intimate contact with the soil, the water inside the porous cup will diffuse into the surrounding soil until a hydraulic equilibrium is reached. The tension developed inside the porous cup as recorded by the vacuum gauze will be equal to the soil moisture tension in the immediate vicinity of the porous cup.

1.17.1.2 Equipments

Tensiometers

1.17.1.3 Procedure

1. Checking of tensiometer prior to its installation. Open the mouth of tensiometer by removing the rubber stopper or else the screw cap. Then add distilled water free from dissolved air through a wash bottle filling the porous cup, connecting single or doubled walled plastic hollow tube and the vacuum gauze. If any air gets entrapped inside the hollow tube during its filling with water, suck from the outer end of the connecting hollow tube, and the air bubble will disappear. Fill the connecting hollow tube completely up to the level of its rim, and then place the rubber stopper into the mouth or put the screw cap on its mouth and tighten it properly.

Now, lay the tensiometer on a table and switch on the fan. Moisture lost from the surface of porous cup by evaporation would result in the development of vacuum inside the connecting tube which will be indicated by the vacuum gauze needle. If the vacuum gauze needle does not deflect, the tensiometer be labelled as defective and it be discarded. Select the tensiometer found satisfactory for the installation into the field.

2. Installation of the tensiometer in the field. Take tensiometer found satisfactory for installation into the field. Fill it completely with distilled water free from dissolved air. Place the screw cap on its mouth and tighten it well. Wrap the porous cup with a wet cotton pad while taking it to the field for installation.

Make a hole in the soil up to the desired depth by a soil auger. The diameter of the hole should be nearly equal to the outer diameter of the porous cup. Drop a handful of loose friable soil into the hole, and then pour some water into the hole to moisten the loose friable soil dropped into the hole.

Remove the wet cotton pad wrapped on the porous cup of tensiometer and lower the porous cup into the hole made in the soil until it reaches the desired depth. Fill in gaps noticed in between the tensiometer outer wall and the hole made in the soil. Tamp the soil to develop good contact between the tensiometer and the surrounding soil. Thereafter, leave the tensiometer for 4-6 hours to enable the porous cup to become in equilibrium with the surrounding soil. The response time is inversely proportional to the product of cup conductance and the gauze sensitivity. When tensiometer is installed in the evening, one can take measurements on soil moisture tension on the following morning.

3. Soil moisture and tension calibration curve. A strip of land uniform in texture, structure, soil reaction, depth, elevation, cropping and management practices be identified. It may be divided into four subplots. Install atleast one tensiometer in each sub-plot at the desired depth. The depth of installation of tensiometer should not vary from one sub-plot to another. Thereafter, irrigate the entire strip uniformly.

• If the soil is of sandy loam texture, take measurements on soil moisture tension beginning from 24 hours after irrigation at 3 days interval for a period during which the tensiometer reading approaches nearly to 0.85 bar value.

Each time measurement is taken on soil moisture tension, three soil samples from adjacent soil in the vicinity of the tensiometer be taken from soil surface up to the depth of porous cup by a soil auger. The moist samples weighing about 25 g be transferred to tared and properly tagged aluminium can. The tagged container lid be immediately placed on the container. The container containing moist soil samples be transferred to laboratory and weighed on a precision balance. Then, the aluminium can filled with moist soil and its lid beneath the can be transferred to an oven at 105°C to dry for 24 hours to a constant weight.

After drying to a constant weight, remove the aluminium can from the oven and transfer the can inside a desiccator, and let it cool before weighment. Find out the loss in weight on oven drying and estimate % moisture in soil sample on oven dry basis. Take the mean of closest values of % moisture in soil sample, and record the mean % moisture in soil sample against the corresponding soil moisture tension in bar.

On a linear scale graph paper, set tensiometer reading scale on the ordinate and set % moisture in soil sample scale on the abscissa. Then plot tensiometer reading corresponding to % moisture in soil sample. Connect the points plotted on the graph paper by a smooth line to obtain the soil moisture and tension calibration curve.

The soil moisture tension increases with the decrease in the soil moisture content vice and versa. The exact relationship vary with the soil texture, mineralogical composition and the content of organic fraction in the soil. A greater fraction of moisture retained by the soil is released relatively at a lower soil moisture tension in sandy loam soil as compared with the fine-textured soil.

1.17.1.4 Calculation

$$\% \text{ moisture in soil} = \frac{\text{Loss in moist weight on drying in g}}{\text{Oven dry mass of soil in g}} \times 100$$

1.17.1.5 Comments

Tensiometer measures the energy status of water in equilibrium with the soil moisture and thus it is a measure of matric potential of soil in bar. The correction due to height of water in the porous cup and the connecting hollow tube is 0.01 bar/ 10 cm height. The correction is usually made when the height measures atleast 50 cm.

When a mercury manometer is fixed into the tensiometer connecting hollow tube instead of a vacuum gauze, the soil moisture tension in terms of height of water column is estimated by the equation $h = 12.55\,x - y - z$ where x is the length of mercury column from the mercury well in cm, y is the height of mercury well from the ground surface in cm, and z is the depth at which porous cup is placed in the soil from the ground surface in cm. The value estimated on substituting the value of x, y and z is the soil moisture tension expressed in cms of water columns. 1 bar soil moisture tension is equal to 1029.8 cm of water column (13.55×76.0). A minus sign is placed before the estimated value to indicate that it is a negative pressure or else there is decrease in the water potential. The logarithm of the soil moisture tension expressed in cms of water column is termed soil pF.

The porus cup is made by making Kaolin suspension of semiliquid consistency. The suspension is poured in wooden mould. The Kaolin mould dries quickly. The porous cup is separated from the wooden encasing, dried first in air, then in an oven and thereafter it is baked in a muffle furnace. Thereafter it is taken out from muffle furnace and allowed to cool.

The porous cup is then placed inside the hollow connecting plastic tubing and fixed with araldite adhesive. The vacuum gauze is also fixed in hole made near the upper end of hollow tube of tensiometer with araldite adhesive. For making a mercury manometer, a thin polythene tubing is fixed into the hole instead of vacuum gauze. Water is filled in porous cup, connecting hollow tube and the polythene tubing. The other end of the polythene tubing filled completely with water is dipped in a mercury well placed above the ground. The porous cup is placed in soil. Water is filled through the upper end of hollow tube which is stoppered after filling.

The porous cup permits in and out diffusion of water and salt but it does not permit diffusion of air through the porous cup inside the water filled connecting tube till soil moisture tension of 0.85 bar. As the soil moisture tension exceed 0.85 bar, the instrument become inoperative due to diffusion of air inside resulting in the elimination of vacuum.

The tensiometers are not used on heavy soils rich in clay exhibiting maximum swelling and shrinkage. The porous cup placed in the soil loose contact with the soil when soil shrink and crack. There is also mechanical damage to porous cup during soil shrinkage. Moreover, the porous cup is so firmly held that if care is not taken while removing the tensiometer, the porous cup may get separated from the tensiometer or else it may develop cracks.

If tensiometer has become inoperative because of entry of air through the other end of connecting tube. Add distilled water to fill the connecting tube completely removing any air bubble. Then tighten the screw cup or stopper making it air tight.

If tensiometer is inoperative and the soil is found to be very dry. It may be the result of diffusion of air through the porous cup inside the connecting tube due to soil moisture tension exceeding 0.85 bar.

Table 1.17.1 Optimum soil moisture tension for irrigation in crops grown on sandy loam soils.

Crop	Yield q/ha	Soil moisture tension	Depth of installation	Place
Maize	55.2	0.25 bar	22.5 cm	New Delhi
Pearl Millet	24.5	0.20 bar	22.0 cm	Anand
Soyabean	35.9	0.50 bar	22.0 cm	New Delhi
Sugar beet	534.0	0.20 bar	25.0 cm	New Delhi
Hookah Tobacco	49.1*	0.40 bar	22.5 cm	New Delhi
Potato	279.7	0.2 – 0.3 bar	15–22 cm	New Delhi
Berseem	652.0**	0.25 bar	22.5 cm	Agra

*Cured leaves, ** green fodder

The scheduling of irrigation in crops should be based on soil moisture regime *i.e.*, on the energy status of soil water rather than on content of moisture in the soil. The experiments on scheduling of irrigation by tensiometer have been particularly undertaken on sandy loam soil in irrigated crops requiring frequent irrigation. The results reveal much better yield of crops irrigated at lower soil moisture tension in the range of 0.2 to 0.4 bar as compared to yield obtained by giving irrigation at higher soil moisture tension (0.5 to 0.7 bar). The optimum soil moisture tension for irrigation in crops grown on sandy loam soils as judged by the maximum yield of crops is summarized in Table 1.17.1.

REFERENCES

Baver, L.D., W.H., Gardner, &. W.R. Gardner, 1976. Soil Physics. Wiley Eastern Limited New Delhi.

Dakshinamurti, C& R.P. Gupta 1968. Practicals in Soil Physics I.A.R.I. New Delhi.

Gardner, W.H. 1965. Water Content In. C.A. Black (ed). Methods of Soil Analysis, Vol. 1. Am. Soc. Agron. Madison. Wisconsin, U.S.A.

WTC. 1977. Water Requirement and Irrigation Management of Crops In India. IARI Monograph 4 (New series), Water Technology Centre IARI.

1.17.2 Soil moisture and tension relationship by use of pressure plate apparatus–Richards (1948).

1.17.2.1 Principle

A saturated soil sample kept on a saturated porous plate is transferred inside a pressure chamber of pressure plate apparatus. The soil is subject to pressure equal to < 1 bar such as 1/10, 1/5, 1/3, 1/2, 3/4 bar for about 48 hours. The moisture retained by the soil against the pressure exerted is determined. Thereafter, a calibration curve is constructed to show % moisture in soil in relation to the pressure exerted in soil sample in bar.

1.17.2.2 Equipment

Pressure plate apparatus, porous plate, air compressor and mercury manometer.

1.17.2.3 Procedure

1. Saturation of the soil sample. Transfer 25g of air dry < 2 mm soil sample inside the rubber ring placed on the porous plate. In case of undisturbed soil sample, the entire soil core with copper cylinder may be placed on the porous plate.

Place the porous plate with disturbed and undisturbed soil sample inside a tray containing water. Let the thin bottom layer of the soil remain submerged in water overnight.

2. Setting the porous plate apparatus. On the following day, if the soil sample is found saturated with water, transfer the porous plate inside the metallic chamber of the pressure plate apparatus.

Connect plate outlet tube leaving the diaphragm to the outlet of the chamber. Then, close the pressure chamber by placing the cover lid. Using special wrenches, tighten the nuts and bolts with the required torque for making the chamber air tight.

3. Development of pressure inside the pressure chamber. Apply pressure from a compressor through control valve in order to maintain desired pressure (1/10 th to < 1 bar) as indicated by the mercury manometer. Also ensure that there is no leakage from the chamber.

Maintain the desired pressure for a period of 48 hours or until the flow through outlet tube continue to trickle water. As the soil become in equilibrium with the applied pressure, the plate outlet tube will stop dripping water.

4. Moisture retention by the soil. When water ceases to flow from outlet tube, close the outlet tube, release the pressure inside, open the cover lid, and transfer the soil from the rubber ring to a tared and numbered aluminium can. Weigh the moist soil sample with aluminium can and transfer the sample with aluminum can into an oven at 105°C to dry for 24 hours to a constant weight. Find out the loss in mass on drying and oven dry mass of the soil.

In case of undisturbed soil sample, remove the soil core with cylinder and weigh. Then, remove it to an oven at 105°C to dry to a constant weight. It takes 3 to 4 days to dry uniformly in the oven. Remove the soil core with cylinder from the oven and let it cool for 3 hours. Measure the oven dry mass of core and its volume. Also determine the loss in mass of sample on drying in oven.

Similarly subject the sample from the same soil after saturating with water to different pressure up to < 1 bar in the pressure chamber of the pressure plate apparatus. Estimate the moisture content and record against the pressure applied in bar.

5. Construction of calibration curve. The moisture content corresponding to each pressure value be plotted on a graph paper to construct soil moisture characteristic curve in relation to the soil moisture tension in bar.

1.17.2.4 Calculations

% moisture in soil on oven dry mass basis

$$= \frac{\text{Loss in mass of sample on drying in oven in g}}{\text{Oven dry mass of the soil in g}} \times 100$$

$$\begin{matrix}\text{\% moisture in soil} \\ \text{on volume basis}\end{matrix} = \begin{matrix}\text{\% moisture in soil} \\ \text{on oven dry basis}\end{matrix} \times \frac{\text{Mass of oven dry soil core in g}}{\text{Volume of the core in cm}^3}$$

1.17.2.5 Comments

Once the calibration curve is constructed from pairs of values on pressure exerted in bar, and the moisture retained by the soil, it would be possible to determine the moisture content of the soil indirectly by measurement of the soil moisture tension.

From the pressure plate apparatus, it is possible to determine moisture content in the soil at field capacity and how much water the soil would release for absorption by plants at soil moisture tension less than 1 bar.

REFERENCES

Dakshinamurti, C. & R.P. Gupta. 1968. Practicals in Soil Physics. I. A.R.I. : Pusa, New Delhi-12.

Gardner, W.H. 1965. Water Content, In. C A. Black (ed.) Methods of Soil Analysis Vol. 1. Am. Soc. Agron., Madison, Wisconsin, U.S.A.

Richards, L.A. 1948. Porous plate apparatus for measuring moisture retention and transmission by soil. Soil Sci, 66 : 105 – 110.

Richards, C.A. (ed). 1954. Diagnosis and Improvement of Saline and Alkali Soils. U.S.D.A. Hand Book 60. Oxford & IBH Publishing Co., New Delhi.

WTC. 1977. Water Requirement and Irrigation Management of Crops in India. I.A.R.I. Monograph 4 (New Series) Water Technology Centre, I.A.R.I. New Delhi.

1.17.3 Soil moisture and tension relationship by use of a Pressure Membrane Apparatus–Richards (1947)

1.17.3.1 Principle

The pressure membrane apparatus consist of a cellulose membrane placed on a wire screen kept on the bottom plate. The chamber below the cellulose membrane remain in equilibrium with air outside through an outlet tube by which water entering into the chamber is drained off. The top plate is provided with a diaphragm to keep the soil samples in contact with the membrane. An "O" ring is provided in between the plates to prevent any leakage after tightening the nuts and bolts.

The cellulose membrane possess fine holes with the high air entry value. With the result it can be subject to air pressure difference from 1 to 15 atmosphere.

Soil is placed inside circular rubber ring on saturated cellulose pressure membrane placed over wire screen of bottom plate is saturated. After putting O ring, the cover lid provided with diaphragm is placed, and tightly held together by tightening with large number of nuts and bolts.

The compressed air is made to enter through the air inlet to maintain desired pressure difference of 1 to 15 atmosphere across the pressure membrane for a period of 24 hours. When the outflow ceases, outlet tube is closed and air pressure is released through exhaust valve. The cover lid of pressure membrane apparatus is opened.

The soil sample is removed and dried in an oven. Moisture retained by the soil is recorded against the corresponding pressure maintained across the cellulose membrane.

1.17.3.2 Equipments

Pressure membrane apparatus and accessories, compressor or compressed air cylinder, rubber rings, rubber tubing, burette, aluminium can, physical balance and oven.

1.17.3.3 Procedure

1. Saturation of soil sample inside the pressure membrane apparatus. Soak the cellulose membrane thoroughly in water and then place it on the wire screen. Then place rubber rings on the membrane. Pour air dry soil passed through 2 mm sieve to fill the rubber ring uniformly. Use separate rubber ring for each soil sample. Add distilled water out side the rubber rings to wet the soil sample from bottom upward. Add enough water around the rubber rings and leave it over night.

2. Placement of top plate and tightening of the nuts and bolts of pressure membrane apparatus. Place O ring over the cellulose membrane and then place the top plate on it. Fit the bolts in the top plate and place the nut on each bolt for tightening. Tighten nuts of bolts opposite each. A torque of 4 to 5 ft.lb. on the clamping bolt will be sufficient to seal the unit for use up to 225 psi (15 atmosphere pressure).

3. Developing and maintaining desired pressure inside the pressure membrane chamber. Connect the out flow tube of the pressure chamber to the bottom of a burette. Connect the compressed air line to the top and side connection of the chamber. Close the exhaust valve and develop the desired air pressure inside the chamber. Maintain the air pressure inside for 24 hours or longer until the outflow of water into the burette ceases.

4. Removal of moist soil. Put a pinch clamp on the outflow tube. Then open the exhaust valve to release the air pressure. Untighten the nuts and bolts. Remove the top plate. Transfer the soil sample from the rubber rings into a preweighed aluminium can and weigh to find the mass of moist soil.

5. Drying and weighment of soil sample. Transfer the aluminium can with lid removed and placed below the bottom of can, inside an oven at 105°C and let it dry for 24 hours.

Then cool the aluminium can + oven dry soil inside a desiccator and weigh to find the mass of oven dry soil.

1.17.3.4 Calculation

$$\% \text{ moisture in soil sample} = \left[\frac{\text{Mass of moist soil}}{\text{Mass of oven dry soil}} - 1.000\right] \times 100$$

Graphical representation. On a graph paper, plot the applied pressure on *y*-axis and the correponding % moisture in the soil sample on the x-axis. There is inverse relationship between % moisture in soil and soil moisture tension in atmosphere or bar.

1.17.3.5 Comments

The soil moisture and tension relationship differs from soil to soil. It is useful particularly in scheduling of irrigation based on the energy status of water in the soil. It is possible to ascertain the percentage of available water in soil and the extent to which the available water is depleted at the time of giving irrigation.

The pressure membrane apparatus enables the development of soil water characteristic curve in the higher range of matric potential (> 1 bar) which is not possible in the pressure plate apparatus.

The pressure values choosen for determining moisture content are normally 1/10, 1/5, 1/3, 1, 3, 5, 7, 10 and 15 bar.

REFERENCES

Dakshinamurti, C. and R.P. Gupta. 1968. Practicals in Soil Physics. Indian Agric. Res. Inst. Pusa, New Delhi-12.

Gardner, W.H. 1965. Water Content. In C.A. Black (ed). Methods of Soil Analysis. Vol. 1. Am. Soc. Agron. Madison, Wisconsin, U.S.A.

Richards, L.A. 1947. Pressure membrane apparatus construction and use. Agric. Engr. 28 : 451-454, 460.

Richards, L.A. 1948. Porous plate apparatus for measuring moisture retention and transmission by soil. Soil Sci. 66 : 105 - 110.

Richards, L.A. 1949. Methods for mounting porous plates used in soil moisture measurements. Agron. J. 41 : 480.

Richards, L.A. 1952. Water conducting and retaining properties of soils in relation to irrigation. Intern. Symp on Desert Res. Proc. pp 1-22. (Res. council in Isarael in Cooperation with UNESCO).

Richards, L.A. 1954. Diagnosis and Improvement of Saline & Alkali Soils. U.S.D.A. Hand Book no. 60. Oxford & IBH Publishing Co., New Delhi.

WTC. 1977. Water Requirement and Irrigation Management of Crops in India. I.A.R.I., New Delhi-12.

Chemical Analysis of Soil

2.1 SOIL SAMPLING FOR CHEMICAL ANALYSIS-ADAPTED FROM JACKSON (1958), MUHR *et al.* (1965)

2.1.1 Principle

A portion of the entire soil mass taken from its natural occurrence for analysis in order to describe the properties of the soil is referred as soil sample. The purpose of the sampling is to estimate the properties of the entire soil mass in question with a certain degree of accuracy at the lowest possible cost. This can be achieved only when the sample is so selected that it represents the entire soil mass.

The soil is usually a heterogenous body of material. It is necessary to delineate units which approach homogeneity within themselves. The intensity of sampling is linked with the size of sampling unit. Normally 20 samples are collected from an area of one hectare. However, in experimental plots, it would be desirable to take atleast two samples from each experimental unit to evaluate the effect of treatments imposed on the soil properties.

2.1.2 Equipments etc : 30 m tape, soil auger or soil sampler

2.1.3 Procedure

1. Sampling Sites. The sampling site may be selected employing different procedure. A sampler may select typical sites which in his judgement represent a representative picture of the entire sampling unit. It is referred as judgement sample. When sampling site is selected by random distances in a coordinate system and using the interactions of the two random distances at the point of sampling. It is referred as random sampling. When the field is divided into a number of sub-plots and random samples are taken from each sub-plot. When samples are collected at regular distances away from each other either in one or two directions, it is referred as systematic sampling.

In sub-sampling procedure, the sampling unit is divided into a number of smaller elements and then sample is drawn at random from these elements. In composite sample procedure, a number of samples taken to represent the sample area is thoroughly mixed to form one composite or bulk sample. The laboratory analysis is done on the composite sample or else on the sub-sample of a composite sample. A valid estimate of the mean of some characteristics of the soil is obtained from the single analysis of the composite sample. The precision increases with the increase in the number of samples forming the composite sample. There is substantial saving in the cost of analysis, but the variance of the mean can not be estimated.

2. Sampling procedure for soil testing. To derive real benefit from soil testing, it is necessary that the soil sample is taken by right and scientific method. Only 60g of the sample brought to the soil testing laboratory is used for soil analysis. On the basis of the tests made, recommendations are made with respect to requirements of fertilizer for the entire field. Therefore, the sample sent to the laboratory should be composite soil sample so that it is representative of the entire field. If the samples are not collected properly, the farmer would be deprived of the benefit which accrue from the result of soil testing. The following procedure may be employed for the collection of soil sample :

(*i*) Weeds, leaves and plant remains on soil surface be removed taking care that while scratching off the extraneous material, the surface soil is not removed by scraping.

(*ii*) After the harvest of a crop, soil sample be immediately collected. Alternatively collect soil sample by driving a soil augar into the soil up to the desired depth. With khurpi or spade one can make a V cut up to the desired depth at each spot and remove a thin slice from the exposed side of each V cut. and sent to soil testing laboratory for obtaining the results prior to sowing of the next crop. When a crop is standing in the field, the samples be collected in between the lines.

(*iii*) If different type of soil is found in the same field or else different crops have been grown in different areas of the field. Then, collect a separate soil sample from each distinct areas.

(*iv*) Select 8-10 spots from one acre field which are quite away from manure heap. Soil samples from roads, depressions and beneath the tree is not taken.

(*v*) At the selected spot dig a rectangular shallow pit of the size of 15 cm long, 10 cm broad and 15 cm deep. Then collect a 2.5 cm slice along the four walls of each wide pit to a depth of 15 cm, collecting the soil excavated in a clean paper or cloth. Mix the soil thoroughly. During mixing of the soil, remove any extraneous materials such as stone, pebbles, charcoal etc which are normally not found in the soil.

(*vi*) After thorough mixing of the soil samples collected, form a rectangular or cone shaped heap from collected soil samples and divide the heap into four quarters. Discard one of the alternate quarters and mix the remaining two quarters. Repeat the mixing, heap for motion, quartering and discarding of the alternate quarters until the size of the soil sample is reduced to half a kg.

(*vii*) The soil sample be air dried before transferring into a cloth bag or thick paper carton. By wet soil the cloth bag as well as thick paper carton gets decomposed and the identity of the sample is lost.

Transfer half a kg of air dried soil sample into thick paper carton or cloth bag. Labed the container giving farmer's name village, identity of field mentioning partientais of Khasra development Khand Tahsil and crops to be grown. The pertinent information may be written on a piece of peper by carbon (lead pencil and it be placed inside the soil container. Close the container and arrange to deliver the container to a soil testing laboratory or soil health laboratory.

(*viii*) In case of saline-alkalie soils, the land has to be levelled, graded and then it should be divided into fields for cropping. Strong broad ridges are made on the periphery of each field to check inflow of bad water from adjacent fields. Then the field is flooded with water to check the levelling of the field. The flooding shall also result in some degree of leaching of salts. When the soil becomes in a workable condition, a composite soil sample for the routine analysis of the soil for pH, EC, available nutrient elements and gypsum requirement be taken. The sample of ground water is also analyzed for its sodicity and salinity hazard.

At the initial stage on representative sites a pit should be dug upto a depth of 1.5 m and the soil samples should be collected from the entire soil profile depthwise *i.e.* one sample per 30 cm soil depth. The sample collected from each depth are analysed for pH, EC, SAR, ESP, GR, hydraulic conductivity for which atleast one kg sample is required. The depth and quality of ground water should also be ascertained in different seasons to assess the feasibility of leaching of soluble salts from the soil and to ascertain the utilization of ground water for arable cropping or else for irrigated forest or orchards. In view of the above, it would be desirable that the Head of the soil testing laboratory is consulted in regard to the collection of the soil sample and subsquent management of the land.

REFERENCES

Jackson, M.L. 1958. Soil Chemical Analysis. Prentice Hall of India Private Limited, New Delhi (1973).

Muhr, G.R., N.P. Datta, H. Sankarasubramoney, V.K. Leley, and R.N. Donahue. 1965. Soil Testing in India, U.S. Agency for Intern. Development Mission to India, New Delhi.

Richards, L.A. (ed). 1954. Diagnosis and Improvement of Saline and Alkali Soils. U.S.D.A. Hand Book 60, Oxford & IBH Publishing Co., New Delhi.

2.2 SOIL PH

2.2.1 Soil pH with a pH meter

2.2.1.1 Principle

Soil pH is defined as the log of the reciprocal of hydrogen ion activity or else as the negative log of hydrogen ion activity.

$$\text{Soil pH} = \log \frac{1}{aH^+} = -\log_{10} aH^+$$

In which the activity of H^+ in the soil-water suspension, a_{H^+}, is expressed as g ions/l.

The soil pH is measured by glass electrode. The glass electrode employs glass membrane. Across the glass membrane develop electric potential that is in proportion to the difference in pH of HCl filled inside the membrane ($a_{H_1^+}$) and the pH of soil water suspension outside the membrane ($a_{H_2^+}$). The KCl diffusing through fine holes of reference calomel electrode into the soil-water suspension form ionic invisible bridge between the electrodes through which current passes. An electric potential develops between the electrodes and the potential difference is measured through a galvanometer. The pH dial is calibrated to the potential difference in relation to the soil pH.

The voltage E, of the glass electrode half cell is given by

$$E = -0.5916 \log 10 \frac{a_{H_2^+}}{a_{H_1^+}}$$

In which 0.5916 is the function of RT/nF, R = 8.313, F = 96500, n = 1 and T = 29.8°K for 25°C). As the pH decreases by one unit, there is decrease in the voltage by 59.16 milli volt. The potential difference between two electrodes at pH 0, 7.0 and 14.0 is approximately +420 mV, 0 mV and –420 mV respectively.

2.2.1.2 Equipments and Reagents

Equipment/Reagents

pH meter with reference and glass electrode or else combined electrode.

(*i*) Buffer solution pH 4.0. Dissolve 11.806 g of citric acid and either 10.9468 g of Na_2HPO_4 or 27.6152 g of $Na_2HPO_4.12H_2O$ in 1l of distilled water.

OR

Add 4 ml of 0.1 N NaOH to 500 ml of 0.1 M potassium acid phthalate and dilute to 1l with distilled water.

OR

Dissolve 10.21 g of potassium hydrogen phthalate in water and dilute to 1l.

OR

Transfer one pH 4.0 buffer tablet into 100 ml vol. flask. Add distilled water to dissolve the tablet and dilute it to 100 ml.

(*ii*) Buffer solution pH 7.0. Dissolve 3.391 g of citric acid and either 23.3844 g of Na_2HPO_4 or 58.9913g of Na_2HPO_4. $12H_2O$ in 1l of distilled water.

OR

Add 296.3 ml of 0.1 N NaOH, 500 ml of 0.1M potassium acid phthalate and dilute to 1l with distilled water.

OR

Transfer one pH 7.0 buffer tablet into 100 ml vol. flask, add distilled water to dissolve the tablet and dilute the solution to 100 ml mark with distilled water.

(*iii*) Buffer solution pH 9.20. Dissolve 3.81 g of $Na_2 B_4O_7$. $10H_2O$ (sodium tetraborate) in distilled water and dilute the solution to 1l with distilled water.

Transfer one pH 9.18 buffer tablet into 100 ml vol. flask and add distilled water to dissolve the buffer tablet and dilute the solution to 100 ml mark.

Prepare the buffer solution with boiled and cooled distilled water and store the buffer solution in polythene or pyrex glass containers.

2.2.1.3 Procedure

1. Preparation of soil water suspension. Transfer 20g air dry 2 mm soil sample into a 100 ml beaker. Add 40 ml distilled water, and stir the suspension with a glass rod intermittently for 30 minutes.

2. Calibration of the pH meter. Connect the leads of the pH meter into a voltage stabilizer. Switch on the pH meter and let it warm for 10 minutes. Adjust the temperature compensator to the temperature of the buffer solution. Remove the glass and reference electrodes from the distilled water and dry with a filter paper. Connect the leads of the glass and reference electrodes into the appropriate sockets of the pH meter as per operation procedure given in the instruction manual. The glass and reference electrodes is dipped in distilled water when not in use so that it does not get dried.

Immerse the two electrodes in pH 9.2 buffer solution and adjust the pH range to pH 7-14. Check the pH recorded by the pH meter. Adjust the pH meter reading to 9.2 with the pH adjustment knob.

Now, turn the pH adjustment range knob to the rest position. Remove the glass and reference electrode from pH 9.2 buffer solution and wash it sufficiently with a jet of distilled water, and then dip the electrodes in dilute acid for half a minute to remove any bases adsorbed on the glass membrane. Wash the electrodes again with a jet of distilled water and then dry the electrodes using a filter paper to absorb the water.

Then immerse the electrodes in pH 7.0 buffer solution and turn the pH range adjustment knob to pH 7.0 - 14.0 range. Note the pH meter reading. If it is found to be exactly 7.0, the pH meter reading has been calibrated and the pH of soil water suspension can be now measured.

3. Measurement of pH of soil-water suspension by pH meter. After calibration of the pH meter, turn the pH range adjustment knob to the rest position, and remove the glass and reference electrode from pH 7.0 buffer solution. Wash the electrodes thoroughly with a jet of distilled water and dry the electrodes by wiping with a filter paper so as to absorb the moisture.

Then stir the soil-water suspension again with a glass rod, and immerse the glass and reference electrodes in the suspension. Turn the pH range adjustment knob to pH 7.0 – 14.0 range. As the galvanometer reading rest at a particular point, note the pH recorded by the pH meter.

Similarly determine the pH of other soil-water suspension or sample solution. After noting the pH, turn the pH range adjustment knob to the rest position. Switch off the pH meter. Remove the glass and reference electrodes from the sample suspension or solution. Wash the electrodes thoroughly by a jet of distilled water. Then immerse the electrodes in distilled water, and let it remain dipped in distilled water until the next determination of pH.

2.2.2 Measurement of soil pH by use of pH indicator/pH paper

2.2.2.1 Principle

A pH indicator is an organic dye, the colour of which is controlled by the hydrogen ion activity in solution. There are about 17 pH indicators each of these have three distinct colours. The centre colour is obtained at the critical pH value.

The use of specific pH indicator is not commonly used for the determination of soil pH as the pH of soil-water suspension or solution may vary from 4–12. Instead a composite indicator solution from BDH which exhibits different colour at different pH in the range of 4 to 11 are used for the measurement of soil pH. The pH papers of the different pH range are also available for the quick determination of pH.

The indicator is either added to clear solution in equilibrium with the soil or dry pH paper soaked in composite pH indicator is treated with the supernant liquid or clear solution. The colour developed is then matched with the colour given corresponding to a particular pH.

2.2.2.2 Reagents

Universal pH indicator along with the pH colour chart

2.2.2.3 Procedure

Transfer 5g air dry 2 mm soil sample into a medium size test tube. Add 10 ml distilled water and a pinch (50 mg) of barium sulphate powder. Stir the suspension with a glass rod carefully and intermittently for 30 minutes and then place the test tube on a test tube stand for the soil particles to settle. As the supernant liquid becomes some what clear, add few drops of the universal pH indicator and gently shake the test tube to mix the indicator with the soil solution. Place the test tube again on the test tube stand for some time so that the supernant liquid become more clear. Then match the colour developed in the soil solution with the pH colour chart and note the pH. Alternatively when the supernant liquid in the test tube become clear, tilt the test tube to let few drops of the soil solution soak the pH paper. Match the colour developed on the pH paper with the colour given for different pH and note the pH with which the colour matches.

2.2.2.4 Calculation : Nil

2.2.2.5 Comments

The pH is an index of soil reaction which has profound effect on soil properties as well as on crop growth because of it's direct and indirect effects.

The crops differ in their pH preferences. The availability of nutrient elements are significantly influenced by the soil pH. The deficiency of Fe, Mn, Zn, and Cu in crops increase with the progressive increase in pH. The availability of N, P, K, Ca, Mg, S and Mo decrease with the decrease in pH below 5.5 to 6.0. At pH below 5.0 boron become less available, and bacterial as well as actinomycetes population decrease. The water soluble phosphorus in soils gets precipitated as tricalcium phosphate in the pH range from 7.5 to 8.5.

One may measure pH of soil saturated paste or else soil-water suspension in the ratio of 1 : 1, 1 : 2, 1 : 2.5, 1 : 5 and 1 : 10. The pH value changes with the soil water ratio. As the soil water ratio increases, the pH increases in both acidic and alkaline soils but to a varying degree. In the soil testing laboratories in India, pH is measured in 1 : 2 soil water suspension (w/v). The pH measurement is done in soil-buffer solution for the estimation of lime requirement. The pH in soil-dilute salt solution minimizes the effect of varying salt concentration in soils on pH. The pH of soil-salt solution suspension is invariably lower than pH in soil water suspension. The soil pH measured in 0.01 M $CaCl_2$ solution using 1 : 2 soil – $CaCl_2$ (w/v) suspension may be interpreted for agronomic purpose as given below :

pH		Interpretation
>7.5	–	Alkali soil
7.5	–	Free lime in soil
7.0	–	100% saturation by cations other than H.
6.5 – 7.0	–	Ideal for alfalfa, satisfactory for most crops
6.0–6.5	–	Ideal for most crops, satisfactory for alfalfa.
5.5 – 6.0	–	Satisfactory for grasses, small grams & corn
5.0 – 5.5	–	Deficient in Ca, should be limed.
4.5 – 5.0	–	Very deficient in Ca, unsatisfactory for almost all crops.

REFERENCES

Jackson, M.L. 1958. Soil Chemical Analysis. Prentice Hall of India. Private Ltd., New Delhi (1973).

Muhr, G. R. N.P. Datta, H. Sankarasubramoney, V.K. Leley, and R.P. Donahue. 1965. Soil Testing in India. U.S. Agency for Intern. Development Mission to India, New Delhi.

Perur, N.G., C.K. Subramaniam, G.R. Muhr, and H.E. Ray. 1965. Soil Fertility Evaluation to Serve the Indian Farmers. Mysore University of Agricultural Sciences. U.S. Agency for International Development. Banglore.

2.3 SOIL Eh

2.3.1 Principle

Eh is a measure of redox potential expressed in milli volts at pH 7.0. It is measured by a potentiometer such as pH meter using a platinum and reference electrode. The platinum electrode is sensitive to diffusion of O_2. As the potential difference is also influenced by the pH, the soil Eh value is adjusted to pH 7.0 on measurement of soil pH by use of glass and reference electrode.

2.3.2 Equipment/Reagents

Equipments

(*i*) pH meter with a glass, reference and platinum electrode. The lower portion of all the three electrodes must remain immersed in distilled water atleast for 24 hours before use.

(*ii*) Experimental pots filled with soil submerged under water in which transplanted rice is being grown. The pots be provided with three holes at the same level but slightly bigger than the diameter of electrodes on one side of the pot 1 to 15 cm below the level of soil surface duly plugged to prevent entry of atmospheric air.

Reagents

Buffer solution of pH 4.0, 7.0 and 9.2 : Transfer one tablet each of above mentioned pH into separate 100 ml volumetric flask and dissolve the buffer tablet in 100 ml distilled water.

2.3.3 Procedure

1. Placement of glass, reference and platinum electrode in the soil-water system. Remove the three plugs employed for plugging the holes on the side wall of experimental pot. Insert a wooden circular stick of the size of electrode in dia and length to soften the mud and then place the three electrodes into the soil layer through the three holes. Allow some time so that the electrodes come in equilibrium with the physico-chemical environment of the soil-water system.

2. Calibration of the pH meter for the measurement of soil Eh. Switch on the pH meter after connecting it's lead to a voltage stabilizer which in turn is connected to the main electric supply. Allow 10 minutes time for the instrument to warm. Adjust the temperature correction knob to the temperature of the soil-water system.

Thereafter, connect socket for reference and glass electrodes in the pH meter by an electric wire fitted with banana plug at both the ends. Insert one end into socket for glass electrode and insert the other end into socket for reference electrode.

Turn the pH range knob to either left or right, and adjust the dial reading to 7.0 *i.e.*, to a potential difference of 0 mV. by use of the pH adjustment knobs. Now take out the banana plugs fitted into the sockets for reference and glass electrode. The pH meter has been calibrated for the Eh measurement.

3. Measurement of soil Eh with reference and platinum electrode. Insert the lead from reference electrode placed in the soil layer of experimental pot into the socket for reference electrode lead, and insert the lead from platinum electrode placed in the soil layer of above experiment pot into the socket for the glass electrode lead in the pH meter.

Then turn the pH range knob first to the right. If the meter needle rest within the specified range. Note the meter reading and designate it as Meter reading with reference and platinum electrode. However, if the meter needle rest out side the specified range turn the pH range knob to the left, and record the meter reading at which the needle rest mentioning the electrodes used.

4. Calibration of the pH meter for measurement of soil pH. Use reference and glass electrode, the lower portion of which have remained immersed in distilled water for atleast 24 hours. Switch on the pH meter 10 minutes before taking the measurement. This may be referred as the warming period of the instrument.

Take out the glass and reference electrodes out of the distilled water, and wipe it dry with a filter paper. Connect their leads into the appropriate sockets provided in the pH meter. Thereafter immerse the lower portion of the two electrodes into buffer of pH 9.2. Turn the pH range knob towards the appropriate range and check the meter reading. If the meter needle is not resting at pH 9.2 adjust the meter needle to pH 9.2. using coarse and fine pH adjustment knob.

Then remove the electrodes from buffer of pH 9.2. Wash it with a jet of distilled water. Collect the washings in a large beaker for discarding. Wipe off the electrodes with a dry filter paper, and then immerse the lower portion of the electrodes in buffer solution of pH 7.0. Turn the pH range knob towards the appropriate range and check the meter reading on the pH dial. If meter needle rest at pH 7.0, the pH meter has been calibrated to measure the soil pH.

5. Measurement of pH of soil layer in experimental pot. Bring the pH range knob to the rest position. Remove the beaker containing buffer of pH 7.0 and wash the two electrodes with a jet of distilled water, and then place them on the electrode stand.

Now take the lead of glass electrode placed inside the soil layer of the experimental pot and fit it into the socket for glass electrode lead provided in the pH meter. Similarly fit the lead of reference electrode placed in the soil layer of experimental pot into the socket for reference electrode lead in the pH meter.

Turn the pH range knob to either left or right and check whether the meter needle rest within the specified range. If the meter needle rest within the specified range, record the meter reading with reference and glass electrode and denote it as pH x.

Switch off the pH meter, take out the leads of reference and glass electrodes from the sockets of pH meter. Then remove the reference, glass and platinum electrode from the soil layer of experimental pot by pulling the electrodes carefully and slowly through their respective hole. Immediately plug the holes again. Wash the three electrodes with a jet of distilled water from the wash bottle. Fix the electrode top in their respective stand immersing lower portion of the electrodes in distilled water.

2.3.4 Calculation

(*i*) Eh of soil layer in mV at pH x $= \left[\text{Meter reading with reference and platinum electrode} - 7.0\right] \times -60$

(*ii*) Eh of soil layer at pH 7.0 $=$ Eh at pHx $\pm$ [(7.0 – pH x) $\times$ –60]

Illustration. If meter reading with reference and platinum electrode is 4.0, and pH x = 8.5, then Eh at pH x = + 180 mv and Eh at pH 7.0 = + 270 mV.

2.3.5 Comments

All the three electrodes are expensive and delicate. The delicate handling of electrodes is necessary. The electrodes should be protected from coming in contact with glass wares. The glass electrode thin membrane bulb may get damaged by rubbing with beaker sides or bottom.

The soil Eh *i.e.* redox potential may vary from –300 mV to + 700 mV. The Eh of soil is interpreted as follows :

Soil Eh at pH 7.0	**Interpretation**
+ 400 mV to + 700 mV	soil is well oxidized or aerated
+ 200 mV to + 400 mV	soil is moderately oxidized
– 100 mV to + 100 mV	soil is in reduced condition
– 300 mV to – 100 mV	soil is in highly reduced condition.

The reduced conditions prevail when the soil is flooded or when soil become water logged for a certain period of time. The physico-chemical environment changes with the biological and chemical reduction of inorganic soil components. The oxidized components are reduced by facultative anaerobes and true anaerobes only on depletion of O_2.

The reduction of NO_3 and higher oxides of Mn occur at high redox potential *i.e.* at Eh in between +200 to +300 mV. The reduction of hydrated ferric oxide occur at intermediate redox potential *i.e.* at Eh around +100 to – 100 mV, and the reduction of sulphate occur at low redox potential *i.e.* at Eh varying from –300 mV to – 100 mV.

The relative amounts of oxidized components control the redox potential. The presence of nitrates and higher oxides of Mn would tend to maintain Eh above + 200 mV whereas significant amount of hydrated ferric oxide would tend to maintain Eh above –100 mV. The redox potential would reach a low level strictly under anaerobic conditions. When native ferric oxide becomes low as in the case of degraded paddy soils in which silica and hydrated ferric oxide become very much depleted. On reduction of sulphates, H_2S is formed which is toxic to semi aquatic plants such as rice.

The reduction of higher oxides of Mn and Fe increase the concentration of soluble Mn and Fe in soil. There is increase in the concentration of Fe and Mn in rice plant to the toxic level in some soils. There is an optimum Fe/Mn ratio for higher yield of rice. A very high Fe/Mn ratio is not conducive for obtaining high yield.

Under the reduced soil conditions, there is also increase in soluble phosphates with the conversion of Ferric phosphate into Ferrous phosphate. The availability of occluded P is also increased.

The moderately reduced conditions reduce the recovery of nitrogen added to rice by encouraging losses of nitrogen by denitrification. However, in some soils in which reduced conditions cause significant reduction in crop yield, nitrates are recommended not as a N-supplement but for preventing the Eh becoming low.

REFERENCES

Patric, W.H. Jr. and I.C. Mahapatro. 1966. Transformation and availability to rice of nitrogen and phosphorus in water logged soils. Advances in Agronomy. 18 : 323-355.

2.4 ELECTRICAL CONDUCTIVITY OF EXTRACT FROM SOIL SATURATED PASTE–WILCOX (1950)

2.4.1 EC of extract from soil saturated paste by non digital conductivity meter

2.4.1.1 Principle

The conductance is the reciprocal of R in the equation E = IR where E is the electric potential in volt, I is the current in amperes, R is the resistance offered to the flow of current in ohms. The unit of conductance in mho.

The conductance measured between a pair of platinized electrode 1 cm^2 in cross sectional area placed 1 cm apart at 25°C is termed specific conductance which is numerically equal to ECt × ft × K, where ECt is EC in mho/cm at 't' temperature, ft is the temperature correction facter to express the conductance at 25°C, and K is the cell constant of conductivity cell which is estimated from the equation $K = 1411.8 \times 10^{-6} \times Rt/ft$ by measuring the resistance of .01N KCl at t temperature, ft is the temperature correction factor, and 1411.8×10^{-6} mho/cm is the specific conductance of .01 N KCl solution at 25°C. The Rt can be also estimated by taking the reciprocal of ECt.

In soil studies, EC is a measure of concentration of soluble salts. The EC of extract from soil saturated paste is measured to ascertain the problem of salinity. The extract from saturated paste of a saline soil has specific conductance greater than 4 milli mho/cm at 25°C. The conductivity is measured by non-digital or digital conductivity meter.

2.4.1.2 Reagents

Normally no reagents are required unless need is felt for either replatinization of the electrodes of conductivity cell or for checking K value of the conductivity cell.

2.4.1.3 Procedure

1. Preparation of soil saturated paste. Transfer 250g of air dry < 2 mm soil sample into 1l capacity plastic beaker. Add distilled water through a burette slowly drop by drop uniformly over the entire soil surface. When the soil appear to have been uniformly wetted, take a stainless steel spatula of large size and work with the soil to form a homogeneous soil saturated paste.

If soil mass appear to be hard, add distilled water in drops and mix the soil with spatula, untill the soil mass begins to glisten with no free water appearing at the soil surface. Besides it should attain a free flowing consistency.

The free flowing consistency of the soil paste may be assessed by resting spatula on one side of soil paste and then moving the spatula across the soil mass towards the opposite side making sufficiently wide straight gap dividing the soil mass into two halves. Now, tap the base of plastic beaker on table, if the gap in soil mass gets readily filled, the paste is free flowing.

If the gap in soil mass remain unfilled on tapping base of the beaker against table top, add some more distilled water in drops and again work the soil mass with spatula. Now, turn the beaker in a slanting position and lift soil mass on the blade of spatula. If the entire soil mass falls *i.e.* slides down rather together on vertical tilting of the spatula blade the soil paste has reached a free flowing consistency.

Again move the spatula blade resting flat on soil mass across the beaker making sufficient gap between the two halves. Tap the base of plastic beaker on table top. Now, the gap in the soil mass should get readily filled to qualify for having reached free flowing consistency.

Remove soil paste adhering to the spatula, and put it back in the beaker. Cover the beaker with a large petri dish or else with a card board and leave it over night so that the composition of soil solution become in equilibrium with the ionic composition of the exchange complex.

On the following day, check whether the soil mass is still retaining the characteristics of a saturated paste. If it has become dry, add some drops of distilled water, and work the soil with spatula until it attains free flowing consistency and it glistens without appearance of free water.

Transfer about 25 g of saturated paste in a tared aluminium can. Weigh aluminium can plus soil saturated paste and then transfer the aluminium can with soil saturated paste in an oven at 105°C to dry the soil for 24 hours to find out percent moisture in soil saturated paste.

2. Drawing of extract from soil saturated paste. Place Whatman no. 44 filter paper in a buchner funnel after cutting the filter paper of the size of buchner funnel. Transfer the entire soil saturated paste on the filter paper. Spread it uniformly sealing the surface, and along the inner wall of the buchner funnel.

Fill straw or glass packing material in a large size suction flask so that it can hold one wide mouth 50 ml capacity pyrex test tube underneath the suction flask mouth. Place a clean 50 ml capacity test tube inside the suction flask and fit the buchner funnel containing soil saturated paste into the mouth of suction flask in such a way that the buchner outlet tube remain inside the mouth of test tube placed in the suction flask. Connect the suction flask to the suction pump.

Now switch on the suction pump for few minutes and then switch it off. Discard any extract collected in the receiving test tube. Again fit the buchner funnel into the mouth of suction flask keeping its outlet tube inside the mouth of test tube placed inside the suction flask.

Switch on the suction pump and check that no air is leaking inside the suction flask by the side of adapter or through the buchner funnel. When the assembly is found air tight, let the suction pump run for several hours until the air begins to leak through the soil mass.

On placing palm tightly on the mouth of buchner funnel, if suction is felt by the palm, air has began to leak through the soil mass and the suction pump be switched off. The buchner funnel be removed from the suction flask. The test tube containing the extract from saturated paste be taken out of the suction flask. A rubber stopper be placed tightly on its mouth. The test tube be labelled properly and safely preserved for the determination of EC, pH and concentration of basic cations.

3. Measurement of EC of extract from soil saturated paste by use of non-digital conductivity meter. Dilute the extract from soil saturated paste 10 times by transferring 5 ml of extract into 50 ml volumetric flask and then add distilled water to make the volume to 50 ml. Stopper, and mix the contents. Then transfer 40 ml of the diluted extract into 100 ml beaker for measuring its electrical conductivity.

4. Operation of the conductivity meter. Follow instructions of the manufacturer of conductivity meter. Following procedure may be followed in case of ELICO conductivity meter provided with cathode ray as null point indicator termed as magic eye.

Before use the conductivity cell should remain immersed in pyrex glass beaker filled with distilled water for atleast 24 hours.

Connect the leads of the conductivity cell in the socket provided in the conductivity meter and provide electrical connection to the conductivity meter through a voltage stabilizer. Then, switch on the conductivity meter, and let it run for 10 minutes time to allow it to warm.

In the mean time, remove the conductivity cell immersed in distilled water, wipe it dry with a blotting paper or else with an ordinary filter paper. Thereafter, rinse the conductivity cell once with diluted extract and discard the rinsing. Then immerse the conductivity cell in diluted extract from soil saturated paste.

5. Measurement of the electrical conductance. Now, rotate the multiplier factor switch one point at a time towards the conductivity range in micro mho/cm until widest gap is attained in the magic eye. Note down the multiplication factor.

Then rotate the meter dial knob towards the left or right which ever widens the gap in the magic eye. Stop rotating the knob when widest gap is attained in the magic eye. Note down the dial reading accurately.

Dip a thermometer in the diluted extract of soil saturated paste and note down its temperature after 1 minute. Find out the ft value *i.e.* the temperature correction factor at the observed temperature of diluted extract from the table giving value of ft at different temperature.

2.4.1.4 Calculation

$$EX_{25} \times 10^3 \text{ of extract from saturated paste} = \frac{DR \times MF \times ft \times K \times dt}{10^3} \text{ milli mho / cm}$$

where

DR is dial reading

MF is multiplication factor in micro mho/cm

ft is the temperature correction factor for expressing EC at 25°C.

K is the cell constant

dt is the dilution times of the extract prior to measurement of EC.

EC_{25} is the specific conductance at 25°C in mho/cm.

2.4.2 Measurement of EC of extract from soil saturated paste by use of digital conductivity meter

2.4.2.1 Principle. Same as in section 2.4.1.1.

2.4.2.2 Reagents

Potassium chloride 0.01N solution. Dissolve 0.7456 g of dry KCl (AR) in distilled water and dilute it to 1000 ml mark with distilled water in 1l volumetric flask. Stopper and mix the contents. This solution has a conductivity of 1411.8×10^{-6} mho/cm at 25°C.

2.4.2.3 Procedure

Take 10 times diluted extract from soil saturated paste and measure its EC in milli mho/cm by use of digital conductivity meter following the procedure described in water analysis (See section 4.3.2).

2.4.2.4 Calculation

$EC_{25} \times 10^3$ of extract from soil saturated paste. $= DMR \times ft \times dt$ milli mho/cm

where

EC_{25} is specific conductance in mho/cm at 25°C

DMR is digital meter reading in milli mho/cm

ft is the temperature correction factor for expressing conductivity at 25°C.

dt is dilution times of the extract from soil saturated paste prior to measurement of its electrical conductivity.

2.4.2.5 Comments

The procedure for the replatinization of electrodes of conductivity and determination of K value of conductivity cell of non-digital conductivity meter are described in water analysis (See section 4.3.1).

% salt in soil may be estimated from the following empirical relationship.

$$\%\text{ salt in soil} = 0.064 \times EC_{25} \times 10^3 \text{ of extract from soil saturated paste} \times \frac{\%\text{ moisture in saturated paste}}{100}$$

The percent moisture in saturated paste can be determined directly by oven dry method or else it may be estimated from the weight of a known volume of saturated soil paste from the following empirical relationship.

$$\%\text{ moisture in saturated paste} = \frac{100\,(2.65\,V \times W)}{2.65\,(W - V)}$$

where V and W denote the volume and mass of saturated paste. The density of water and soil is assumed to be 1.00 and 2.65 g/ml.

If $EC_{25} \times 10^3$ of extract from 1 : 2 soil water suspension (w/v) has been determined, an estimate of specific conductance of extract from soil saturated paste can be calculated from the following empirical relationship.

$$\text{EC}_{25} \times 10^3 \text{ of extract from soil saturated paste} = \text{EC}_{25} \times 10^3 \text{ of extract from 1 : 2 soil water suspension} \times F$$

$$\text{Where } F = \frac{200}{\%\text{ moisture in saturated paste}}$$

When EC_e *i.e.* EC of extract from soil saturated paste is greater than 4 millimho/cm at 25°C, the soil is grouped as saline and when EC_c is less than 4 millimho/cm at 25°C, the soil is grouped as non-saline. The salinity scale for rating salt effected soils are given below :

Specific Conductance EC_e mmho/cm	% salt in soil	Salinity Rating	Remarks
0–2	0–0.1	Non-saline	Salinity effects mostly neligible
2–4	0.1 – 0.3	Very slightly Saline	Yield of very sensitive crops restricted
4–8	0.3 – 0.5	Moderately saline	Yield of many crops affected
8–16	0.5–1.0	Strongly saline	Only tolerant crops yield satisfactory
>16	>1.0	Very strongly saline	Only a few very tolerant crops yield satisfactory.

REFERENCES

Jackson, M.L. 1958. Soil Chemical Analysis, Prentice-Hall of India, Private Limited, New Delhi (1973).

Richard, L.A. et. al. (ed). 1968. Diagnosis and Improvement of Saline And Alkali Soils. USDA Hand Book no. 60, Oxford & IBH, Publishing Co. New Delhi.

Wilcox L.V. 1950. Electrical conductivity, Amer. Water Works Assoc. J. 42 : 775–776

2.5 DECOMPOSITION OF SOIL SAMPLE FOR TOTAL ELEMENTAL ANALYSIS

2.5.1 HCl extract of soil

2.5.1.1 Principle

Soil is digested with relatively concentrated hydrochloric acid over a hot sand bath for an hour. Soil gets decomposed and mineral elements other than silica are solubilized.

2.5.1.2 Reagents

Approx. 8 N HCl : Dilute 675 ml of AR grade hydrochloric acid to 1l with distilled water.

2.5.1.3 Procedure

1. Digestion of soil sample with hydrochloric acid. Transfer 10 g air dry 2 mm soil sample into a 250 ml conical flask. Add 25 ml of approx. 8N hydrochloric acid. Keep a funnel over the flask and keep the flask on a hot sand bath. Digest the contents for one hour.

2. Filtration of the acid digest of soil. Thereafter, cool the digest and add 50 ml of distilled water. Place the funnel on the mouth of 250 ml vol. flask. Set dry Whatman no. 42. filter paper in the funnel. Transfer the digest in small portions into the funnel lined with filter paper and collect the filtrate. Transfer the digest in conical flask quantitatively into the funnel and collect the filtrate. Wash the residue on filter paper with hot distilled water until the filtrate becomes free from chloride. Make the volume of the filtrate to the 250 ml mark with distilled water. Stopper the flask and shake the contents.

2.5.1.4 Comments

As mineral elements in soil other than silica are solubilized in the HCl digest, the filtrate may be preserved for the estimation of total P, sesquioxides and basic cations in the soil.

The soil sample for the preparation of HCl extract may be preferably ignited in a muffle furnace at 600°C for 6 hours in a platinum dish.

For igniting soil sample, transfer 15g of oven dry soil sample into a platinum dish. Cover it with a mica sheet, and transfer it into a muffle furnace. Then raise the temperature to 600°C and maintain the temperature for 6 hours. Cool the ignited sample and weigh to determine % loss on ignition. Deduct % moisture estimated earlier.

The HCl soil extract may be used for the estimation of Si, R_2O_3, Fe, Al, Ca, Mg and P.

REFERENCES

Amma, M.K. 1989. Plant and Soil Analysis, Rubber Research Institute, Rubber Board, Kottayam-686009, KERAL.

Chopra, S.L. and J.S.K. Kanwar, 1982. Analytical Agricultural Chemistry. Kalyani Publishers, B.O. 4862 2B. Bharat Ram Road, 24-Daryaganj, New Delhi–110002.

2.5.2 Decomposition of soil by Na_2CO_3 fusion

2.5.2.1 Principle

The elements in the soil sample is brought into solution by Na_2CO_3 fusion. The finely ground soil sample is mixed with Na_2CO_3 – 7.5 times the mass of soil sample in a 30 ml platinum crucible. The mixture is covered with Na_2CO_3 2.5 times of the sample mass. The crucible is covered with lid, the top remaining partially opened.

After placing the crucible on a silica covered triangle, the mixture is heated by burner flame from the crucible sides until the melt is liquified. Small curds of precipitate float. Otherwise the solution is clear and evolution of gas ceases. Subsequently, the melt is heated with full flame for few minutes and the platinum crucible is swirled to spread the melt in thin layers for easy solution.

2.5.2.2 Equipment/Reagents

Equipments. 30 ml capacity platinum crucible with lid, silica covered triangle or Nichrome triangle, Pure nickle tongs.

Reagents. Sodium carbonate anhydrous (Analar)

2.5.2.3 Procedure

1. Treatment of the soil sample with anhydrous Na_2CO_3 powder. Weigh 0.1000 g of finely ground soil sample < 0.2 mm dried at 100°C for 2 hours or else air dry soil sample a portion of which is being dried in an oven for moisture determination.

Transfer the weighed sample carefully into 30 ml capacity platinum crucible. Add 0.75 g of anhydrous Na_2CO_3 into the platinum crucible. Rotate the crucible with the finger so that added Na_2CO_3 get mixed with the soil sample. Thereafter, place 0.25 g of anhydrous Na_2CO_3on top of the soil plus Na_2CO_3 mixture.

2. Heating the mixture for Na_2CO_3 fusion. On a silica covered triangle, place the platinum crucible in a standing position and place a lid to cover 7/10 th of top. Heat one side of the platinum crucible with a burner. Gradually increase the heat until the melt is liquified.

The fusion is usually completed in 1-2 minutes time when the sample is finely ground. At this time, the flux appears quiet with small curds of precipitate floating. The solution is otherwise clear and the evolution of gas also ceases.

3. Spreading the flux and melt in thin layer. Now grasp the crucible with nickle tongs and swirl the platinum crucible to spread the flux in a thin layer on the sides of lower half of the crucible so that any silica adhering to the sides is fused.

Then, heat the contents with full flame for 2-3 minutes. Thereafter, remove the burner, and swirl the platinum crucible to spread the melt in thin easily dissolved layer.

2.5.2.4 Calculation : Nil

2.5.2.5 Comments

Fusion in the electric furnace at 1200°C avoids transfer of Fe from the sample to the platinum metal which occurs if very good oxidizing conditions are not maintained.

The Na_2CO_3 fused sample is used for the determination of Se, Al, Ca and Mg. One can also determine Fe and Ti in the fused sample. However, for Fe and Ti determination more quickly and accurately the sample should be decomposed with HF.

For bringing elements of sample into solution, sample is either fused with Na_2CO_3 or else the sample is heated in acids such as HCl, HF, $HClO_4$, H_3PO_4 and H_2SO_4.

The platinum crucible is expensive. It should be carefully handled to prevent any deformation. It should be stored in cotton packing to avoid denting. A porcelain plate or asbestos pad or beaker be employed to set the porcelain utensil on.

Keep the platinum crucible chemically clean by soaking in hot water, boiling in hot HCl, washing with ethanol to remove manganate, fusion with potassium pyrosulfate $K_2S_2O_7$ and pouring out the melt in waste sand.

Do not use chemicals that react with platinum or cause solution or volatization of platinum. Chlorine, free bromine, and iodine attack platinum. Do not digest with aqua-regia in which Cl is liberated. Hydroxides, oxides, peroxides, nitrites and cyanides of alkalies attack platinum. Reducing conditions result in alloying the platinum. Silica and borate react with platinum under reducing conditions. $FeCl_3$ in the presence of HCl attacks platinum with the partial reduction of iron to ferrous and the simultaneous appearance of platinum in solution.

Platinum melts at 1770°C but it is volatalized at 1000°C and above. Heating of platinum at a very high temperature for a longer period will cause more loss of platinum by volatalization.

As platinum becomes soft on heating, it should be allowed to cool sufficiently so as to become stiff before grasping with tongs.

2.5.3 Decomposition of soil with HF

2.5.3.1 Principle

For bringing elements present in soil sample in solution, finely ground soil sample < 0.2 mm oven dried for two hours weighing 0.1 g is decomposed with HF in presence of $HClO_4$ in a platinum crucible by heating on a sand bath at a temperature of 200 to 225°C. Condensed organic fraction adhering to cover lid and upper sides of crucible is decomposed by directing burner flame to sides of the crucible.

The constituents remaining in crucible are treated with HCl and then diluted to 2/3 rd of the crucible volume and subsequently it is heated slowly by placing it on a crucible radiator to boil the contents so that the residue get completely dissolved.

If the residue is not completely dissolved, the suspension is evaporated to dryness and then the entire treatment beginning with the addition of HF and $HClO_4$ is repeated until clear solution with no insoluble constitutent are obtained. The extract is then diluted to 100 ml to prepare the test solution for the elemental analysis.

2.5.3.2 Equipments and Reagents

Equipments

(*i*) Platinum crucible-30 ml capacity with cover lid

(*ii*) Crucible Radiator.

It may be made from 100 ml crucible of nickel, iron, or porcelain over which silica covered nichrome triangle is fitted to hold a platinum crucible.

Reagents

(*i*) Perchloric acid - 60% $HClO_4$ (Analar)

(*ii*) Hydrogen fluoride - 48% HF (LR)

(*iii*) Hydrochloric acid - 6N

Add 50 ml conc. HCl into a 100 ml volumetric flask containing 50 ml of distilled water.

2.5.3.3 Procedure

1. Decomposition of soil sample with acids– HF and $HClO_4$. Weigh 0.1000g of finely ground soil sample dried at 100°C for 2 hours or else air dry soil sample–a portion of which is being dried in oven for moisture determination. Transfer the weighed sample into a 30 ml capacity platinum crucible.

Wet the sample with few drops of water and then with 0.5 ml of $HClO_4$ and 5 ml of 48% HF. With a lid, cover 9/10 th of the crucible top after placing the crucible on a sand bath at a temperature of 200-225°C to evaporate the acids to dryness.

2. Decomposition of condensation products of organic fraction. Organic matter which condense on to the cover and upper sides of the crucible are often not attacked completely by $HClO_4$. The presence of dark colour on the cover and inner sides of crucible indicate condensation of organic fraction.

Direct the burner flame on external sides of the crucible to a faint red heat for sufficiently long time to dispel organic matter.

3. Solution of mineral elements. Remove the crucible from sand bath. Allow it to cool. Then, add 5 ml of 6N HCl. Dilute the suspension to 2/3rd of the volume of crucible with distilled water.

Place the platinum crucible on silica covered nichrome triangle fitted on to a crucible radiator. Cover the crucible with the lid. Now heat the radiator crucible with relatively low flame so that the solution boils gently without bumping. The residue should get completely dissolved after 5 minutes of boiling.

4. Repetition of the decomposition process. If the residue is not completely dissolved for reason of presence of coarse grained particles and Al_2O_3, evaporate the solution remaining in the platinum crucible to dryness, and then repeat the entire decomposition process described above beginning with the addition of few drops of water, 0.5 ml of $HClO_4$ and 5 ml of 48% HF. The other alternative is to take fresh sample and proceed further following the entire procedure and continue repetition of the decomposition process until all the residue is dissolved.

5. Preparation of the test solution. When no residue remains in the platinum crucible on completion of the process of decomposition of soil sample with HF - $HClO_4$ treatment, transfer the solution into a 100 ml volumetric flask quantitatively using portions of distilled water for rinsing of the platinum crucible. Transfer all the rinsings to the same 100 ml volumetric flask and make the volume to 100 ml mark with distilled water. Stopper and preserve it not too long for elemental analysis.

2.5.3.4 Calculation : Nil

2.5.3.5 Comments

All the F must be driven off by heating with $HClO_4$. When F is present in the test solution, there is interference in Fe & Al determination. Ordinarily it is possible to determine total K, Na, Ca, Mg and Ti in the test solution. If F has been completely driven off by fuming $HClO_4$, Fe and Al can be also determined in the test solution.

The test solution be preferably stored in a polythene bottle as borosilicate glass ware is attacked by F.

The 30 ml capacity platinum crucible is expensive. The care needed to prevent physical deformation and chemical decomposition of platinum crucible are dealt under the heading comments in section on decomposition of soil sample by Na_2CO_3 fusion (Section 2.5.2.5).

REFERENCES

Amma, M.K. 1989. Plant and Soil Analysis. Rubber Research Institute Rubber Board, Kottayam, Keral, India.

Chopra, S. L and J. S.K. Kanwar, 1982. Analytical Agricultural Chemistry, Kalyani Publishers, B.O. 4862/2B, Bharat Ram Road. 24 Daryaganj, New Delhi-110002.

Jackson, M.L. 1958. Soil Chemical Analysis. Prentice Hall of India Private Limited, M-97 Connaught Circus, New Delhi (1973).

Muthuval P., C. Udayasoorian, R. Natesan, P.P. Ramaswami, 1992. Introduction to Soil Analysis, Tamil Nadu Agricultural University, Coimbatore-641002.

2.6 DETERMINATION OF SiO_2, SESQUIOXIDE AND PHOSPHORUS IN HCL SOIL EXTRACT

2.6.1 Acid insoluble matter (SiO_2) in the soil

2.6.1.1 Principle

The residue obtained on filtration of HCl digest is SiO_2. SiO_2 is determined by igniting filter paper containing SiO_2 inside a tared crucible.

2.6.1.2 Apparatus and Reagents

Crucible, gas burner, desiccator balance.

2.6.1.3 Procedure

Transfer the residue made free from chloride along with the filter paper into a previously weighed silica crucible. Dry the filter paper and residue in an oven and ignite the residue on a strong flame for about an hour. Occasionally stir the mass in the silica crucible with a glass rod. Cool the residue in a desiccator and weigh.

2.6.1.4 Calculation

$$\%\ SiO_2 \text{ in soil sample} = \frac{(\text{Mass of ignited residue in g} - \text{Mass of filter paper ash}) \times 100}{\text{mass of soil samnple in g}}$$

2.6.1.5 Comments

The soils derived from highly weathered parent material under intensive leaching are low in silica. The ratio of SiO_2 to sesquioxides in soil minerals is an index of weathering.

2.6.2 Phosphorus in acid extract of soil A.O.A.C (1950)

2.6.2.1 Principle

The phosphorus extracted from soil by hydrochloric acid is first made alkaline by adding ammonium hydroxide. Thereafter it is made acidic by adding nitric acid. In nitric acid medium, ammonium molybdate is added to precipitate phosphorus as yellow compound of ammonium phosphomolybdate.

The precipitate is collected on a filter paper and washed to make it free from acid. Thereafter, the precipitate is dissolved by adding standard sodium hydroxide in excess. The alkali added in excess is back titrated by standard acid to the phenolphthalein end point. The phosphorus extracted from soil is calculated by multiplying meq of standard alkali used in dissolving precipitated phosphorus by the mass of phosphorus in g dissolved by one meq of alkali used.

Reactions

(*i*) $H_3PO_4 + 12\,(NH_4)_2\,MoO_4 + 21\,HNO_3 \rightarrow$
$\rightarrow (NH_4)_3\,PO_4\,.\,12\,MoO_3 + 21\,NH_4\,NO_3 + 12\,H_2O$

(*ii*) $(NH_4)_3\,PO_4\ 12\,MoO_3 + 23\,NaOH \rightarrow NaNH_4\,HPO_4 + (NH_4)_2\,MoO_4$
$+ 11Na_2\,MoO_4 + 11H_2O$

(*iii*) $2\,NaOH + H_2SO_4 \rightarrow Na_2\,SO_4 + 2H_2O.$

2.6.2.2 Apparatus/Reagents

Reagents

(*i*) conc. NH_4OH - reagent grade

(*ii*) conc. HNO_3 - reagent grade

(*iii*) 5.0% Ammonium molybdate solution. Dissolve 12.5 g of ammonium molybdate in 200 ml of hot distilled water, cool and dilute the solution to 250 ml with distilled water.

(*iv*) 3.0% KNO_3 solution. Dissolve 15.0 g potassium nitrate in distilled water and make the volume to 500 ml mark with distilled water.

(*v*) 4.0 N NaOH solution. Dissolve 40g of AR sodium hydroxide in distilled water and dilute the solution to 250 ml with distilled water.

(*vi*) 1.0N NaOH solution. Dilute 25 ml of 4N sodium hydroxide solution to 100 ml with distilled water.

(*vii*) 0.1N NaOH solution. Dilute 25 ml of 1.0 N sodium hydroxide solution to 250 ml with distilled water and standardize it against standard sulphuric acid to the phenolphthalein end point.

(*viii*) 0.5% phenolphthalein indicator. Dissolve 0.5g phenolphthalein indicator in 100 mL of ethyl alcohol.

(*ix*) 0.1N H_2SO_4 solution. Dilute 3.0 ml of conc. sulphuric acid (AR) to 1 l with distilled water. Standardize the acid against 0.1 N sodium carbonate to the methyl orange end point.

(*x*) 0.1 N Na_2CO_3 solution. Dissolve 0.530 g AR anhydrous dry sodium carbonate in distilled water and make the volume to 100 ml with distilled water.

(*xi*) 0.1% Methyl orange indicator. Dissolve 0.100 g of methyl orange indicator in distilled water and make the volume to 100 ml with distilled water.

2.6.2.3 Procedure

1. Precipitation of phosphorus in hydrochloric acid extract as ammonium phosphomolybdate. Pipette 50 ml of acid extract into 250 ml beaker. Make the solution distinctly alkaline by adding ammonium hydroxide solution. Then, add conc. nitric acid till the solution turns acidic. Now, add 20 ml of 5% ammonium molybdate solution to form ammonium phosphomolybdate (a yellow precipitate). Rub inner side of beaker with a glass rod to promote formation of granular precipitate.

Transfer the beaker on a hot water bath at 40-50°C and let it remain on water bath for 30 minutes. Add little more solution of 5% ammonium molybdate along the inner wall of the beaker to ensure complete precipitation of phosphorus as ammonium phosphomolybdate. Then let the precipitate to settle for 1/2 to 1 hour.

2. Washing of precipitated phosphorus to make it free from acid. Filter the contents through a funnel lined with dry Whatman No. 42 filter paper. Wash the precipitate atleast 5 times with cold 3% potassium nitrate solution to make the precipitate free from acid. Thereafter, wash the precipitate twice with cold water to remove potassium nitrate.

3. Dissolving of precipitated phosphorus in standard sodium hydroxide solution. Now, transfer the precipitate along with the filter paper into the beaker used for the precipitation of phosphorus. Add 0.1 N sodium hydroxide solution from a burette till the precipitate is completely dissolved as indicated by the disappearance of the yellow colour. Add few ml of 0.1 N sodium hydroxide in excess and note down the sodium hydroxide titre used in dissolving the precipitate.

4. Back titration of standard alkali added in excess by standard acid solution. Add few drops of 0.5% phenolphthalein indicator and back titrate the excess standard sodium hydroxide added with 0.1 N H_2SO_4 through a burette until the colour of solution changes from pink to colourless. Note down the 0.1 N sulphuric acid titre.

2.6.2.4 Calculation

g phosphorus per meq of sodium hydroxide

$\because$ 23 g mol of NaOH dissolve 1 g mol. of P

$\therefore$ 1g mol of ,, ,, $\frac{1}{23}$ g mol of P

$\therefore$ 1 g equivalent of NaOH will dissolve $= \frac{1}{23} \times 31$ g of P

1 meq of NaOH will ,, $= \frac{1 \times 31}{23 \times 1000} = .001348$ g P

$$\% \text{ Total P in soil} = \frac{(A - B) \times C \times T \times 100}{A \times W}$$

$$\% \text{ Total } P_2O_5 \text{ in soil} = \% \text{ Total P in soil} \times 2.29.$$

where

S is meq of NaOH added

B is meq of H_2SO_4 used for back titration

C is gP/meq of NaOH

T is total volume of HCl extract prepared in ml.

A is ml extract taken for P precipitation

W is mass of soil in g taken for the preparation of the HCl extract.

2.6.2.5 Comments

The forms of P found in soils are organic P, soluble P, adsorbed P and insoluble P. These forms of P remain in equilibrium with each other. That form of P which readily enters into soil solution to maintain the equilibrium concentration is termed labile P. Thus labile P is of significance for plant nutrition. The labile P is estimated as isotopically exchangeable P or P extracted from soil by Olsen's or by Bray's extractant. The sum of the four forms of P is the total P in soil which is an index of total P reserve in soil.

REFERENCES

A.O.A.C. 1950. Official and Tentative methods of Analysis of the Association of Official Agricultural chemists Ed. 7. 910 pp. Washington.

2.6.3 Sesquioxides (Fe_2O_3+Al_2O_3) in acid extract of soil–A.O.A.C. (1950)

2.6.3.1 Principle

Fe in the extract is first oxidized to Fe^{+++} by adding nitric acid and then boiling it for a minute. Then Fe & Al is precipitated by making the solution distinctly alkaline with conc. ammonium hydroxide. The contents are then boiled on a hot sand bath, and the precipitate is allowed to settle.

The precipitate is collected on a filter paper by filtration of the contents. The precipitate is then washed with distilled water to make it free from chloride. The residue is dried, ignited and weighed after cooling to determine the content of elements as oxides.

2.6.3.2 Reagents

1 Conc. HNO_3 - reagent grade

2. Conc. NH_4OH - reagent grade

2.6.3.3 Procedure

1. Precipitation of Fe & Al as hydroxides. Pipette 50 ml of HCl extract in a 500 ml beaker. Add 1 ml of conc. nitric acid to convert Fe^{++} to Fe^{+++}. Boil the contents for a minute after placing a watch glass on the beaker. Thereafter cool the contents.

Then add conc. ammonium hydroxide drop by drop till the contents become distinctly alkaline to litmus. Boil for a minute after placing watch glass on the beaker. Then, place the beaker with the contents on a hot sand bath for about half an hour so as to allow the precipitate to settle.

2. Collection of the precipitate on filter paper and making it free from chloride by washing with distilled water. Filter the contents through a funnel lined with Whatman No. 42. filter paper. Wash the residue on filter paper with hot water until the filtrate (washing) becomes free from chloride.

3. Drying and igniting the residue collected on filter paper and its weighment after cooling. Weigh a dry clean silica crucible and note its tare weight. Then transfer the precipitate together with the filter paper into the silica crucible. Dry it in an oven and then ignite it on a strong flame until the residue become red hot. Cool inside a desiccator and weigh accurately. The residue is composed of sesquioxide + P_2O_5 + filter paper ash.

2.6.3.4 Calculations

$$\% \text{ Sesquioxide} + P_2O_5 = \frac{(S - B) \times T \times 100}{A \times W}$$

% sesquioxide = % sesquioxide + P_2O_5 – % P_2O_5 in soil.

where

S is the mass of sesquioxide plus P_2O_5 plus ash in g.

B is the mass of ash in g

T is the volume of total extract.

A is ml extract used for the precipitation

W is the mass of soil in g.

2.6.3.5 Comments

Soils formed under tropical climate are high in sesquioxides as compared to the content of SiO_2.

The filtrate together with the washings for making the residue free from chloride in the above described procedure for the estimation of sesquioxide may be collected and evaporated over a water bath to a volume less than 250 ml. It may be preserved after making the volume to 250 ml for the estimation of Ca, Mg and K.

REFERENCES

A.O.A.C. 1950. Official and Tentative Methods of Analysis of the Association of Official Agricultural Chemists Ed. 7. 910 pp. Washington.

Muthuval, P.C. Udayasoorian, R. Natesan, and P.P. Ramaswami, 1992. Introduction to Soil Analysis, Tamil Nadu Agricultural University, Coimbatore 641002.

Jackson, M.L. 1958. Soil Chemical Analysis Prentice Hall of India, New Delhi.

2.7 DETERMINATION OF ORGANIC CARBON–WALKLEY (1947)

2.7.1 Principle

The organic carbon of soil is usually determined by wet oxidation method of Walkley and Black (1947). In this method organic carbon is oxidized by $K_2Cr_2O_7$ in the presence of H_2SO_4 leading to the formation of CO_2 according to the equation given below :

$$2\,K_2\,Cr_2O_7 + 8H_2SO_4 \rightarrow 2\,K_2SO_4 + 2\,Cr_2\,(SO_4)_3 + 8\,H_2O + 3O_2$$

$$3C + 3O_2 \rightarrow 3CO_2$$

While organic carbon is oxidized, $K_2Cr_2O_7$ is reduced. The amount of oxygen consumed during the oxidation of organic carbon is calculated from the difference between the amount of $K_2Cr_2O_7$ taken and the amount remaining after oxidation which is determined by titration with 0.5N ferrous sulphate or ferrous ammonium sulphate in the presence of redox indicator such as diphenylamine, ortho-phenanthroline –ferrous complex or phenylanthranilic acid.

2.7.2 Reagents

(*i*) 1N Potassium dichromate. Dissolve 49.04 g $K_2Cr_2O_7$ in distilled water and dilute it to 1 l. The molecular weight of $K_2Cr_2O_7$ is devided by 6 to obtain the equivalent weight as one gram mole $K_2Cr_2O_7$ furnishes three gram mole oxygen, and three gram mole oxygen is equivalent to six gram mole hydrogen.

(*ii*) 0.5N Ferrous ammonium sulphate or Ferrous sulphate. Dissolve 196.1 g $Fe\,(NH_4)_2\,(SO_4)_2 \cdot 6H_2O$ in 800 ml distilled water containing 20 ml conc. H_2SO_4 and dilute with distilled water to 1 l.

Alternatively, dissolve 140g $FeSO_4 \cdot 7H_2O$ in distilled water, add 40 ml conc. H_2SO_4, cool and dilute with distilled water to 1 l. The reagents should be standardized every day from the blank titre *i.e.* ml of $Fe\,(NH_4)_2\,(SO_4)_2 \cdot 6H_2O$ or $FeSO_4 \cdot 7H_2O$ used to reduce 10 ml of 1N K_2CrO_7.

(*iii*) Conc. H_2SO_4 - 96 percent, sp. gr. 1.84 containing preferably 1.25% Ag_2SO_4.

(*iv*) 0.025 M. Orthophenanthroline Indicator. Dissolve 1.485 g Orthophenanthroline monohydrate and 0.695 g $FeSO_4 \cdot 7H_2O$ in distilled water and dilute the contents with distilled water to 100 ml.

OR

0.5% Diphenylamine indicator. Dissolve 0.5 g reagent grade diphenylamine in 100 ml conc. H_2SO_4. (Analytical reagent). Discard the reagent if it turns brown or blue.

(*v*) Orthophosphoric acid - 85% pure.

(*vi*) 2% Sodium fluoride. Dissolve 2g NaF in distilled water and dilute it to 100 ml.

2.7.3 Procedure

1. Weigh accurately 1.0 g of soil passed through 2 mm sieve and transfer it to a 500 ml conical flask. Add 10 ml of 1 N $K_2Cr_2O_7$ by a pipette and mix the contents

by swirling the conical flask gently. Then add 20 ml of conc. H_2SO_4 containing preferably 1.25% of Ag_2SO_4 slowly along the inner wall of the flask. Rotate the flask gently to ensure complete contact of the reagents with the soil. Keep the flask on a wire gauze for 30 minutes for allowing the completion of the oxidation of organic carbon.

2. After 30 minutes of oxidation of organic carbon by the chromic acid, add 100 ml of distilled water, 10 ml of 85% H_3PO_4, 10 ml of 2% NaF or 0.2 g NaF and 2 ml of diphenylamine indicator into the flask.

Titrate the solution by adding ferrous ammonium sulphate or ferrous sulphate in small portions until the solution colour changes from blue violet to green or brilliant green. Swirl the flask after each addition of the titre. Note down the burette reading, add 0.5 ml of 1 N $K_2Cr_2O_7$ and titrate the solution again by ferrous ammonium sulphate or ferrous sulphate to the same end point. Note down the second burette reading. Similarly run a blank without soil sample.

3. When 0-phenanthroline ferrous complex is to be used as redox indicator instead of diphenylamine, add 200 ml of distilled water to the reaction mixture after 30 minutes of oxidation of organic carbon by the chromic acid.

Then add 5 to 10 drops of o-phenanthroline-ferrous complex indicator and titrate the reaction mixture by adding ferrous ammonium sulphate or ferrous sulphate in small portions until the colour of the reaction mixture changes from dull green to chocolate red. Swirl the flask after each addition of the titre. Note down the burette reading. Add 0.5 ml of 1 N $K_2Cr_2O_7$ and 3 drops of o-phenanthroline ferrous complex indicator, and titrate the reaction mixture again by ferrous ammonium sulphate or ferrous sulphate solution to the same end point. Note down the second burette reading. Similarly run a blank without soil sample.

2.7.4 Calculations

$$\% \text{ Organic Carbon in soil} = \frac{(B - S) \times N \times \text{meq. wt. of C} \times 100}{\text{mass of soil in g}}$$

$$\% \text{ Organic matter} = \% \text{ Organic Carbon} \times 1.724$$

where

B is blank titre i.e. ml of $Fe(NH_4)_2(SO_4)_2$ used for titration of 10.5 ml of 1.0 N $K_2Cr_2O_7$.

S is sample titre i.e. ml of $Fe(NH_4)_2(SO_4)_2$ used for titration of soil sample + 10.5 ml of 1.0 N K_2CrO_7.

$$\text{N is the normality of } Fe(NH_4)_2(SO_4)_2 = \frac{10.5 \times 1.00}{\text{blank titre}}$$

meq wt. of carbon = 0.003

2.7.5 Comments

1. The % organic carbon in soil is used to assess the nitrogen status of soil. The carbon and nitrogen content of organic matter is 58 and 5 percent respectively. The C : N ratio in soils tend to become stabilized. The soil health labs in U.P. interpret the results of % organic carbon as follows :

Very Low	–	Low	–	Medium	–	High
< 0.20		0.21-0.40		0.41-0.80		>0.80%

2. There are other methods of organic carbon determination such as dry combustion method, loss on ignition, oxidation with hydrogen peroxide. However, the wet oxidation method is satisfactory even though there is recovery of only 89% of organic carbon. The recovery of carbon in soil sample increases by raising the temperature of the reaction mixture to 155-160°C in 20-30 minutes time and maintaining the temperature attained for an additional 5 minutes.

3. The size of the soil sample is inversely related to its humus content. The soil sample may be preferably grinded in a pestle-mortar to pass through 0.2 mm sieve. The sample size should be such that the reaction mixture after the oxidation of organic carbon remains orange yellow to yellowish brown. If the reaction mixture turns green, it should be discarded and the determination be repeated with a smaller soil sample. The size of the soil sample be such that more than 80% of $K_2Cr_2O_7$ added is not reduced in the oxidation of organic carbon.

4. The interference from chloride in organic carbon determination is eliminated by use of conc. H_2SO_4 containing 1.25% of Ag_2SO_4. However, Ag_2SO_4 is costly and its use is not feasible for regular soil testing of large number of samples. Ag_2SO_4 powder (a pinch) may be added to the soil sample prior to adding $K_2Cr_2O_7$.

5. The glasswares used in organic carbon determination be degreased by use of detergents such as Labwash or Teepol (0.2% solution). After washing with tap water, it should be rinsed with chromic acid ($K_2Cr_2O_7$ dissolved in conc. H_2SO_4 and diluted with distilled water) and again washed with tap water. The stop cock of burette should also be degreased.

6. The redox indicator diphenylamine is oxidized into blue violet compound of diphenyl benzidine by $K_2Cr_2O_7$. The oxidized diphenylamine is reduced back to colourless on complete reduction of $K_2Cr_2O_7$by the addition of Fe^{+++} ions, and the reaction mixture turns green due to formation of chromic sulphate. As diphenylamine is oxidized at a lower redox potential, the concentration of Fe^{+++} ions formed during titration by Fe $(NH_4I_2$ $(SO_4)_2$in the reaction mixture must be kept low by the incorporation of H_3PO_4 and NaF prior to titration.

7. The redox indicator O–phenanthroline forms a colourless solution on being dissolved in distilled water. However, it combines with Fe^{++} ions to form a red complex. When O-phenanthroline is used as a redox indicator in the organic carbon

determination, the colour of the reaction mixture changes from dull green to chocolate red at the end point. The end point is sharp and readily attained even without the incorporation of 85% H_3PO_4 in the reaction mixture. Some soils adsorb O-phenanthroline and so near the end point some (2-3 drops) more indicator may be added with advantage. Alternatively, the reaction mixture may be filtered using a rapid filter paper on a Buchner funnel before adding the indicator. This would improve titration by ensuring sufficient redox indicator at the end point.

8. In Russia, chromic acid for the wet oxidation of organic carbon is prepared by mixing 0.4 N $K_2Cr_2O_7$ and conc. H_2SO_4 in equal volume. 0.3 to 0.5 g soil sample containing 2 to 4% organic matter is digested with 10 ml chromic acid with a pinch of Ag_2SO_4 in a 100 ml Erlenmeyer flask. The reaction mixture is heated to boil for 5 minutes after placing a funnel into the mouth of Erlenmeyer flask .The funnel on cooling is rinsed with 10 ml distilled water. The rinsings together with the reaction mixture is titrated with 0.1 N ferrous ammonium sulphate after adding 3 & 4 drops of 0.2% CH_3 $H_{11}O_2N$ (phenylanthranilic acid) in 0.2% Na_2CO_3. The phenylanthranilic acid (0.2 g) is grinded after moistenring with 0.2% $Na_2CO_3$with a glass rod in a porecelain crucible. When a liquid paste is formed it is diluted to 100 ml with 0.2% Na_2CO_3. At the end point, the colour of the reaction mixture changes from cherry-violet to bright green. The recovery of organic carbon in the above method is claimed to be 95%. Simakov (1957) recommended the use of N phenylanthranillic acid as a redox indicator in organic carbon determination.

9. In colorimetric determination of organic carbon, titration is omitted after wet oxidation and the clear reaction mixture is decanted to determine its optical density at 420 mμ wave length. The standards are prepared with dextrose after wet oxidation with chromic acid. The optical density of standard is determined and a calibration curve is drawn by plotting optical density on the ordinate and the corresponding mg of organic carbon in the abscissa of a linear scale graph paper. The content of organic carbon in soil sample is estimated by reference to the calibration curve.

10. The humus is composed of 58% of organic carbon. Thus, % organic carbon multiplied by 1.724 give an estimate of % organic matter. The organic matter contains 5% N, but in soil 98% of total N remains in organic combination.

The C : N ratio in most soils is usually stabilized to 10 : 1 in sub-tropical region. Consequently if % organic C is known, % N in soil can be estimated. One can estimate organic carbon from total N, but the determination of total N is laborious, time consuming and expensive and so total N determination is practically not feasible for regular testing of large number of soil samples. The organic carbon give a measure of nitrogen status of the soil and the need of fertilizer nitrogen of different crops in arable cropping is based on % organic carbon in soil.

REFERENCES

Black, C.A. (ed). 1965. Methods of Soil Analysis. Vol. 2. Am. Soc., Agron. Madison, Wisconsin, U.S.A.

Jackson, M.L. 1973. Soil Chemical Analysis. Prentice-Hall of India New Delhi-12.

Muhr, G.R., N.P. Datta, H. Sankarasubramoney, V.K. Leley, and R.L. Donahue. 1965. Soil Testing in India ; U.S. Agency for Indian. Development Mission to India, New Delhi.

Walkley, A. 1947. Critical examination of rapid method for determining organic carbon in soils, effect of variation in digestion conditions and of inorganic soil constituents. Soil Sci. 632 : 251.

2.8 TOTAL N IN SOIL SAMPLE MODIFIED KJELDAHL METHOD

2.8.1 Principle

Soil sample is digested with sulphuric acid-salicylic acid mixture in the presence of sodium thiosulphate and the catalyst mixture viz. anhydrous sodium sulphate and metallic mercury or mercuric oxide with external heat. During the digestion, C of organic fraction is oxidized to CO_2, and organic N is initially converted into amino form (R–$COONH_2$) which on oxidation by sulphuric acid is converted into ammonium sulphate. Nitrates in the soil form nitric acid in the presence of sulphuric acid. The nitric acid react with salicylic acid ($C_6H_4OHCOOH$) to form nitrosalicylic acid ($C_6H_3OHCOOHNO_2$) which is reduced by sulphurous acid (H_2SO_3) produced by the action of sulphuric acid on sodium thiosulphate to form amino salicylic acid. The amino salicylic acid ($C_6H_3NH_2OHCOOH$) is oxidized by sulphuric acid to convert amino nitrogen into ammonium sulphate.

Then, the digested material is distilled using Kjeldahl nitrogen distillation assembly after making it strongly alkaline. The ammonium distilled is collected into boric acid solution containing mixed indicator. Ammonia reacts with boric acid to form ammonium borate (NH_4. B $(OH)_4$) which is titrated by a standard acid.

2.8.2 Reagents

(*i*) Conc. H_2SO_4 (sp. gr. 1.84. N.-free)

(*ii*) $C_6H_4OHCOOH$ (reagent grade)

(*iii*) $Na_2S_2O_3$ (reagent grade)

(*iv*) Anhydrous Na_2SO_4or K_2SO_4 (reagent grade)

(*v*) Hg or HgO (reagent grade)

(*vi*) Zinc granules (reagent grade)

(*vii*) Paraffin

(*viii*) 40% NaOH. Dissolve 400g sodium hydroxide pellets in 600 ml distilled water and let it remain under a fume hood until it cools down. Store the solution in a screw top bottle.

(*ix*) 10.0 N HCl. Dilute 218.2 ml of conc. hydrochloric acid (sp. gr. 1.18, Minimum Assay 35.4%) to 250 ml with distilled water.

(*x*) 2.0 N HCl. Dilute 20 ml of 10 N hydrochloric acid to 100 ml with distilled water.

(*xi*) 0.05 N HCl. Dilute 25.0 ml of 2N hydrochloric acid to 1000 ml in a volumetric flask with distilled water.

(*xii*) 0.1 N Na_2CO_3. Dissolve 0.530 g AR sodium carbonate in distilled water and make the volume to 100 ml. It is desireable to transfer 10-20 g AR grade Na_2CO_3 in pyrex beaker and then heat it to 270°C for 2 hours. Thereafter cool the beaker in a desiccator before weighing desired amount of the chemical.

(*xiii*) 0.1% Methyl orange. Dissolve 0.100g methyl orange in distilled water and make the volume to 100 ml.

(*xiv*) Methyl red-bromcresol green mixed indicator. Dissolve 0.066 g methyl red and 0.099 g bromcresol green in 100 ml of ethyl alcohol.

(*xv*) 2% H_3BO_3 solution containing mixed indicator. Dissolve 20g boric acid in distilled water and make the volume to 1000 ml. Then add 20 ml mixed indicator into 1 l volumetric flask.

2.8.3 Procedure

1. Digestion of the soil sample in a 800 ml Kjeldahl flask. Weigh 5g of loam to clayey soil on a butter paper. Fold the sample with the butter paper and drop it at the bottom of Kjeldahl flask. Add 25 ml of 96% conc. sulphuric acid containing 5% salicylic acid wetting the sample thoroughly. Let it stand for 30 minutes and then add 5g sodium thiosulphate, swirl the contents and again let it stand for 30 minutes.

After 30 minutes, add 10 g sodium sulphate or potassium sulphate, 0.65 g Hg or 0.70 HgO and 1 g paraffin. Heat gently till frothing continues and then heat briskly till the contents become clear. Thereafter, digest it for 60 minutes, and then let it cool to room temperature.

2. Distillation of the digested material. Add 200 ml of distilled water, some glass beads, few zinc granules and 1 g paraffin into the Kjeldahl flask containing sample digest. Then fit the Kjeldahl flask into the Kjeldahl distillation assembly.

Transfer 50 ml of 2% boric acid solution containing the mixed indicator into 250 ml Erlenmeyer flask. Place the Erlenmeyer flask below the condenser, and dip the condenser outlet tube in the boric acid solution. Fill the feeding funnel of distillation assembly with about 10 ml distilled water. Lift rod shaped stopper slightly to let a portion of distilled water flow into the Kjeldahl flask. If boric acid solution give off air bubbles due to displacement of air by introduction of distilled water into the Kjeldahl flask, the glass joints are air tight and ammonia shall not leak through the joints during the distillation.

Then add 120 ml of 40% sodium hydroxide solution into the Kjeldahl flask through the feeding funnel. As the last portion of sodium hydroxide solution in the feeding funnel enters into the drain hole, plug the drain hole with glass stopper and add about 50 ml distilled water in the feeding funnel. Let about 20 ml distilled water flow into the Kjeldehl flask so that the alkali from the drain hole get washed into the Kjeldahl flask.

Thereafter, regulate the flame or heating unit so that the contents of Kjeldahl flask begin to boil in 5 minutes time. Continue heating until 120 ml distillate is collected. Lift the condenser outlet tubing from the boric acid solution and wash the tip of glass tubing with 5 ml distilled water and collect the washing also in the distillate receiving flask. Place a red litmus paper below the condenser outlet tubing collecting few drops of distillate. If the red litmus paper does not turn blue, the distillate is free from ammonia, stop the distillation by switching of the heating unit and by stopping the flow of gas into the burner. Similarly run a blank determination using 0.2 g sucrose and following the above described procedure for the digestion of sample and for the distillation of the digested material.

3. Standardization of acid. Pipette 5.0 ml of 0.1 N sodium carbonate solution into 125 ml conical flask. Add 2 drops of 0.1% methyl orange indicator, and titrate with the acid in small portions until the colour of solution changes from yellow to orange. Heat the contents to boil gently just for 1 minute and cool.

If the orange colour is retained by the solution on cooling, the end point has reached and note down the titre. Repeat the titration taking another 5.0 ml of 0.1 N sodium carbonate solution into another conical flask. Estimate the normality of acid by taking the product of volume and normality of sodium carbonate solution and divide it by acid used in the titration of alkali.

4. Titration of sample and blank distillate collected in 2% boric acid containing mixed indicator. Fill in the burette with the standard acid and titrate first sample distillate collected in boric acid by adding the standard acid in small portions. Continue titration with the acid until ammonium borate (NH_4. $B(OH)_4$) is completely neutralized. At the end point the colour of solution changes from green to distinct pink. Note down the acid titre, and fill the burette again with the standard acid. Then titrate the blank distillate with the standard acid until the colour of the solution changes from green to distinct pink as obtained on titration of sample distillate. Note down the acid titre for the blank.

2.8.4 Calculation

% N in soil

$$= \frac{\text{Acid titre (Sample–blank)} \times \text{Normality of acid} \times \text{meq. wt. of N} \times 100}{\text{mass of soil in g}}$$

where meq. wt. of N = 0.014 g

2.8.5 Comments

The Kjeldahl method of total N determination is satisfactory provided procedure described is strictly followed and it does not differ from sample to sample. It is, however, time consuming and it is not practically feasible to undertake the analysis for large number of samples.

Plants absorb ammonical and nitrate forms of nitrogen which in soil is not greater than 2.5% of total N, and yet total N in soil is important as soil productivity increases with the increase in total N. The total N can be a measure of potential fertility of the soil. In alluvial soils of Allahabad, sewage irrigated soil containing 2 to 3% organic C give good yield of all crops grown without use of any fertilizer but adjoining fields irrigated with tube well water contain 0.3 to 0.4% organic C and all crops grown require liberal application of N fertilizer for satisfactory yield. The soil tests show that sewage irrigated soils are high in primary nutrient elements as well as high in cation exchange capacity.

REFERENCES

Jackson, M.L. 1973. Soil Chemical Analysis, Prentice–Hall of India Private Limited, New Delhi.

2.9 READILY MINERALIZABLE NITROGEN IN SOIL–ALKALINE PERMANGANATE METHOD OF SUBBIAH AND ASIJA (1956)

2.9.1 Principle

Soil is distilled in the presence of excess of alkaline permanganate. The nascent oxygen liberated by $KMnO_4$in the presence of NaOH oxidizes readily oxidizable fraction of organic matter. Organic carbon is lost as CO_2 and organic N is transformed into NH_3.

The NH_3 formed remains unstable in alkaline medium and it is readily distilled at boiling temperature in the distillation unit. The distillate containing NH_3 is collected in 2% boric acid solution. The ammonia reacts with boric acid forming ammonium borate which is neutralized by titration with a standard acid using mixed indicator.

2.9.2 Reagents

(*i*) 0.32% $KMnO_4$. Dissolve 0.80 g of potassium permanganate in distilled water and make the volume to 250 ml.

(*ii*) 2.5 % NaOH. Dissolve 6.25 g sodium hydroxide pellets in distilled water and make the volume to 250 ml.

(*iii*) Mixed indicator. Dissolve 0.066 g methyl red and 0.099 g brom cresol green in 100 ml of ethyl alcohol.

(*iv*) 2% H_3BO_3. Dissolve 20 g boric acid in distilled water and make the volume to 1000 ml. Then add 20 ml of mixed indicator into 1000 ml of 2% boric acid solution. If the solution colour is faint pink, add 1 or 2 drop of 2N hydrochloric acid.

(*v*) Paraffin.

(*vi*) Glass beads.

(*vii*) 10.0 N HCl. Dilute 218.2 ml of conc. hydrochloric acid (Sp. Gr. 1.18, Minimum Assay 35.4%) to 250 ml with distilled water.

(*viii*) 2.0 N HCl. Dilute 20 ml of 10 N hydrochloric acid to 100 ml with distilled water.

(*ix*) 0.02 N HCl. Dilute 10 ml of 2N hydrochloric acid to 1000 ml with distilled water. Standardize the dilute acid against 0.1 N sodium carbonate using 0.1% methyl orange indicator.

(*x*) 0.1% Methyl Orange Indicator.. Dissolve 0.1 g of methyl orange in distilled water and make the volume to 100 ml.

(*xi*) 0.1N Na_2CO_3. Dissolve 0.530g AR sodium carbonate in distilled water and make the volume to 100 ml.

2.9.3 Procedure

1. Distillation of readily mineralizable nitrogen from the soil sample. Set the Kjeldahl nitrogen distillation assembly after placing 1g paraffin, few glass beads and 20 g air dry 2 mm soil sample in the 800 ml kjeldehl flask. Transfer 40 ml of 2% boric acid solution containing the mixed indicator into a 250 ml Erlenmeyer flask. Dip the outlet of condenser tube into the boric acid solution. Then fill the feeding funnel with about 10 ml distilled water. Lift the rod shaped stopper to let the distilled water into the 800 ml Kjeldehl flask. If boric acid solution give off air bubbles, lower the stopper to plug the drain hole. The appearance of air bubbles suggest that all the glass joints are air tight, and the ammonia will not escape through the joints during the distillation.

Thereafter, transfer first 100 ml of freshly prepared 0.32% potassium permaganate solution into the feeding funnel. Lift the funnel stopper and let the potassium permanganate solution flow into the 800 ml Kjeldahl flask. Similarly transfer 100 ml of 2.5% sodium hydroxide solution into the 800 ml Kjeldahl flask. Then add about 25 ml distilled water into the feeding funnel and lift the funnel stopper slightly to allow a portion of distilled water to flow in the kjeldahl flask. This would remove the alkali from the drain hole.

Then switch on the heating unit or light the burner for heating the Kjeldahl flask to boil in 5 minutes. Continue heating until 120 ml distillate is collected in the receiving conical flask filled with 40 ml of 2% boric acid containing mixed indicator.

During the distillation if there is violent bumping and vigorous frothing, switch off the heating unit or reduce the burner flame for few seconds so that frothing does not reach the inlet of the condenser. Switch on the heater or raise the burner flame, the moment bumping and frothing calms down.

After collecting 120 ml distillate, lift the condenser outlet tubing out of the boric acid solution. Wash the outlet tubing with 10 ml distillated water and collect the washing into the receiving flask containing the distillate. Similarly run a blank without soil sample and collect 120 ml distillate in 40 ml of 2% boric acid containing mixed indicator.

2. Standardization of the acid. Transfer 5 ml of 0.1 N sodium carbonate solution into 125 ml capacity conical flask. Add 1 drop of 0.1% methyl orange indicator and titrate with the acid until the colour of the solution changes from yellow to orange. Heat the contents to boil gently and cool. If orange colour is retained by the solution, the end point has reached and note down the acid titre. Repeat the titration taking another aliquot of 5.0 ml of 0.1 N sodium carbonate. Estimate the normality of acid

by taking the product of volume and normality of sodium carbonate and divide the product by the acid used in the neutralisation of the alkali.

3. Titration of sample and blank distillate collected in 40 ml of 2% boric acid. Titrate sample distillate collected in boric acid by adding standard acid in small portions. Continue titration with the acid until the ammonium borate is completely neutralized. At the end point colour of the solution changes from green to distinct pink. Note down the acid titre. Then titrate the blank distillate with the acid to the same end point.

2.9.4 Calculation

N in lbs/acre

$$= \frac{\text{HCl (Sample–blank) titre} \times \text{Normality of acid} \times \text{meq. wt. of N} \times 2 \times 10^{6}}{\text{mass of soil in g}}$$

$$\text{N in kg/ha} = \frac{\text{N in lbs/ acre} \times 2.47}{2.2}$$

meq. wt. of N = 0.014 g

Assumption – mass of one acre furrow slice weighs 2×10^6 lbs.

2.9.5 Comments

If the distillate get sucked from the receiving flask into the Kjeldahl distillation flask during the distillation process, take fresh 40 ml of 2% H_3BO_3solution in the receiving flask, and redistill the contents of Kjeldahl distillation flask. Such occurrences can be, however prevented by deputing a person with proper instruction to immediately remove the extension tube of condenser outlet dipped in the contents of receiving flask when burner estinguishes, or electricity goes off or else on noticing sucking back of the distillate from the receiving flask.

The readily mineralizable N is a measure of available N in soil. The available N < 250 kg/ha may be interpreted as low, 250-500 kg N/ha as medium and >500 kg N/ha as high. The mineralizable N is likely to vary with the substrate, moisture content, soil aeration and soil temperature.

The total N in temperate soils may be taken to be high as the soils normally contain 5 to 10% organic matter. However, the crops grown on these soils do require starter nitrogen to boost their initial growth when the rate of mineralizable N is low for reason of low temperatures. The mineralizable N can serve as an index of active fertility of the soil.

REFERENCES

Black, C.A. (ed) 1965. Methods of Soil Analysis. Vol. 2 Am. Soc. Agron. Madison, Wisconsin, U.S.A.

Subbiah B.V. and C.L. Asija 1956. A rapid procedure for the estimation of available nitrogen in soils. Current Sci. 25 : 259-260.

2.10 AMMONICAL AND NITRATE NITROGEN IN SOIL EXTRACTS

2.10.1 NH_4-N in soil extract by Kjeldahl distillation method Jackson (1958)

2.10.1.1 Principle

NH_4 - N in soil may be extracted by acidified NaCl or by morgan's reagent: Acidified NaCl extractable NH_4. N is conveniently determined by Kjeldahl distillation method.

Some paraffin wax, few glass beads, and the extract from soil is added to 800 ml Kjeldahl flask. The kjeldahl flask is fitted into the distillation apparatus.

The soil extract is made strongly alkaline by adding 40% NaOH. Then, it is heated to boil continuously. The condenser outlet tube is placed inside a receiving flask immersed in sufficient 2% H_3BO_3containing mixed indicator. NH_3 distilled into H_3BO_3 is entrapped as ammonium borate which is indicated by change in colour from pink to green. The ammonium borate is titrated by standard mineral acid.

2.10.1.2 Reagents

(*i*) Sodium chloride - 10% (pH adjusted to 2.5). Dissolve 100g NaCl in distilled water and dilute the solution to 900 ml. Acidify with HCl. Adjust the pH to 2.5 and make the volume to 1000 ml with distilled water.

(*ii*) Boric acid - 2% solution containing mixed indicator. Dissolve 20g H_3BO_3 in distilled water and dilute it to 900 ml. Then, add 20 ml of brom cresol green-methyl red mixed indicator and make the volume to 1000 ml with distilled water.

(*iii*) Sodium hydroxide 40%. Transfer 400g NaOH pellets into 1.5 l beaker and add 600 ml of distilled water keeping the beaker inside a fume hood. Allow the beaker to cool and then transfer the NaOH solution into a glass reagent bottle with screw cup.

(*iv*) Brom cresol green – methyl red indicator. Dissolve 0.1g of brom cresol green and 0.07 g of methyl red in 100 ml of 95% ethyl alcohol.

(*v*) Hydrochloric acid – .02 N. Dilute 4.0 ml of 5.0 N HCl to 1000 ml with distilled water. Check the normality of the acid against 0.1 N Na_2CO_3 using methyl orange indicator.

2.10.1.3 Procedure

1. Extraction of NH_4 from soil sample. From a freshly taken soil sample, transfer about 20 g moist soil into a tared aluminium can, weigh aluminium can + moist soil and transfer the container to an oven at 105°C to dry for 24 hours in order to determine the moisture extent.

Weigh 100 g soil sample from the same lot and transfer into 500 ml capacity Erlenmeyer flask. Add 200 ml of 10% NaCl (acidified) into the flask containing 100 g soil. Stopper the flask and shake for 30 minutes on a mechanical shaker.

Filter the suspension through a buchner funnel lined with 11 cm dia. Whatman 42 filter paper. The buchner funnel be fitted into the mouth of a clean suction flask connected to a vacuum pump to accelerate the process of filtration. The initial filtrate 5-10 ml be discarded after rinsing the suction flask and the sample test solution container. Continue filtration until most of the soil extract is obtained.

2. Distillation of NH_4-N from soil extract. Add few glass beads, 1 g of paraffin wax and 120 ml of acidified NaCl soil extract into 800 ml Kjeldhal flask. Attach the Kjeldahl flask carefully into the Kjeldahl distillation assembly. Immerse the condenser outlet tube in 40 ml of 2% H_3BO_3 added to 300 ml Erlenmeyer flask. Now add 5 ml of distilled water into the feeding funnel, and allow it to pass into the Kjeldahl flask. If air bubbles through H_3BO_3solution in which condenser tube is immersed, the distillation unit fittings at the joints are air tight.

Now pour 80 ml of 40% NaOH solution through the feeding funnel into the Kjeldahl flask. Add 10 ml of distilled water into the feeding funnel and lift the glass rod stopper to let distilled water pass into the Kjeldahl flask washing away the alkali from the outlet. Place the stopper back and fill the feeding funnel with 25 ml distilled water.

Switch on the electric heater, or light the gas burner. Heat the contents to boil and let it boil until 120 ml of distillate is collected in the receiving flask. Similarly run a blank with 120 ml of acidified NaCl instead of soil extract following the entire procedure as described for the sample extract.

3. Titration of NH_3 collected in 2% H_3BO_3 by a standard acid. Add standard acid through a burette in small portions into sample distillate, and then swirl the contents. Continue titration of the sample distillate with standard acid until the distillate changes its colour from green to distinct pink. Note down the sample titre. Fill the burette again with the standard acid and then titrate the blank distillate similarly with standard acid more slowly adding only few drops of the acid at a time. Continue titration until the distillate colour changes from green to distinct pink of same intensity as in sample distillate. Note down the blank titre.

2.10.1.4 Calculation

$$\text{ppm } NH_4 \text{ in soil} = \frac{(S - B) \times T \times N \times ME \times 10^6}{A \times W}$$

$$\text{Kg } NH_4/\text{ha} = \frac{\text{ppm } NH_4 \text{ in soil} \times 2 \times 2.47}{2.2}$$

where,

S is standard acid titre for the sample distillate in ml.

B is the standard acid titre for the blank distillate in ml.

T is the total volume of extractant used for the extraction of NH_4 from the soil.

N is the normality of standard acid.

ME is the milligram equivalent weight of NH_4 *i.e.* 0.018 g

A is aliquot of soil extract in ml taken for the distillation of NH_4 in Kjeldahl flask.

W is the mass of oven dry soil sample actually used for the extraction of NH_4.

2.10.2 NH_4 in soil extract by Nessler's method. Peech *et. al.* (1947)

2.10.2.1 Principle

NH_4 react with potassium mercuric iodide (Nessler's reagent) forming an orange coloured complex, the intensity of which is proportional to the concentration of NH_4upto a certain range. The optical density of the coloured complex is measured at 420 mμ wave length using a spectrophotometer or photoelectric colorimeter. The concentration of NH_4 in soil extract is estimated by reference to the calibration curve prepared with the standards.

$$2K_2Hg\,I_4 + 3KOH + NH_4 \rightarrow \underset{\text{Orange}}{Hg_2O\,(NH_4)_3} + 7\,KI + 2H_2O$$

2.10.2.2 Reagents

(*i*) Sodium acetate-acetic acid solution. Transfer 100g of sodium acetate trihydrate in 1 l beaker. Dissolve the reagent in about 500 ml distilled water by stirring with a glass rod. Add 30 ml of 99.5% acetic acid, and transfer the solution to 1 l volumetric flask quantitatively and dilute the solution to 1000 ml mark with distilled water. This is identical with the Morgan's extracting solution.

(*ii*) Activated charcoal or Darco G-60. The powdered charcoal be washed with 0.1N HCl and with 0.5 M $NaHCO_3$, and thereafter, with distilled water until free from chloride.

(*iii*) 20 ppm NH_4 solution. Dissolve 0.0733 g of $(NH_4)_2\,SO_4$ in 1 : 1 diluted sodium acetate-acetic acid and dilute the solution to 1000 ml mark.

(*iv*) 5 ppm NH_4 solution. Transfer 25 ml of 20 ppm NH_4 solution, and dilute it to 100 ml with 1 : 1 diluted morgan's solution (*i.e.* sodium acetate-acetic acid).

(*v*) Nessler's reagent (stock solution). Dissolve 50g of HgI_2 and 40 g of KI in about 100 ml distilled water separately. Allow the solution to cool, and then transfer the two reagents into a 500 ml volumetric flask and dilute the solution to 500 ml with distilled water. Store in an amber coloured bottle.

(*vi*) Dilute Nessler's reagent (a working solution). Mix 1 part of Nessler's stock solution by volume with 9 parts of distilled water. Store in an amber coloured bottle. This reagent be prepared every week.

(*vii*) Gum arabica solution. Dissolve 10g of gum arabica in about 150 ml distilled water. Add 5 ml of Nessler's reagent (stock solution) and make the volume to 200 ml. Keep in the dark for 14 days. Thereafter, siphon off the supernant opalescent liquid in an amber coloured bottle.

(*viii*) Tartrate reagent. Dissolve 80g of sodium tartrate in distilled water, and also dissolve 20 g NaOH in distilled water. Transfer the two solutions into 1 l volumetric flask and dilute the contents to 500 ml with distilled water. Cool and add 50 ml of gum arabica solution and dilute the solution to 1000 ml mark with distilled water.

2.10.2.3 Procedure

1. Extraction of mineral nitrogen from the soil. Transfer 10g of soil sample into a 150 ml Erlenmeyer flask. Then, add 0.25 g of activated charcoal or Darco

G-60 and 50 ml of sodium acetate-acetic acid extracting solution. Stopper the flask and shake for 30 minutes on a mechanical shaker. Filter the solution through a funnel lined with Whatman no. 1 filter paper. Collect the filtrate in a 150 ml Erlenmyer flask. Discard the initial filtrate and preserve the subsequent filtrate for nutrient analysis.

2. Development of the coloured complex with Nessler's reagent. Transfer 0, 2, 4 and 6 ml of 5 ppm NH_4 solution into separate 50 ml volumetric flasks. Add 2 ml distilled water, 3 ml tartrate reagent, 7 ml of Nessler's working solution in each of the four 50 ml volumetric flasks. Shake immediately to mix the contents to avoid trouble from turbidity. Dilute the contents to 50 ml with distilled water. Stopper and turn the flask upside down holding the stopper firmly, and allow the colour to develop for 5 minutes.

3. Measurement of the optical density of coloured complex developed with standards. Switch on the photoelectric colorimeter after placing 420 mμ blue filter in the filter socket and after placing colorimeter test tubes filled with distilled water in the socket provided. Allow the instrument to warm for 10 minutes before taking measurement of the optical density of the standards. Optical density measurements be done after 5 minutes but within 20 minutes of development of coloured complex.

Set the instrument reading to 0 optical density on placing colorimeter test tube filled with coloured complex developed with 0 ml of 5 ppm NH_4 solution and then measure the optical density of coloured complex developed with known volume of 5 ppm NH_4 solution using test tubes of identical optical properties. Record the optical density corresponding to the volume of 5 ppm NH_4 solution taken.

4. Preparation of calibration curve. Set optical density scale on the ordinate of linear graph paper and μg NH_4/50 ml of coloured complex on the abscissa. Plot the optical density corresponding to amount of NH_4 taken and connect the points to obtain a calibration curve.

5. Concentration of NH_4 in sample and blank test solution. Transfer 2 ml of sample and blank test solution into 50 ml vol. flask and proceed as for standard omitting transfer of aliquot of 5 ppm NH_4 solution.

Set 0 optical density with colour complex developed with blank test solution and measure the optical density of colour complex developed with the sample test solution. Find the concentration of NH_4in sample test solution by reference to the standard calibration curve.

2.10.2.4 Calculations

$$\text{ppm } NH_4 \text{ in soil} = \frac{S \times T}{A \times W}$$

$$\text{Kg } NH_4 \text{ in soil/ha} = \frac{\text{ppm } NH_4 \text{ in soil} \times 2 \times 2.47}{2.2}$$

where

S is microgram NH_4 in aliquot of soil extract *i.e.* sample test solution by reference to the calibration curve.

T is the total volume of the extractant in ml used for equilibration with the soil.

A is ml aliquot of soil extract *i.e.* sample test solution taken for the development of the coloured complex.

W is the mass of soil sample taken in g for the extraction of NH_4.

2.10.3 NO_3 N in soil extracts by phenoldisulphonic acid method adapted from Harper (1924), and Prince (1945)

2.10.3.1 Principle

Nitration of position 6 of 2, 4- phenol-disulphonic acid in fuming H_2SO_4 yields a product which is colourless in acid medium, and yellow in neutral and alkaline medium. The intensity of colour developed is proportional to the concentration of NO_3.

The calibration curve is obtained by plotting the optical densities of standards determined at 420 mμ wave length on the ordinate of a linear scale graph paper against the corresponding concentration of NO_3 plotted on the abscissa. The concentration of NO_3 in the soil extract is determined by taking the optical density of colour complex developed on reacting soil extract with phenoldisulphonic acid and then estimating NO_3 concentration in the soil extract by reference to the calibration curve.

$$C_6H_3OH\ (HSO_3)_2 + HNO_3 \rightarrow C_5H_2OH\ (HSO_3)_2\ NO_2 + H_2O$$

2.10.3.2 Reagents

(*i*) Sodium acetate-acetic acid. As described in section 2.10.2.2.

(*ii*) Darco G-60 or activated charcoal

(*iii*) 100 ppm NO_3–N solution. Transfer 0.7221 g of KNO_3 in 1 l volumetric flask and dissolve in extracting solution diluted 1 : 1 with distilled water and make the volume to 1l.

(*iv*) 20 ppm NO_3–N solution. Transfer 20 ml of 100 ppm NO_3 solution into a 100 ml volumetric flask and dilute the solution to 100 ml with 1 : 1 diluted extracting solution.

(*v*) Phenoldisulphonic acid. Dissolve 25g of crystal clear pure phenol in 150 ml of conc. H_2SO_4(AR) in a 500 ml capacity flat bottom round flask. Add 75 ml of fuming H_2SO_4 (minimum 20% SO_3) under a fume hood. Immerse the flask in a hot water bath and keep it immersed for 2 hours. Store in a amber coloured bottle.

(*vi*) Sodium hydroxide-EDTA solution. Dissolve 360g NaOH pellets in 1 l distilled water. After cooling add 15g EDTA and mix to dissolve EDTA thoroughly. Let the solution stand over night, and filter out the precipitated sodium carbonate through Whatman No. 1 filter paper using a vacuum pump.

(*vii*) Calcium carbonate powder (AR)

2.10.3.3 Procedure

1. Extraction of nitrate from the soil sample-as described in 1st step of section 2.10.2.3.

2. Development of coloured complex by reaction of NO_3 standard with phenol disulphonic acid. Transfer 0, 2, 4, and 6 ml of 20 ppm NO_3–N solution into separate porcelain evaporating dish and add a pinch of $CaCO_3$ powder to each evaporating dish. Then evaporate the contents to dryness on a water bath.

Add 2 ml of phenol disulphonic acid and rotate the porcelain evaporating dish so that all the residues come in contact with phenol disulphonic acid. After 10 minutes, add 15 ml of distilled water to each evaporating dish. Allow to cool and then add 10 ml of NaOH– EDTA solution.

Now transfer the contents of each evaporating dish into separate 100 ml volumetric flask, and dilute the solution to 100 ml mark with distilled water. Stopper and mix the contents by turning the flask upside down holding the stopper firmly. Allow ten minute time for the maximum development of the colour complex.

3. Measurement of the optical density of the coloured complex developed with the NO_3–N standard. Place blue filter in the filter holder or set the wave length knob at 420 mμ wave length. Then place cuvette/colorimeter test tube filled with distilled water in the socket provided. Switch on the instrument as well as the tungsten lamp and allow 10 minutes time for the instrument to warm up before taking measurements on optical density of the coloured complex.

Take a pair of cuvettes/colorimeter test tubes with identical optical properties. Fill one cuvette with coloured complex developed with 0 μg NO_3–N/100 ml and fill the other curvette with coloured complex developed with 40 μg NO_3–N/100 ml. From the instrument take out the cuvette filled with distilled water and place the cuvette filled with colour complex developed with 0 μg NO_3–N/100 ml. Adjust the optical density to zero. Then replace the cuvette immediately by the cuvette containing coloured complex developed with 40 μg NO_3–N/100 ml. Note down the optical density recorded by the instrument. Similarly take the optical density of coloured complex developed with 80 and 120 μg NO_3–N/100 ml adjusting optical density with colour complex developed with 0 μg NO_3–N/100 ml.

4. Standard calibration curve. On a linear scale graph paper, set optical density range on the ordinate and the NO_3–N concentration in μg /100 ml coloured complex on the abscissa. Then plot the recorded values of optical density corresponding to the concentration of NO_3–N. Connect the points of optical densities by a straight line to obtain the standard calibration curve.

5. Concentration of NO_3–N in soil extract solution. Take 10 ml aliquot of soil extract and the blank test solution into separate porcelain evaporating dish and add a pinch of $CaCO_3$and proceed following the procedure as for the standard omitting taking aliquot of standard 20 ppm NO_3 – N solution. Transfer the coloured complex developed with sample and blank extract to separate 100 ml volumetric flask and then dilute each to 100 ml mark with distilled water. Allow 10 minutes time for full development of the coloured complex.

Now place cuvette filled with coloured complex developed with blank extract and adjust the optical density to zero. Then replace the cuvette by an identical curvette filled with coloured complex developed with soil extract and note down the optical density recorded by the instrument.

Then refer to the calibration curve, locate the optical density of soil extract on the ordinate and draw an imaginary line parallel to the abscissa intersecting the calibration curve. From the point of intersection draw a perpendicular to the abscissa and note down the concentration of NO_3 given at the base of the perpendicular.

2.10.3.4 Calculation

$$\text{ppm } NO_3\text{–N in soil} = \frac{S \times T}{A \times W}$$

$$\text{kg } NO_3\text{–N/ha} = \frac{\text{ppm } NO_3 - \text{N in soil} \times 2 \times 2.47}{2.2}$$

where

S is μg NO_3–N/100 ml of coloured complex in aliquot of sample test solution by reference to the standard calibration curve.

T is the total volume of the extracting solution equilibrated with the soil.

A is the ml aliquot of soil extract taken for the development of coloured complex.

W is the mass of soil sample in g equilibrated with the extracting solution.

2.10.4 NO_3-N determination in soil extract by Kjeldahl distillation in presence of Devarda alloy. A.O.A.C. (1950)

2.10.4.1 Principle

NO_3 being water soluble is extracted from soil after equilibration with distilled water by filtration and leaching. The filtrate plus leachate is transferred to 800 ml kjeldahl flask. The volume of distillate is reduced after adding MgO. Thereafter, Devarda alloy is added for the reduction of nitrates. The ammonium formed is distilled in the presence of alkali and the distillate is collected in boric acid. The ammonium borate formed is titrated back by standard mineral acid using brom cresol green-methyl red indicator.

2.10.4.2 Reagents

(*i*) Acidified distilled water. Add 1 ml of conc. H_2SO_4 into 250 ml volumetric flask and make the volume to 250 ml with distilled water.

(*ii*) conc. H_2SO_4.

(*iii*) MgO

(*iv*) Devarda alloy

(*v*) Sodium hydroxide-40%.

(*vi*) Boric acid -2% solution.

(*vii*) Brom cresol green-methyl red indicator

(*viii*) Hydrochloric acid. 0.02 N.

The preparation of reagents (*v*) to (*viii*) are described in section 2.10.1.2.

2.10.4.3 Procedure

1. Extraction of nitrates from soil. Spread 200 g of soil sample brought from the field on a large flat dish and dry in an oven at 55°C for 12-16 hours. Simultaneously take a soil sample of about 20 g in a tared aluminium can and after weighment transfer the aluminium can plus soil into an oven at 105°C to determine the moisture content in soil sample on oven dry basis.

Transfer the partially dried soil sample at 55°C on to a 9 cm dia buchner funnel lined with Whatman no. 50 filter paper. Place the buchner funnel adapter tightly into the mouth of suction flask connected to a vacuum pump.

Wet the soil sample with acidified distilled water giving a thin film of free water over the soil surface. Run the vacuum pump for 5 minutes. After initial filtration continue leaching of the soil mass with distilled water until the volume of filtrate plus leachate measure 600 ml.

2. Concentration of filtrate cum leachate from the soil. Transfer 600 ml of filtrate cum leachate collected from the soil into a 800 ml Kjeldahl flask. Then, heat the Kjeldahl flask to reduce the volume of contents to 500 ml. Then, add 1 g of MgO, and continue heating of Kjeldahl flask until the volume is reduced to about 200 ml.

3. Treatment of residual filtrate cum leachate with Devarda alloy and distillation of ammonium formed. Add few glass beads and 1 g paraffin wax into the Kjeldahl flask. Place the glass adapter on the Kjeldahl flask and connect it to condenser inlet and also fit the feeding funnel into the grounded joint.

Now lower the condenser outlet tube inside a 300 ml Erlenmeyer flask, and let it dip into 40 ml of 2% H_3BO_3 solution containing brom cresol green-methyl red mixed indicator. Add 5 ml of distilled water into the Kjeldahl flask through the feeding funnel. If air bubbles through H_3BO_3 solution, the distillation apparatus is air tight and free from escape of gas.

Then, add 2g of Devarda alloy, and 50 ml of 40% NaOH solution through the feeding funnel. Fill the feeding funnel with distilled water and allow 10 ml portion of distilled water to pass through the feeding funnel to remove the alkali from the feeding funnel outlet.

Thereafter, heat the contents to boil, and let it boil until 125 ml of distillate is collected in the receiving flask. Preserve the sample distillate for titration of ammonium borate formed. Similarly prepare a blank distillate following the entire procedure without taking the soil sample.

4. Titration of sample and blank distillate with a standard mineral acid. Titrate sample distillate by adding standard acid through a burette. Swirl the contents after adding standard acid in small portions. Continue titration until the sample distillate in H_3BO_3 in the presence of mixed indicator changes its colour from green to distinct pink. Note down the sample titre. Fill the burette again with standard acid. Titrate the blank distillate collected in H_3BO_3containing mixed indicator slowly adding few drops of acid at a time. Swirl the contents after each addition of acid. Continue

titration until the colour of blank distillate changes from green to distinct pink of the same intensity as attained in the titration of sample distillate. Note down the blank titre.

2.10.4.4 Calculations

$$\text{ppm } NO_3\text{–N in soil} = \frac{(S - B) \times N \times ME \times 10^6}{W}$$

$$\text{kg } NO_3\text{/ha} = \frac{\text{ppm } NO_3 - \text{N in soil} \times 2 \times 2.47}{2.2}$$

where

S is acid titre in ml for titration of ammonium borate formed on distillation of soil extract in the presence of Devarda alloy.

B is acid titre in ml for titration of ammonium borate formed on distillation of blank extract in the presence of Devarda alloy.

N the normality of acid taken for the titration of distillate collected in H_3BO_3.

ME is the milligram equivalent weight of NO_3- N *i.e.* 0.014 g

W is the oven dry mass of soil sample taken for NO_3 - N determination.

2.10.5 Comments

In the determination of NO_3-N by phenol disulphonic acid method, Cl concentration greater than 0.03 % cause loss of NO_3 during the reaction between phenol disulphonic acid and NO_3. In order to prevent the loss of NO_3 due to Cl, powdered Ag_2SO_4 be mixed with 10 g of soil @ 0.225 g for each 1% Cl in the extracting solution. When Ag_2SO_4 is added in excess, Ag be precipiated by addition of $Ca(OH)_2$and $MgCO_3$ prior to filtration of soil suspension (Jackson, 1973). The loss of NO_3 due to Cl can also be avoided by reacting $NO_3$with phenol disulphonic acid in aqueous medium (Prasad and Lancaster, 1963).

Devarda alloy is composed of 50% Cu, 45% Al and 5% Zn. In the presence of Devarda alloy, NO_3 is reduced which makes it possible to determine NO_3 - N by Kjeldahl distillation method.

In the Nessler's method of NH_4 determination, at concentration greater than 5000 ppm Cu, 500 ppm Mg, 0.5 ppm Mn and 0.2 ppm Fe, turbidity is encountered. In such cases more diluted soil extract be used.

Exchangeable NH_4 account for nearly 1% of the exchange capacity and NO_3 concentration in soil is not more than 100 ppm when soil contain 0.10%N. However, Russel (1953) reported NO_3 concentration in arable soils range from 3 to 60 ppm. The NO_3 concentration in soil vary with the season and from hour to hour. Usually plant test is a better index of NO_3 level than soil analysis. However, NO_3 in plant tissues accumulate with the deficiency of Mo. Both NH_4 and NO_3 are readily absorbed by the plants, but these forms constitute a small fraction of total N, and do not serve as a good index of N supply to plants.

REFERENCES

A.O.A.C. 1950. Official and Tentative Methods of Analysis of the Association of Official Agricultural Chemical. Ed. 7,910 pp. Washington.

Harper 1924. Ind Eng. Chem 16 : 180.

Jackson M.L. 1958. Soil Chemical Analysis Prentice Hall of India Private Limited, New Delhi (1973).

James, W.W. 1975. Nitrogen pp. 310-323. In. Chapman, H.D. Diagnostic Criteria. For Plants And Soils. Eurasia Publishing House (P) Ltd. Ram Nagar, New Delhi 110055.

Pecch et. al. 1947. U.S.D.A. Cir. 757 p10.

Prasad R. 1970. A Practial Manual for Soil Fertility. Division of Agronomy, Indian Agricultural Research Institute, New Delhi.

Prasad R., and J.D. Lancaster, 1963. Proc. 50th Indian Sci. Congress PTM : 599

Prince 1945. Soil Sci. 59 : 47

Russel, E.J. 1953. Soil Conditions And Plant Growth. Longman Green & Co. NY.

2.11 AVAILABLE PHOSPHORUS IN SOIL

2.11.1 0.5 M $NaHCO_3$ (pH 8.5) extractable phosphorus in soils in neutral, alkaline and calcareous soils–Olsen *et. al.* (1954)

2.11.1.1 Principle

P is extracted from soil by 0.5 M $NaHCO_3$ (pH 8.5). P in the soil extract is determined colorimetrically using a photoelectric colorimeter after developing molybdenum blue colour, the intensity of which varies with the P concentration. The orthophosphate ion in an acid molybdate solution form phosphomolybate complex which is reduced by stannous chloride to produce molybdenum blue colour.

2.11.1.2 Reagents

(*i*) 0.5 M $NaHCO_3$ (pH 8.5). Dissolve 42 g $NaHCO_3$ in distilled water preferably in 500 ml beaker and transfer the solution into 1 l volmetric flask. Add more distilled water into the beaker in small portions and transfer the washings into the same volumetric flask. Dilute the solution to the 1 l mark. If the pH is slightly less than 8.5, add 1-2 pellets of NaOH and shake to dissolve. Check the pH.

(*ii*) Activated carbon. Use Darco G 60 or any suitable decolorizing carbon made free from soluble phosphorus by repeated washings with 0.5 M $NaHCO_3$ (pH 8.5) - solution.

(*iii*) 10 N HCl. Dilute 218.2 ml of conc. HCl (sp. gr. 1.18, Minimum assay 35.4) to 250 ml mark in a volumetric flask with distilled water.

(*iv*) 1.5% Ammonium-molybdate in 4 N HCl. Transfer 7.5 g of $(NH_4)_6 MO_7O_{24}$. 4 H_2O into 500 ml volumetric flask. Add about 150 ml of warm distilled water and shake to dissolve the reagent. Filter the solution if necessary and allow it to cool. Then add 200 ml of 10 N HCl and dilute the contents to the 500 ml mark with distilled water. Store the reagent in a black or amber coloured glass bottle. Prepare fresh reagents every two months.

(*v*) Concentrated $SnCl_2 . 2H_2O$ solution. Dissolve 5g crystals of stannous chloride in 12.5 ml of conc. HCl in a small beaker and transfer the contents immediately into a 25 ml volumetric flask. Add 1-2 small balls of metallic tin. Stopper and place an inverted glass funnel carefully over the stopper so that it does not get blown away.

(*vi*) Dilute $SnCl_2 . 2H_2O$ solution. Transfer 66 ml distilled water into a 100 ml volumetric flask. Then add 0.5 ml of concentrated stannous chloride solution. Place the stopper and mix the contents. Preferably make a fresh solution every two hours.

(*vii*) 100 ppm P solution. Transfer 0.4434 g dry KH_2PO_4 analytical reagent (AR) into 1 l volumetric flask. Add suficient distilled water and shake to dissolve the reagent. Add 4.8 ml of conc. H_2SO_4 (AR) and dilute the solution to 1 l mark.

(*viii*) 2 ppm P solution. Pipette 5 ml of 100 ppm P solution into 250 ml volumetric flask. Add distilled water so as to make the volume to 250 ml mark.

2.11.1.3 Procedure

1. Preparation of standard graph for phosphorus. Pipette 0, 1, 2, 3, and 4 ml of 2 ppm P solution containing 0, 2, 4, 6 & 8 μg P into separate 25 ml volumetric flasks. Add to each flask 5 ml of 0.5 M $NaHCO_3$ (pH 8.5). Then add slowly 5 ml of 1.5% ammonium molybdate in 4 N HCl. When CO_2 evolution ceases, mix the contents by gentle shaking of the flask. Fill the flask upto its neck with distilled water to make the volume to about 22 ml and do not mix the contents. Add 1 ml of dilute $SnCl_2$ solution and make the volume up to the 25 ml mark with distilled water. Stopper the flask. Then turn the flask upside down and immediately remove the stopper for a moment to let the CO_2 escape. Allow just 10 minutes time for the full development of molybdenum blue colour. Simultaneously, place the colorimeter test tube (cuvette) filled with distilled water upto the mark in the appropriate cuvette holder of the instrument. Then fit the red filter or adjust the wavelength knob to 660 mμ. Switch on the instrument and let it warm for 10 minutes.

Thereafter, remove the cuvette filled with distilled water from the instrument and place the cuvette filled with coloured complex containing 0 μg P/25 ml. Set the instrument reading to 100% transmittance or zero optical density by rotating the adjustment knob gently. Fill the second cuvette with the coloured complex containing 2 μg P/25 ml after rinsing the cuvette with the coloured complex. Place the cuvette inside the cuvette holder removing the cuvette filled with coloured complex containing 0 μg P/25 ml. Note down the % transmittance or the optical density recorded by the instrument. Again set the instrument to 100% transmission/or zero optical density on placing cuvette filled with coloured complex pertaining to 0 μg P/25 ml. Then read the % transmittance or optical density of the subsequent standard coloured complex containing graded amount of P.

Plot the percent transmittance against P concentration (μg P/25 ml coloured complex) on single cycle semi log graph paper. The P concentration be plotted on a linear scale and the corresponding % transmittance on a log scale. Alternatively, the optical density be plotted on "Y" axis and the corresponding P concentration on the "X" axis on ordinary graph paper. The numerical value of Klett photoelectric colorimeter reading is equal to optical density multiplied by 500. Repeat the entire procedure described until all points fall very nearly on a straight line.

2. Extraction of 0.5 M $NaHCO_3$ (pH 8.5) soluble P from soils. Transfer 2.5 g air dry passed through 2 mm sieve, 1 g activated charcoal free from soluble phosphorus and 50 ml of 0.5 M $NaHCO_3$ (pH 8.5) into a 250 ml Erlenmeyer flask (conical). Place a rubber stopper on the mouth of the conical flask and fit the flask firmly on a power operated mechanical shaker. Switch on the mechanical shaker at a moderate speed and let it run for 30 minutes. When the shaking period is over, switch off the mechanical shaker, remove the conical flask and filter the suspension through a dry Whatman 42 paper collecting the filtrate in a clean receiving flask after descarding 3-5 ml of initial filtrate. Whatman 42 paper has very fine pores which does not permit colloidal particles of soil and charcoal to pass through it. Besides, it is free from phosphorus. If the filtrate is not clear, add more activated charcoal to obtain clear filtrate. Shake the flask immediately before pouring the suspension into the funnel lined with filter paper. Similarly run a blank.

3. Determination of P concentration in the soil extract. Transfer 5 ml aliquot of the sample and blank extract filtrate into seperate 25 ml volumetric flasks. In another 25 ml volumetric flask, transfer 5 ml of 0.5 M $NaHCO_3$ (pH 8.5) solution. Then add 5 ml of 1.5% ammonium molybdate solution to each flask slowly with gentle shaking to let escape of CO_2 formed. Mix the contents by shaking the flask with mouth open. Then fill the flask upto its neck with distilled water without shaking. Add 1 ml of dilute $SnCl_2$ (freshly prepared) in each volumetric flask and then make the volume up to the mark with distilled water. Place the stopper on the mouth of vol. flask and turn it upside down once. Immediately remove the stopper for a moment to let the CO_2 escape. Let the molybdenum blue colour develop for 10 minutes.

Then, adjust wave length of incident light to 660 mμ in the spectrophotometer or else fit the red filter in the colorimeter. Thereafter, place cuvette filled with distilled water in the cuvette holder. Then, switch on the instrument to warm up for 10 minutes. Just after 10 minutes time allowed for the full development of molybdenum blue colour, determine the % transmittance or optical density of the coloured complex developed with sample and blank extract adjusting 100% transmittance or zero optical density with coloured complex containing 0 μg P/25 ml. Alternatively adjust zero optical density with coloured complex developed with blank extract and measure the optical density of coloured complex developed with sample extract. Find out microgram P present in 5 ml of sample and blank extract or only sample extract if zero optical density is adjusted with blank extract by reference to the standard graph for phosphorus.

2.11.1.4 Calculations

ppm P in soil =

$$\frac{\mu g\ P / 25\ ml\ \text{coloured complex (Sample - blank)} \times \begin{matrix}\text{Volume of extractant}\\ \text{used for preparing extract}\end{matrix}}{\text{ml of extract taken for colour development} \times \text{mass of soil in g}}$$

lbs P/acre = ppm P in soil × 2

$$\text{Kg P/ha} = \frac{\text{Lbs P acre} \times 2.47}{2.2}$$

Kg P_2O_5/ha = Kg P/ha × 2.29

2.11.1.5 Comments

P in soil solution is replenished from organic P, P adsorbed on positively charged surfaces, sparingly soluble Fe, Al and Ca phosphates and P occluded inside particles with oxide coating. The amount of P which can dissolve to replenish P in soil solution is termed labile P. The P extracted from soil by 0.5 M $NaHCO_3$(pH 8.5) is a measure of labile p in soil. The P from Ca, Fe & Al phosphates is brought into solution at high pH because of decrease in the concentration of Fe, Al and Ca. The extractant exchanges also surface bound phosphates from different sources.

An increase in temperature and shaking speed result in an increase in extractable P. The intensity of molybdenum blue colour is also influenced by the temperature. There is thus need for doing extraction of P from soil at moderate speed with minimum variation in temperature. Some standards need to be prepared from time to time to check whether the optical density of the coloured complex fall on the line of standard curve.

A number of ions such as As^{5+}, Si^{4+}, Ge^{4+}, Mo^{6+} and B^{3+} form complexes with molybdate in a fashion analogous to phosphate ions and thus the interference from such ions may be significant when their concentrations are high.

The molybdenum blue complex is sensitive to acidity, the concentration of molybdate and the reductant. It is, therefore, necessary that the recommended procedure be strictly followed.

The double distilled water be preferably used for colorimetric determination as it gives lower blank value in comparison with the blank value obtained from the use of ordinary distilled water.

The soils containing more than 10 ppm of bicarbonate soluble P are not considered to be deficient in plant available P. However, the critical level of available P in India differ from soil to soil and from crop to crop. In U.P. the results of soil test for bicarbonate soluble P expressed in Kg P_2O_5/ha is interpreted in soil health lab as follows :

Very low	Low	Medium	High
0 – 10	11 – 20	21 – 40	> 40

REFERENCES

Bingham, F.T. 1949. Soil test for phosphate, Calif. Agr. 3 : 11 – 14

Olsen, S.R., Cole, C.V., Watnahe, F.S. and Dean, L.A. 1954. Estimation of available phosphorus in soils by extraction with sodium bicarbonate U.S. Deptt. Agr. Circ. 939.

Jackson M.L. 1958. Soil Chemical Analysis Prentice- Hall, Inc, Englewood Cliffe, N.J.

2.11.2 0.03 N NH_4F + 0.025 N HCl extractable phosphorus in acid soils by the method of Bray and Kurtz (1945)

2.11.2.1 Principle

The combination of HCl and NH_4F extracts acid-soluble forms of P-largely calcium phosphate and a portion of Al and Fe phosphates. The fluoride ions complex trivalent Al and Fe in acid solution and thereby release P held by Al and Fe. The P extracted from soil is a measure of labile P.

2.11.2.2 Reagents

(*i*) 1.0 N NH_4F. Dissolve 37g NH_4F in distilled water, and dilute the solution to 1 l. Store the solution in a polythene bottle.

(*ii*) 0.5 N HCl. Dilute 21.8 ml of concentration HCl (sp. gr. 1.18, minimum assay 35.4%) to a volume of 500 ml with distilled water.

(*iii*) 0.03 N NH_4 F + 0.025 N HCl. Transfer 15 ml of 1.0 N NH_4 F and 25 ml of 0.5 N HCl into 500 ml capacity volumetric flask and add distilled water to make the volume up to the 500 ml mark.

(*iv*) Other reagents. As listed in the determination of available phosphorus by the method of Olsen *et. al.* (1954) excluding activated carbon and 0.5 M $NaHCO_3$ (pH 8.5). In preparation of 500 ml of 1.5% ammonium molybdate use only 175 ml of 10 N HCl instead of 200 ml of 10 N HCl required for the preparation of the reagent in the method of Olsen *et. al.* (1954).

2.11.2.3 Procedure

1. Preparation of standard graph for phosphorus. Pipette 0, 1, 2, 3 and 4 ml of 2 ppm P solution into separate 25 ml volumetric flask. Add to each flask 5 ml of 0.03 N NH_4F in 0.025 N HCl. Then proceed following the procedure described in the Olsen's method of determination of available phosphoprus omitting the addition of 2 ppm P solution and 5 ml of 0.5 M $NaHCO_3$ (pH 8.5) to each flask.

2. Extraction of 0.03 N NH_4 in 0.025 N HCl soluble P from acid soils. Transfer 5g acid air dry soil passed through 2 mm sieve into 250 ml Erlenmeyer flask (conical). Add 50 ml of 0.03 N NH_4F in 0.025 N HCl into the flask, stopper and fit it on a mechanical shaker. Switch on the mechanical shaker and run it at a moderate speed for 5 minutes. Thereafter remove the flask and pour the suspension after shaking in small portions into a funnel lined with dry Whatman No. 42 paper. Collect the

filtrate in a clean receiving flask after discarding 3-5 ml of initial filtrate. If the filtrate is not clear, pour it back into the flask. Shake the flask and transfer the suspension into a funnel lined with Whatman No. 42 paper. Preserve the filtrate for the determination of P concentration in the extract.

3. Determination of P concentration in the soil extract. Transfer 5 ml aliquot of the sample and blank filtrate into separate 25 ml volumetric flask. Then proceed as described in the Olsen's method of determination of available P omitting the addition of sample and blank filtrate prepared from extraction of P from soil by 0.5 M $NaHCO_3$ (pH 8.5).

2.11.2.4 Calculation

Use the same formula as given in the Olsen's method of determination of available P. It may, however, be kept in mind that the mass of soil taken in Bray and Kurtz method is 5 g whereas in the Olsen's method the mass of soil used for the extraction of P is 2.5 g.

2.11.2.5 Comments

On some acid sandy soils, fluorides may interfere in the colour development. The fluoride interference can be eliminated by use of boric acid. The soil test values expressed in ppm are interpreted in general as follows :

Very low	Low	Medium	High
< 3	3–7	7–20	>20

REFERENCES

Bray, R.H. and Kurtz, L.T. 1945. Determination of total, organic and available forms of phosphorus in soils. Soil Sci. 59 : 39 – 45.

Kurtz, L.T. 1942. Elimination of fluoride interference in the molybdenum blue reaction. Ind. Eng. Chem. Anal. Ed. 14,855.

Jackson, M.L. 1958. Soil Chemical Analysis, Prentice Hall, Inc. Englewood Cliffe, N.J.

2.12 1N NH_4OAC (PH 7.0) EXTRACTABLE POTASSIUM IN SOILS– TOTH & PRINCE (1949)

2.12.1 Principle

Ammonium acetate is used to extract exchangeable K in soil as ionic diameter of NH_4^+ is such that exchangeable K in soil is readily displaced by NH_4^+. K in the extract is atomized into blue flame of a flame photometer so that it gets excited on gaining energy and emits radiation of a certain wave length in proportion to the concentration of K. The light emitted strikes a photocell after passing through a filter. The photocell converts light energy into electrical energy which is measured by a galvanometer. The galvanometer is first calibrated by atomizing solution varying in K concentration from 0 to 20 ppm.

2.12.2 Reagents

(*i*) 1.0 N CH_3 $COONH_4$ (pH 7.0). Dissolve 77.0 g of ammonium acetate in distilled water and transfer the solution into 1 l volumetric flask. Dilute the contents to 1 l mark with distilled water. Check the reaction of the solution, and if necessary add few drops of acetic acid or NH_4OH to adjust the reaction of the solution to pH 7.0.

(*ii*) 1000 ppm K solution. Weigh $1.90\overline{66}$ g of pure dry KCl equivalent to 1.0 g K and transfer it to 1 l volumetric flask. Add distilled water to dissolve the KCl and dilute the solution to the 1000 ml mark with distilled water. Stopper the flask and shake to mix the contents.

(*iii*) 250 ppm K solution. Transfer 25 ml of 1000 ppm K solution into a 100 ml volumetric flask. Add distilled water to dilute the solution to the 100 ml mark. Stopper and mix the contents by shaking.

(*iv*) 0 to 20 ppm K solution. Transfer 0, 2, 4, 6 and 8 ml of 250 ppm K solution into separate 100 ml volumetric flask. Add neutral normal ammonium acetate solution to each volumetric flask and make the volume to the 100 ml mark to obtain 0, 5, 10, 15 and 20 ppm K solution respectively. Add a drop of butyl alcohol to each volumetric flask to improve the spraying property of the solution.

2.12.3 Procedure

1. Preparation of standard graph for potassium. Attach the tube from the Indane gas cylinder to the gas inlet of the built in burner. Attach the thin plastic tubing to the atomizer hollow pin and introduce the other end of plastic tubing into 10 ml beaker or vial filled with distilled water. Start the compressor and adjust the air pressure to the specified level. Light a match stick, and place it on the burner through the window and immediately open the gas. Adjust the gas regulator to obtain a central bluish cone shaped flame.

Thereafter fill the vial with 0 ppm K solution and immerse the tubing from the atomizer pin into it. Adjust the flame photometer reading to zero. Then place the vial filled with 20 ppm K solution below the atomizer pin and dip its tubing end inside the solution and adjust the flame photometer reading to 100. Repeat the process to check whether the flame photometer reading remains 0 and 100 with the atomization of 0 and 20 ppm K solution. When the flame photometer reading is found to be satisfactory, atomize the intermediate standards *i.e.* 5, 10 and 15 ppm K solutions and note down the reading recorded by the flame photometer. Atomize distilled water in between two standards and before switching off the instrument.

Plot the flame photometer reading on y-axis and the corresponding concentration of K in ppm (μg/ml) on the x-axis of an ordinary graph paper so as to prepare a standard graph for potassium. The points should fall on a straight line. In case of significant departure from the straight line, the standards be prepared again and their corresponding flame photometer reading be determined and plotted on the graph until all the points fall in a straight line.

2. Extraction of 1.0 N ammonium acetate (pH 7.0) extractable K in soil. Transfer 5g of air dry soil passed through 2 mm sieve into a 125 ml Erlenmeyer flask. Add exactly 25 ml of 1 N ammonium acetate (pH 7.0) solution into the conical flask. Stopper and place it on a mechanical shaker firmly. Switch on the mechanical shaker and run it for 5 minutes at a moderate speed. Remove the flask from the shaker. After shaking the suspension, pour it in small portions in a funnel lined with dry Whatman No. 1 paper. Collect the filtrate in a clean receiving flask after discarding about 3 ml of the initial filtrate.

3. Determination of the K concentration in the soil extract. Transfer 5 ml aliquot of the neutral normal ammonium acetate soil extract into 25 ml volumetric flask and dilute it to the 25 ml mark with the extractant. Add 2 drops of butyl alcohol and turn the flask up side down to mix the contents.

Set the flame photometer for K determination as described under the heading "Preparation of standard graph for potassium". Then atomize 0 ppm K solution and adjust the flame photometer reading to zero. There after, atomize 20 ppm K solution and set the flame photometer reading to 100. Recheck the flame photometer reading with the above two standards. If the readings are found satisfactory, atomize the undiluted extract first and then the diluted extract, and record the corresponding flame photometer reading. By reference to the standard graph for potassium, find out the concentration of K in the diluted and undiluted ammonium acetate extract.

2.12.4 Calculations

ppm K in soil

$$= \frac{\dfrac{\mu g\ K / ml \text{ in undiluted extract}}{(\text{Sample} - \text{Blank})} \times \dfrac{\text{Volume of extractant used}}{\text{in preparation of extract}}}{\text{mass of soil in g}}$$

$$\text{lbs K/acre} = \text{ppm K in soil} \times 2$$

$$\text{Kg K/ha} = \frac{\text{lbs K / acre} \times 2.47}{2.2}$$

$$\text{Kg } K_2O\text{/ha} = \text{Kg K/ha} \times 1.20$$

$$\text{meq K/100g soil} = \frac{\text{ppm K in soil}}{\text{Eq. wt. of K} \times 10}$$

*As flame photometer reading by atomizing extractant is set at zero, the blank value be taken as zero.

2.12.5 Comments

In the extraction of K from soils with ammonium acetate, both exchangeable and water soluble forms of K is extracted. However, water soluble K in non-saline soils form only few ppm of the soil dry weight. Other forms of K in soils are non exchangeable and non-exchangeable but useful K. The K in soil not extracted by ammonium acetate is termed non-exchangeable. It is the K which is held deep within the crystal lattices of clay minerals. The non-exchangeable but useful K is extracted from soil by boiling HNO_3. All these forms remain in equilibrium in soil. However,

K status of soil is primarily evaluated on the basis of exchangeable K. In U.P. the results of soil testing for ammonium acetate extractable K expressed in Kg K_2O/ha is rated in soil health lab as follows :

Very low	Low	Medium	High
0-50	51-100	101-250	>250

The available K may be determined on field moist soil, but the results be expressed either on air dry or oven– dry soil basis. It is, therefore, desirable to determine moisture content of the soil. The drying in air or oven may increase or decrease the amount of K extracted by ammonium acetate depending upon whether the exchangeable K present prior to drying is below or above the equilibrium level. If the exchangeable K is lower than the equilibrium level, the non-exchangeable K is released during drying. If the exchangeable K is greater than the equilibrium level, certain amount of K is fixed or transformed into non-exchangeable form during the drying of the soil.

The German scientist (Herrn Kemmler) engaged in Potash Agency considered desirable level of K in extract from saturated paste to be 1.0 meq per l *i.e.* 39.1 ppm K. He also emphasized that the lack of significant response from added K does not imply that the soil is adequate in K supply. In some soils, significant response from added K have been obtained after regular application of K each year for a period of 7-12 years. Presumably, K reserve in soil was gradually build up to a level that the added K to the soil remained available for absorption by the crop plants.

REFERENCES

Muhr, G.R., N.P. Datta, H. Sankaran subramoney, V.K. Leley and K.L. Donahue. 1965. Soil Testing in India. U.S. Agency for Intern. Mission to India., New Delhi.

Stanford, S. and L. English 1949. Use of the flame photometer in rapid soil tests for K and Ca. Agron. J. 41 : 446–445.

Toth, S.J. and A.L. Prince (1949). Estimation of cation exchange capacity and exchangeable Ca, K and Na content of soil by flame photometer technique. Soil Sci. 67 : 439-445.

2.13 CATION EXCHANGE CAPACITY OF SOIL

2.13.1 CEC by saturation with Na^+–Bower *et. al.* (1952)

2.13.1.1 Principle

Soil colloids are negatively as well as positively charged. With the result, cations and anions are adsorbed by the soil colloids. Cations are adsorbed on negatively charged colloids. The capacity of the soil to adsorb cations in an exchangeable form is termed the cation exchange capacity and it is expressed in milligram equivalent per 100 g soil. The CEC is usually determined by saturating the exchange complex of the soil with a cation such as Na^+, NH_4^+, Ca^{++}or Ba^{++} and then determining the amount of adsorbed cation by appropriate method.

Soil is treated with 1 N sodium acetate (pH 8.2) to saturate the exchange sites with Na. Soluble Na in soil is then removed by washing with 99% isopropyl alcohol. Thereafter, exchangeable Na is extracted with 1 N ammonium acetate (pH 7.0). The Na extracted by ammonium acetate is determined by flame photometer.

2.13.1.2 Reagents

(*i*) 1.0 N CH_3COONa (pH 8.2). Dissolve 136 g of sodium acetate in distilled water and transfer the solution into 1 l volumetric flask. Make the volume up to the 1 l mark with distilled water. Add few drops of acetic acid or few pellets of NaOH to adjust the reaction of the solution to pH 8.2.

(*ii*) 1.0 N CH_3 COO NH_4 (pH 7.0). Dissolve 77.0 g of ammonium acetate in distilled water, and transfer the solution into 1 l volumetric flask. Make the volume up to 1 l mark with distilled water. Add few drops of acetic acid or NH_4OH to adjust the reaction of the solution to pH 7.0.

(*iii*) 99% Isopropyl alcohol.

(*iv*) 1000 ppm Na solution. Transfer 2.5412g of pure dry NaCl equivalent to 1.0 g Na into 1 l volumetric flask. Add distilled water to dissolve sodium chloride and dilute the solution to the 1 l mark with distilled water. Mix the contents by shaking.

(*v*) 250 ppm Na solution. Transfer 25 ml of 1000 ppm Na solution into 100 ml volumetric flask and dilute the solution with distilled water up to the 100 ml mark.

(*vi*) 0 to 20 ppm Na solution. Transfer 0, 2, 4, 6 and 8 ml of 250 ppm Na solution into separate 100 ml volumetric flask. Add 1 N ammonium acetate (pH 7.0) solution to each volumetric flask to make the volume to 100 ml mark for obtaining 0, 5, 10, 15 and 20 ppm Na solution. Add few drops of butyl alcohol in each flask to improve the spraying property of the solution.

2.13.1.3 Procedure

1. Preparation of standard graph for sodium. Operate the flame photometer and calibrate the instrument using sodium standards (0 to 20 ppm) and sodium filter in place of K standards and K filter as described under the heading "preparation of standard graph for potassium. After obtaining the flame photometer reading corresponding to 0, 5, 10, 15 and 20 ppm Na solution, plot the flame photometer reading on y-axis of ordinary graph paper against the corresponding Na concentration in ppm on the x-axis. It is desirable that all the points fall on a straight line.

2. Preparation of soil extract for CEC determination. Weigh 6g of coarse textured or 4.5 g of medium to fine textured air dry soil passed through 2 mm sieve into 50 ml capacity centrifuge plastic tubes. Add 33 ml of 1.0 N sodium acetate (pH 8.2) solution into the centrifuge tube, and triturate with a glass rod for 5 minutes. Then place the centrifuge tube in the centrifuge in pairs opposite to each other for a proper balance. Close the centrifuge lid, switch on the centrifuge and gradually increase the speed. Run the centrifuge at 2000–2440 rpm for 10 minutes or until the supernant liquid become clear. Decant the supernant liquid as completely as possible and discard it. Treat the sample in the same manner two times more each time using 33 ml portions

of 1 N sodium acetate solution and discard the supernant liquids each time. This would complete the process of saturation of the exchange complex with sodium.

Now soil saturated with Na be washed in an identical manner with three 33 ml portions of 99% Isopropyl alcohol using the procedure described above. This would remove soluble sodium acetate in soil without causing any hydrolysis of exchangeable sodium.

Then extract Na adsorbed by the soil with three 33 ml portions of 1.0 N ammonium acetate (pH 7.0) solution. Decant each washing with ammonium acetate from one sample into a 100 ml volumetric flask. Add few ml of ammonium acetate solution directly into the flask to make the volume up to 100 ml mark.

3. Determination of concentration of sodium in the CEC extract. Transfer 1.0 ml of the CEC extract into a 100 ml volumetric flask. Add 1N ammonium acetate (pH 7.0) to dilute the contents to the 100 ml mark. Then add 2 drops of butyl alcohol and mix the contents by turning the flask upside down. Now the CEC extract has been diluted 100 times.

Switch on the flame photometer again. Adjust the air pressure and gas regulator to obtain central blue cone shaped flame on lighting the burner. Then atomize first 0 ppm Na solution and adjust the flame photometer reading to zero. Thereafter atomize 20 ppm Na solution into the blue flame and adjust the flame photometer reading to 100. Recheck the flame photometer reading after atomizing 0 and 20 ppm Na solution in sequence. If the flame photometer reading is found satisfactory atomize the CEC extract diluted 100 times and note down the flame photometer reading. Find out the concentration of Na in the diluted CEC extract by reference to the standard graph for Na and multiply the concentration of Na by 100 to obtain the concentration of Na in undiluted CEC extract.

2.13.1.4 Calculations

$$\text{meq. of Na / l of CEC extract} = \frac{\text{ppm Na in CEC undiluted extract}}{\text{Equivalent Weight of Sodium}}$$

$$\text{meq. of Na / 100 g soil} = \frac{\text{meq. Na / l of CEC extract} \times 100}{10 \times \text{mass of soil in g taken for preparation of CEC extract}}$$

CEC of soil = meq. of Na/100 g soil

2.13.1.5 Comments

It is desirable to use isopropyl alcohol instead of ethyl alcohol or methyl alcohol for the removal of residual electrolyte, because it will cause less loss of organic materials (Peach *et. al.*, 1962).

The standard solution be kept in polythene bottles to avoid contamination. The flame photometer should be set up in a room free from drafts, dust or fumes, and smoking should not be permitted.

In case centrifuge tubes of only 10 ml capacity are available, the soil sample be apportioned to four centrifuge tubes, and the volume of sodium acetate, isopropyl alcohol and ammonium acetate be also accordingly apportioned. However, the sample extract be combined together into a 100 ml volumetric flask and the volume be made to 100 ml with 1N CH_3COONH_4 (pH 7.0).

REFERENCES

Bower, C.A., R.F. Reitemeir, and M. Fireman. 1952. Exchangeable cation analysis of saline and alkali soils. Soil Sci. 73 : 251-261.

Jackson, M.L. 1973. Soil Chemical Analysis. Prentice-Hall of India. Private Limited.

Toth, S.J. and A.L. Prince. 1949. Estimation of cation exchange capacity and exchangeable Ca, K and Na content of soils by flame photometer technique. Soil Sci. 67 : 439-445.

2.13.2 CEC by saturation with NH_4^+ ions–Amma (1989)

2.13.2.1 Principle

The exchange sites in soil is first saturated with NH_4^+ by treatment of the soil with 1N ammonium acetate (pH 7.0). The residual electrolytes are then removed by leaching with 99% isopropyl alcohol. Thereafter, NH_4^+adsorbed by soil as exchangeable ion is extracted by leaching with acidified 10% NaCl solution. The leachate containing NH_4^+ is distilled by Kjeldahl distillation method into 2% H_3BO_3. The boric acid containing ammonium borate is titrated by standard acid in the presence of brom cresol green-methyl red mixed indicator to find out the meq. of exchangeable NH_4^+ adsorbed by the soil sample. It forms the basis of estimation of the cation exchange capacity of soil.

2.13.2.2 Reagents

(*i*) 1.0N CH_3COONH_4 (pH 7.0). Dissolve 77.0g of ammonium acetate in distilled water and transfer the solution into 1 l vol. flask. Add distilled water to make the volume to 1 l mark. Add few drops of acetic acid or NH_4OH to adjust the reaction of the solution to pH 7.0

(*ii*) 1.0 N NH_4Cl. Dissolve 53.453 g of ammonium chloride in distilled water and dilute the solution to 1 l mark with distilled water. Adjust the pH to 7.0 by NH_4OH or HCl.

(*iii*) 0.25 N NH_4Cl. Dissolve 13.3632 g of ammonium chloride in distilled water and dilute the solution to 1 l mark with distilled water. Adjust the pH to 7.0 with NH_4OH or HCl.

(*iv*) 99% Isopropyl alcohol

(*v*) 10% $(NH_4)_2 C_2O_4.H_2O$. Dissolve 5g of ammonium oxalate in distilled water and make the volume to 50 ml with distilled water in a 50 ml vol. flask.

(*vi*) Approx. 7.5N NH_4OH. Transfer 25 ml of conc. NH_4OH into a 50 ml vol. flask and dilute the solution with distilled water upto the 50 ml mark.

(*vii*) 0.10N $AgNO_3$. Dissolve 0.4254 g silver nitrate in distilled water and make the volume to 25 ml with distilled water.

(*viii*) 10% NaCl in 0.005 N HCl. Dissolve 100g sodium chloride (ammonium free) in distilled water and transfer the solution into 1l vol. flask. Add 10 ml of 0.5 N HCl and dilute the solution to 1l mark with distilled water.

(*ix*) 0.5N HCl. Dilute 21.8 ml of concentrated hydrochloric acid (specific gravity 1.18 and minimum assay 35.4%) to a volume of 500 ml with distilled water.

(*x*) 1.0 N NaOH. Dissolve 40g sodium hydroxide in distilled water and make the volume to 1 l mark with distilled water in vol. flask.

(*xi*) Brom cresol green-methyl red indicator. Dissolve 0.066 g methyl red and 0.099 g bromcresol green indicator in 100 ml of ethyl alcohol.

(*xii*) 2% H_3BO_3 containing brom cresol green-methyl red indicator. Transfer 10g boric acid into a 500 ml volumetric flask. Add distilled water to dissolve the boric acid. Then add 10 ml of brom cresol green-methyl red indicator and dilute the solution to 500 ml mark with distilled water. If the colour of the solution is found to be faint pink, add few drops of dilute acid to make the boric acid solution distinct pink.

(*xiii*) 0.1N H_2SO_4. Dilute 2.8 ml of conc. H_2SO_4 (sp. gr. 1.84, Minimum assay 96%) to 1000 ml with distilled water. Standardize the normality of acid against 0.1 N Na_2CO_3 using methyl orange indicator.

2.13.2.3 Procedure

1. Saturation of exchangeable sites with ammonium ions. Transfer 10g soil passed through 2mm sieve into 500 ml conical flask. Add 250 ml of 1N ammonium acetate (pH 7.0). Stopper the flask and shake it thoroughly. Then keep the suspension over night after closing the flask mouth with a rubber stopper.

Set a Buchner funnel lined with Whatman 42 paper on a suction flask connected to a suction pump. Pour the suspension inside the Buchner funnel for filtration. Leach the soil remaining in the Buchner funnel with 25 ml portions of 1N ammonium acetate (pH 7.0) until the leachate is free from calcium.

For checking the presence of Ca in the leachate, take 10 ml of the last leachate, add few drops of 1N NH_4Cl solution and few drops of 10% ammonium oxalate solution and enough approx. 7.5N ammonium hydroxide solution to make the leachate alkaline. Then heat the solution. If on heating no turbidity develops in a few minutes time, the leachate is free from calcium.

When the leachate become free from calcium, leach the soil with 25 ml portions of 1N NH_4Cl four times. Thereafter, leach the soil once with 25 ml of 0.25 N HCl.

2. Removal of residual electrolytes from the soil. Wash the soil on filter paper inside Buchner funnel with 150-200 ml of 99% isopropyl alcohol in small portions until the leachate is free from chloride. Test chloride in leachate by adding few drops of 0.1 N $AgNO_3$ solution. If the turbidity does not develop in the leachate, the leachate is free from chloride. Then allow the soil to get drained thoroughly.

3. Extraction of adsorbed ammonium ions from the soil. Now, leach the ammonium saturated soil with 225 ml of acidified 10% NaCl in small portions at a time. Let sodium chloride solution to pass through the soil before adding the next portion. Collect the leachate containing ammonium ions in a conical flask and stopper.

4. Distillation of ammonium ions in leachate by Kjeldahl distillation apparatus. Set the Kjeldahl nitrogen distillation unit with 800 ml Kjeldahl flask containing few glass beads, 1 ml paraffin and few zinc granules. Transfer the leachate containing ammonium ions from the soil sample quantitatively into 800 ml Kjeldahl flask through the feeding funnel of Kjeldahl distillation unit. Thereafter, stopper the drain hole of the feeding funnel.

Then take a 250 ml capacity conical flask and fill it with 50 ml of 2% boric acid solution containing brom cresol green-methyl red indicator. Dip the end of condenser outlet tubing into the boric acid solution. Fill the feeding funnel with 5 ml distilled water and lift the stopper for a while to let the distilled water pass into the Kjeldahl flask. If boric acid solution begins to give off air bubbles, the distillation assembly is air tight and the ammonia will not escape through the joints during the distillation.

Now, add 25 ml of 1N sodium hydroxide solution into the Kjeldahl flask through the feeding funnel. Stopper the feeding funnel and fill it with 20 ml distilled water allowing a portion to pass into the Kjeldahl flask. Then regulate the flame or heating unit so that the contents begin to boil within about 5 minutes time. Continue heating until 60 ml distillate is collected into the receiving flask containing 50 ml of 2% boric acid solution. Test the fresh distillate with a red litmus paper. If it does not turn blue, the distillate is free from ammonia. Stop the distillation and immediately remove the condenser outlet tube dipped in 2% boric acid solution of the receiving flask. Wash it with 5 ml of distilled water and collect the washing in the same receiving flask. Similarly run a blank by distilling 225 ml of acidified 10% NaCl solution in a Kjeldahl nitrogen distillation unit after adding 25 ml of 1.0 N sodium hydroxide solution into the Kjeldahl flask through the feeding funnel. Collect 60 ml distillate in 50 ml of 2% boric acid containing mixed indicator.

5. Titration of ammonia collected in 2% H_3BO_3 by 0.1H H_2SO_4. Titrate 5ml of 0.1N Na_2CO_3 into 100 ml conical flask. Add one drop of 0.1% methyl orange indicator and titrate with approx. 0.1 N H_2SO_4 until the colour change from yellow to orange which does not disappear on gentle boiling. From the acid titre, estimate the normality of the acid.

Then titrate the distillate collected in 2% boric acid containing the mixed indicator with the standard sulphuric acid. Note down the titre as the colour of the solution changes from green to distinct pink. The blank distillate should also be titrated to the same end point.

2.13.2.4 Calculations

$$\text{meq of } NH_4^+/100\text{g soil} = \frac{\text{(Sample} - \text{Blank) titre} \times \text{Normality of acid} \times 100}{\text{mass of soil in g}}$$

CEC in meq/100g soil = meq of NH_4^+/100 g soil

2.13.2.5 Comments

The removal of excess electrolyte by isopropyl alcohol cause less loss of ammoniated organic constituents than leaching by 95% ethyl alcohol or 100% methyl alcohol (Peach *et. al.*, 1962).

REFERENCES

Amma, M.K. 1989. Plant and Soil Analysis. Rubber Research Institute, Rubber Board, Kottayom, Keral, India.

Jackson, M.L. 1973. Soil Chemical Analysis. Prentice–Hall of India Private Limited, New Delhi.

Muthuval, P., C. Udayasoorian, R. Natesan, and R.P. Ramaswami 1952. Introduction to Soil Analysis. Tamil Nadu Agricultural University, Coimbatore-641002.

Peach, M.R., L. Cowan, and J.H. Baker (1962). A critical study of the $BaCl_2$-triethanolamine, and ammonium acetate methods for determining the exchangeable hydrogen content of the soils. Soil Sci. Soc. Am. Proc. 26 : 37-40.

2.13.3 Cation exchange capacity of the soil by saturation with Ca^{++}–Jackson (1958) and Prasad (1970)

2.13.3.1 Principle

The exchangeable sites of the soil is saturated with Ca after displacing adsorbed cations by Na. The soil is then washed with alcohol for the removal of residual electrolytes. Thereafter exchangeable Ca is extracted by 1.0N sodium acetate (pH 5.0). Ca extracted is estimated by titration with EDTA using Eriochrome Black T indicator on introducing few drops of dilute $MgCl_2$.

2.13.3.2 Reagents

(*i*) 1.0 N CH_3COONa (pH 5.0). Dissolve 136 g of sodium acetate in distilled water and dilute the solution to 1 l mark with distilled water. Add acetic acid to adjust the reaction to pH 5.0.

(*ii*) 1.0 N $CaCl_2$. Dissolve 73g of $CaCl_2.2H_2O$ in CO_2 free distilled water and dilute it to 1 l mark with distilled water.

(*iii*) 0.01N $CaCl_2$. Dissolve 0.5g pure dry $CaCO_3$ in 10 ml of approximately 3N (1 + 3) HCl and dilute the solution to 1l mark.

(*iv*) NH_4Cl - NH_4OH Buffer. Transfer 33.75 g NH_4Cl into 500 ml volumetric flask. Add 285 ml of conc. NH_4OH into the flask. Then make the volume to 500 ml mark with distilled water.

(*v*) Erichrome Black T Indicator. Mix 0.5 g Eriochrome Black T and 4.5 g Hydroxylamine hydrochloride thoroughly.

(*vi*) 50% Triethanolamine. Transfer 50 ml triethanolamine into 100 ml volumetric flask. Then add 40 ml distilled water and mix the solution thoroughly. Dilute the solution to 100 ml mark with distilled water and shake to mix the contents.

(*vii*) 0.1% Methyl red indicator. Dissolve 0.1g methyl red indicator in 95% ethanol and dilute it to 100 ml with 95% ethanol.

(*viii*) 0.01N EDTA. Transfer 2g disodium salt of ethylene diamine tetracetic acid and 0.05 g $MgCl_2$ into 1l vol. flask. Add distilled water to dissolve the salt and dilute the solution to 1l mark with distilled water. Standardize EDTA against 0.01N $CaCl_2$ using EBT indicator.

(*ix*) 2% KCN. Dissolve 2g KCN in distilled water and dilute the solution to 100 ml mark with distilled water. Label the solution as deadly poison with red ink. Fill a plastic indicator bottle with a portion of the solution for dispensing when required.

(*x*) 0.01N Mg Cl_2. Dissolve 0.5081 g $MgCl_2.6H_2O$ in distilled water and make the volume to 500 ml with distilled water.

(*xi*) 99% Isopropyl alcohol.

2.13.3.3 Procedure

1. Saturation of the soil with calcium. Transfer 2g soil into 50 ml centrifuge tube and add 20 ml of 1.0 N sodium acetate (pH 5.0) into the centrifuge tube. Stir the solution + soil with policeman tipped glass rod for 5 minutes and then place the tubes inside the centrifuge in pairs opposite each other for proper balance. Close the centrifuge lid and run the centrifuge at 2000–2440 rpm for 10 minutes. Switch off the centrifuge after lowering the speed gradually. When the centrifuge drum stops, open the centrifuge lid, remove the centrifuge tube and decant off the supernant liquid and discard the solution. Similarly give two more washings with 1N sodium acetate (pH 5.0). Each washing operation includes adding of washing solution, stirring with glass rod, centrifuging and decanting off the supernant liquid. By the washings adsorbed cations shall be displaced by sodium.

Thereafter add 20 ml of 1N $CaCl_2$ solution into the centrifuge tube, stir the soil plus solution with policeman tipped glass rod for 5 minutes. Place the centrifuge tube into the centrifuge, and run the centrifuge at 2000–2440 rpm for 10 minutes. Then decant off the supernant liquid and discard it. Similarly give four more washing with 1N $CaCl_2$decanting the supernant liquid each time. This would effect displacement of exchangeable sodium by calcium resulting in the saturation of exchangeable sites by calcium.

2. Removal of the residual electrolytes. Give 3 or more washings to the soil with isopropyl alcohol for removing the residual electrolyte. The washings be continued until the leachate *i.e.* washing becomes free from chloride as indicated by the silver nitrate test.

3. Extraction of exchangeable calcium from the soil. In order to extract Ca adsorbed on the exchange sites, add 20 ml of 1N sodium acetate to each centrifuge tube after removal of the residual electrolyte.

Then stir the contents with policeman tipped glass rod for 5 minutes. Place the centrifuge tube inside the centrifuge in pairs opposite each other. Run the centrifuge at 2000–2440 rpm for 10 minutes. When the centrifuge revolving drum stops, remove the centrifuge tube and decant off the supernant clear Ca-extract into a 250 ml Erlenmeyer flask. Similarly give four more washings with 1N sodium acetate (pH 7.0) each time using 20 ml of the extracting solution for each centrifuge tube and decanting off the supernant clear Ca extract from the same soil into the same 250 ml Erlenmeyer flask.

4. Estimation of exchangeable Ca in the extract from soil. Add 5 ml of 50% triethanolamine and 0.5 ml of 2% KCN solution into 250 ml Erlenmeyer flask containing adsorbed Ca from soil extracted by 100 ml of 1N sodium acetate (pH 5.0). Then add 10 ml of NH_4Cl - NH_4OH Buffer solution and a pinch of Eriochrome Black T indicator mixed with hydroxyl amine hydrochloride, and 2 drops of 0.01N $MgCl_2$. Titrate the solution by 0.01N ethylene diamine tetra acetic acid (sodium salt) using a micro burette until the colour of solution changes from wine red to purple. At this point add 5 drops of 0.1% methyl red indicator, and continue titration by EDTA until the colour of the solution changes to bright green colour. Similarly titrate 100 ml of 1N sodium acetate (pH 5.0) by 0.01N EDTA solution to the same end point for obtaining the EDTA titre for the blank.

2.13.3.4 Calculations

Cation exchange capacity meq/100 g soil

$$= \frac{\text{EDTA titre (Sample} - \text{Blank)} \times \text{Normality of EDTA} \times 100}{\text{mass of soil in g}}$$

2.13.3.5 Comments

Two drops of 0.01N $MgCl_2$ is introduced into the solution prior to titration by EDTA as EBT is a Mg indicator, and in EDTA titration the end point is sharp with Mg indicator than with Ca indicator.

The introduction of one drop of 0.1% methyl red indicator near the end point changes the end point from purplish blue to green. The colour change from wine red to green makes the end point sharp.

If centrifuge tubes of only 10 ml capacity fit into the centrifuge. The soil sample may be equally apportioned into four centrifuge tubes. The reagents for washing of the soil and for the extraction of adsorbed Ca from exchange sites be also apportioned equally into four portions so that it gets equally divided for the four centrifuge tubes. However the sample extract containing adsorbed Ca from the four centrifuge tubes be combined together into a 100 ml volumetric flask and the volume be made to the 100 ml mark with the extractant.

REFERENCES

Barrows, H.L. and F.C. Simpson 1962. An EDTA method for the direct routine determination of Ca and Mg in soils and plant tissues. Soil Sci. Soc. Am. Proc. 26 : 440-445.

Black, C.A. (ed.). 1965. Methods of Soil Analysis. Vol. 2. Am. Soc. Agron., Madison, Wisconsin, U.S.A.

Cheng, K.L. and R.H. Bray. 1951. Determination of Calcium and Magnesium in Soil and Plant material. Soil. Sci. 72 : 449-458.

Jackson, M.L. 1958. Soil Chemical Analysis. Prentice-Hall of India Private Limited, New Delhi (1973).

Prasad, R. 1970. Practical Manual For Soil Fertility. Div. Agron. Indian Agric. Res. Inst., New Delhi-12.

Tucker, B.B. and L.T. Kurtz (1961). Calcium and Magnesium determination by EDTA titration. Soil Sci. Soc. Am. Proc. 25: 27-29.

2.14 EXCHANGEABLE BASES IN SOIL

2.14.1 Estimation of exchangeable bases in soil after removal of soluble salts-Black (1965)

2.14.1.1 Principle

The soluble salts from soil sample is first removed by leaching with 99% isopropyl alcohol. Then the exchangeable cations are extracted by 1.0 N $NH_4C_2H_3O_2$ (pH 7.0). Na and K extracted is determined directly by flame photometer. For Ca and Mg determination, an aliquot of soil extract is evaporated to dryness and after treatment of the residue with aqua-regia, the solution is again evaporated to dryness to remove dissolved organic constituents. As $NH_4C_2H_3O_2$ is volatile at lower temperature it is removed during the heating. The residue remaining is dissolved in distilled water and is titrated by 0.01N disodium ethylene diamine tetra acetic acid solution (EDTA) using Ca and Mg indicator.

2.14.1.2 Apparatus/Reagents

Apparatus

Volumetric flasks-1000, 500, 250 and 100 ml capacity, Beakers 500, 250 and 100 ml capacity preferably tall form, suction flasks, funnel, rubber rings, 50 ml test tubes, suction pump, centrifuge with conical heavy duty 40 ml capacity centrifuge tubes calibrated at 5 ml, graduated pipettes, Erlenmeyer flasks-250 and 150 ml capacity, burette 50 ml cap ; micro burette -10 ml capacity and flame photometer.

Reagents

1. Reagents for the extraction of exchangeable bases from the soil

(*i*) 99% Isopropyl alcohol (reagent grade)

(*ii*) 1.0 N Ammonium acetate solution (pH 7.0) : Dissolve 77.0 g of $NH_4C_2H_3O_2$ (AR) in 1000 ml distilled water. Adjust the pH to 7.0 with acetic acid or ammonium hydroxide.

(*iii*) Approx. 0.1N Ammonium chloride solution (pH 7.0) : Dissolve 0.5350 g of NH_4Cl (AR) in 100 ml of distilled water. Adjust its pH to 7.0 with NH_4OH (AR).

(*iv*) Approx. 0.1N Silver nitrate solution : Dissolve 1.6989 g of $AgNO_3$ (AR) in 100 ml distilled water

(*v*) Approx. 1.0N Ammonium Chloride solution (pH 7.0) : Dissolve 5.3500 g of NH_4Cl (AR) in 100 ml distilled water. Adjust its pH to 7.0 with NH_4OH (AR).

(*vi*) 10% Ammonium oxalate solution : Dissolve 10g of $NH_4C_2O_4.H_2O$ in 100 ml distilled water (*vii*) 50% Ammonium hydroxide solution (v/v). Transfer 50 ml of conc. NH_4OH (AR) into 100 ml. volumetric flask and dilute it to 100 ml with distilled water.

2. Reagents for the determination of K and Na in soil extract

(*i*) 1000 ppm K solution. Dissolve 1.9066 g. of pure dry KCl (AR) in 1000 ml distilled water.

(*ii*) 250 ppm K solution. Dilute 25 ml of 1000 ppm K solution to 100 ml with distilled water.

(*iii*) 0 to 20 ppm K solution. Transfer 0, 2, 4, 6 and 8 ml of 250 ppm K solution into separate 100 ml volumetric flasks and then dilute the solution to 100 ml with 1.0N $NH_4C_2H_3O_2$ (pH 7.0) to obtain 0, 5, 10, 15 and 20 ppm K solution. Add one drop of butyl alcohol to each in order to improve spraying property.

(*iv*) 1000 ppm Na solution. Dissolve 2.5412 g of pure dry NaCl in 1000 ml distilled water.

(*v*) 250 ppm Na solution. Dilute 25.0 ml of 1000 ppm Na solution to 100 ml with distilled water.

(*vi*) 0 to 20 ppm Na solution. Transfer 0, 2, 4, 6 and 8 ml of 250 ppm Na solution into separate 100 ml volumetric flasks and then dilute the solution to 100 ml with 1.0N $NH_4C_2H_3O_2$ (pH 7.0) solution to obtain 0, 5, 10, 15 and 20 ppm Na solution. Add a drop of butyl alcohol to improve spraying property of the solution.

3. Reagents for the determination of Ca and Mg in the soil extract

(*i*) 0.01N Calcium chloride solution. Dissolve 0.5000 g. of pure dry $CaCO_3$ (AR) in 10 ml of (1 + 3) HCl, and dilute it to 1000 ml mark with distilled water.

(*ii*) Approx. 0.01N EDTA solution. Transfer 2.0 g of disodium dihydrogen ethylene diamine tetracetic acid and 0.05 g of $MgCl_2 . 2H_2O$ into 1l volumetric flask. Add distilled water to dissolve the reagents and dilute it to 1000 ml mark with distilled water. Standardize the solution against 0.01N $CaCl_2$ with Ca and Mg indicator separately following the procedure as described for the sample titration ommiting the treatment with aqua-regia or conc. nitric acid for the removal of interference from ammoniated organic constituents.

(*iii*) approx. 4.0N sodium hydroxide solution. Dissolve 16 g of NaOH pellets (AR) in 100 ml distilled water.

(*iv*) Ca-indicator. Mix 0.5 g of ammonium purpurate and 100 g of powdered K_2SO_4 (AR) in a pestle mortar.

(*v*) Mg. Indicator. Dissolve 0.5 g of Eriochrome Black T (EBT) and 4.5 g. of $NH_2OH.HCl$ (hydroxylamine hydrochloride) in 100 ml of 95% ethanol. Store in a refrigerator. Prepare fresh reagent every three weeks.

(*vi*) 1.0% Potassium cyanide solution. Dissolve 1.0 g of KCN in 100 ml. distilled water and label it as deadly poison.

(*vii*) 5.0% Hydroxylamine hydrochloride solution. Dissolve 5.0 g of $NH_2OH.HCl$ in 100 ml distilled water.

(*viii*) 4.0% Potassium ferrocyanide solution Dissolve 4.0g of $K_4Fe(CN)_6.3H_2O$ (reagent grade) in 100 ml. distilled water.

(*ix*) 50% Triethanolamine. Transfer 50 ml of triethanolamine (reagent grade) into 100 ml volumetric flask and dilute it to 100 ml mark with distilled water.

(*x*) Ammonium chloride. Ammonium hydroxide buffer solution : Dissolve 67.5g of NH_4Cl in 570 ml concentrated NH_4OH and make the volume to 1l.

(*xi*) 20% Sodium tungstate solution. Dissolve 20g. of $Na_2WO_4.2H_2O$ in 100 ml. of distilled water.

2.14.1.3 Procedure

1. Leaching of soil sample with 99% isopropyl alcohol for the removal of soluble salts. Set a suction flask filled with straw to provide support to a 50 ml. test tube. Place a funnel lined with Whatman No. 42 filter paper on the rubber rings placed inside the mouth of suction flask.

Transfer 10.0g of air dry <2 mm soil sample into a 400 ml beaker. Add 50 ml of 99% isopropyl alcohol to the beaker. Stir the contents with a glass rod for 5 minutes and transfer the contents of the beaker into the funnel in small portions. Apply a light suction to accelerate the filtration. When the soil sample has been completely transferred from the beaker into the funnel, add a drop of 0.1N NH_4Cl (pH 7.0) on the soil sample.

Continue leaching the soil with 25 ml portions of 99% isopropyl alcohol. Rinse the beaker with the 99% isopropyl alcohol, and then transfer the rinsings into the funnel containing the soil sample. Allow each portion to pass through the soil sample before adding the next portion. Use a total of about 150 to 200 ml of 99% isopropyl alcohol. Continue leaching the soil sample until the leachate is free from chloride.

For testing chloride in leachate, add 1 ml of 0.1N $AgNO_3$ solution to 10 ml of leachate. The absence of white precipitate is indicative of leachate being free from chloride.

2. Extraction of exchangeable basic cations by 1.0N $NH_4C_2H_3O_2$ (pH 7.0). When the leachate is free from chloride, allow the soil to drain off for some time, and then transfer the soil from filter paper into the same 400 ml beaker.

Add 100 ml of 1.0 N $NH_4C_2H_3O_2$ (pH 7.0), stir the contents for 10 minutes, and let it stand overnight at room temperature after covering the beaker with a watch glass. This would permit displacement of exchangeable cations by ammonium ions.

On the following day, transfer the contents of the beaker in small portions into a funnel lined with Whatman No. 42 filter paper, and placed on suction flask connected to a suction pump. Collect the filtrate (extract) into a 50 ml test tube placed below the funnel end inside the suction flask filled with straw. Transfer the extract into 250 ml volumetric flask.

Rinse the beaker with 25 ml portions of 1.0N $NH_4C_2H_3O_2$, and transfer the rinsings into the funnel. Allow each portion to pass through the funnel before adding the next portion. Transfer the extract collected in each test tube into 250 ml volumetric flask. Continue washing of the soil with 25 ml portions of 1.0N $NH_4C_2H_3O_2$ until the extract collected reaches a volume of 250 ml. Similarly prepare a blank test solution.

At this stage the leachate be free from Ca. Take 10 ml of fresh leachate, add few drops of neutral 1.0N NH_4Cl, and 10% ammonium oxalate. Then, add enough dilute NH_4OH solution to make it alkaline. Heat the contents. If no turbidity develops in a few minutes time, the leachate is free of calcium.

3. Determination of basic cations in soil extract.

2.14.1.3.1 Estimation of Exchangeable K in soil Extract by Flame Photometer

Switch on the flame photometer and place K filter disc against the built in gas burner. Place a candle flame just above the gas burner. Adjust the gas regulator pin to let indane gas flow into the burner. When the burner is lighted, remove the candle. Adjust the air inlet knob to attain the desired pressure and a clear blue flame of sufficient length.

Now atomize 0 ppm K solution and adjust the flame photometer reading to zero by regulating the reference knob. Then, atomize 20 ppm K solution and adjust the flame photometer reading to 100 by regulating the knob provided. Thereafter, atomize the intermediate K standards and record the corresponding flame photometer readings. Plot flame photometer reading on the ordinate of a linear scale graph paper and the corresponding concentration of K standards on the abscissa. Connect points for 0 and 100 flame photometer reading by a straight line to obtain the calibration curve.

Check the flame photometer reading by atomizing 0 and 20 ppm K solution. Then, atomize sample and blank extract and record the flame photometer reading. Determine the concentration of K in sample and blank extract by reference to the calibration curve or else by dividing the flame photometer reading of the extracts by the slope of the calibration curve *i.e.* flame photometer reading per ppm K solution.

Calculation

$$\text{Exchangeable K is soil (meq / 100 g soil)} = \frac{(A - B) \times C \times 100}{10^6 \times D \times E}$$

$$= \frac{(A - B) \times 250 \times 100}{10^6 \times 0.0391 \times 10}$$

where A and B denote ppm K in sample and blank extract respectively as estimated by reference to the standard calibration curve for K. C is total volume of 1N $NH_4C_2H_3O_2$ (pH 7.0) used for the extraction of K from the soil sample. D is the milligram equivalent weight of potassium. E is the mass of soil sample in gram taken for the estimation of exchangeable K from the soil.

2.14.1.3.2 Estimation of Exchangeable Na in Soil Extract by Flame Photometer

Rotate the filter disc to place the Na filter against the burner. Switch on the flame photometer and the compressor. Place candle flame just above the gas burner. Adjust the gas regulator pin to let Indane gas flow into the burner. When the burner is lighted, remove the candle. Adjust the air inlet knob to attain the desired air pressure and a clear blue flame of sufficient length.

Atomize 0 ppm Na solution and adjust the flame photometer reading to zero. Then, atomize 20 ppm Na solution and adjust the flame photometer reading to 100. Thereafter, atomize the intermediate Na standards and record the corresponding flame photometer reading. Prepare a straight line calibration curve by plotting flame photometer readings corresponding to Na concentration in ppm on a linear scale graph paper.

Recheck the flame photometer reading on atomizing 0 and 20 ppm Na solutions. Then, atomize blank and sample extract, and note the corresponding flame photometer reading. Estimate the concentration of Na in blank and soil extract in ppm by reference to the calibration curve for Na or else by dividing flame photometer reading by the slope of the calibration curve *i.e.* flame photometer reading per ppm Na in the standard

Calculation

$$\text{Exchangeable Na in soil (meq Na/100g soil)} = \frac{(A - B) \times C \times 100}{10^6 \times D \times E}$$

$$= \frac{(A - B) \times 250 \times 100}{10^6 \times 0.023 \times 10}$$

where A and B denote ppm Na in sample and blank extract respectively as estimated by reference to the standard calibration curve for Na.

C is total volume of 1N $NH_4C_2O_3O_2$ (pH 7.0) used for the extraction of exchangeable Na from the soil.

D is the milligram equivalent weight of sodium.

E is the mass of soil sample in gram taken for the extraction of exchangeable Na from the soil.

2.14.1.3.3 Ca and Mg Determination in Soil Extract

1. Normality of EDTA solution against standard 0.01N $CaCl_2$ solution using Ca-indicator. Transfer 5.0 ml of 0.01N $CaCl_2$ solution into a 125 ml Erlenmeyer flask. Then add 20 ml of distilled water, and 2 drops each of 1.0% KCN, 4.0% K_4 Fe $(CN)_6$, 5% NH_2OH-HCl and 50% triethanolamine. Alternatively one may add a pinch of disodium diethyl dithio carbamate crystals to complex heavy metals. Thereafter, add 4N NaOH solution to raise the pH to 12.0 or more (normally 0.5 ml of 4N NaOH is required). Then add a pinch of Ca-indicator (ammonium purpurate

grinded with K_2SO_4 in the ratio of 1 : 200 (w/w) and titrate by EDTA solution until the colour changes from wine red to purple. Swirl the contents after adding EDTA solution. Near the end point, add 1 or 2 drop of EDTA and then swirl the contents. Perform the titration in triplicate and take the mean of two closest titre. Similarly run a blank and note the blank titre. Divide 0.05 by (sample- blank) titre to obtain the normality of EDTA.

Alternatively transfer 5.0 ml of approx. 0.01N EDTA solution into 125 ml Erlenmeyer flask. Add 20 ml of distilled water and 2 drops each of the 1.0% KCN, 4% $K_4Fe(CN)_6$, 5% $NH_2OH.HCl$ and 50% triethanolamine. Then gradually add 4N NaOH to raise the solution pH to 12 or more. Thereafter, add a pinch of Ca-indicator and titrate the solution by 0.01 N $CaCl_2$ until the colour changes from purple to wine red. Perform the titration in triplicate and take the mean of two closeest titre. Simillarly run blank and note down the blank titre. Estimate the normality of EDTA by taking the product of $CaCl_2$titre (sample-blank) and its normality and then dividing the product by the aliquot of EDTA solution taken for titration. This procedure is considered to be preferable for solutions in which interference from high phosphate concentration is likely.

2. Normality of EDTA solution against standard 0.01N $CaCl_2$ using Mg-indicator (EBT). Transfer 5.0 ml of 0.01N $CaCl_2$ solution into three separate 125 ml. Erlenmeyer flasks. Add 20 ml of distilled water and one drop of 0.01N $MgCl_2$ to each flask. Then add 2 drops each of 1.0% KCN, 4% $K_4Fe\ (CN)_6$, 5% $NH_2OH.HCl$ and 50% triethanolamine solution. Mix the contents and add NH_4Cl-NH_4OH buffer to raise the solution pH to 10.0. Normally 2.0 ml NH_4Cl-NH_4OH buffer is sufficient for attaining the desired pH. Thereafter, add 2 drops of Mg-indicator (EBT) and immediately titrate the solution by adding EDTA solution in small portions using a microburette. Swirl the contents after each addition of EDTA. Continue titration by EDTA solution until the colour changes from wine red to purple. Now add EDTA drop by drop swirling the contents after each addition until the solution colour changes from purple to clear blue or green. Note down the titre and similarly perform the other two titrations of solution containing 5 ml of .01N $CaCl_2$, 20 ml distilled water and 1 drop of .01N $MgCl_2$. Similarly run blank titration beginning with the addition of 20 ml of distilled water. Divide 0.05 by EDTA titre (sample-blank) to estimate the normality of EDTA.

3. Treatment of the soil extract with aqua-regia. Transfer 5.0 ml of soil extract into separate six 150 ml Erlenmeyer flask (pyrex or corning). Evaporate the soil extract to dryness on a sand bath. Remove the Erlenmeyer flask from sand bath and let it cool. Add 10 ml of distilled water along the inner side of the flask and again evaporate the contents to nearly dryness on the sand bath. Remove the flask from the sand bath and let it cool. Then add 4 ml of aqua-regia (3 volume conc. HCl + 1 volume conc. HNO_3) freshly prepared and heat the contents to dryness. This would make the residue free from dissolved organic constituents. The residue will be also free from ammonium acetate as it is volatile at low temperature.

4. Dissolution of residue after treatment with aqua regia in distilled water and then titration by EDTA using Mg indicator. When the residue after treatment with

aqua-regia has been evaporated to dryness, allow the flask to cool and then add 25 ml of distilled water to dissolve the residue. Thereafter, proceed following the procedure described in step 2 of section 2.14.1.3.3, beginning with the addition of 1 drop of 0.01N $MgCl_2$. Perform the titration in triplicate and take the mean of two closest readings. Similarly run a blank taking 5 ml of blank extract and following the entire procedure given in step 2 of section 2.14.1.3.3.

5. Back titration of .01N EDTA added in excess by standard $CaCl_2$ using Ca indicator. On dissolving the residue in 25 ml of distilled water. Add 2 drops each of 1% KCN, 4% $K_4Fe\ (CN)_6$, 5% $NH_2OH.HCl$, and 50% triethanolamine. Then add 20 ml of 0.01N EDTA solution through a microburette. Then, add 4N NaOH to raise the pH to 12 or more. Add a pinch of Ca-indicator (ammonium purpurate plus K_2SO_4 in the ratio of 1:200 grinded in a mortar) and swirl the contents. If the solution develops a purple colour, EDTA is present in excess. Back titrate the EDTA added in excess by titration with standard 0.01N $CaCl_2$ until the colour of the solution changes from purple to wine red . Perform the titration in triplicate and take the mean of two closest titre. Similarly run a blank with 5 ml of blank extract following the entire procedure described in step 3 and 5 of section 2.14.1.3.3 for the soil extract.

Calculation

$$\text{Exchangeable Ca + Mg in soil (meq / 100g soil)} = \frac{(A-B)\times C\times D \times 100}{E \times F}$$

$$= \frac{(A-B)\times C\times 250\times 100}{5\times 10}$$

where

A and B denote EDTA titre for the aliquot of sample and blank extract respectively.

C is the normality of EDTA solution against .01N $CaCl_2$ using Mg indicator.

D is the total volume of neutral normal ammonium acetate used for the extraction of exchangeable bases from the soil.

E is an aliquot of soil extract taken for the determination of Ca + Mg *i.e.* before treatment of the extract with aqua-regia.

F is the mass of soil sample taken for the extraction of exchangeable bases.

$$\text{Exchangeable Ca in soil (meq / 100 g soil)} = \frac{(A - B) \times C \times 100}{D \times E}$$

$$= \frac{(A - B) \times 250 \times 100}{5 \times 100}$$

where

A is meq of EDTA used in complexing Ca present in aliquot of soil extract which is equal to the meq of EDTA added *i.e.*, ml of EDTA added $\times$ normality minus meq of EDTA back titrated with standard .01N $CaCl_2$ *i.e.* ml of $CaCl_2$ used × normality.

B is meq of EDTA used in complexing Ca present in the aliquot of blank extract which is equal to the meq of EDTA added to the blank extract after treatment with aqua-regia minus meq of EDTA back titrated with standard 0.01N $CaCl_2$.

C, is the total volume of 1.0N $NH_4C_2H_3O_2$ (pH 7.0) used for the extraction of exchangeable bases from the soil.

D, is the ml aliquot of soil extract taken for the determination of exchangeable Ca.

E, is the mass of soil in gram taken for the extraction of exchangeable bases by 1.0N $NH_4C_2H_3O_2$ (pH 7.0).

Exchangeable Mg in soil = Exchangeable Ca + Mg – Exchangeable Ca in soil
(meq/100 g soil) (meq/100 g soil) (meq/100 g soil)

2.14.1.3.4 Determination of Exchangeable Mg in the Soil Extract after Separation of Mg by Precipitation of Ca

Method 1. Soil extract is evaporated to dryness and treated with aqua regia. Then Ca is precipitated as Ca tungstate and separated by filtration. The filtrate and washing of residue with dilute NH_4Cl–NH_4OH buffer is collected and titrated with EDTA using Mg indicator.

1. Treatment of the soil extract with aqua-regia. Transfer 10 ml of soil extract into 250 ml tall beaker tall form and give the extract the same treatment as described in step-3 of section 3.3 of the procedure (2.14.1.3) using 8 ml instead of 4 ml of aqua-regia.

2. Precipitation of Ca in residue dissolved in distilled water, and filtration of the suspension. Dissolve the residue obtained on treatment with aqua-regia in 25 ml of distilled water. Add 5 ml of 20% sodium tungstate solution to precipitate Ca in solution. Add NH_4Cl-NH_4OH buffer to raise the pH to 10.0. Heat the contents on a hot plate and filter the suspension through a funnel lined with Whatman No. 42 filter paper. Collect the filtrate in a 150 ml Erlenmeyer flask. Wash the precipitate and filter paper with 20 ml of dilute NH_4Cl-NH_4OH buffer (1 ml buffer diluted to 20 ml with distilled water). Collect the washings in the 150 ml Erlenmeyer flask containing the filtrate.

3. Titration of filtrate cum washings containing Mg by EDTA using Mg indicator. Adjust the pH of filtrate cum washings in the Erlenmeyer flask to 10.0 by adding NH_4Cl–NH_4OH buffer. Then add 4 drops each of 1% KCN, 4% $K_4Fe(CN)_6$, 5% NH_2OH-HCl and 50% triethanolamine. Add 2 to 3 drops of Mg indicator (EBT). Heat to warm the contents and titrate by 0.01N EDTA solution until the colour of reaction mixture changes from wine red to clear blue or green. Perform the titration in triplicate, and take the mean of the closest titre. Similarly run a blank using blank extract.

2.14.1.4 Calculation (Method 1)

Exchangeable Mg is soil (meq. / 100 g soil)

$$= \frac{(A-B)\times C\times D\times 100}{E\times F}$$

$$= \frac{(A-B)\times C\times 250\times 100}{10\times 10}$$

Where,

A is the EDTA titre for complexing Mg in the aliquot of the soil extract.

B is the EDTA titre for complexing Mg in the aliquot of blank extract.

C is the nomality of EDTA determined against 0.01N $CaCl_2$ using Mg-indicator (EBT)

D is the total volume of 1.0N $NH_4C2H_3O_2$ (pH 7.0) used for the extraction of exchangeable bases from the soil.

E is ml aliquot of soil extract taken for the determination of exchangeable Mg in the soil.

F is the mass of soil sample in gram taken for the extraction of exchangeable bases by 1.0 N $NH_4C_2H_3O_2$ (pH 7.0).

Method 2. Soil extract is evaporated to dryness and the residue is treated with HNO_3, H_2SO_4 and ethyl alcohol. Alcoholic solution containing Mg is separated by centrifugation. Alcohol is removed by evaporation on hot plate. The residual solution is titrated by EDTA using Mg-indicator.

1. Evaporation of soil extract to dryness. Transfer 10.0 ml of soil extract into a 250 ml tall form beaker (pyrex or corning). Place the beaker on a sand bath and evaporate the contents to dryness. Cool the beaker, and then add 10 ml of distilled water along the inner side of beaker and again evaporate the contents to dryness. Remove the beaker from sand bath and let it cool.

2. Treatment of the residue with HNO_3, H_2SO_4 and ethyl alcohol and separation of solution containing Mg. When the beaker containing soil extract residue has cooled, add 3 ml of approx. 1N HNO_3 to dissolve the residue. Transfer the solution to a 40 ml heavy duty centrifuge tube. Rinse the beaker with distilled water, and transfer the rinsings to the same 40 ml heavy duty centrifuge tube. Now place the centrifuge tube in a water bath to evaporate the solution in excess of 5 ml.

Thereafter, add 1 ml of 6NH_2SO_4. Swirl the centrifuge tube to mix the contents. Then, add 33 ml of 95% ethanol, and cover the centrifuge tube with a piece of aluminium foil and let it stand over night. On the following day, remove the aluminum foil cover and centrifuge the tube at 2440 rpm for 10-15 minutes. Decant the alcoholic supernant clear solution containing Mg into a 250 ml tall form beaker.

3. Removal of alcohol by heating and titration of Mg in residue after dissolving in distilled water by EDTA titration using Mg indicator. Place the tall form beaker containing Mg in alcoholic supernant clear solution on a hot plate to evaporate the contents at moderate heat until the volume is reduced to less than 1 ml.

Let the beaker cool. Then, add 25 ml of distilled water and about 2 ml of NH_4Cl-NH_4OH buffer to raise the solution pH to 10.0. Thereafter, add 2 drops each of 1% KCN, 4% $K_4Fe(CN)_6$, 5% $NH_2OH.HCl$ and 50% triethanolamine, and 2 drops of Mg indicator (EBT). Titrate the solution containing Mg by 0.01N EDTA solution until the colour changes from wine red to clear blue or green. Similarly run a blank taking an aliquot of 10 ml blank extract carried through the entire procedure.

2.14.1.4 Calculation (Method 2)

$$\text{Exchangeable Mg in soil (meq/100 g soil)} = \frac{(A - B) \times C \times D \times 100}{E \times F}$$

$$= \frac{(A - B) \times C \times 250 \times 100}{10 \times 10}$$

where A, B, C., D, E & F denote the same as described under calculation of section 2.14.1.3.3.

2.14.1.5 Comments

The exchangeable bases are usually extracted by 1.N $NH_4C_2H_3O_2$ (pH 7.0). When soils contain soluble salts in appreciable amounts, the soluble salts are removed by leaching with alcohol prior to extraction of exchangeable bases. As ethyl alcohol removes ammoniated organic constituents also, leaching of soluble salts with 99% isopropyl alcohol is preferable. Leaching with alcohol removes salts other than sulphates as sulphates are not readily soluble in alcohol. Leaching with water is avoided as it causes hydrolysis of adsorbed ions. In soils containing free carbonates, gypsum or excess soluble salts, the extraction of exchangeable Ca or Mg is not recommended. If exchangeable Ca/Mg are determined, the results be interpreted with great care.

The concentration of Na and K in the extract is directly determined by flame photometer after preparation of the calibration curve with Na and K standards using appropriate filter.

The concentration of Ca and Mg is determined conveniently by ethylene diamine tetra acetic acid (EDTA) titration using Ca and Mg indicator. However, the organic constituents extracted has to be decomposed by treating the residue remaining after evaporation with concentrated nitric acid or else with aqua-regia (3 volumes HCl plus 1 volume HNO_3) and then heating again to dryness. The dark coloured extract would require second treatment with conc. nitric acid or with aqua-regia. The residue obtained is dissolved in distilled water or HCl prior to determination of Ca and Mg by EDTA titration. As $NH_4C_2H_3O_2$ is volatile at lower temperature, it also gets removed during heating of the extract.

The EDTA acts as a chelate. The end point is indicated by metal sensitive indicator which are also chelate. The Ca indicator is murexide (ammonium purpurate mixed with potassium sulphate) and the Mg indicator is Eriochrome Black T (EBT). In the presence of free Ca, ammonium purpurate gives a wine red colour which turns to purple when free Ca has been completely complexed by EDTA. When EBT is used as an indicator, EBT gives a wine red colour in the presence of free Mg, and when Mg is completely complexed by EDTA, EBT gives a clear blue or green colour. As Ca EDTA complex is more stable than Mg EDTA complex, it is possible to determine Ca + Mg by EDTA titration using Mg indicator. The colour developed at the end point is the result of reaction of Mg with EBT, but it is indicative that the reaction of Ca with EDTA has been completed.

The pH of solution is adjusted to 12 or more by adding NaOH when Ca-indicator is employed. At this pH, Mg become precipitated as $Mg(OH)_2$. However, when Mg-indicator is used, the pH of the solution is adjusted to 10.0 by adding NH_4Cl-NH_4OH buffer which permits the formation of both Ca-EDTA and Mg-EDTA complex. As ammonia is volatalized in alkaline medium, the titration by EDTA be completed immediately after adding the NH_4Cl-NH_4OH buffer.

The EDTA forms chelate with Ca, Mg and other metals. It is therefore desirable to remove interferences from other metals either by chelation or precipitation. Some metal ions need to be transferred to a reduced state prior to chelation.

The KCN forms very stable complexes with Cu, Zn, Ni and Fe^{++} whereas triethanolamine chelates Al and Fe^{++} effectively. The $NH_2OH.HCl$ reduce Fe^{+++} to Fe^{++}. The $K_4Fe(CN)_6$ precipitate Mn as $Mn_2Fe(CN)_6$. The interference from Cu and other metals is also removed by incorporation of sodium diethyl dithiocarbamate reagent.

In EDTA titration, Ca recovery is not influenced by the amount of Mg present. However, the presence of 15 mg of Ca prevent complete recovery of Mg. Consequently, it is preferable to determine Ca and Mg separately. In Ca determination, Mg gets precipitated as $Mg(OH)_2$on adjusting the pH of the solution to 12 or more by the addition of 4N NaOH. The interference from ammonium salts and dissolved organic constituents is removed by evaporating the contents to dryness and by decomposition of organic constituents with nitric acid or else with aqua-regia. In Mg determination, the interference from Ca can be removed by precipitation of Ca by sodium tungstate or by adding alcohol on introducing 1 ml of 6N H_2SO_4which effect flocculation of $CaSO_4$ formed. The sedimentation is accelerated by centrifugation and decantation of the supernant clear solution free from Ca. The alcohol is dispelled by evaporation of the contents to dryness prior to Mg determination.

The presence of phosphate in concentration greater than 10 ppm also interfere in Ca recovery by EDTA titration. It becomes difficult to maintain the clear blue colour at the end point as a result of dissolution of Ca-phosphate by EDTA. In soil extract, high P concentration rarely occur unless soil is analysed for total Ca. The soluble P may be removed by passing the solution through anion resin or by precipitation as Zirconium phosphate. When sample is high in soluble P, EDTA be added in excess prior to adjustment of solution pH for back titration of excess EDTA with standard $CaCl_2$ solution. This procedure avoids any risk of precipitation of Ca while raising the pH of the solution.

The sum of exchangeable Ca, Mg, K and Na is referred as exchangeable bases. The exchangeable bases expressed as a percentage of cation exchange capacity is termed % base saturation. The soil pH is very much related to % base saturation, but the relationship differs a great deal depending upon the clay mineral.

The availability of basic cations to plants increases with the increase in % base saturation. The 1 : 1 type of soil colloids adsorb monovalents with greater bonding energy than 2 : 1 type of clay mineral. Consequently a higher percent K saturation is desirable in 1 : 1 type of clay mineral than in 2 : 1 type of clay mineral.

In most soils, the basic cations in exchangeable form occur in the order of Ca > Mg > K > Na. However, plants absorb basic cations in the order of K > Ca > Mg. Na is essential for sugar beet and marigolds only. The exchangeable Na exceed the amount of exchangeable K, and the exchangeable Ca + Mg decrease in proportion to increase in exchangeable Na in saline-sodic soils. The increase in exchangeable Na is associated with the increase in soil pH and soil dispersibility but decrease in the hydraulic conductivity.

REFERENCES

Barrows, H.L. and F.C. Simpson (1962). An EDTA method for the direct routine determination of Ca and Mg in soils and plant tissues. Soil Sci. Soc. Amer. Proc. 26 : 440 - 445.

Black, C.A. ed. (1965). Methods of Soil Analysis (Vol. 2). Am. Soc. Agron, Madison, Wisconsin, USA.

Cheng, K.L. and R.H. Bray (1951). Determination of calcium and magnesium in soil and plant material. Soil Sci. 72 : 449-458.

Diehl, H., C.A. Goetz, C.C. Hach (1950). The versenate titration for total hardness. Amer. Water Works Assoc. Journ. 42 : 40-48.

Jackson, M.L. (1968). Soil Chemical Analysis. Prentice-Hall India, New Delhi.

Toth, S.J., A.L. Prince, A. Wallace and D.S. Mikkelsen (1948). Rapid quantitative determination of eight mineral elements in plant tissue by a systematic procedure involving use of flame photometer. Soil. Sci. 66 : 459-466.

Tucker, B.B. and L.T. Kurtz (1961). Calcium and Magnesium determination by EDTA titration. Soil Sci. Soc. Amer. Proc. 25 : 27-29.

2.14.2 Estimation of exchangeable bases in soil without removal of soluble salts–Bower *et. al.* (1952)

2.14.2.1 Principle

Ammonium acetate extractable bases *i.e.*, basic cations without the removal of soluble salts give an estimate of exchangeable plus water soluble bases. Water soluble basic cations is estimated by the chemical analysis of extract from saturated paste. Exchangeable basic cations in soil is estimated by subtracting water soluble basic cations in soil from ammonium acetate extractable basic cations in soil.

2.14.2.2 Apparatus/Reagents etc.

Apparatus and Equipmentss

Centrifuge with 50 ml heavy duty centrifuge tubes, volumetric flasks, Erlenmeyer flasks, graduated pipette, water bath/sand bath, hot plate, oven, Buchner funnel, suction flask with suction pump, and flame photometer.

Reagents

1. Reagents for the extraction of exchangeable plus water soluble cations in soil.

1.0N ammonium acetate solution (pH 7.0). Dissolve 77.0 g of $NH_4C_2H_3O_2$ (AR) in about 900 ml distilled water. Adjust the pH to 7.0 with acetic acid or ammonium hydroxide and dilute the solution to 1000 ml mark with distilled water.

2. Reagents for flame photometric determination of potassium and sodium.

As listed in section 2.14.1.2.

3. Reagents for the determination of Ca and Mg by EDTA titration.

As listed in section 2.14.1.2.

2.14.2.3 Procedure

1. Extraction of exchangeable plus water soluble cations from soil sample by 1.0N $NH_4C_2H_3O_2$ (pH 7.0). Follow the step 2 of procedure 2.14.1.3 omitting the leaching of soluble salts by 99% isopropyl alcohol.

Alternatively weigh 4.0 g of loam to clay soil or 6.0g of sandy loam soil (air dry < 2 mm size). Determine the content of moisture in air dry soil to make correction for the moisture content.

Transfer the soil sample to 50 ml heavy duty bottom centrifuge tube. Add 33.0 ml of 1N $NH_4C_2H_3O_2$ (pH 7.0) solution. Triturate the contents with a glass rod for 5 minutes or else stopper and shake the contents for 5 minutes. Remove the stopper and place the centrifuge tube in centrifuge in pairs opposite to each other for a proper balance. Close the centrifuge lid, and gradually increase the speed of centrifuge drum to 2440 rpm or RCF = 1000, and let it run for 10 minutes. Thereafter, gradually lower the centrifuge speed until it comes to rest.

Then remove the centrifuge tube, and decant off the supernant clear liquid into a 100 ml volumetric flask. Repeat the extraction of exchangeable plus water soluble bases with two more 33 ml portions of 1.0N $NH_4C_2H_3O_2$ (pH 7.0) and after centrifugation decant off the supernant clear liquid into the same 100 ml volumetric flask. Thereafter dilute the $NH_4C_2H_3O_2$ extract to 100 ml with the extracting solution. Preserve the extract for the determination of basic cations.

2. Estimation of $NH_4C_2H_3O_2$-extractable K and Na in soil by flame photometer. Follow step 3.1 and 3.2 of procedure described in section 2.14.1.3.3 for the determination of K and Na concentration in $NH_4C_2H_3O_2$ –soil extract in ppm.

Thereafter, substitute the values in the appropriate equations to estimate exchangeable K and Na expressed in meq/100 g soil. The results obtained would give an estimate of exchangeable plus water soluble K and Na in soil.

3. Estimation of $NH_4C_2H_3O_2$ –extracble Ca in soil by EDTA titration using Ca indicator. Follow steps 1 and or 5 of section 2.14.1.3.3 and determine EDTA titre for complexing Ca in sample and blank extract. Thereafter, substitute the values in the given equation for exchangeable Ca expressed in meq/100 g soil. The result will give an estimate of exchangeable plus water soluble Ca in soil.

4. Estimation of $NH_4C_2H_3O_2$ –extractable Ca + Mg in soil by EDTA titration using Mg indicator. Follow steps 3 and 4 of section 2.14.1.3.3 given in chapter 2.14.1 to determine EDTA titre for complexing Ca plus Mg in sample and blank extract. Thereafter, determine exchageable plus water soluble Ca + Mg expressed in meq/100 g soil from which substract the value of exchangeable plus water soluble Ca in meq/100 g soil to obtain exchangeable plus water soluble Mg in soil.

It may be noted that in step 3 of section 2.14.1.3.3, the volume of total extractant employed is 250 ml, and the mass of soil sample taken for extracting exchangeable bases is 10.0 g. On the other hand, in step 1 of procedure 2.14.2.3 the volume of total extractant employed is 100 ml and the mass of soil taken for extracting exchangeable plus water soluble bases is either 4.0 g or 6.0 g depending upon the soil texture. The above must be kept in mind in doing the calculation.

5. Preparation of extract from saturated paste for the determination of water soluble cations in the soil. Transfer 200 to 300 g of air dry < 2 mm soil sample into one litre capacity plastic beaker. Add distilled water drop by drop until the soil become almost saturated without the appearance of free water. Work the soil mass with a stainless steel spatula until the soil mass begins to glisten without appearance of free water. It should be at the same time free flowing. Run the spatula across the soil paste making gap dividing the soil paste into two halves. If the gap formed gets readily filled on tapping the beaker gently on the table. The soil paste may be considered to have reached the saturation point. At this consistency, if the soil paste is lifted with the spatula the entire soil paste above spatula will slide and fall down with the tilting of the spatula. Cover the mouth of the plastic beaker with a large size petri dish and allow it to stand over night.

On the following day, remove the petri dish and make the paste again saturated fulfilling the criteria described above. Transfer a portion of soil saturated paste (about 25 g) into a tared aluminium can. Take the mass of aluminium can plus saturated paste, and then place it in an oven at 105°C to dry for 24 hours to determine % moisture in saturated paste.

Thereafter, transfer the remaining soil paste into a Buchner funnel lined with Whatman No. 42 filter paper. Fill the Buchner funnel uniformly with the saturated paste and smoothen the surface of the soil paste with fingers so that the air can not leak through the paste. Fill suction flask with straw to hold a 50 ml test tube. Fit Buchner funnel adapter into the mouth of suction flask firmly placing lower portion of Buchner funnel outlet inside the 50 ml test tube. Now, run the suction pump for 4 to 6 hours until air begins to leak through the soil mass. The initial 4-5 drops of the extract be used for rinsing the test tube before discarding. The subsequent extract collected in the test tube be stoppered, labeled and preserved for the determination of water soluble K, Na, Ca and Mg.

6. Flame Photometric determination of Na and K in the extract from saturated paste. For the determination of K in the extract from saturated paste, an aliquot of the extract may be diluted 25 to 50 times. However, for the determination of Na, the saturated paste extract from normal soil may require 50 to 100 times dilution whereas the extract from sodic soil may require 2000 to 4000 times dilution.

The volume of original extract from saturated paste may not exceed 10 ml. Normally 5 ml extract is initially diluted 10 times to enable measurement of the specific conductance of the saturated past extract. The 10 times diluted extract may be further diluted as required to estimate Na and K concentration in the undiluted extract.

Follow procedure given in section 2.14.1.3.1 and 2.14.1.3.2 for the flame photometric determination of K and Na respectively in the diluted extract in ppm. Multiply the concentration by the dilution factor to obtain the concentration of the element in undiluted extract. Thereafter, divide the concentration of the element in ppm by its equivalent weight to express the concentration of the element in meq/1.

7. Determination of Ca and Mg in extract from saturated paste by EDTA titration using Ca and Mg indicator.

1. Determination of Ca by EDTA titration using Ca–indicator. Transfer 1.0 ml of undiluted extract or 10.0 ml of 10 times diluted extract into 150 ml of Erlenmeyer flask and follow steps given in chapter 2.14.1.3.3 to obtain EDTA titre for complexing Ca in the sample extract. Similarly run a blank with an aliquot of 1.0 or 10.0 ml of distilled water. EDTA titre (sample-blank) × normality of EDTA × 1000 would give Ca concentration in meq/l.

2. Determination of Ca plus Mg by EDTA titration using Mg-indicator. Transfer 1.0 ml of undiluted extract or 10.0 ml of ten times diluted extract into 150 ml of Erlenmeyer flask and follow steps 3 and 4 of section 2.14.1.3.3 given to obtain EDTA titre for complexing Ca plus Mg in an aliquot of sample extract. Similarly run a blank titration with an aliquot of 1.0 or 10.0 ml of distilled water used for the dilution of sample extract. EDTA titre (sample-blank) × normality of EDTA × 1000 would give concentration of Ca + Mg in meq/l of undiluted extract. Substract the concentration of Ca from the concentration of Ca + Mg to obtain the concentration of Mg in undiluted extract from saturated paste.

2.14.2.4 Calculations

Water soluble cations (meq/l) :

$$K = \frac{\text{ppm K in undiluted extract from saturated paste}}{\text{Equivalent Weight of potassium}}$$

$$Na = \frac{\text{ppm Na in undiluted extract from saturated paste}}{\text{Equivalent Weight of sodium}}$$

$$\text{Ca or Mg} = \frac{\text{EDTA titre (sample} - \text{blank)} \times \text{normality of EDTA} \times 1000}{\text{ml undiluted extract taken for EDTA titration}}$$

Water soluble cations (meq/100 g soil)

$$= \frac{\text{Concentration of cations in meq}/l \times \text{saturation percentage i.e.\% moisture in saturated paste}}{1000}$$

$$\text{Saturation percentage or \% moisture in saturated paste} = \left[\frac{\text{Moist mass of saturated paste}}{\text{Oven dry mass of saturated paste}} - 1.00\right] \times 100$$

$$\begin{array}{c}\text{Exchangeable plus}\\ \text{Water soluble cations}\\ (\text{meq./100 g soil})\end{array} = \frac{\begin{array}{c}1\text{N } NH_4C_2H_3O_2\\ \text{extractable cation in (meq/}l\text{)}\end{array} \times \begin{array}{c}\text{Total Volume of}\\ \text{extractant}\end{array} \times 100}{1000 \times \text{mass of soil sample in g}}$$

$$\begin{array}{c}\text{Exchangeable}\\ \text{cations}\\ (\text{meq./100 g soil})\end{array} = \begin{array}{c}\text{Exchangeable +}\\ \text{Water soluble cations}\\ (\text{meq./100 g soil})\end{array} - \begin{array}{c}\text{Water soluble}\\ \text{cations}\\ (\text{meq./100 g soil})\end{array}$$

2.14.2.5 Comments

If soil contains soluble salts in significant amounts, ammonium acetate extraction would extract water soluble cations in addition to exchangeable cations. It would be, therefore, desirable to substract soluble cations from extractable cations to estimate exchangeable cations.

REFERENCES

Bower, C.A., R.F. Reitemeiyer, and M. Fireman (1952). Exchangeable cation analysis of saline and alkaline solis. Soil Sci. 73 : 251-261.

2.15 EXCHANGEABLE H^+ IN SOIL BY EXTRACTION WITH $BaCl_2$ - TRIETHANOLAMINE REAGENT–PEECH *et al.* (1962)

2.15.1 Principle

The soil sample is equilibrated with 0.5N $BaCl_2$ –0.055 N triethanolamine in the ratio of 1 : 10 (w/v) overnight. The exchangeable H^+extracted by $BaCl_2$ is neutralized by triethanolamine. The suspension is filtered through Buchner funnel lined with filter paper of fine porosity and the soil is further leached with 125 ml of extracting solution. The leachate is collected and combined with the filtrate and the volume is made to 250 ml with the extracting solution. Thereafter, the entire soil extract and leachate is titrated with 0.2 N HCl using a mixed indicator. Similarly a blank is run. The (blank-sample) titre gives a measure of exchange acidity.

2.15.2 Apparatus and reagents/Apparatus

Erlenmeyer flasks –500 ml and 250 ml capacity, volumetric flasks-1000 ml, 250 ml and 100 ml capacity, burette, graduated pipette 10, 5, 2 and 1 ml capacity, Pyrex Buchner funnel with suction and adapter, suction pump.

Reagents

(*i*) 0.50 N Barium chloride -0.055N Triethanolamine. Dissolve 61.07g of $BaCl_2$ $2H_2O$ in carbon dioxide free distilled water.

Then add 27.8 ml of 2.0N Triethanolamine and 4 ml of 6NHCl and dilute it to 1000 ml with carbon dioxide free distilled water. Mix well and adjust the pH to 8.0 ± 0.02 with HCl or triethanolamine. This solution is 0.03N with respect to free base. Prevent absorption of CO_2 by the solution by providing a siphon and connecting it to soda lime to the air inlet.

(*ii*) 2.0 N Triethanolamine solution. Dilute 26.6 ml of conc. triethanolamine (approx. 7.5 N) to 100 ml with distilled water. Standardize the base by titration with 0.2N HCl using mixed indicator.

(*iii*) 10.0 N Hydrochloric acid. Dilute 218.2 ml of conc. HCl (AR) to 250 ml with distilled water.

(*iv*) 6.0N Hydrochloric acid. Dilute 60 ml of 10 N HCl to 100 ml with distilled water.

(*v*) 0.2N Hydrochloric acid. Dilute 5.0 ml of 10 N HCl to 250 ml with distilled water

(*vi*) Mixed indicator. Dissolve 0.22g of brom cresol green and 0.075 g of methyl red in 96 ml of 95% ethanol and add 3-5 ml of 0.1N NaOH and mix the contents.

(*vii*) 1.0N Sodium hydroxide solution. Dissolve 4.0 g of NaOH (AR) in distilled water and dilute the solution to 100 ml with distilled water.

(*viii*) 0.1N Sodium hydroxide solution. Dilute 10.0 ml of 1N NaOH to 100 ml with distilled water.

2.15.3 Procedure

1. Equilibration of the soil with the extractant. Transfer 10g of soil into a 250 ml Erlenmeyer flask. Then, add 100 ml of 0.05N $BaCl_2$-0.055N Triethanolamine solution (pH 8.0). Stopper the flask and mix well. Thereafter, allow it to stand over night.

2. Filtration of the soil extractant suspension and thereafter leaching of the soil with the extractant. Transfer the soil extractant suspension into a pyrex Buchner funnel (size 40) fitted with 4.25 cm Whatman No. 32 filter paper placed on a suction flask connected to a suction pump. Thereafter rinse the flask with small portions of the extractant several times and transfer each rinsing to the same Buchner funnel. Collect the filtrate as well as the leachate in a 250 ml volumetric flask. Continue leaching the soil by adding small portions of the extractant at a time until the volume of filtrate cum leachate reaches about 240 ml. Each portion of solution added be allowed to drain before the addition of next portion. Care be taken to prevent the soil from cracking before prolonged drying. Dilute the filtrate cum leachate collected to 250 ml mark with the extractant. Similarly run a blank.

3. Titration of sample and blank filtrate cum leachate with standard 0.2N HCl. Transfer 250 ml of sample filtrate cum leachate into a 500 ml Erlenmeyer flask. Add 5 drops of the mixed indicator and titrate the solution with 0.2N HCl until the colour changes from green to pink end point (pH 5.1). Rinse the volumetric flask with some of the titrated solution and transfer it back to the same Erlenmeyer flask and complete the titration by reaching to the previous end point. Similarly run a blank. The blank and sample filtrate cum leachate be titrated to the same end point.

2.15.4 Calculation

Exchangeable H^+ (acidity) in meq/100 g soil

$$= \frac{(\text{Blank titre} - \text{Sample titre}) \times \text{Normality of acid} \times 100}{\text{mass of soil is g}}$$

2.15.5 Comments

With some soils, the end point will fade (pH will increase) upon standing owing to the slow dissolution of aluminium hydroxide, but this colour fading should be ignored.

The small amount of aluminate that may be present in the soil extract will be converted into aluminium hydroxide. On completion of the titration it introduces no error in the determination of exchangeable H^+. On standing, however, the precipitated aluminium hydroxide will continue to react with the free acid and any further addition of acid to restore the end point would thus lead to a small negative error.

REFERENCES

Peech, M.,R.L. Cowan, and J.H. Baker (1962). A critical study of the $BaCl_2$–triethanolamine and ammonium acetate methods for determining the exchangeable hydrogen content of the soils. Soil Sci. Soc. Amer. Proc. 26 : 37–40.

2.16 LIME REQUIREMENT

2.16.1 Lime requirement by neutralization of soil acidity with $Ca(OH)_2$.PNP solution–Schofield (1933)

2.16.1.1 Principle

A measured volume of $Ca(OH)_2$.PNP solution is made to react with a certain mass of air-dry acid soil for several hours preferably overnight. The unreacted $Ca(OH)_2$. PNP is determined by back titration of an aliquot of soil–$Ca(OH)_2$ PNP suspension. The filtrate is titrated with standard mineral acid in presence of brom cresol green indicator. An aliquot of the same volume of $Ca(OH)_2$. PNP solution not equilibrated with the soil sample is similarly titrated with the standard mineral acid to obtain the blank titre. The (blank–sample) titre give a measure of lime requirement.

2.16.1.2 Reagents

(*i*) Brom cresol green indicator, .04%. Dissolve 20 mg of brom cresol green in 50 ml of ethyl alcohol.

(*ii*) Hydrochloric acid, .05 N. Transfer 12.5 ml of 2.0 N HCl into 500 ml volumetric flask and dilute the solution to 500 ml mark with distilled water. Standardize the acid against 0.1 N Na_2CO_3 using methyl orange indicator.

(*iii*) Hydrochloric acid, 2.0 N. Dilute 20.0 ml of 10 N HCl to 100 ml with distilled water.

(*iv*) Hydrochloric acid, 10.0 N. Dilute 218.2 ml of conc. HCl (AR, Sp. gr. 1.18, percent purity 35.4) to 250 ml in a 250 ml volumetric flask with distilled water.

(*v*) Methyl orange indicator, 0.1%. Dissolve 0.1g of methyl orange indicater in 100 ml distilled water.

(*vi*) Sodium carbonate, 0.1N. Dissolve 0.530 g of Na_2CO_3 anhydrous (AR) in 100 ml distilled water.

(*vii*) Calcium hydroxide–P–nitrophenol (PNP) solution. Dissolve 2.667 g of $Ca(OH)_2$ (analytical reagent) in 900 ml of freshly prepared distilled water in 1l volumetric flask. Add 8.35 g P–nitrophenol (PNP) and shake or stir intermittently until dissolved. Then make the volume to 1l with distilled water.

2.16.1.3 Procedure

1. Equilibration of the soil–$Ca(OH)_2$–PNP solution. Weigh air dry < 2 mm soil sample equal to twice the reciprocal of CEC expressed in meq/g soil (*i.e.* 10g soil sample if CEC is 0.2 meq/g soil) and transfer the weighed sample into 250 ml Erlenmeyer flask. Then, add 100 ml of $Ca(OH)_2$–PNP solution into the Erlenmeyer flask. Stopper and shake the contents for 2 hours and then leave it overnight.

2. Filtration of the soil –$Ca(OH)_2$–PNP suspension. On the following day take a clean Erlenmeyer flask and place on its mouth a funnel lined with dry Whatman No. 2 filter paper. Pour the soil –$Ca(OH)_2$– PNP suspension in small portions using a glass rod into the funnel about 1 cm below the rim of the filter paper. Discard the initial 5 ml filtrate after rinsing the receiving flask. Collect the subsequent filtrate for chemical analysis.

3. Tiration of the sample and blank filtrate with mineral acid. Transfer 20.0 ml aliquot of the sample filtrate into 150 ml Erlenmeyer flask. Add 2-4 drops of brom cresol green indicator and titrate with 0.05N HCl until a faint green in yellow of PNP is attained at a pH of 5.2. Similarly take 20 ml aliquot of blank filtrate *i.e.* $Ca(OH)_2$–PNP solution not equilibrated with the soil sample and titrate with 0.05N HCl after adding few drops of brom cresol green indicator to the same end point and designate it as blank titre. Perform each titration atleast in duplicate.

2.16.1.4 Calculations

$$\text{Lime requirement LR meq/100g soil} = \frac{(B-S) \times N \times T \times 100}{A \times W}$$

$$\text{Lime requirement in } CaCO_3 \text{ - t/ha} = \frac{\text{LR in meq/100 g soil} \times 20 \times EW \times 2.47}{2.2 \times 1000}$$

where,

B is ml of standard acid used in titration of aliquot of $Ca(OH)_2$–PNP solution not equilibrated with the soil sample.

S is ml of standard acid used in titration of aliquot of $Ca(OH)_2$–PNP solution equilibrated with the soil sample *i.e.* sample filtrate.

N is the normality of standard acid used for the titrations.

T is the total volume of $Ca(OH)_2$–PNP equilibrated with the soil sample

A is ml of blank or sample filtrate taken for the titration with the standard acid.

W is the mass of air dry < 2 mm soil sample actually equilibrated with 100 ml of $Ca(OH)_2$–PNP solution.

EW is the equivalent weight of liming material. The equivalent weight of $CaCO_3$ is 50 whereas the equivalent weight of CaO is 28.

LR is Lime required in most soils to raise the soil pH to 6.5

2.16.2 Estimation of lime requirement of soil–Woodruff (1948)

2.16.2.1 Principle

The soil is equilibrated with a buffer mixture of calcium acetate and p–nitro phenol adjusted to pH 7.0. The drop in the pH of buffer after equilibration with the soil form the basis of LR as the buffer give a linear change of pH in relation to exchangeable H (meq /100 g soil) in the pH range 6 to 7.

2.16.2.2 Reagents

(*i*) Buffer solution (pH 7.0) : Dissolve 8g p–nitrophenol, 40g calcium acetate and 0.625 g magnesium oxide in distilled water and make the volume to 1l with distilled water. Adjust the pH of the solution to 7.0 with HCl or MgO.

2.16.2.3 Procedure

Transfer 10g soil, 10 ml distilled water and 20 ml buffer solution into a 100 ml beaker. Stir the suspension with a glass rod and let it stand for 30 minutes. Then measure the pH of soil-water-buffer suspension by a pH meter. Note the pH as pHb. If the pH become less than 6.0, then repeat the entire procedure taking only 5g soil and note the pH as pHb′.

2.16.2.4 Calculations

$$\text{LR in meq/100 g soil} = \frac{\text{Buffer pH} - \text{pHb}}{0.1} \text{ or } \frac{(\text{Buffer pH} - \text{pHb}') \times 2}{0.1}$$

$$\text{LR in lbs/acre} = \text{LR in meq/100g soil} \times 20 \times \text{Eq Wt of } CaCO_3$$

$$\text{LR in t/ha} = \frac{\text{LR the lbs/acre} \times 2.47}{2.2 \times 1000}$$

2.16.3 Estimation of lime requirement of soil–Woodruff (1948)

2.16.3.1 Principle

The soil is equilibrated with a buffer (pH 7.0) prepared by dissolving calcium acetate, p–nitrophenol, salicylic acid and calcium hydroxide. The drop in the pH of buffer on equilibration with the soil form the basis for the estimation of LR.

2.16.3.2 Reagents

(*i*) 1.0 M $CaCl_2$. Dissolve 14.7 g of $CaCl_2 \cdot 2H_2O$ in distilled water and dilute it to 100 ml.

(*ii*) Buffer solution. Dissolve 10 g Ca acetate, 12 g p–nitrophenol, 10 g salicylic acid and 4 g calcium hydroxide in 900 ml distilled water and then dilute the buffer solution to 1000 ml with distilled water. Adjust buffer pH to 7.0 using pH meter by adding sodium hydroxide solution.

2.16.3.3 Procedure

Transfer 20g air dry 2 mm soil sample into 100 ml beaker, add 20 ml distilled water and stir the suspension with a glass rod for 30 minutes and measure the pH with a pH meter. Note the pH as pH_W.

Then add 4 drops (0.2 ml) of 1M calcium chloride solution, Stir with a glass rod for 10 minutes and measure the pH of the suspension with a pH meter and note it as pH_S.

Now pipette 20 ml of buffer solution into the beaker containing soil-water-calcium chloride suspension. Stir with a glass rod thoroughly immediately after adding buffer and then at 10 minutes interval until 30 minutes. Thereafter measure the suspension pH and note it as pHb. If the pH becomes less than 6.0, then repeat the entire procedure taking only 10g soil and note the pH as pHb′.

2.16.3.4 Calculations

$$\text{LR in meq/100g soil} = = \frac{\text{Buffer pH} - pH_b}{0.1} \text{ or } \left[\frac{\text{Buffer} - \text{pH b}'}{0.1}\right] \times 2$$

$$\text{LR in lbs/acre} = \text{LR in meq/100 g soil} \times 20 \times \text{Eq. wt of } CaCO_3$$

$$\text{LR in t/ha} = \frac{\text{LR in lbs/acre} \times 2.47}{2.2 \times 1000}$$

The LR as estimated by the above equation gives lime requirement to attain a pHs of 7.0. In practice liming is done to attain a pH's (desired pHs) of 6.0 to 6.5. Consequently LR' (LR or desired pHs) is estimated assuming a linear relationship between pHs and the lime requirement by the following equation

$$LR' = \left[\frac{pH's - pHs}{7.0 - pHs}\right] \times LR = \frac{6.5-5.2}{7.0-5.2} = \frac{1.3}{1.8} \times 4$$

where,

LR' is lime required to attain a pHs less than 7.0

$PH's$ is the pHs desired after adding lime.

pHs is the observed pH of the soil water calcium chloride suspension.

LR is the lime required for attaining a pHs of 7.0

Illustration. If LR is 4 meq/100 g soil, pHs 5.2 and pH's is 6.5, then LR′ will be 2.89 meq per 100 g soil.

2.16.4 Estimation of lime requirement-Shoemaker *et. al.* (1961)

2.16.4.1 Principle

A buffer solution (pH 7.5) is prepared by dissolving p–nitrophenol triethanolamine, potassium chromate, calcium acetate and calcium chloride in distilled water. The soil is equilibrated with buffer solution, and the pH of soil buffer suspension is determined by a pH meter. The drop in the pH of buffer on equilibration with the soil depend on the extractable acidity which form the basis of estimation of lime requirement to bring the soil to a desired pH. The lime requirement is noted from a table giving pH of soil buffer suspension and the corresponding lime requirement to bring the soil to a pH of 6.0, 6.4 and 6.8.

2.16.4.2 Reagents

Buffer solution. Dissolve 1.8 g p–nitrophenol, 2.5 ml triethanolamine, 3.0 g potassium chromate, 2.0 g calcium acetate and 53.1 g calcium chloride dihydrate in distilled water and make the volume to 1l. Adjust the pH of the buffer solution to 7.5 with sodium hydroxide.

2.16.4.3 Procedure

Transfer 10 g air dry 2 mm soil sample having a pH of 6.3 or below into a 150 ml conical flask. Add 10 ml distilled water and 20 ml buffer solution. Stopper and shake the conical flask on a mechanical shaker at moderate speed for 10 minutes. Remove the conical flask from the shaker and pour the suspension immediately after shaking into a 100 ml beaker. Measure the pH of soil water buffer suspension by a pH meter. Note down the pH and record it as pHb. Refer to the table giving the requirement of pure $CaCO_3$ in t/acre at desired pH of 6.0, 6.4 and 6.8 against the pH of soil water buffer *i.e.* pHb. Note down the lime requirement (LR) from the table 2.16.1. The amount may be converted in terms of locally available lime stone on the basis of % $CaCO_3$ equivalence using the following equation

Quantity of Liming Material

$$= \frac{\text{Quantity of Calcium Carbonate needed} \times 100}{\%\ CaCO_3 \text{ equivalence}}$$

Table 2.16.1 Lime required to bring the soil to an indicated pH according to pH of the soil buffer suspension (Shoemaker *et al.* 1961).

pH of soil buffer	**Pure $CaCO_3$ in t/acre**		
	Indicated pH		
suspension	**pH 6.0**	**pH 6.4**	**pH 6.8**
6.7	1.0	1.2	1.4
6.6	1.4	1.7	1.9
6.5	1.8	2.2	2.5
6.4	2.3	2.7	3.1
6.3	2.7	3.2	3.7
6.2	3.1	3.7	4.2
6.1	3.5	4.2	4.8
6.0	3.9	4.7	5.4
5.9	4.4	5.2	6.0
5.8	4.8	5.7	6.5

5.7	5.2	6.2	7.1
5.6	5.6	6.7	7.7
5.5	6.0	7.2	8.3
5.4	6.5	7.7	8.9
5.3	6.9	8.2	9.4
5.2	7.4	8.6	10.0
5.1	7.8	9.1	10.6
5.0	8.2	9.6	11.2
4.9	8.6	10.1	11.8
4.8	9.1	10.6	12.4

Multiply t/acre by 2.47 to obtain t/ha.

2.16.5 Lime Requirement–Pratt and Blair (1962)

2.16.5.1 Principle

The soil is equilibrated with a buffer solution of pH 7.5 in the ratio of 1 : 2 soil buffer suspension (w/v) by shaking in a mechanical shaker for 10 minutes. The buffer is prepared with nitrophenol, triethanolamine, potassium chromate, calcium acetate and calcium chloride. The pH of soil buffer suspension is determined by a pH meter.

The lime requirement expressed in pure $CaCO_3$ per acre is determined by reference to a Table giving Lime requirement in t/acre of pure $CaCO_3$corresponding to pH of soil-buffer suspension or else LR may be estimated from the decrease in pH of soil buffer suspension. A pH drop of 0.8 unit gave LR of 1.6 t $CaCO_3$/acre and further decrease in pH of soil-buffer suspension by 0.1 unit increased LR by 0.6 t/acre.

2.16.5.2 Reagents

(*i*) Buffer solution pH 7.5. Dissolve 1.8 g of nitrophenol, 2.5 ml of triethanol amine, 3.0 g of potassium chromate, 2.0g of calcium acetate and 53.1 g of calcium chloride in about 800 ml of distilled water. Adjust the solution pH to 7.5 with NaOH or HCl, and then dilute the solution to 1000 ml mark with distilled water in 1l volumetric flask. Stopper and mix the contents by turning the flask upside down holding the stopper firmly.

(*ii*) Buffer solution of pH 7.0. Dissolve one Buffer tablet of pH 7.0 in 100 ml distilled water.

(*iii*) Buffer solution of pH 4.0. Dissolve one buffer tablet of pH 4.0 in 100 ml distilled water.

2.16.5.3 Procedure

1. Equilibration of the soil with buffer of pH 7.5. Transfer 10.0 g of air dry < 2 mm soil sample into a 100 ml Erlenmeyer flask. Then add 20 ml of buffer

solution of pH 7.5. Stopper the flask and fit it on a mechanical shaker firmly. Switch on the mechanical shaker and run it at moderate speed for 10 minutes. Switch off the shaker after reducing the speed when the shaker is at rest position, remove the Erlenmeyer flask.

2. Measurement of pH of the soil-buffer suspension by a pH meter. Let the glass and reference electrodes of the pH meter remain dipped in distilled water for a period of atleast 24 hours. Connect the leads of the glass and reference electrodes into the appropriate sockets of the pH meter. Connect the leads of the pH meter to a voltage stabilizer which in turn is connected to the main line of the electricity. Let the pH meter run for 10 minutes to allow the instrument to warm up. Measure the temperature of the soil-buffer suspension and accordingly adjust the temperature correction knob.

Remove the electrodes from distilled water and wipe it dry with a Whatman No 1 filter paper. Then immerse the electrode in buffer of pH 7.0. Adjust the pH meter reading to the pH of buffer by adjusting with coarse and fine adjustment knob provided in the pH meter.

Thereafter, remove the pH electrodes from the buffer solution. Rinse the electrodes with distilled water and again wipe it dry with Whatman No. 1 filter paper. Then dip a portion of lower end of the electrodes in buffer of pH 4.0. Check the pH meter reading. If the pH recorded by pH meter is exactly 4.0, the pH meter is is calibrated for the measurement of pH of soil-buffer suspension.

Shake soil–buffer suspension, and then pour it into 50 ml beaker. Remove the pH electrodes from buffer of pH 4.0, rinse it with distilled water and wipe it dry with a filter paper and immerse the lower end of glass and reference electrodes into the soil-buffer suspension. After 1 minute, note down the pH reading recorded by the instrument.

Refer to the Table 2.16.5 giving lime requirement corresponding to the pH of soil–buffer suspension (1 : 2 w/v) or else estimate lime requirement by substituting the value of pH of soil-buffer suspension in the equation given under the heading calculations.

2.16.5.4 Calculations

$$\text{Lime Requirement } (CaCO_3 - \text{t/acre}) = \{(6.7 - \text{pH of soil} - \text{buffer suspension}) \times 0.6 + 1.6\}$$

$$\text{Lime Requirement } (CaCO_3 - \text{t/ha}) = \text{LR in t/acre} \times 2.47$$

Table 2.16.2 Soil–buffer suspension (1 : 2 w/v) pH and Lime Requirement by Pratt & Blair (1962)

pH Soil–buffer suspension	Lime Requirement $CaCO_3$–t/acre	pH Soil–buffer suspension	Lime requirement $CaCO_3$–t/acre
6.7	1.6	5.7	7.6
6.6	2.2	5.6	8.2
6.5	2.8	5.5	8.9
6.4	3.4	5.4	9.5
6.3	4.0	5.3	10.1
6.2	4.6	5.2	11.0
6.1	5.2	5.1	11.7
6.0	5.8	5.0	12.4
5.9	6.4	4.9	13.2
5.8	7.0	4.8	14.0

2.16.6 Lime requirement of soil by estimation of extractable acidity (exchangeable H + Al) Hunter (1971)

2.16.6.1 Principle

The soil is equilibrated with 1N potassium chloride in order to extract exchangeable H + Al. The suspension is filtered and a portion of the filtrate is titrated by standard sodium hydroxide to the phenolphthalein end point to estimate the extractable acidity which give a measure of lime requirement.

2.16.6.2 Reagents

(*i*) 1.0 N KCl. Dissolve 74.56 g of AR potassium chloride in distilled water and dilute it to 1000 ml.

(*ii*) 1.0 N NaOH. Dissolve 40g AR sodium hydroxide pellets in distilled water and dilute it to 1000 ml.

(*iii*) 0.1N NaOH. Dilute 25 ml of 1N sodium hydroxide to 250 ml with distilled water.

(*iv*) 0.01N NaOH. Dilute 50 ml of 0.1N sodium hydroxide to 500 ml with distilled water.

(*v*) 0.5% Phenolphthalein Indicator. Transfer 0.5 g phenolphthalein into 100 ml vol. flask. Add 50 ml of alcohol and 50 ml distilled water. Shake the contents. If there is any precipitate filter through Whatman No. 1 filter paper to obtain a clear solution.

(*vi*) Distilled water containing phenolphthalein indicator. Add 2 ml (20 drops) of phenolphthalein indicator into 100 ml volumetric flask and dilute the solution to 100 ml mark with distilled water.

2.16.6.3 Procedure

1. Preparation of soil extract. Transfer 5.0g of air dry 2 mm soil sample into 125 ml conical flask. Then add 50 ml of 1N potassium chloride solution. Stopper and fit the conical flask on a mechanical shaker. Switch on the shaker and run it at moderate speed for 5 minutes. Switch off the shaker and filter the suspension through Whatman No. 1 filter paper. Discard the initial filtrate and preserve the subsequent filtrate.

2. Standardization of dilute sodium hydroxide solution. Pipette 5.0 ml of 0.02 N hydrochloric acid into a 125 ml conical flask, and 10 ml of distilled water containing 2 drops of the phenolphthalein indicator. Then titrate with approx. 0.01N sodium hydroxide in small portions until the solution changes from colourless to persistent pink. Repeat the titration taking another 5.0 ml aliquot of the standard acid.

Then run a blank titration taking 10 ml of distilled water containing 2 drops of phenolphthalein indicator. Note down the blank titre. Substract blank titre from alkali titre used in titration of 5 ml of standard acid. Estimate the normality of dilute sodium hydroxide by taking the product of volume and normality of standard acid and then dividing it by the alkali titre.

3. Titration of soil extract by standard sodium hydroxide solution. Pipette 10.0 ml soil extract in duplicate in separate 125 ml conical flask. Add 10.0 ml distilled water containing 2 drops of phenolphthalein indicator to each conical flask. Fill a 10 ml micro burette with 0.01N sodium hydroxide and titrate until the solution changes it's colour to persistent pink. Thereafter, titrate the second aliquot of the soil extract with the standard alkali to the same end point. Similarly run a blank titration taking 10.0 ml of 1N potassium chloride solution in place of soil extract.

2.16.6.4 Calculations

LR in meq/100g soil

$$= \frac{\text{Alkali titre (Sample} - \text{Blank)} \times \text{Normality of Alkali} \times \text{Vol of total extractant} \times 100}{\text{ml extract taken for titration} \times \text{mass of soil in g}}$$

$$\text{LR in lbs/acre} = \text{LR in meq/100 g soil} \times 20 \times \text{Equivalent wt of } CaCO_3$$

$$\text{LR in t/ha} = \frac{\text{LR in lbs/acre} \times 2.47}{2.2 \times 1000}$$

2.16.7 Lime requirement of arable acid soils by measurement of soil pH in N/10 KCl–Bergmann *et al.* (1966)

2.16.7.1 Principle

Lime requirement is based on soil pH in N/10 KCl. The strong electrolyte such as KCl have been employed for the displacement of exchangeable Al^{3+} and H^+. The pH of soil-N/10 KCl suspension should decrease in proportion to the amount of exchangeable $Al^{3+} + H^+$ displaced from the solid phase and released into the solution phase. It will at the same time eliminate differences in soil pH which normally result as a consequence of differences in the concentration of salt.

2.16.7.2 Reagents

(*i*) Potassium Chloride, 0.1 N. Dissolve 7.456 g of KCl (AR) in distilled water and dilute the solution to 1000 ml mark with distilled water.

(*ii*) Buffer solution of pH 7.0. Dissolve one buffer tablet of pH 7.0 in 100 ml distilled water.

(*iii*) Buffer solution of pH 4.0. Dissolve one buffer tablet of pH 4.0 in 100 ml distilled water.

2.16.7.3 Procedure

1. Equilibration of the soil with N/10 KCl solution. Transfer 20g of air dry < 2 mm soil sample into a 150 ml Erlenmeyer flask, and then add 40 ml of 0.1N KCl solution. Stopper and shake the flask atleast for 2 hours and allow it to stand over night.

2. Calibration of the pH meter. Let the glass and reference electrodes of the pH meter remain dipped in distilled water for atleast 24 hours. Connect the leads of the electrodes into the sockets provided in the pH meter. Connect the leads of the pH meter to a voltage stabilizer connected to the main. Switch on the pH meter and let it warm for 10 minutes.

After 10 minutes remove the reference and glass electrode from distilled water and wipe it dry with an ordinary filter paper. Then, immerse the lower part of both reference and glass electrodes in buffer solution of pH 7.0. Adjust the pH meter reading after setting the temperature correction knob to the temperature of the buffer solution. Rotate coarse and fine pH adjustment knob gently to adjust pH meter reading to 7.0.

Thereafter remove the electrodes from buffer of pH 7.0. Wash the electrodes with distilled water and wipe it dry with a filter paper, and then immerse the electrodes in buffer solution of pH 4.0. If the instrument also records a pH of 4.0, the pH meter has been calibrated for pH measurement.

3. Measurement of pH of soil–N/10 KCl suspension. Shake the flask containing the suspension and pour the contents into a 100 ml beaker after rinsing the beaker with 5 ml of suspension.

Thereafter, remove the glass and reference electrodes from the buffer solution of pH 4.0, wash it with a jet of distilled water from the wash bottle. Wipe it dry with a ordinary filter paper. Then, immerse the lower part of glass and reference electrodes into the soil-KCl suspension transferred to a 100 ml beaker. Note the pH recorded by the pH meter.

4. Assessment of Lime requirement. Ascertain the soil texture by feel method and refer to the pH and lime requirement chart corresponding to the textural class of the soil.

Table 2.16.3 Lime Requirement in soils of different texture under arable farming based on soil pH with N/10 KCL.

pH n/10 KCl	(CaO in quintals per hectare) Sandy to Sandy Loam	Loamy Sand to Silty Loam	Loam	Heavy Loam to clayey	pH n/10 KCl
3.5	10/15	20/50	40/85	40/100	3.5
3.6					3.6
3.7		20/45	40/75	40/90	3.7
3.8					3.8
3.9		20/40	40/70	40/85	3.9
4.0	10				4.0
4.1		20/35	40/65	40/80	4.1
4.2					4.2
4.3			40/60	40/75	4.3
4.4		20/30			4.4
4.5			40/55	40/70	4.5
4.6					4.6
4.7			30/45	40/60	4.7
4.8	5	20/25			4.8
4.9			30/40	30/50	4.9
5.0					5.0
5.1		20	30/35	30/40	5.1
5.2					5.2
5.3			30	30/35	5.3
5.4	3–4	15			5.4
5.5				30	5.5
5.6		10	25		5.6
5.7	2–3				5.7
5.8		5	20	25	5.8
5.9	0				5.9
6.0		3/4	15	20	6.0
6.1	HIGH				6.0
					6.1

Source. Bergmann *et. al.* (1966)

2.16.8 Estimation of $CaCO_3$ equivalence of Liming material A.O.A.C. (1950)

2.16.8.1 Principle

A small dried and ground lime stone sample passed through 0.26 mm sieve is made to react with known volume of standard 1N HCl and the unreacted acid is titrated back with standard 1N sodium hydroxide to a phenolphthalein end point. The meq of base neutralized by the acid in 100 g sample is estimated and is multiplied by the molecular weight of $CaCO_3$ and the product is divided by the meq of the base present is one mole of $CaCO_3$ to obtain the % $CaCO_3$ equivalence.

2.16.8.2 Reagents

(*i*) 10.0 N HCl. Dilute 218.2 ml of AR conc. hydrochloric acid (sp. gr. 1.18, Minimum Assay 35.4%) to 250 ml with distilled water.

(*ii*) 1.0 N HCl. Dilute 25 ml of 10 N Hydrochloric acid to 250 ml with distilled water. Standardize the reagent against 0.1N sodium carbonate using 0.1% methyl orange indicator.

(*iii*) 0.1% Methyl orange. Dissolve 0.100 g methyl orange in distilled water and make the volume to 100 ml.

(*iv*) 1.0% Phenolphthalein indicator. Dissolve 1.0g phenolphthalein indicator in 100 ml of 95% ethyl alcohol. If there is precipitate, filter the solution.

(*v*) 0.1N Na_2CO_3. Dissolve 0.530 g AR anhydrous sodium carbonate and make the volume to 100 ml with distilled water.

(*vi*) 1.0 N NaOH. Dissolve 40.0 g AR sodium hydroxide pellets in distilled water and dilute the solution to 1000 ml with distilled water.

2.16.8.3 Procedure

1. Neutralization of sample of limestone by standard hydrochloric acid. Transfer 1.0 g of air dried and ground lime stone passed through 0.26 mm sieve into a 500 ml conical flask. Then add 25 ml standard 1.0N hydrochloric acid. Swirl the suspension to mix thoroughly and then heat nearly to boiling. Thereafter, keep the flask on a steam bath for about 5 to 45 minutes to complete the reaction. Then add 100 ml distilled water and boil the solution on a small burner exactly for 1 minute and then cool it to room temperature.

2. Back titration of standard hydrochloric acid unreacted with the sample of limestone. Add 5 drops of 1% phenolphthalein indicator solution into the 500 ml conical flask containing standard hydrochloric acid unreacted with the sample of limestone. Then, back titrate the unneutralized standard acid by standard 1N sodium hydroxide solution to the phenolphthalein end point *i e.* pink colour which persists for 15 seconds as the solution is shaken.

2.16.8.4 Calculation

% $CaCO_3$ equivalence

$$= \frac{\text{ml of reacted acid} \times \text{normality of acid} \times 100 \times \text{mol. wt. of } CaCO_3}{\text{mass of sample in g} \times \text{meq of base in one mol of } CaCO_3}$$

$$= \frac{(\text{ml of 1N HCl added} - \text{ml of 1N NaOH used in back titration}) \times 1.0 \times 100 \times 100}{1 \times 2000}$$

$$= \left[\text{ml of 1NHCl added} - \text{ml of 1NNaOH used in back titration}\right] \times 5$$

2.16.5 Comments

The neutralizing equivalence of compounds having lower molecular weight than $CaCO_3$ may exceed 100%. For example, the $CaCO_3$ equivalence of $MgCO_3$ and CaO is 117% and 179% respectively.

1. Forms of soil acidity. The forms of soil acidity in the pH range of <4.0 to 7.0 as stated by Black (1968) is summarized below :–

Group 1. Strong acids (soil pH 4.2 and below).

The acidity is due to the presence of sulphuric acids basic ferric sulphate, exchangeable hydrogen and active hydronium ion.

Group 2. Week acids (soil pH 4.2 to 5.2)

The acidity is due to accumulation of monomeric hydrated aluminium ion in large quantities.

Group 3. Very weak acids (soil pH 5.2 to 7.0)

The acidity is due to carboxyl group of organic matter, edge group of hydroxy aluminium polymers in inter layer positions and edge of silicate clay particles.

The exchange acidity is the acidity that is replaced by unbuffered neutral salts such as KCl, $CaCl_2$, or NaCl. It is almost entirely due to monomeric Al^{3+}ion in strongly acid soil. Moderately acidic soils do not have exchangeable Al. The exchange acidity is due to H^+ in soils in which organic matter contribution to acidity is important. The titratable, but not displaceable acidity is estimated by using buffered $BaCl_2$ triethanolamine (pH 8.2).

2. Lime requirement. The lime requirement of acid soils is determined by three methods :

1. Titration with a base or moist incubation in soil with liming materials.

2. Equilibration of the soil with strongly buffered system or else with strong electrolytes followed by the determination of exchange acidity by direct titration or by measuring change in pH of the extract or soil buffer suspension.

3. Substracting the exchangeable bases from cation exchange capacity both expressed in meq/100 g soil.

Method 1. Tamboli and Tulsi (1967) determined LR by titration with $Ca(OH)_2$ and expressed the result in amount of basic slag containing 30% CaO and 7.2% MgO.

Schofield (1933) equilibrated a sample of air dry soil with 100 ml of $Ca(OH)_2$. PNP for atleast 2 hours preferably overnight. After filtration of the suspension, an aliquot of the filtrate was titrated with 0.05 N HCl to brom cresol green end point *i.e.* a faint green in the yellow of PNP (pH about 5.2). Similarly an aliquot of $Ca(OH)_2$. PNP was titrated as Blank. The blank-sample titre gave a measure of LR for liming to attain a pH of 6.5.

Ricra (1946) equilibrated 100 g soil samples with 0–300 mg of $CaCO_3$in increments of 25 mg or else with 0–600 mg of $CaCO_3$ in increments of 50 mg. After mixing added $CaCO_3$ with soil sample, the mixture was wetted to field capacity and incubated for 45 days at 33°C with periodic wetting. Then, the change in soil pH was recorded against the amount of liming material added. From the above relationship, LR for raising the soil pH to the desired value was estimated.

Some scientists equilibrated 100g soil samples with 0-120 mg paper mill sludge Ghosh (1966) or else with powdered basic slag considered to be an excellent liming material in Bihar.

The moist incubation of soil sample with liming material is considered to give a reliable estimate of Lime Requirement.

Method 2. Hunter (1971) determined exchangeable H + Al in soil by extraction with 1 N KCl solution which formed the basis for the estimation of LR. However, the soils in the pH range have little KCl exchangeable Al and yet respond to liming.

The LR determination in eastern region of Germany is based on pH of soil equilibrated with 0.1N KCl. In pH range of 5.2 to 6.5 humus functional groups become important, if organic matter content of soil is high. However, Jackson (1958) reported that the use of strong electrolyte such as KCl largely excludes the replacement of pH dependent charge.

Several researchers have equilibrated soil with buffer solution, and have estimated LR from the decrease in pH of the buffer. The reagents used in the preparation of buffer is summarized below :

1. $BaCl_2$-triethanolamine by Mehlich (1939)
2. Ca–acetate–triethanolamine by Rich (1955)
3. Ca–acetate–p–nitrophenol by Woodruff (1948)
4. Ca–acetate–p–nitrophenol–triethanolamine by Pratt and Blair (1962)
5. Ca acetate–p–nitrophenol–triethanolamine–potassium chromate–calcium acetate and calcium chloride by Shoemaker *et. al.* (1961).

The observations and inferences drawn from equilibration of soil with different buffer are briefly described below :–

Both Woodruff (1946) and Mehlich (1939) buffer appear to be too strong to indicate any change in pH in relatively weaker acidity of soils rich in alumina.

Reeve and Summer (1970) also reported that strongly buffered solution are unable to indicate appreciable change in pH of weakly buffered soils high in extractable aluminum.

Mclean *et. al* (1960) reported Woodruff method indicated only half the amount of acidity as indicated by moist incubation with $CaCO_3$, but it indicated too much acidity in partially limed soil.

Nye *et al.* (1961) found that Al is held quite strongly in comparison with other cations, and that in soils with low salt concentration, appreciable amount of Al in soil solution are not found until Al saturation is greater than 60 percent.

Mclean *et al.* (1960) reported that the LR method of Shoemaker. *et al.* (1960) indicated more correctly the acidity of soil rich in Al.

Solanky (1965) tested several methods of LR determination. The method of Pratt and Blair (1962) was comparable to the method of Riera (1946).

Method 3. Bhumla and Nijhawan (1965) found the method of substraction of exchangeable cation from CEC both expressed in meq/100 g soil as a measure of LR more dependable than the titrimetric method of LR determination.

2.16.4 Lime requirement as influenced by soil texture and pH

The effect of soil texture and pH on lime requirement in acid soils of Bihar and Punjab are presented in Table 1 and 2.

Table 2.16.4 Lime requirement of acid red soils of Bihar $CaCO_3$ t/ha

Soil pH	Soil Texture		Clay Loam
	Sandy Loam	Loam	
5.5	3.6	4.0	4.5
5.6	3.2	3.6	4.0
5.7	2.7	3.2	3.6
5.8	2.2	2.7	3.2
5.9	1.8	2.2	2.7
6.0	1.4	1.8	2.2

Table 2.16.5 Lime requirement of acid soils of Punjab–CaO t/ha

Soil pH	Soil Texture		Clay Loam
	Sandy	Loam Loam	
5.0	1.26	1.89	2.95
5.2	1.09	1.64	2.55
5.4	0.93	1.39	2.16
5.6	0.76	1.14	1.77
5.8	0.59	0.88	1.38
6.0	0.42	0.63	0.98
6.2	0.25	0.38	0.59
6.4	0.08	0.13	0.20

Note : To convert CaO into $CaCO_3$ multiply by 1.786 **Source** : Kanwar and Joshi (1964)

Table 2.16.6 Lime requirement for raising soil pH by one unit in acid soils of Germany ($CaCO_3$ - t/ha)

Sandy	Sandy Loam	Loam Silty	Loam Clay	Loam	Clayey
2.47	3.70	4.94	6.18	7.41	8.64

The lime requirement for raising the soil pH by one unit in acid soils of different texture in Germany are presented in Table 2.16.6.

5.4 FINENESS OF THE LIMING MATERIAL

The liming material be in a powdered state for application to the field. The standard fixed in USA demand that 70–100 percent of lime should pass through 3–10 mesh sieve, 20–80% should pass through 80–100 mesh sieve.

The relative efficiency of various sized limestone particles are rated as given below :

Table 2.16.7.

Particles Size Relative	Efficiency
Particles passing through 80–100 mesh	100
Particles passing through 20–60 mesh	60
Particles passing through 3–10 mesh	20

In the first year, pH rise is more with 80–100 mesh fine particles than with other particles. However, the difference become negligible in the subsequent year. It is desireable that liming material comprises of fine, medium and coarse particles so that it is effective for a longer period when applied to field.

5.5 LIMING FACTOR

The lime requirement determined in the laboratory is multiplied by a factor varying from 1 to 3 to determine the amount of lime to be applied to the field. This factor is determined on the responsiveness of the crops grown under the field conditions over a period of few years after the applications of lime in multiples of LR.

The liming factor depend on the clay minerals, solubility of liming materials, uptake by crops grown and leaching losses during the reaction period.

5.6 FREQUENCY OF LIMING

The lime applied to the field become depleted in 2–5 years due to absorption by crops, leaching and soil erosion. At chotanagpur, Bihar the average loss of Ca/ year amounted to 1.3 meq/100 g soil. This resulted in annual decreases in soil pH also to the extent of 0.25 unit (Sinha *et. al* 1955). Consequently, the residual effect is considerably reduced within a certain period and the need for lime application is again felt. Motiramani *et al* (1964) reported that the residual effect of lime applied to acid soils of MP lasted for 2–4 years. When annual depletion of applied lime is high, it may be desireable to apply a certain amount of lime each year.

5.7 RATE OF LIME APPLICATION IN DIFFERENT STATES

The general rate of lime application on acid soils of different states are given below :

Table 2.16.8 Rate of-lime application in different states

State	Particulars	Rate of Lime application	Source
Tamilnadu	General	1.10 to 3.36 t/ha	Subramoney (1971)
Mysore	Kanara District	0.13 to 6.50 t/ha	Perur (1971)
Bihar	General	2 t/ha first year and 6t/ha/ year in subsequent year with lime containing 30% CaO.	Mandel *et.al* (1955)
Keral	Lateritic with pH 4.5	3t/ha of slaked lime or 6t/ha of $CaCO_3$ slurry	
	Karpodon with pH 3.6 to 4.5	1.1t/ha fully burnt lime.	

5.8 CROP RESPONSES TO LIMING

Liming did not increase the yield of rice in kari soils of Kuttanad tract of Keral with pH varying from 3.5 to 4.5. Rice crop in some districts of Tamilnadu, West Bengal and Bihar did not respond significantly to liming.

However, Paddy crop responded significantly to liming in Karpodon soils of Kuttanad district of Keral and at Pattandi Res Sta of Palaghat district in Keral, at Ponnapet in Coorg, at Manglore in Kannara district of Mysore. At Palaghat liming @ 5 to 7 t/ha increased the rice yield by 2.4 to 4.8 q/ha. In Assam significant increase in rice yield was obtained when liming was done in conjunction with NPK.

The significant increase in rice yield on liming in some areas of Keral seem to be due to neutralization of sulphates in addition to reduction in toxicity of Fe, Al and Mn. In laterites, Fe, Al and Mn toxicity depress rice yield and lime application lowers the active concentration of these elements.

The other crops which have responded significantly to lime application when applied in conjunction with NPK are Wheat, Maize and Jowar in Bihar (Mandal, 1962), wheat, maize and potatoes in Himachal Pardesh (Kanwar and Joshi, 1962), pigeon pea, soyabean, lentil, groundnut, peas and cotton in Bihar (Mandal *et. al* 1966), Mustard and Phaseolus mungo on alluvial acidic soil of Assam. (Barua, 1962) and Jute in Nowgong. Liming alone without organic manure and/or NPK give a slight increase in crop yield. In some acidic soils there is acute deficiency of Molybdenum which may be applied as ammonium or sodium molybdate.

Mandal *et. al* (1966) grouped upland crops into three groups–

1. Highly responsive to liming – Pigeon Pea, Soyabean, Cotton

2. Medium responsive to liming – Gram, Lentil, Peas, ground nut, maize and Jowar.

3. Low or non responsive to liming–Minor millet, paddy and potatoes.

When liming is done, crops highly responsive to liming be grown. The acidic soils are not invariably limed. The plantation crops such as tea, coffee, rubber and spices grow well on laterite soil without liming.

REFERENCES

A.O.A.C. 1950. Official and Tentative Methods of Analysis of the Association of Official Agricultural Chemists. Ed. 7, 910 pp. Washington.

Barua, J.N. 1964. Bull. Natn. Inst. Sci. India. 26 : 135.

Bergmann, W., A. Günther, and B.Witter 1966. Anleitung für die Aufstellung von Düngungsplañen. Institüte fur pflanzenernährung, Jena, Deutschland (Germany).

Bhumla, D.R., and S.D. Nighawan. 1965. J. Res. Punjab Agric. Univ. 2 : 1–5.

Black C.A. 1968. Soil Plant Relationship. Wiley and Sons, Inc., New York.

Ghosh, P.C. 1966. Personnel Communication.

Jackson, M.L. 1958. Soil Chemical Analysis Prentice Hall Inc. Engleward, Cliff's N.J.

Kanwar, J.S. and J.S. Grewal, 1959. J. Indian Soc. Soil Sci. 7 : 135.

Mandal, S.C., S.K. Das, H.N. Mukherjee, 1955. J. Indian Soc. Soil Sci. 3 : 71.

Mandal, S.C. 1962. J. Ranchi agric. Coll. and Res. Insti. 2 : 39.

Mandal, S.C., N. Sinha, C.R. Prasad and M.A. Ali, 1966. J Indian Soc. Soil Sci. 14 : 127.

Mandal, S.C. and K.K. Jha. 1969. Tech. Bull No. 1. Agric Res. Inst. Kanke Ranchi, Bihar.

Mandal, S.C., M.K. Sinha and H. Sinha. 1975. Acid Soils of India and Liming. I. C.A.R., New Delhi.

McLean, E.O., H.E. Shoemaker and W.R. Hourigan. 1960. Trans. 7th Int. Congr. Soil Sci.

Mehlich, A. 1933. Proc. Soil Sci. Soc. Am. 3 : 162.

Motiramani, D.P., K.K. Vyas, and N.K. Sharma, 1964. J. Indian Soc. Soil Sci. 12 : 157.

P.M. Tamboli, and V.K. Mishra. 1965. Tech. Bull. J.L.N.K.V.V., Jabalpur 1 : 10.

Nye, P. D. Craig, N.T. Coleman, and J.H. Ragland. 1961. Proc. Soil Sci. Soc. Amer. 25 : 14.

Perur, N.G. 1971. Personal Communication.

Pratt P.R and F.L. Blair. 1962. Calif. Agric. 15 : 13.

Riera, A. 1946. J Agric. Univ. R.R.. 30 : 184–185.

Riera, N.G. and M.E. Sumer. 1970. Proc. Soil Sci Soc. Am. 34 : 595.

Rich, C.I. 1955. Bull. Verginia Agric. Exp. Sta. 475 : 2.

Roy, H.K. 1967. M.Sc. Thesis, Ranchi University.

Schofield, R.K. 1933. J. Agric. Sci. 23 : 252.

Shoemaker, H.H., E.O. Mclean, and P.F. Pratt 1960. Proc. Soil Sci. Soc. Amer. 24 : 115.

Sinha, N.P. and S.C. Mandal. 1955. Proc. Bihar Acad. Agric Sci. 4 : 5–8.

Solanky, R.S. 1965. M. Sc Thesis, J.N. Agric. Univ.

Subramoney, N. 1971. Tech. Bull. I.C.A.R., New Delhi 22 : 123.

Tamboli, P.N. and M.J.S. Tulsi. 1967. J. Indian Soc. Soil Sci. 15 : 229

Woodruff, C.M. 1948. Soil Sci. 66 : 33.

2.17 GYPSUM REQUIREMENT–SCHOONOVER (1952)

2.17.1 Principle

5 g of a saline-sodic or sodic soil is treated with 100 ml of nearly saturated gypsum solution containing about 28 meq Ca/l. The Ca is adsorbed by the soil in proportion to the Na displaced from the exchange sites. Thus, the decrease in calcium concentration of saturated gypsum solution on being equilibrated with the saline–sodic or sodic soil gives a measure of gypsum requirement. The Ca concentration in the gypsum solution before and after equilibration with the soil is determined by titration with standard EDTA using a Mg indicator (Eriochrome Black T).

2.17.2 Reagents

(*i*) Saturated $CaSO_4.2H_2O$ solution. Dissolve 5 g of reagent grade gypsum in 1 l distilled water by shaking in a flask for 10 minutes on a mechanical shaker or for one hour by hand. Filter the suspension through a funnel lined with Whatman No. 1 filter paper. Collect the clear filtrate in a clean plastic bottle.

(*ii*) .01 N $CaCl_2$. Dissolve 0.500 g of pure dry calcium carbonate in 10 ml of approx. 3 N (1 + 3) hydrochloric acid and dilute the solution to 1 l.

(*iii*) 0.01 N EDTA. Dissolve 2.00 g disodium dihydrogen ethylene diamine tetra acetic acid and 0.05 g magnesium chloride hexahydrate in distilled water and dilute it to 1 l mark with distilled water.

(*iv*) NH_4Cl–NH_4OH Buffer. Dissolve 67.5 g of ammonium chloride in 570 ml of concentrated ammonia solution and dilute the solution to 1 l with distilled water.

(*v*) Eriochrome Black T indicator. Dissolve 0.5 g of EBT and 4.5 g of hyroxylamine hydrochloride in 100 ml of 95 % ethanol.

(*vi*) 0.01 N $MgCl_2$. Dissolve 0.5081 g of magnesium chloride hexahydrate in distilled water and make the volume to 500 ml with distilled water.

(*vii*) Complexing reagents for heavy metals

1% KCN. Dissolve 1.0 g potassium cyanide in distilled water and make the volume to 100 ml with distilled water.

5 % NH_2OH–HCl. Dissolve 5.0 g of hydroxylamine hydrochloride in distilled water and dilute it to 100 ml with distilled water.

4 % $K_4Fe(CN)_6.3H_2O$. Dissolve 4.0 g potassium ferrocyanide in distilled water and dilute it to 100 ml with distilled water.

Triethanolamine–reagent grade :

20 % $Na_2WO_4.2H_2O$. Dissolve 20 g of sodium tungstate in distilled water and dilute it to 100 ml with distilled water.

2.17.3 Procedure

1. Preparation of G.R. filtrate. Transfer 5.0 g of air dry 2 mm sample of a saline–sodic or sodic soil into a 250 ml Erlenmeyer flask. Then add 100 ml of saturated gypsum solution. Place rubber stopper into the flask mouth and fit the flask on a power driven mechanical shaker. Switch on the mechanical shaker and run it at a moderate speed for 5 minutes. Thereafter, remove the Erlenmeyer flask from the mechanical shaker and pour the suspension in small portions inside a funnel lined with Whatman No. 1 filter paper. Discard the initial 5 ml filtrate after rinsing the receiving flask with it, and collect the subsequent filtrate. Preserve the filtrate for the determination of calcium concentration by titration with standard EDTA.

2. Titration of 0.01 N $CaCl_2$ solution by approx. 0.01 N EDTA. Pipette 5.0 ml aliquots of 0.01 N $CaCl_2$ solution into three 125–150 ml conical flasks. Add 20 ml distilled water in each flask. Then add one drop of 0.01 N magnesium chloride and two drops each of the 5 complexing reagents for heavy metals in each flask. Now add 2 ml of NH_4Cl–NH_4OH buffer and one drop of EBT indicator and titrate by approx. 0.01 N EDTA solution filled in a 10 ml micro–burette.

Add EDTA solution in small portions and then shake the flask gently observing any change in the colour of the solution. Carefully continue to add EDTA solution in small portions until the colour of the solution in flask turn purple. Then add EDTA solution drop by drop and shake the flask after each addition. The end point is reached when the colour of the solution changes from purple to clear blue or green. Note down the EDTA titre. Similarly titrate the remaining two aliquots of 0.01 N $CaCl_2$ by EDTA solution to the same end point. Note down the titre and take the mean of two more reliable titrations. Use the EDTA titre for estimating the normality of EDTA solution.

3. Titration of saturated gypsum solution by standard EDTA solution. Pipette 5.0 ml aliquot of saturated gypsum solution into three 125 conical flasks. Then add 20 ml distilled water, 1 drop 0.01N $MgCl_2$, 2 drops each of the 5 complexing reagents for heavy metals, 2 ml NH_4Cl–NH_4OH and 1 drop EBT indicator to one of the three conical flask and titrate the solution by standard EDTA as described above. The colour of the solution changes from wine red to purple and usually with few drops of EDTA from purple to clear blue or green. Note down the EDTA titre on reaching the end point. Fill the micro burette again with EDTA solution and titrate the remaining two aliquots of 5 ml saturated gypsum solution in separate conical flasks. Take the mean of the nearest two reliable titrations by standard EDTA and note it as a blank titre.

4. Titration of G.R.–filtrate by standard EDTA solution. Pipette 5.0 ml aliquots of G.R. filtrate into 125–150 ml conical flask in triplicate and titrate the aliquots of G.R. filtrate one by one following the procedure described for the titration of saturated gypsum solution by the standard EDTA solution. Note down the three EDTA titre and take the mean of the nearest two reliable titrations by standard EDTA and note it as sample titre.

2.17.4 Calculations

Normality of EDTA solution

$$= \frac{\text{ml of standard } CaCl_2 \text{ taken for titration} \times \text{normality}}{\text{EDTA titre for the standard } CaCl_2}$$

GR in meq/100g soil

$$= \frac{\begin{array}{l}\text{(Blank} - \text{Sample) titre} \times \text{Normality of EDTA} \\ \times \text{ Volume of Saturated gypsum solution equilibrated with soil sample}\end{array}}{\text{Volume of aliquot taken for titration} \times \text{mass of soil in g}} \times 100$$

GR in lbs/acre = GR in meq/100 g soil × 20 × Equivalent wt. of $CaSO_4 \cdot 2H_2O$

$$\text{GR in t/ha} = \frac{\text{GR in lbs/acre} \times 2.47}{2.2 \times 1000}$$

2.17.5 Comments

The soil samples for estimating G.R. be taken after grading, levelling and flooding of the fields. The composite soil samples be collected from the plough depth when the fields are not wet. Gypsum is required as soil amendment only on soils with pH more than 9.0 in 1:2 soil water suspension (w/v).

The glassware be washed with detergent such as Lab Wash or Teepol. It should be also rinsed with approx. 6 N (1 + 1) HCl to make glasswares free from calcium. After rinsing with HCl, the glasswares be thoroughly washed with tap water and then it should be rinsed twice with distilled water.

The use of complexing reagents for heavy metals is desirable as EDTA complexes Ca, Mg as well as heavy metals. The NH_4Cl–NH_4OH buffer is added to raise the solution reaction to pH 10.0. The pH of the solution may be checked after adding 2 ml of the buffer.

It is desirable to do the titrations in triplicate. Therefore, the steps upto the addition of complexing reagents and one drop of .01 N $MgCl_2$ be done simultaneously in the three conical flasks containing the aliquot of the sample, but the subsequent procedure be undertaken one by one. Besides, the microburette be filled with EDTA and be kept ready for titration before introducing buffer into the solution so that the titration may be performed immediately after the addition of buffer and the EBT indicator. If near the end point the colour of the solution being titrated become quite faint, then one more drop of the EBT indicator may be added with advantage.

The gypsum requirement of soils vary with the soil texture and pH. Gypsum recommended to the farmer is normally 50% of the laboratory estimation of gypsum requirement. The effectiveness of the gypsum is also linked with its fineness and the crop management after its application.

REFERENCES

Schoonover, W.R. 1952. Examination of soils for alkali. Univ. Calif. Ext Ser. Berkely, California (unpublished).

Richards, L.A. (ed.). 1954. Diagnosis and Improvement of Saline and Alkali Soils. U.S.D.A. Hand Book 60. Oxford and IBH Publishing Co. New Delhi.

2.18 CARBONATES IN SOIL BY NEUTRALIZATION WITH AN ACID–SCHOLLENBERGER (1945)

2.18.1 Principle

The soil sample is treated with an standard acid in excess of carbonates present. The acid decomposes the carbonates. The unreacted acid is back titrated with a standard alkali to the phenolphthalein end point.

2.18.2 Reagents

(*i*) Standard hydrochloric acid, 0.5 N. Dilute 21.8 ml of conc. HCl(sp.gr. 1.18, Assay 35.4 %) to a volume of 500 ml. Standardize the acid against 0.1 N Na_2CO_3 using methyl orange indicator.

(*ii*) Sodium carbonate, 0.1 N. Dissolve 0.530 g of Na_2CO_3 in distilled water and then dilute the solution to 100 ml mark with distilled water.

(*iii*) Methyl Orange, 0.1 %. Dissolve 0.1 g of methyl orange in distilled water and then dilute the solution to 100 ml mark in a volumetric flask.

(*iv*) Sodium Hydroxide Solution 0.25 N. Dissolve 10.0 g Na OH pellets (AR) in distilled water and dilute the solution to 1000 ml in a volumetric flask. Standardize the reagent against standard acid to the phenolphthalein end point.

(*v*) Phenolphthalein solution, 1.0 %. Transfer 1.0 g of phenolphthalein indicator into 100 ml volumetric flask. Add 60 ml of ethanol or absolute alcohol. Shake to dissolve phenolphthalein for 5–10 minutes. Then make the volume to 100 ml with distilled water. Filter the solution through Whatman No. 1 filter paper.

2.18.3 Procedure

Transfer 5 to 10 g of finely grinded soil sample into a 300 ml tall beaker. Then add 50.0 ml of 0.5 N HCl through a burette. Place a watch glass on the beaker. Heat to boil the contents gently for 5 minutes. Allow it to cool.

Filter the contents through a funnel lined with Whatman No. [illegible] ter paper. Collect the filtrate in a 250 ml Erlenmeyer flask. Transfer the soil from the be[illegible] er to the funnel with the small portions of distilled water . Then wash the soil on filter paper with distilled water until the filtrate become free from acid. Collect the filtrate and the washing in the same 250 ml Erlenmeyer flask. Use a blue litmus paper to test whether leachate (washing) is free from acid.

Now add 3 drop of the phenolphthalein indicator and titrate with 0.25 N NaOH solution until the solution colour chages from colourless to persistent pink. Note down the NaOH titre.

2.18.4 Calculations

$$\text{meq } CO_3 / 100 \text{ g soil} = \frac{(\text{meq HCl added} - \text{meq NaOH used}) \times 100}{\text{oven dry mass of soil in g}}$$

$CaCO_3$ equivalent percent = meq CO_3/100g soil × meq wt. of $CaCO_3$

CO_3- Carbon percent = meq CO_3/100 g soil × meq wt. of CO_3 - carbon

Note. Titre × Normality = meq of acid or base ; meq weight of $CaCO_3$ is .05 g and meq weight of CO_3-C is .006 g

2.18.5 Comments

The carbonates are found in soils chiefly as $CaCO_3$ (calcite) and as $CaCO_3.MgCO_3$(dolomite). These minerals are sparingly soluble and are seldom distributed uniformly in soils. In sodic soil, carbonates occur chiefly as $Na_2CO_3$and $NaHCO_3$. Precipitated $CaCO_3$occur also in alkali soils particularly at lower depth. When precipitated $CaCO_3$ occur in upper layer, it would be preferable to use acidifying amendments such as sulphur or pyrite. Otherwise it would be preferable to use gypsum.

In soil there is equilibrium between HCO_3^- and CO_2 as well as in between HCO_3^- and CO_3^{--}. These ions are found in soil solution at pH 8.5 or higher.

The soils saturated with $CaCO_3$ are termed as calcareous soils. The calcareous soils have pH around 8.2. These soils also show $CaCO_3$ concretions and give strong effervescence when treated with 1N HCl. On these soils there is lime induced deficiency of micronutrient elements particularly in citrus orchards. Calcareous soils also adsorb phosphates. With ageing, soluble phosphates are precipitated as tricalcium phosphate. $CO_3^=$ associated with Na cause Fe chlorosis in plants.

Several methods are employed for the determination of carbonates in soil. The carbonate minerals other than dolomite are rapidly decomposed by dilute acid with an evolution of CO_2. As an index of CO_3carbon, the CO_2 evolved could be measured directly employing manometer apparatus, vacuum distillation or by gravimetric method. The CO_2evolved could be also measured indirectly by loss of weight in an open system or by increase in volume or pressure using appropriate calcimeter in a closed system. The vacuum distillation procedure of Schollenberger (1945), and the rapid gravimetric procedure of Allison (1960) are useful for more precise determination of CO_3^- carbon. However, a rapid acid neutralization method which provide a rough index of soil carbonates is quite satisfactory.

REFERENCES

Allison, L.E., 1960. Wet combustion apparatus and procedure for organic and inorganic carbon in soil. Soil Sci. Soc. Amer. Proc. 24 : 36–40.

Jackson, M.L. 1973. Soil Chemical Analysis, Prentice Hall of India Private Limited, New Delhi.

Richards, L.A. (ed). 1954. Diagnosis and Improvement of Saline and Alkali Soils. U.S.D.A. Hand Book 60. Oxford & IBH Publishing Co. New Delhi.

Schollenberger C.J., 1945. Determination of carbonates in soil. Soil Sci. 59 : 57–63.

2.19 GYPSUM IN SOIL

2.19.1 (Method 1) Gypsum in soil extract flocculated by acetone and estimated after dissolving in distilled water by measuring the electrical conductance–Bower and Huss (1948)

2.19.1.1 Principle

Gypsum from soil is extracted by distilled water using a soil-extractant ratio of 1 : 5 or 1 : 10 (w/v) with 30 minutes shaking on a mechanical shaker. A soil extractant ratio of 1 : 10 is desirable for soils containing more than 15 meq of gypsum /100 g of soil.

A portion of the clear extract is treated with equal volume of acetone in a conical centrifuge tube to flocculate the gypsum. The solution is centrifuged for sedimentation of the gypsum floccules, and the supernant clear solution is decanted off.

Then, the gypsum flocceules are completely dissolved in measured volume of distilled water and the electrical conductance of the solution is measured by a conductivity bridge using a conductivity cell. The specific electrical conductance of the gypsum solution expressed in milli mho/cm at 25°C multiplied by 12.5 give an estimate of gypsum in meq/l of solution.

2.19.1.2 Apparatus/Reagents

Apparatus

Erlenmeyer flasks. 250 ml capacity, rubber stopper wrapped in thin polythene, 50 ml burette, mechanical shaker, pipette 10 and 20 ml capacity, funnel, thermometer, conductivity'bridge with conductivity cell.

Reagents

(*i*) Distilled water free from sulphur. Use deionized water or water distilled from a completely glass distill.

(*ii*) Acetone (AR)

(*iii*) 0.01N Potassium Chloride Solution. Dissolve 0.7456 g dry KCl (AR) in distilled water and dilute it to 1000 ml mark with distilled water. This is to be used as a standard reference solution as it has an electrical conductivity of

1411.8×10^{-6}mho/cm at 25°C. The electrical conductivity increases by 2% for each degree increase in temperature of the solution.

2.19.1.3 Procedure

1. Extraction of gypsum from the solution. Weigh 10.0 g of air dry soil passed through 2 mm sieve having round holes in duplicate and transfer the samples into separate 250 ml Erlenmeyer flasks. Thereafter, add 50 ml of distilled water to one of the extraction flask, and 100 ml of distilled water into other extraction flask. Place the stoppers, and shake the contents on a mechanical shaker for 30 minutes or else by hand 6 times at 15 minutes interval. Then, filter the suspension separately into a clean flask using a funnel lined with dry Whatman No. 42. filter paper. Discard the initial filtrate and collect the subsequent filtrate.

2. Flocculation of $CaSO_4$ and removal of other soluble salts. Transfer 20 ml aliquot of the filtered soil extract into 50 ml conical centrifuge tubes. Then add 20 ml of acetone. Mix the contents and allow the suspension to stand for 5 to 10 minutes until $CaSO_4$flocculates. Place the centrifuge tubes into the sockets provided opposite each other in the centrifuge. Gradually increase the speed of centrifuge drum to 2440 rpm and let it run for 10 minutes. Thereafter gradually decrease the speed until the centrifuge drum stops moving.

Then, remove the centrifuge tube and decant off the clear supernant solution. Thereafter, invert the centrifuge tube on a filter paper for 5 minutes to drain. Disperse the $CaSO_4$ flocculates with a glass rod, and then add fresh 10 ml portion of acetone through a pipette washing down the walls of the tube. Triturate with a glass rod for 5 minutes and then repeat centrifugation, decantation and drainage on a filter paper.

3. Dissolving of flocculated $CaSO_4$ in distilled water. Thereafter add excactly 40.0 ml of distilled water in centrifuge tube containing floccules of $CaSO_4$. Triturate the contents with a glass rod for 5 minutes or shake the contents after putting stopper until $CaSO_4$ floccules become completely dissolved.

4. Measurement of the specific conductance of solution containing dissolved $CaSO_4$. Connect the terminals of conductivity cell placed in distilled water into the appropriate electric points of the conductivity bridge. Switch on the instrument, and let it run for 10 minutes to warm. For checking the cell constant, remove the conductivity cell from distilled water, and place a blotting paper to soak the water adhering to it and then dip the conductivity cell in 0.01N KCl solution. Measure the temperature of the solution in °C, and note down the temperature correction factor from the physical tables.

Rotate the multiplier knob in the resistance range and let it rest at the point of widest gap in the magic eye. Then, rotate the dial knob until widest gap in the magic eye is attained. Note the main dial reading and the multiplier range, and multiply the two values to obtain Rt in ohms/cm. Then estimate K by substituting the values in the equation $K = 1411.8 \times 10^{-6} \times Rt/ft$. Thereafter, remove the conductivity cell from 0.01N KCl solution, and wash it with a jet of distilled water and dry it with a blotting paper. Thereafter, place it in solution containing dissolved gypsum.

Now rotate the multiplier knob towards conductance range in micro mho/cm and let it rest at the point giving the widest gap in the magic eye. Thereafter, rotate the dial knob until widest gap is attained in the magic eye. Note the multiplication factor and the main dial reading. Multiply the two values to obtain ECt in micro mho/cm. Measure the temperatures of the solution in °C, and note the temperature correction factor (ft) from the physical tables. Thereafter, estimate the specific conductance of the solution in milli mho/cm at 25°C by solving the following equation after substituting the values :

$$EC_{25} \times 10^3 = \frac{ECt \times ft \times K}{10^3}$$

In case of digital conductivity meter, follow the procedure given in the instruction manual. Regulate the temperature of the 0.01N KCl solution to the temperature of the sample solution. Then note the conductance of the reference solution (.01N KCl) at the observed temperature.

Prior to measurement of the electrical conductance, switch on the conductivity meter placing the conductivity cell in distilled water and let it warm for 10 minutes. Thereafter, set the CAL/READ knob at the CAL point, and adjust the meter reading by the CAL knob to the specified value (maximum value) given in the instruction manual. Then, remove the conductivity cell from the distilled water and wipe it dry with a blotting paper. Now, dip the conductivity cell in 0.01 N KCl solution. Set the CAL/READ knob at the READ point and adjust the conductivity range knob and the cell constant knob until the conductivity meter records the desired conductance value of the reference solution (.01 N KCl). Thereafter, remove the conductance cell from the reference solution, wash it with a jet of distilled water, dry it with a blotting paper and place the conductance cell in the sample solution. Note the specific conductance recorded by the conductivity meter in millimhos/cm. Note the temperature of the sample solution and read the temperature correction factor (ft) from the physical table. Multiply ft by the specific conductance in millimho/cm to express specific conductance in millimho/cm at 25°C.

5. Estimation of concentration of $CaSO_4$ in solution prepared by dissolving $CaSO_4$ flocculated from aliquot of soil extract in 40 ml distilled water.

The relationship between the concentration of $CaSO_4$ and the specific conductance of the solution at 25°C is presented in the under given Table 2.19.1.

Table 2.19.1 Specific conductance of the solution as influenced by the concentration of $CaSO_4$

Specific Conductance	**Concentration of $CaSO_4$ in meq/l**					
at 25°C	1	2	5	10	20	30.5
milli mho/cm	0.121	0.226	0.500	0.900	1.584	2.205

A close approximation of the above mentioned relationship may be obtained by the following equation.

$CaSO_4$ meq/l = Specific conductance at 25°C in milli mho/cm × 12.5

The concentration of $CaSO_4$ in solution may be estimated by reference to the above mentioned relationship between $CaSO_4$ concentration and specific conductance of the solution.

2.19.1.4 Calculations

$$CaSO_4 \text{ in meq/100 g soil} = \frac{A \times B \times C \times 100}{1000 \times D \times E} = \frac{A \times 40 \times C \times 100}{1000 \times 20 \times 10}$$

$$\% \text{ gypsum in soil sample} = CaSO_4 \text{ meq/100 g soil} \times \text{meq wt. of gypsum}$$

$$= CaSO_4 \text{ meq/100 g soil} \times 0.0861$$

where

A is meq of $CaSO_4$/l in solution as estimated from the relationship between concentration of $CaSO_4$ in solution and its specific conductance at 25°C expressed in milli mho/cm.

B is volume of distilled water used for dissolving $CaSO_4$ flocceules formed on adding acetone to aliquot of soil extract.

C is volume of distilled water used for the extraction of gypsum from the soil in ml.

D is ml aliquot of soil extract taken for flocceulation of $CaSO_4$ in solution by adding acetone.

E is mass of air dry < 2 mm soil sample taken for the extraction of gypsum by distilled water.

2.19.1.5 Comments

The soil should not be oven dried because heating promotes the conversion of $CaSO_4.2H_2O$ to $CaSO_4 0.5H_2O$. The latter hydrate has a higher solubility in water for an indefinite period following solution.

Sodium and potassium sulphate when present in sufficiently high concentration are also precipitated by acetone. The maximum concentration of sodium sulphate and of potassium sulphate that may be tolerated are 50 and 10 meq/l respectively.

At 1 : 5 soil-water ratio, 50 ml of distilled water will dissolve 0.1 g $CaSO_4$ or 0.13 g $CaSO_4.2H_2O$ *i.e.* approximately 15 meq of the salt which is approximately 1.5 meq of gypsum per 100 g of soil.

Three factors besides the solution of gypsum may influence the amount of Ca or SO_4 extracted from the soil.

(*a*) Solution of Ca from sources other than gypsum *e.g.* from $CaCO_3$.

(*b*) Exchange reactions by which the dissolved Ca replaces some ions such as Na and Mg.

(*c*) The solution of SO_4 from sources other than gypsum.

The above mentioned errors preclude a highly accurate determination of gypsum.

The separate determination of Ca and SO_4 is fairly lengthy procedure. A rapid conductance method of gypsum determination developed by Bower and Huss (1948) gave good concordance.

REFERENCES

Bower, C.A. and R.B. Huss (1948). Rapid conductometric method for estimating gypsum in soil. Soil Sci. 66 : 199–204.

Gypsum in Soil (Method 2)

2.19.2 Gypsum content in soil by increase in soluble calcium and magnesium content on dilution–Richards ed. (1954)

2.19.2.1 Principle

The extraction of saturation extract removes $CaSO_4$ only to the extent that it normally contributes to the osmotic pressure of the soil solution in the field. As 50 ml of distilled water dissolves only 0.1g of $CaSO_4$or 0.13 g of $CaSO_4 . 2H_2O$, only a very small fraction of gypsum present in soil will be solublized in saturation extract. There is need to increase the soil extractant ratio to solubilize greater fraction of gypsum. The use of 100 ml distill water will enable dissolving of gypsum up to 2.6 g from the soil. This will increase the concentration of gypsum in the equilibrium solution which can serve as a basis for the estimation of gypsum in soil.

2.19.2.2 Apparatus/Reagents

Apparatus. 2 mm sieve preferably with round holes, Erlenmeyer flasks–250 ml capacity, conical flask with stoppers, volumetric flasks, graduated pipettes, 1, 2, 5 & 10 ml capacity, Bulb pipette 20 ml capacity, Erlenmeyer flasks–125 ml capacity, 50 ml burette, 10 ml micro burette.

Reagents. All the reagents needed for Ca + Mg determination by EDTA titration as listed in the determination of gypsum requirement in sodic soil (See section 2.17.2).

2.19.2.3 Procedure

1. Preparation of saturated paste, estimation of saturation percentage and collection of extract from saturated paste. Transfer 200 g of air dry < 2 mm soil sample into a 500 ml or 1000 ml plastic beaker. Add distilled water slowly drop by drop to make the soil sufficiently moist. Work the moist soil with a stainless steel spatula. When the soil paste become free flowing without the presence of free water and the soil paste begins to glisten also, the soil has reached the saturation point. Run the spatula across the soil paste. If the gap formed get filled quickly on tapping the beaker on table, the soil paste may be referred as free flowing.

Transfer a small portion of saturated paste into a tared aluminium can. Take the mass of moist soil and place it in an oven set at 105°C placing the can lid below the can. Let it dry in the oven for 24 hours. Then take the mass of dry soil and estimate the percent moisture on oven dry basis for determining the saturation percentage.

Allow the saturated paste to stand overnight after covering the beaker with a large size petri dish. On the following day make the paste saturated again and transfer the entire soil saturated paste to a Buchner funnel lined with Whatman No. 42 filter paper. Place the Buchner funnel on a suction flask filled with straw to hold a large 20–30 ml test tube below the funnel. Connect the suction flask to a suction pump, and let it run until air begins to leak through the soil mass. Preserve the extract collected in the test tube for estimating the concentration of Ca + Mg by EDTA titration using EBT as Mg indicator.

2. Estimation of Ca + Mg concentration in soil saturated extract by EDTA titration. Transfer 2.0 ml of original soil saturated extract into a 125 ml Erlenmeyer flask, add 20 ml of distilled water, 2 drops each of 1% KCN, 4% solution of potassium ferrocyanide, 5% hydroxylamine hydrochloride, and 50% triethanolamine. Thereafter, add 1 drop of 0.01N $MgCl_2$ and 2 ml of NH_4Cl-NH_4OH Buffer, 2 drops of Eriochrome Black T indicator, and titrate with standard 0.01N EDTA solution until the colour of the solution changes from pink to purple and then to clear blue or green. Perform the titration in duplicate and similarly run a blank taking 2 + 20 ml of distilled water. Follow step 3 of section 2.20.3 when extract is coloured.

3. Extraction of soluble Ca + Mg from soil using a soil extract ratio of 1 : 5 and 1 : 10 with 10 g of air dry < 2 mm soil sample. Transfer 10g of air dry 2 mm soil sample in duplicate into separate 250 ml Erlenmeyer flasks. Then, add 50 ml distilled water to one of the Erlenmeyer flask and 100 ml of distilled water to the other Erlenmeyer flask. Stopper the flask, and place the flask firmly on a mechanical shaker. Run the mechanical shaker for 30 minutes at a moderate speed.

Thereafter, filter the suspension from the two flasks separately through separate funnels lined with Whatman No. 42 filter paper. Discard about 4 ml of the initial filtrate, and collect the subsequent filtrate. Label the flask mentioning the soil extractant ratio and preserve it for complexometric titration by EDTA.

4. Estimation of Ca + Mg in filtrate from 1 : 5 and 1 : 10 soil-water suspension (w/v). Pipette 5.0 ml of filtrate in triplicate into separate 125 ml Erlenmeyer flask and then proceed as described in step 2 of procedure (section 2.19.2.3) for titration of Ca + Mg by EDTA beginning with the addition of 20 ml distilled water.

2.19.2.4 Calculations

(*a*) Saturation Extract

(*i*) Ca + Mg meq/l

$$= \frac{(\text{Sample titre} - \text{Blank titre}) \times \text{Normality of EDTA} \times 1000}{\text{ml aliquot of saturation extract taken}}$$

(*ii*) Ca+Mg meq/100g soil

$$= \frac{\text{Ca} + \text{Mg in saturation extract (meq / l)} \times \text{saturation percentage}}{1000}$$

(*b*) 1 : 5 or 1 : 10 soil-water suspension extract

(*i*) Ca + Mg in meq/l

$$= \frac{(\text{Sample titre} - \text{blank titre}) \times \text{Normality of EDTA} \times 1000}{\text{ml aliquot of soil extract taken}}$$

(*ii*) meq/100g soil

$$= \frac{\text{Ca} + \text{Mg in solution meq / l} \times \text{total volume of extractant} \times 100}{1000 \times \text{mass of} < 2 \text{ mm soil sample in gram}}$$

(*c*) Gypsum content in soil meq/100 g soil

$$= \frac{\text{Ca+Mg in soil-water (1:10) extract}}{(\text{meq/100 gsoil})} - \frac{\text{Ca+Mg in saturation extract}}{(\text{meq/100 g soil})}$$

Note. If soluble Ca + Mg in 1 : 5 soil-water extract approaches 15 meq/100 g soil, determine Ca + Mg in 1 : 10 soil- water extract and if Ca + Mg in soil is less than 30 meq per 100 g soil, use the data for estimating gypsum content in soil.

2.19.2.5 Comments

The method deserve to be tested and compared with the result obtained by the method of Bower and Huss (1948).

REFERENCES

Richards I.A. (ed.) (1954). Diagnosis and Improvement of Saline and Alkali Soils. U.S.D.A. Hand Book 60, Oxford and IBH Publishing Co., New Delhi, 1968.

Bower, C.A. and R.B. Huss. 1948. Rapid conductometric method for estimating gypsum in soils. Soil Sci. 66 : 199–204.

2.20 SODIUM ADSORPTION RATIO AND EXCHANGEABLE SODIUM PERCENTAGE–RICHARDS ed. (1954)

2.20.1 Principle

The sodium adsorption ratio (SAR) is defined as $Na^+ / \sqrt{\frac{Ca^{++} + Mg^{++}}{2}}$ where Na^+, Ca^{++}, Mg^{++}refer to the concentration of the designated soluble cations expressed in milliequivalents per litre. The ratio designated as SAR is employed for describing equilibrium relationship between soluble and exchangeable cations in soil.

The SAR of soil extract at the saturation moisture percentage in 59 soil samples of 9 western states of USA showed linear relationship with exchangeable sodium divided by exchange capacity *i.e.* CEC minus exchangeable sodium. The regression equation worked out to be y = – 0.0126 + 0.01475 SAR with a regression coefficient r = 0.923. An empirical relationship has been also worked out to define the relationship between exchangeable sodium percentage (ESP), and the sodium-adsorption ratio (SAR) which is given below :

$$\text{ESP} = \frac{100\,(-.0126 + .01475\ \text{SAR})}{1 + (-.0126 + .01475\ \text{SAR})}$$

In view of the above mentioned relationship, it is possible to determine ESP indirectly by estimating SAR from the ionic composition of extract from the soil saturated paste.

2.20.2 Reagents

(*i*) 0–20 ppm Na solution

(*ii*) EDTA–.01N, and Eriochrome Black T as described in section 2.17.2.

(*iii*) $MgCl_2$–.01N

2.20.3 Procedure

1. Preparation of soil saturated paste and extract from soil saturated paste. As described in step 5 of procedure 2.14.2.3 given in chapter 2.14.2

2. Sodium concentration in extract from soil saturated paste. The soil saturated paste extract from sodic soil has high concentration of Na, and so the extract require 1000 to 4000 times dilution for the determination of Na by the flame photometer technique.

Dilute 1.0 ml extract to 100 ml to obtain 100 times dilution. Transfer 10.0, 5.0 and 2.5 ml of 100 times diluted extract into separate 100 ml volumetric flasks, and dilute each to 100 ml with distilled water to obtain 1000, 2000 and 4000 times dilution of the original extract.

Determine Na concentration in diluted extract by flame photometer as given below :

First switch on the flame photometer and then switch on the air compressor. Set the air pressure to the desired level. Then, connect the gas cylinder tube to the gas inlet of the flame photometer. Place a lighted match stick above built in gas burner of the flame photometer just before opening the gas valve of the gas cylinder. When the flame gets lighted, regulate the gas valve knob of the flame photometer to obtain a blue flame of sufficient length.

Now, atomize 20 ppm Na solution and adjust the flame photometer reading to 100. Then atomize 0 ppm Na solution and adjust the flame photometer reading to 0. Repeat atomization of 20 ppm and 0 ppm Na solution setting the flame photometer reading to 100 and 0 respectively. Thereafter, atomize the intermediate standards of Na solution and note down the flame photometer reading. Plot the flame photometer reading on the ordinate of a linear scale graph paper against the corresponding Na concentration on the abscissa. Connect flame photometer reading of 0 ppm and 20 ppm Na by a straight line.

When the flame photometer reading has been set with the standard Na solution, atomize 1000 to 4000 times diluted extract and record the corresponding flame photometer reading. Find out the Na concentration in diluted extract by reference to the standard graph for Na or else divide the flame photometer reading by 5 to obtain the concentration of Na in diluted extract in ppm.

Estimate Na concentration in meq/l of original extract by substituting the values in the following equation

$$\begin{array}{c}\text{Na concentration (meq / l)} \\ \text{in original extract}\end{array} = \frac{\text{ppm Na in diluted extract} \times \text{dilution time}}{\text{equivalent weight of sodium}}$$

3. Ca + Mg concentration in extract from soil saturated paste. Transfer 2.0 ml of extract from soil saturated paste into a 150 ml of conical flask. Evaporate the contents to dryness on a hot plate set at low heat or else on a water bath. Remove the

conical flask and allow it to cool. Then, add 2 ml of conc. HNO_3 and place the conical flask on a sand bath. Heat to evaporate the contents to dryness. Allow the flask to cool. Add 2 ml of conc. HCl, swirl the flask and then evaporate the contents to dryness on a sand bath. Now, add 3 ml of distilled water and again evaporate the contents to dryness. Remove the flask from sand bath and let it cool.

Thereafter, add 20 ml distilled water, 1 drop of 0.01N $MgCl_2$ solution, 2 drops each of four heavy metal complexing reagents such as 1% KCN solution, 4% potassium ferrocyanide solution, 5% hydroxylamine hydrochloride solution, triethanolamine (1 : 1) solution. Add sufficient ammonium chloride-ammonium hydroxide buffer (2 ml or more) to raise the solution pH to 10.0. Add 1 or 2 drop of Eriochrome Black T indicator and immediately titrate with 0.01N EDTA solution drop by drop swirling the contents until the colour changes from pink to clear blue or green. Note down the titre. Similarly run a blank without taking the soil extract. Both sample and blank titration be done in duplicate drop by drop as extract from sodic soil are very low in soluble Ca. Estimate the concentration of Ca + Mg in meq/l of extract by substituting the values in the following equation.

Ca + Mg (meq/l)

$$= \frac{\text{EDTA titre (Sample - blank)} \times \text{normality of EDTA} \times 1000}{\text{Extract taken for titration in ml}}$$

2.20.4 Calculations

Estimation of sodium-adsorption-ratio (SAR)

Substitute the value of Na^+, Ca^{++}, Mg^{++} concentration expressed in meq/l in the following equation and simplify

$$\text{SAR} = \frac{Na^+}{\sqrt{\frac{Ca^{++} + Mg^{++}}{2}}}$$

Estimation of exchangeable sodium percentage (ESP) from sodium-adsorption-ratio (SAR)

Substitute the value of SAR of extract from soil saturated paste in the following equation and simplify to obtain ESP.

$$\text{ESP} = \frac{100\,(-.0126 + .01475\ \text{SAR})}{1+(-.0126 + .01475\ \text{SAR})}$$

2.20.5 Comments

The sodic soils have ESP greater than 15. There is a close relationship between SAR and ESP at relatively lower value of SAR, but at higher values of SAR, ESP value falls behind SAR considerably. The ESP value is always lower than SAR. The emperical relationship between ESP and SAR may vary depending upon the mineralogical composition of the soil. The following emperical relationship between ESP and SAR has been worked out on study of soils of Rajasthan by Paliwal and Maliwal (1970).

$$ESP = \frac{100\,(0.23 + .0042\ SAR)}{1+(0.23+ .0042\ SAR)}$$

In flame photometric determination, direct readings without an internal standard give satisfactory results if an air-propane flame is used.

When optimum concentration of lithium chloride (5–10 meq/l) is required as an internal standard for the flame photometric determination. The calculated volume of lithium chloride solution of a particular strength (.05N) be added to the 100 ml volumetric flask containing an aliquot of 100 times diluted soil saturated extract for further dilution. In preparation of Na standard also, lithium chloride be added to obtain its desired concentration as internal standard.

Besides, the interference from other ions such as K, Ca and Mg may be compensated for by determining the concentration of interfering elements in the unknown and by adding the desired quantity of the interfering elements to the Na standards using KCl (12.5 meq/l), $CaCl_2$(50 meq/l) and $MgCl_2$ (50 meq/l). K interferes when K/Na ratio is 5 or greater, Ca interfere when Ca/Na ratio is 10 or greater, and Mg interferes when Mg/Na ratio is in excess of 100. In soil saturation extract of a sodic soil, Na concentration is so high that interference from K, Ca and Mg is ruled out.

If $LiCl_2$ concentration of 5 meq/l is desired as internal standard then add 10 ml of .05 N $LiCl_2$ into 100 ml volumetric flask containing 20 ml of 100 ppm Na solution, and dilute the contents to 100 ml with distilled water to obtain 20 ppm Na solution in .005 N $LiCl_2$ having Li concentration of 5 meq/l. To prepare 1000 times diluted soil extract in .005 N $LiCl_2$. Add 10 ml of .05 N $LiCl_2$ into 100 ml volumetric flask, containing 10 ml of 100 times diluted soil extract and dilute the contents to 100 ml with distilled water.

REFERENCES

Richards, L.A. (ed). 1954. Diagnosis and Improvement of Saline Alkali Soils. U.S.D.A. Hand Book 60. Oxford and IBH Publishing Co., New Delhi 1968.

2.21 SULPHUR IN SOIL

2.21.1 Total S in soil after oxidation to sulphates by the reduction method –Johnson and Nishita (1952)

2.21.1.1 Principle

Total S in soil is first oxidized to sulphates by acid digestion or alkaline digestion or else by dry ashing. The residue from dry ashing is transferred quantitatively to 50 ml digestion flask and then it is treated with 2 ml distilled water. The digestion by alkaline oxidation is done in 50 ml digestion flask. The residue obtained on cooling is treated with 1 ml of formic acid to decompose remaining NaOBr. If soil sample is digested with HNO_3 and $HClO_4$, 2 ml aliquot or more of the acid digest is transferred to 50 ml digestion flask and the contents are evaporated on a hot plate to reduce the volume to 2 ml.

The solution in the flask is treated with a reduction mixture and the flask is quickly attached to a distillation apparatus. The purified nitrogen gas is passed through the reduction mixture and the digestion flask is heated by a micro burner to distill H_2S liberated. The H_2S distilled is made to pass through a gas washing column before its absorption by zinc acetate-sodium acetate buffer solution. The H_2S absorbed is treated with acid solution of p–aminodimethylaniline sulphate and ferric ammonium sulphate solution to form methylene blue colour. The optical density (absorbance) of the methylene blue colour is measured at 670 mμ wave length using a spectrophotometer or colorimeter.

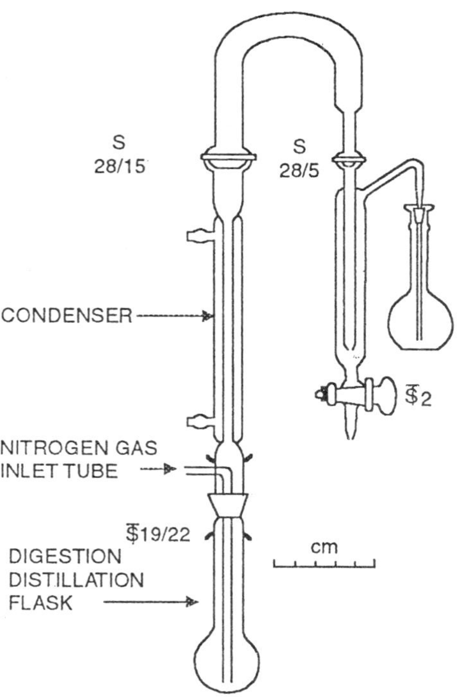

Fig. 2.21.1 Dgestion-distillation apparatus.

2.21.1.2 Apparatus/Reagents

Apparatus. Digestion distillation apparatus (Fig. 2.21.1), Apparatus for preparation of reducing mixture (Fig. 2.12.2), Fume hood, hot plate, steam plate, porccelain crucible, Philip's beakers, volumetric flasks, pipettes, thermometers, tall beakers, cylinder of water pumped N_2, photoelectric colorimeter or spectrophotometer.

Reagents

(*i*) Reagents 1 to 12 as described in the determination of total S in plant by the reduction method (Section 3.9.3.2).

(*ii*) Phosphoric acid (AR)
(*iii*) Sodium bicarbonate (AR)
(*iv*) Silver oxide
(*v*) Sodium hypobromate
(*vi*) Formic acid

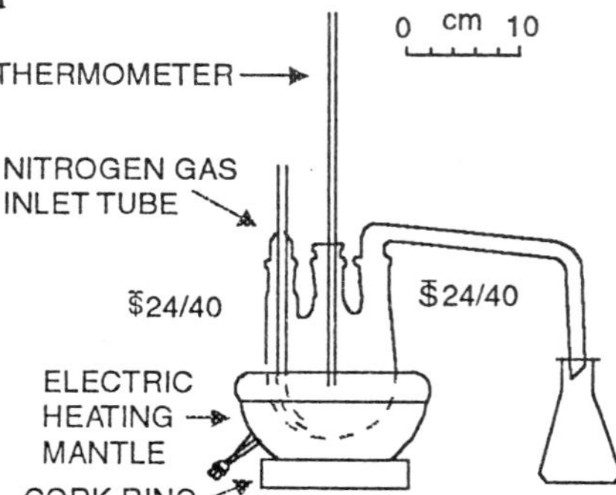

Fig. 2.21.2 Apparatus for preparation of the reducing mixture.

2.21.1.3 Procedure

1. Oxidation of total sulphur in soil sample to sulphate.

Acid digestion. Transfer 2.0 g of finely ground < 40 mesh soil sample into 125 ml Philip's beaker. Then add 3 ml of 69% HNO_3 and swirl the beaker to mix the contents. Cover the beaker with a watch glass and heat on water bath or steam plate for 1 hour. Remove the beaker from the water or steam plate, uncover and add 3 ml of 60% $HClO_4$, and 7 ml of conc. H_3PO_4. Thereafter heat the beaker on a sand bath or hot plate under fume hood. Regulate the temperature of the sand to about 190° to 210°C until heavy white fumes of $HClO_4$ are visible. Place the watch glass and continue heating for 30 min.

Remove the beaker from the sand bath, cool, uncover, and add 2 ml of 37% HCl. Heat again until white fumes of $HClO_4$ is again visible. Cool and transfer the digest quantitatively into a 100 ml volumetric flask using 1N HCl solution and adjust the volume to 100 ml mark with 1N HCl.

Alkaline oxidation

Transfer 0.1 to 0.2 g of air dry finely ground < 100 mesh soil containing 10 to 50 μg S into dry 50 ml digestion-distillation flask. Add 3 ml of NaOBr solution. Swirl the flask for a few seconds to mix the contents and allow it to stand for 5 minutes. Then, swirl the flask again for a few seconds, and place the flask upright in a sand bath under a fume hood. Regulate the temperature of sand to 250–260°C. Heat the flask until the contents are evaporated to dryness, and continue heating for an additional 30 minutes.

Remove the flask from the sand bath, allow it to cool for about 5 minutes and add 1 ml water and heat the flask on the sand bath for few seconds to bring the residue into suspension. Remove the flask from the sand bath, swirl it for a few seconds, and allow it to cool to room temperature.

Dry Ashing

Transfer 0.1 to 0.2 g air dry finely ground < 100 mesh soil sample containing 10 to 50 μg S in a small porcelain crucible (1.5 cm in dia. by 1.5 cm in height). Mix it thoroughly with 0.25 g of $NaHCO_3$, and add 0.01 g of Ag_2O. Place the crucible in cold electric muffle furnace. Raise the temperature to 550°C and heat for 3 hours. Cool and transfer the mixture quantitatively to a 50 ml digestion flask.

2. Preparation of the soil digest for further treatment. If the soil sample is digested by the acid digestion procedure, transfer 1 to 5 ml aliquot of the soil digest into a 50 ml digestion-distillation flask. With aliquot greater than 2 ml, reduce the volume to about 2 ml by heating the flask on a sand bath.

If the alkaline oxidation procedure is used, add 1 ml of formic acid to the flask containing the oxidized sample, swirl the flask for a few seconds to mix the contents and allow it to stand for about 10 minutes.

If the dry ashing procedure is used for the oxidation of S, add 2 ml of distilled or deionized water, and swirl the flasks to mix the contents.

3. Reduction of SO_4–S and distillation of the H_2S formed for absorption into zinc acetate-sodium acetate buffer solution. Add 4 ml of reduction mixture into the 50 ml digestion-distillation flask containing soil digest. Immediately attach the reduction-distillation flask to the condenser assembly, and follow steps 2 and 3 of procedure described in total S determination in plant sample by the reduction method (Section 3..9.3.3).

4. Development of methylene blue colour, measurement of its absorbance (optical density), preparation of standard calibration curve, and estimation of SO_4–S in the soil digest. Follow the steps 4 to 9 of procedure described in total S determination in plant sample by the reduction method (Section 3.9.3.3).

2.21.1.4 Calculation

$$\text{Total S in soil (ppm)} = \frac{A}{B}$$

Where A is μg SO_4–S in soil digest as estimated by reference to the standard calibration curve. In case of acid digest of soil, μg SO_4–S in x ml aliquot of acid digest be multiplied by $\frac{100}{x}$ to obtain μg SO_4–S in 100 ml acid digest prepared from 2g soil sample.

B is the mass of soil sample taken for the preparation of soil digest.

2.21.1.5 Comments

The total S in Indian soils has been summarized by Takkar (1988). The mean concentration of total S in soils of Punjab and U.P. is reported to be about 140 ppm when the soils of Gujrat and West Bengal contained only 74 ppm of total S. The reddish brown soils of Maharastra contained 2.5 times more sulphur than the alluvial soil of U.P. whereas the soils of Tamilnadu recorded 9 to 10 times more sulphur than the alluvial soils of U.P. There is regional differences in respect of Total S. There exist positive correlation between organic S and total S as well as in between organic carbon and total S.

REFERENCES

Johnson, C.M. and Hideo Nishita (1952). Micro estimation of Sulphur in plant materials, soils and irrigation water. Anal. Chem. 24 : 736–742.

Takkar, P.N. (1988). Sulphur states of Indian soils pp S 1/2–1 to 31. In TSI–FAI Sulphur in Indian Agriculture, Sulphur Institute, Washington D.C. and FAI, New Delhi.

2.21.2 Reserve Sulphur–Bardsley and Lancaster (1960)

2.21.2.1 Principle

Organic and reduced inorganic sulphur which form reserve sulphur in soils is converted to sulphates by ignition of a soil-sodium bicarbonate mixture at 500°C. Converted and soluble sulphates forming soil sulphur are then extracted with a solution that is 2 N to acetic acid and 0.1 N to sodium dihydrogen phosphate and determined turbidimetrically as $BaSO_4$ after acidifying a portion of the filtrate with 1 : 1 HCl.

Soluble sulphates are extracted from a separate sample with 0.5 N ammonium acetate + 0.25 N acetic acid. Activated charcoal is added to decolorize the extract. A portion of the filtrate is acidified with 1 : 1 HCl containing a "seed" solution of sulphate and sulphur is determined turbidimetrically as $BaSO_4$.

Reserve sulphur is the difference between soil sulphur and soluble sulphates

2.21.2.2 Apparatus/Reagents

Apparatus

High form porcelain crucible, electric muffle furnace, Erlenmeyer flasks, colorimeter or spectrophotometer.

Reagents

(*i*) 1 : 1 Hydrochloric acid solution. Dilute 125 ml conc. HCl (AR) to 250 ml with distilled water

(*ii*) 0.1 N sodium dihydrogen phosphate in 2 N acetic acid. Dissolve 4.6 g of $NaH_2PO_4.H_2O$ in 1l of 2N CH_3COOH.

(*iii*) Sodium bicarbonate (AR) finely divided with no lumps.

(*iv*) Barium chloride (AR) crystals. Ground to pass a 20-mesh screen but retained on a 60 mesh screen

(*v*) 5% Acacia solution in 50% glacial acetic acid. Dissolve 0.5 g acacia in 50 ml of distilled water, and dilute the solution to 100 ml with glacial acetic acid.

(*vi*) 100 ppm Sulphur solution. Dissolve 0.7703 g of $MgSO_4.7H_2O$ in 1 l of extracting solution (0.1 N $NaH_2PO_4.H_2O$ in 2N CH_3COOH)

(*vii*) 0-25 ppm SO_4–S solution. Transfer 0, 5, 10, 15, 20 and 25 ml of 100 ppm SO_4–S solution into separate 100 ml volumetric flasks and dilute the solution to 100 ml mark with the extracting solution to obtain 0, 5, 10, 15, 20 and 25 ppm SO_4–S solution.

2.21.2.3 Procedure

1. Dry ashing of the soil sample. Transfer 2.5 g of 20-mesh soil into a high form porcelain crucible (Coors OA, 18 ml). Then add 0.5 g of $NaHCO_3$ and mix it with the soil thoroughly. Thereafter, add an additional 0.5 g of $NaHCO_3$uniformly as a surface layer. Ignite at 500°C in an electric muffle furnace for 3 hours. Allow the sample to cool.

2. Preparation of the test solution after dry ashing. Transfer the contents of the crucible after dry ashing into 50ml Erlenmeyer flask. Add 25 ml of the extracting solution slowly to avoid loss by spattering. Shake the contents of the flask for 30 minutes and filter the suspension through a funnel lined with Whatman No. 1 filter paper. Similarly prepare a blank test solution.

3. Development of $BaSO_4$ turbidity in aliquot of sample and blank test solution. Transfer 10 ml of the sample and blank test solution to separate 50 ml Erlenmeyer

flask with a pipette. Add 1 ml of acacia solution, and 1 ml of 1 : 1 HCl to each flask. Swirl the contents of the flask. Then add 0.5 g of $BaCl_2$ crystals to each flask and allow the flask to stand for 1 minute, then swirl the contents frequently until the crystals are dissolved.

4. Measurement of the absorbance of the $BaSO_4$ turbidity using a photoelectric colorimeter or a spectrophotometer. Place cuvette or colorimeter test tube filled with distilled water in the socket provided and fit the appropriate filter or adjust the wavelength to 420 mμ. Then switch on the instrument, and let it run for 10 minutes before measurement of the absorbance of turbidity developed with the aliquot of blank and sample test solution. Set 0 absorbance (optical density) with the turbidity developed with aliquot of blank test solution, and then read the absorption of turbidity developed with the aliquot of sample test solution. The absorbance be read within 2 to 12 minutes of developing the turbidity.

5. Development of $BaSO_4$ turbidity with aliquots of sulphur standards containing 0 to 25 ppm of SO_4–S, and measurement of absorbance of the turbidity. Transfer 25.0 ml of each sulphur standards into separate 50 ml Erlenmeyer flask, add 1.0 g $Na_2CO_3$to each flask. When the reaction has subsided, pipette 10.0 ml of each concentration into 50 ml of separate Erlenmeyer flask. Then add 1 ml of acacia solution and 1 ml of 1 : 1 HCl. Thereafter, add 0.5 g of $BaCl_2$ crystals, and allow it to stand for 1 minute. Swirl the flask frequently to completely dissolve the crystals.

Within 2 to 12 minutues of developing $BaSO_4$ turbidity, measure the absorbance of turbidity developed with aliquot of each SO_4–S standards adjusting 0 absorbance with the turbidity developed with 0 ppm SO_4–S solution. Prior to measurement of the absorbance of standards, switch on the instrument for 10 minutes to warm. Record the value of absorbance corresponding to the μg SO_4–S taken for development of turbidity.

6. Preparation of standard calibration curve. On a linear scale graph paper, plot the value of absorbance corresponding to the μg of SO_4–S taken for developing turbidity by dissolving $BaCl_2$crystals. Let ordinate be the scale for absorbance and abscissa be the scale for μg SO_4–S taken for developing the turbidity. Connect the point of 0 absorbance with the point of maximum absorbance by a straight line to obtain the standard calibration curve.

7. Estimation of the content of μg SO_4–S in the turbid solution prepared with aliquot of sample test solution. Locate the value of absorbance recorded for the sample test solution on the ordinate of the standard calibration curve and draw an imaginary line parallel to the abscissa intersecting the straight line calibration curve. From the point of intersection, draw a line perpendicular to the abscissa. The point at which the perpendicular line intersect the abscissa, read the μg SO_4–S in the turbid solution.

2.21.2.4 Calculation

$$\text{Soil sulphur (ppm)} = \frac{A \times B}{C \times D}$$

Where A is μg SO_4–S recorded in the turbid solution prepared with the aliquot of sample test solution by reference to the standard calibration curve.

B is total volume of the sample test solution prepared.

C is ml aliquot of sample test solution taken for developing turbidity.

D is mass of soil sample taken for dry ashing by igniting sample after treatment with $NaHCO_3$ in a electric furnace.

REFERENCES

Bardsley, C.E. and J.D. Lancaster. 1960. Determination of reserve sulphur and soluble sulphates in soils. Soil Sci. Soc. Am. Proc. 24 : 265–268.

2.21.3 Soluble sulphates in soil by turbidimetric method–Bardsley and Lancaster (1960)

2.21.3.1 Principle

Sulphates in soil is extracted by 0.5 N ammonium acetate in 0.025 N acetic acid solution employing a 1 : 2.5 soil extractant (w/v) ratio in the presence of decolorizing agent. The soil extract is treated with acid seed solution and crystals of barium chloride to develop $BaSO_4$ turbidity for the measurement of absorbance by a colorimeter or spectrophotometer at 420 mμ wave length.

2.21.3.2 Apparatus/Reagents

Apparatus

Erlenmeyer flasks, mechanical shaker, funnels, volumetric flasks, pipettes, photoelectric colorimeter or a spectrophotometer.

Reagents

(*i*) Distilled water. Use deionized water or water from a completely glass distill.

(*ii*) 0.5 N ammonium acetate in 0.025 N acetic acid. Dissolve 38.5 g of $NH_4C_2H_3O_2$ (anhydrous) in 1 l of 0.025 N CH_3COOH.

(*iii*) Darco–G-60 : (activated charcoal free from sulphur)

(*iv*) Acid "seed" solution. Transfer 50 ml of conc. HCl (AR) into a 100 ml volumetric flask. Then add 20 ml of 100 ppm SO_4–S solution, and dilute the solution to 100 ml with distilled water to obtain 1 : 1 HCl containing 20 ppm of SO_4–S.

(*v*) Barium Chloride crystals. Grind $BaCl_2$ crystals (AR) to pass a 20 mesh. sieve, but which is retained by a 60 mesh sieve.

(*vi*) 100 ppm SO_4–S solution. Dissolve 0.7703 g of $MgSO_4 \cdot 7\,H_2O$ in 1 l of $NH_4C_2H_3O_2$ in 0.025 N CH_3 COOH solution.

(*vii*) 0–25 ppm SO_4–S solution. Transfer 0, 5, 10, 15, 20, and 25 ml of 100 ppm SO_4–S solution into separate 100 ml. volumetric flasks and dilute the solution to 100 ml with the extracting solution to obtain 0, 5, 10, 15, 20 and 25 ppm SO_4–S solution.

2.21.3.3 Procedure

1. Extraction of soluble SO_4–S from the soil. Transfer 10.0 g of air dry soil which has been passed through a 20 mesh sieve into a 100 ml Erlenmeyer flask. Add

25.0 ml of the extracting solution, and shake the contents for 30 minutes on a mechanical shaker. Then add 0.25 g of activated charcoal (Darco–G–60) and resume shaking for 5 minutes. Filter the suspension through a funnel lined with sulphur free Whatman No. 42 filter paper. Similarly run a blank.

2. Development of $BaSO_4$ turbidity in aliquot of sample and blank extract and measurement of the absorbance. Pipette 10.0 ml of soil and blank extract into separate 50 ml Erlenmeyer flasks. Add 1 ml of acid "seed" solution to each. Swirl the contents and add 0.50 g of $BaCl_2$ crystals into each flask. Let the mixture stand for 1 minute. Then swirl the flasks frequently until the crystals are dissolved.

Measure the absorbance of the $BaSO_4$ turbidity at 420 mμ wave length by a photoelectric colorimeter or spectrophotometer within the time interval of 2–8 minutes of developing turbidity. Prior to the measurement of the absorbance, place the cuvette or colorimeter tube filled with distilled water in the socket provided and set the desired wave length or filter. Then switch on the instrument and allow it to warm for 10 minutes. Adjust 0 absorbance with the $BaSO_4$turbidity developed with aliquot of blank extract, and read the absorbance of turbidity developed with aliquot of the sample extract.

3. Development of the $BaSO_4$ turbidity with standard SO_4–S solution and measurement of the absorbance (optical density). Transfer 25.0 ml of 0 to 25 ppm SO_4–S solution into separate 50 ml of Erlenmeyer flasks. Then add 0.25 g of activated charcoal (Darco–G–60) to each Erlenmeyer flask. Shake for 3 minutes and filter through Whatman No. 42 filter paper, and collect the filtrate.

Pipette 10.0 ml of each concentration of the standard SO_4–S solution into 50 ml Erlenmeyer flask. Then add 1 ml of acid "seed" solution. Mix the contents and add 0.5 g $BaCl_2$ crystals. Allow it to stand for one minute. Then swirl the contents of the flasks frequently until the crystals are dissolved.

Measure the absorbance of the $BaSO_4$ turbidity at 420 mμ wave length by a photoelectric colorimeter or spectrophotometer within the time interval of 2–8 minutes of developing turbidity. Prior to the measurement of the absorbance, place the cuvette or colorimeter tube filled with distilled water in the socket provided and set the desired wave length or filter. Then switch on the instrument and allow it to warm for 10 minutes. Adjust 0 absorbance with $BaSO_4$turbidity developed with 0 μg SO_4–S and read the absorbance of $BaSO_4$ turbidity developed with known amount of SO_4–S.

4. Preparation of standard calibration curve. On a linear scale graph paper, plot the value of absorbance corresponding to the μg of SO_4–S taken for developing $BaSO_4$ turbidity. Let ordinate be the scale for absorbance and abscissa be the scale for μg SO_4–S taken for developing the turbidity. Connect the point of 0 abrsorbance with the point of maximum absorbance by a straight line to obtain the standard calibration curve.

5. Estimation of μg SO_4–S in aliquot of the sample test solution. Locate the value of absorbance recorded for the sample test solution on the ordinate of the standard calibration curve, and draw an imaginary line parallel to the abscissa

intersecting the straight line calibration curve. From the point of intersection, draw a line perpendicular to the abscissa. The point at which the perpendicular line intersect the abscissa, read the μg SO_4–S in the turbid solution.

2.21.3.4 Calculation

$$SO_4\text{–S in soil (ppm)} = \frac{A \times B}{C \times D}$$

Where A is μg SO_4–S recorded in the turbid solution prepared with the aliquot of soil extract by reference to the standard calibration curve.

B is the total volume of 0.5 N NH_4 $C_2H_3O_2$ in 0.025 N CH_3COOH used for extracting soluble SO_4–S from the soil sample.

C is ml aliquot of soil extract taken for developing $BaSO_4$ turbidity

D is mass of air dry 20 mesh soil sample taken for the extraction of soluble SO_4–S.

2.21.3.5 Comments

The filter paper used for filtration be free from sulphur. If the filter paper is contaminated, the same volume of sample and blank extracts be filtered through filter paper of the same weight.

If SO_4–S concentration in soil extract or sample test solution is >20 ppm, use 2 ml of 0.5% acacia solution as stabilizer and accordingly the development of the turbidity with the standard SO_4–S solution be also modified.

For soil samples containing fairly large amount of sulphates as expected in sub-soils, a wider soil/extractant ratio may be required to extract the sulphates effectively.

The determination of SO_4–S in soil extract after ignition of the sample with $NaHCO_3$indicated a recovery of 103 per cent as against AOAC method (1955).

In surface soils of humid region, sulfur occur in organic combination. Consequently organic matter serve as reserve of both N and S. The N/S ratio in organic matter remains nearly constant indicating the mineralization of N and S in soil occur in proportion to the amount present in organic fraction. It is, therefore, desired to assess organic S in evaluating S status of soil.

Organic S compounds are easily volatalized in wet oxidation procedure. However organic S is oxidized and retained in alkaline system containing sodium carbonate, sodium oxides or mixtures of the two.

The determination of S in soil extracts after oxidation is done successfully by the turbidimetric method. However, the determination of sulphates by the turbidimetric method following the procedure of Chesnin and Yien (1950) generally give low results for surface soils. The difficulty occur in coloured extracts which can be overcome by carrying a duplicate sample with each determination to which no $BaCl_2$ is added in an effort to compensate for colour extracts.

SO_4–S added to the soil sample and extracted by the method was completely recovered by the method employed.

There existed a highly significant positive correlation between total N in soil and reserved sulphur, and the SO_4–S represented about 10% of the soil sulphur. The yield of sulphur by growing white clover on surface soils gave significant curvilinear relationship with the reserve sulphur. However, the yield of sulphur did not show significant positive correlation with the SO_4–S. The content of SO_4–S in soils was found to be in order of sandy loam < loam < clay loam < clay. It is proposed that reserve sulphur be also determined apart from SO_4–S in evaluating sulphur states of soil.

REFERENCES

Bardsley, C.E. and J.D. Lancaster1960. Determination of reserve sulphur and soluble sulphates in soils. Soil Sci. Soc. Amer. Proc. 24 : 265–268.

Chesnin, L. and C.H. Yien. 1950. Turbidimetric determination of available sulphate Soil. Sci. Amer. Proc. 15 : 149–151.

2.21.4 Soluble and adsorbed sulphate S in soil as extracted by 0.01 M Ca $(H_2PO_4)_2$–Chesnin and Yien (1950), and Johnson and Nishita (1952)

2.21.4.1 Principle

The soluble and adsorbed sulphate S from soil is extracted by 0.01M $Ca(H_2PO_4)_2$ employing soil extractant ratio of 1 : 10 (w/v) with 30 minutes shaking on a mechanical shaker. The suspension is filtered through Whatman No. 42 filter paper.

The SO_4–S in the extract is determined by either the turbidimetric or by the reduction method.

2.21.4.2 Apparatus/Reagents. (See section 2.21.5.1.2 and 3.9.3.2).

2.21.4.3 Procedure

1. Turbidimetric determination of soluble and adsorbed SO_4–S in the 0.01M $Ca(H_2PO_4)_2$ soil extract – Chesnin and Yien (1950). Transfer 5g of air dry < 2mm soil sample into a 250 ml Erlenmeyer flask. Add 50 ml of 0.01M $Ca(H_2PO_4)_2$ solution. Fit a rubber stopper wrapped in thin polythene into mouth of the Erlenmeyer flask. Shake the contents on a mechanical shaker for 30 minutes. Filter the suspension through Whatman No. 42 filter paper. Similarly run a blank.

Transfer 10 ml of the sample and blank extract into separate 25 ml volumetric flask. Thereafter follow steps 2 to 7 of procedure given in section 2.21.5.1.3 for determination of soluble SO_4–S in soil by extraction with 0.15% $CaCl_2$. Estimate SO_4–S in soil by the equation given in section 2.21.5.1.4.

2. Reduction method of determination of soluble and adsorbed SO_4–S in soil–Johnson and Nishita (1952). Transfer 5 g of air dry < 2 mm soil sample into a 250 ml Erlenmeyer flask. Add 50 ml of 0.01M $Ca(H_2PO_4)_2$ solution. Fit a rubber stopper wrapped in thin polythene into mouth of the Erlenmeyer flask. Shake the contents on a mechanical shaker for 30 minutes. Filter the suspension through Whatman No. 42 filter paper. Similarly run a blank.

Transfer 2–20 ml of soil extract into 50 ml of digestion distillation flask. Heat the flask on a hot plate until the extract is reduced to a volume of about 2 ml.

Thereafter, follow the reduction method of S determination developed by Johnson and Nishita (1952) as described in sectiion 3.9.3 following steps 2&3–9 beginning with the addition of 4 ml of reduction mixture into 50 ml digestion distillation flask.

2.21.4.4 Calculation

Estimate SO_4–S in soil by equation given in section 2.21.5.2.4

2.21.4.5 Comments

The reagent 0.01M $Ca(H_2PO_4)_2$ contains sufficient phosphates to displace most of the adsorbed sulphates. Thus, it extracts soluble and adsorbed sulphates from the soil. Besides, the soil extract is clear. However, adsorption of sulphates by soil is negligible above pH 6.5 and increases with decreasing pH below 6.5.

Glass wares must be carefully cleaned to remove any potential nuclei for the formation of irregular sized barium sulphate crystals and the cuvetts or colorimeter test tubes must be cleaned regularly with the alkaline EDTA to remove adhering barium sulphate and maintain sensitivity.

REFERENCES

Chesnin. L and C.H. Yien (1950). Turbidimetric determination of available sulphates. Soil Sci. Soc. Amer. Proc. 15 : 149–151.

Fox. R.L., R.A. Olson, and H.F. Rhoades (1964). Evaluating the sulfur status of soils by plant and soil tests. Soil Sci. Soc. Amer. Proc. 28 : 243–246.

Johnson, C.M. and Hideo Nishita (1952). Micro estimation of sulfur in plant materials, soils and irrigation water. Anal. Chem. 24 : 736–742.

Williams. C.H. and A. Steinberg (1959). Soil sulphur fractions as chemical indices of available sulphur in some Australian soils. Australian J. Agric. Res. 10 : 342–352.

2.21.5 Calcium chloride extractable sulphate sulphur determined by turbidimetric and reduction method

2.21.5.1 Method 1. Turbidimetric method–Chesnin and Yien (1950)

2.21.5.1.1 Principle

The soil is extracted with 0.15% $CaCl_2$ using a soil extractant ratio of 1 : 5. The SO_4–S in soil extract is determined colorimetrically by developing $BaSO_4$ turbidity in the presence of sodium acetate–acetic acid buffer.

2.21.5.1.2 Apparatus/Reagents

Apparatus

250 ml Erlenmeyer flasks, 25 ml and 50 ml volumetric flasks, pipettes, rubber stopper wrapped with polythene sheet, photoeletric colorimeter or spectrophotometer

Reagents

(*i*) 0.15% Calcium chloride. Dissolve 1.5 g $CaCl_2$. $2H_2O$ (AR) in distilled water, and dilute it to 1 l with distilled water.

(*ii*) 0.25% gum acacia solution. Dissolve 0.25 g gum acacia in 100 ml distilled water.

(*iii*) Sodium acetate-acetic acid buffer (pH 4.8). Dissolve 100g of CH_3COONa (AR) in about 500 ml of distilled water. Add 30 ml of 99.5% CH_3COOH (AR) and dilute the solution to 1 l with distilled water.

(*iv*) Barium Chloride (AR) crystals. Collect crystals finer than 30 mesh sieve but larger than 60 mesh sieve.

(*v*) 100 ppm SO_4–S solution. Dissolve 0.5434 g of K_2SO_4 (AR) in 0.15% $CaCl_2$ solution and dilute the solution to 1 l with distilled water.

(*vi*) 0–20 ppm SO_4–S solution. Transfer 0, 2.5, 5.0, 7.5 and 10.0 ml of 100 ppm SO_4–S solution into separate 50 ml volumetric flasks and dilute the solution to 50 ml mark with 0.15% $CaCl_2$ solution to obtain 0, 5, 10, 15 and 20 ppm SO_4–S solution.

2.21.5.1.3 Procedure

1. Extraction of SO_4–S from soil by 0.15% $CaCl_2$ solution. Transfer 10g of air dry 2 mm soil sample into a 250 ml Erlenmeyer flask. Then add 50 ml of 0.15% $CaCl_2$ solution. Place rubber stopper wrapped with thin polythene and shake the contents on a mechanical shaker for 30 minutes. Remove the Erlenmeyer flask from the shaker and filter the soil-$CaCl_2$ suspension through funnel lined with Whatman No. 42 paper. Discard few ml of initial filtrate, and collect the subsequent filtrate in a clean vessel.

2. Development of $BaSO_4$ turbidity in the sample and blank extract. Transfer 10 ml of sample and blank extract into separate 25 ml volumetric flasks. Add 10 ml sodium acetate–acetic acid buffer and 1 g $BaCl_2$ crystals (60 to 30 mesh fraction) into each vol. flask and mix the contents for 1 minute. Thereafter, if the SO_4–S concentration in the soil extract is less than 20 ppm, add 1 ml of 0.25% gum acacia, and dilute the solution to 25 ml with distilled water. Shake the contents until $BaCl_2$ crystals are dissolved. If the concentration of SO_4–S in the soil extract is in between 20 to 40 ppm, use 2 ml of 0.25% gum acacia solution instead of 1 ml for stabilizing the turbidity or else dilute the soil extract two times with 0.15% $CaCl_2$ solution and then follow the entire procedure for the development of $BaSO_4$ turbidity.

3. Measurement of the absorbance of $BaSO_4$ turbidity developed with sample and blank extract. Within the time interval of 2–8 minutes after $BaCl_2$ crystals are dissolved, measure the absorbance (optical density) of the $BaSO_4$ turbidity developed with aliquot of sample extract on a photoelectric colorimeter with 420 mμ filter or else on a spectrophotometer at 420 mμ wavelength adjusting 0 absorbance with the $BaSO_4$ turbidity developed with the aliquot of blank extract. Prior to taking the absorbance, the instrument should be switched on after placing cuvette or colorimeter tube filled with distilled water to warm for 10 minutes.

4. Development of $BaSO_4$ turbidity with standard SO_4–S solution. Transfer 10.0 ml of 0, 5, 10, 15 and 20 ppm SO_4–S solution into separate 25 ml volumetric flasks. Then add 10 ml of sodium acetate–acetic acid buffer solution and 1g of $BaCl_2$ crystals to each flask. Mix the contents for 1 minute. Then add 1 ml of 0.25% gum acacia. Dilute the solution to 25 ml with distilled water. Mix the contents well until the crystals are dissolved. If it is desired to prepare calibration curve with standards containing 20-40 ppm SO_4–S, use 2.0 ml of 0.25% gum acacia for stabilizing the turbidity.

5. Measurement of the absorbance of $BaSO_4$ turbidity with standard SO_4–S solution. Within the time interval of 2–8 minutes after $BaCl_2$ crystals are dissolved, measure the absorbance (optical density) of the $BaSO_4$ turbidity developed with aliquots of SO_4–S standards–on a photoelectric colorimeter fitted with 420 mμ filter or else on a spectrophotometer at 420 mμ wave length adjusting 0 absorbance with the $BaSO_4$turbidity developed with 0 μg SO_4–S. Prior to taking the absorbance, the instrument should be switched on after placing cuvette or colorimeter tube filled with distilled water to warm for 10 minutes. Record the value of absorbance corresponding to aliquot of each concentration of the SO_4–S standards.

6. Preparation of standard calibration curve. On a linear scale graph paper plot the value of absorbance corresponding to the μg SO_4–S taken for developing $BaSO_4$ turbidity. Let ordinate be the scale for absorbance and abscissa be the scale for μg SO_4–S taken for developing the turbidity. Connect the point of 0 absorbance with the point of maximum absorbance by a straight line to obtain the standard calibration curve.

7. Estimation of the μg SO_4–S in aliquot of soil sample extract. Locate the value of absorbance recorded for the aliquot of soil extract on the ordinate of the standard calibration curve and draw an imaginary line parallel to the abscissa intersecting the straight line calibration curve. From the point of intersection, draw a line perpendicular to the abscissa. The point at which the perpendicular line intersect the abscissa, read μg SO_4–S in the turbid solution.

2.21.5.1.4 Calculations

$$SO_4\text{–S in soil} = \frac{A \times B}{C \times D} = \frac{A \times 50}{10 \times 10}$$

Where A is the μg SO_4–S recorded in the turbid solution prepared with aliquot of soil extract by reference to the standard calibration curve.

B is the total volume of 0.15% $CaCl_2$ used for extracting SO_4–S from the soil sample.

C is ml aliquot of the soil extract taken for developing $BaSO_4$ turbidity.

D is mass of air dry 2 mm soil sample taken for the extraction of soluble SO_4–S.

2.21.5.2 (Method 2). Reduction method developed by Johnson and Nishita (1952)

2.21.5.2.1 Principle

SO_4–S from soil is extracted by 0.15% $CaCl_2$ using soil extractant ratio of 1 : 10. Then 2–20 ml extract containing 10–20 μg SO_4–S is transferred to a reduction-distillation flask. The content is evaporated to dryness in an oven if the volume is more than 2 ml and then the residue is dissolved in 2 ml of distilled water.

The solution in the flask is treated with a reduction mixture and the flask is attached to a distillation apparatus. The purified nitrogen gas is passed through the reduction mixture, and the digestion flask is heated by a micro burner to distill H_2S liberated. The H_2S distilled is made to pass through a gas washing column before its absorption by zinc acetate-sodium acetate buffer solution. The H_2S absorbed is treated with acid solution of p–aminodimethylaniline sulphate and ferric ammonium sulphate solution to form methylene blue colour. The optical density (absorbance) of the methylene blue colour is measured at 670 mμ wave length using a spectrophotometer or colorimeter.

2.21.5.2.2 Apparatus/Reagents

Apparatus

Digestion distillation apparatus (Fig. 2.21.1), Apparatus for preparation of reducing mixture (Fig. 2.21.2), oven, fume hood, microburner, mechanical shaker, spectrophotometer or photoelectric colorimeter, cylinder of water pumped N_2, conical flasks, tall beakers, volumetric flasks, pipettes and thermometers.

Reagents

(*i*) 0.15% Calcium chloride dihydrate solution. Dissolve 1.5 g $CaCl_2$. $2H_2O$ (AR) in about 700 ml of distilled water and make the volume to 1 l with distilled water.

(*ii*) Reagents 3 to 12 as described in the determination of total S in plant by the reduction method (See section 3.9.3.2).

2.21.5.2.3 Procedure

1. Extraction of soluble SO_4–S by 0.15% $CaCl_2$ solution. Transfer 5g of air dry < 2 mm soil sample into a 250 ml Erlenmeyer flask. Add 50 ml of 0.15% $CaCl_2$ solution. Fit a rubber stopper wrapped in thin polythene into the flask mouth and shake the Erlenmeyer flask on a mechanical shaker for 30 minutes. Filter the suspension through a funnel lined with Whatman No. 42. filter paper into a clean Erlenmeyer flask. Discard the initial filtrate (2–3 ml) and collect the subsequent filtrate.

2. Treatment of the aliquot of soil extract with the reduction mixture and distillation of the H_2S formed for absorption into zinc acetate-sodium acetate buffer solution. Transfer 2 to 20 ml of soil extract containing 10 to 50 μg SO_4–S into 50 ml capacity reduction-distillation flask. If the soil extract aliquot taken is more than 2 ml, evaporate the contents to dryness in an oven at 100°C, cool the flask and add 2 ml distilled water and dissolve the residue. Then add 4 ml of reducing mixture. Immediately attach the reduction- distillation flask to the condenser assembly and follow steps 2 and 3 of procedure described in total S determination in plant sample by the reduction method. (See section 3.9.3.3).

3. Development of methylene blue colour, measurement of its absorbance (optical density), preparation of standard calibration curve, and estimation of

SO_4–S in the aliquot of soil extract. Follow the steps 4 to 9 of procedure described in total S determination in plant sample by the reduction method. (See section 3.9.3.3).

2.21.5.4 Calculations

$$\text{Soluble } SO_4\text{–S in soil sample (ppm)} = \frac{A \times B}{C \times D} = \frac{A \times 50}{2 - 20^* \times 5}$$

*depending on volume of aliquot

Where

A is μg SO_4–S in aliquot of soil extract as estimated by reference to the standard calibration curve for SO_4–S by the methylene blue method.

B is the total volume of 0.15% $CaCl_2$ used for the extraction of SO_4–S from the soil.

C is ml aliquot of soil extract taken for reduction and subsequent distillation of H_2S for the development of methylene blue colour.

D is the mass of air dry < 2 mm soil sample taken in gram for extraction of SO_4–S.

2.21.5.2.4 Calculations

The 0.15% $CaCl_2$ solution is a satisfactory extractant for soluble sulphate sulphur in soils. The presence of Ca in the extractant flocculate the soil particles during the extraction period giving rise to a clear filtrate. The extract is not coloured and it does not pose any interference in the determination of sulphate by the turbidity method. Besides, 0.15% $CaCl_3$extractable SO_4–S gives a good measure of plant available sulphur in coarse textured neutral to alkaline soil.

The critical level of 0.15% $CaCl_2$ extractable SO_4–S in alluvial brown soils of Punjab has been reported by Takkar (1988) to be 9 ppm for Mustard (Brsassica juncea L.), 10 ppm for groundnut, 15 ppm for wheat and 19 ppm for sunflower. The sulphur status of soil in ppm as extracted by 0.15% $CaCl_2$ is interpreted as follows :

Highly deficient	Deficient	Marginal	High	Very High
0–5	5–10	10–15	15–20	20–25 ppm

There is a gradual oxidation of sulphur compounds in the rubber stopper and so the rubber stopper be wrapped in thin polythene before placing it on mouth of Erlenmeyer flasks or centrifuge tube.

In turbidimetric determination of SO_4–S, the glasswares must be carefully cleaned to remove any potential nuclei for the formation of irregular sized barium sulphate crystals, and the measurement cell i.e. cuvette or colorimeter test tubes must be cleaned regularly with alkaline EDTA to remove adhering barium sulphate and maintain sensitivity.

A small amount of sulphate is generally added to the aliquot of the sample test solution before measurement of the absorbance of $BaSO_4$turbidity to improve the sensitivity of the method, and linearity of the standard calibration curve.

Apart from gum arabic, one can also use gelatin, Tween 80, polyvinyl alcohol and sorbitol as a stabilizer of $BaSO_4$ turbidity.

REFERENCES

Chesnin, L. and C.H. Yien (1950). Turbidimetric determination of available sulphates. Soil. Sci. Soc. Amer. Proc. 15 : 149–151.

Johnson C.M. and Hideo Nishita (1952). Micro estimation of sulfur in plant materials, soils and irrigation water. Anal. Chem. 24 : 736–742.

Randall, P.J. (1988). Evaluation of the sulphur status of soils and plants-techniques and interpretation. S1/3–1 to 15 In. TSI–FAI. Sulphur in Indian Agriculture. Sulphur Institute, Washington D.C. and FAI New Delhi.

Takkar, P.N. (1988). Sulphur status of Indian soils. pp S1/2 – 1 to 31. In TSI–FAI Sulphur in Indian Agriculture, Sulphur Institute, Washington, D.C. and FAI, New Delhi.

William, C.M. and A. Steinbergs, (1959). Soil sulphur fractions as chemical indices of available sulphur in some Australian soils. Australian J. Agric. Res. 10 : 340–352.

2.21.6 Sodium acetate-acetic acid (Morgan's solution) extractable SO_4–S in soil by turbidimetric method of Chesnin and Yien (1950)

2.21.6.1 Principle

The SO_4–S from the soil is extracted by CH_3COONa–CH_3COOH (pH 4.8) using soil extractant ratio of 1:5 (w/v) with 30 minutes shaking on a mechanical shaker. An aliquot of the extract is reacted with $BaCl_2$crystals to form $BaSO_4$ turbidity. Gum acacia solution is added to stabilize the turbidity. The absorbance of the $BaSO_4$ turbidity is measured at 420 mμ wave length on a spectrophotometer or else on a Klett photoelectric colorimeter with 420 mμ filter.

2.21.6.2 Apparatus/Reagents

Apparatus

Erlenmeyer flask 250 ml cap. with rubber stopper wrapped in thin polythene, Burette, funnels, volumetric flasks, pipettes, mechanical shaker, 0.25, 0.50 and 2.00 mm sieve, spectrophotometer or Klett photoelectric colorimeter with 420 mμ filter.

Reagents

(*i*) Sodium acetate–acetic acid buffer (pH 4.8). Dissolve 100 g of CH_3COONa (AR) in about 500 ml of distilled water. Then add 31 ml of 99.5% CH_3COOH (AR). Mix the content and add 450 ml of distilled water. Adjust the pH of solution to 4.8 and make the volume to 1 l with distilled water.

(*ii*) Barium chloride crystals (AR). Grind $BaCl_2$ crystals to pass through 0.50 mm sieve but which is retained by 0.25 mm sieve.

(*iii*) 0.25% Gum acacia solution. Transfer 0.25 g of finely ground gum acacia into a 100 ml vol. flask. Add distilled water and shake to dissolve the contents. Dilute the solution to 100 ml mark and let it stand over night. Thereafter, filter the solution and collect the filtrate.

(*iv*) 100 ppm SO_4–S solution. Dissolve 0.5370 g of $CaSO_4.2H_2O$ in sodium acetate–acetic acid buffer, and dilute the solution to 1000 ml with the buffer solution.

(*v*) 1.0 ppm SO_4–S solution. Dilute 1.0 ml of 100 ppm SO_4–S solution to 100 ml with sodium acetate-acetic acid buffer.

2.21.6.3 Procedure

1. Extraction of SO_4–S from soil by sodium acetate-acetic acid (pH 4.0) solution. Transfer 20.0 g of air dry < 2 mm soil sample into 250 ml Erlenmeyer flask. Then, add 100 ml of sodium acetate-acetic acid solution. Stopper the flask with rubber stopper wrapped in thin polythene and shake the contents on a mechanical shaker for 30 minutes. Thereafter, filter the suspension through Whatman No. 42 filter paper. Discard about 4 ml of initial filtrate and collect the subsequent filtrate in a clean Erlenmeyer flask rinsed with the initial filtrate. Similarly prepare a blank filtrate.

2. Development of $BaSO_4$ turbidity with aliquot of sample and blank extract. Transfer 10 or 20 ml of soil extract containing SO_4–S extracted from the soil into 25 ml volumetric flask. Add 1 g of $BaCl_2$ crystal (0.25 mm to 0.50 mm size). Allow it to stand for 1 minute and then swirl the contents until the crystals are dissolved. Add 1 ml of 0.25% gum acacia solution if the concentration of SO_4–S is < 20 ppm. If the concentration of SO_4–S is between 20–40 ppm, add 2 ml of 0.25% gum acacia solution. Mix the contents and dilute the solution to 25 ml mark with distilled water. Similarly run a blank.

3. Measurement of absorbance of $BaSO_4$ turbidity developed with the sample and blank extract. Within 5 to 30 minutes of developing the $BaSO_4$ turbidity, measure the absorbance of $BaSO_4$ turbidity developed with sample extract aliquot on a Klett photoelectric colorimeter fitted with 420 mμ (blue) filter or else on a spectrophotometer at 420 mμ wave length adjusting 0 absorbance with $BaSO_4$ turbidity developed with the aliquot of the blank extract. Prior to measurremeent of the absorbance, place cuvette or colorimeter test tube filled with distilled water in the socket provided, and switch on the instrument to allow it to warm for 10 minutes.

4. Development of $BaSO_4$ turbidity with SO_4–S standards. Transfer 0, 2, 4, 6, 8 and 10 ml of 1.0 ppm SO_4–S solution into separate 25 ml volumetric flasks and then follow step 2 of the procedure beginning with the addition of sodium acetate-acetic acid buffer instead of soil extract.

5. Measurement of absorbance of $BaSO_4$ turbidity developed with SO_4–S standards. Within 5–30 minutes of development of the $BaSO_4$ turbidity adjust 0 absorbance with $BaSO_4$ turbidity developed with 0 μg SO_4–S at 420 mμ wavelength on a spectrophotometer or on a Klett photoelectric colorimeter fitted with 420 mμ (blue) filter.

Thereafter, measure the absorbance of $BaSO_4$ turbidity developed with 2 to 10 μg SO_4–S adjusting 0 absorbance each time with $BaSO_4$ turbidity developed with 0 μg SO_4–S. Prior to measurement of the absorbance, the instrument be switched on to warm for 10 minutes.

6. Preparation of standard calibration curve. Record the value of absorbance corresponding to the μg SO_4–S taken for development of $BaSO_4$ turbidity. On a linear scale graph paper, let the ordinate be the scale of absorbance, and the abscissa be the scale for μg SO_4–S taken for developing $BaSO_4$ turbidity. Plot the values of the absorbance recorded, and connect the point of 0 absorbance with the point of maximum absorbance by a straight line to obtain the standard calibration curve.

7. Estimation of SO_4–S in aliquot of the soil extract. Locate the value of absorbance of $BaSO_4$ turbidity developed with the aliquot of soil extract on the ordinate of the standard calibration curve. Draw a line from the point parallel to the abscissa intersecting the straight line calibration curve. From the abscissa, draw a perpendicular line that would meet the point of intersection. Read the content of SO_4–S on the abscissa at the origin of the perpendicular line.

2.21.6.4 Calculation

$$SO_4\text{–S in soil (ppm)} = \frac{A \times B}{C \times D} = \frac{A \times 100}{10 \text{ or } 20 \times 20}$$

Where

A is μg SO_4S recorded in the turbid solution prepared with the aliquot of soil extract by reference to the standard calibration curve.

B is the total volume of Morgan's solution used for the extraction of SO_4–S from the soil.

C is ml aliquot of soil extract taken for developing $BaSO_4$ turbidity *i.e.* 10 or 20 ml.

D is mass of air dry < 2 mm soil sample in gram taken for the extraction of SO_4–S by the Morgan's solution.

2.21.6.5 Comments

Glasswares must be carefully cleaned to remove any potential nuclei for the formation of irregular sized barium sulphate crystals, and the measuring cell must be cleaned regularly with alkaline EDTA to remove adhering barium sulphate, and maintain sensitivity.

A small amount of sulphate is generally added to the aliquot of the sample test solution before adding crystals of $BaCl_2$ in order to increase the absorbance of $BaSO_4$ turbidity so as to improve the sensitivity of the method and linearity of the standard calibration curve.

Apart from gum arabic, one can also use gelatin, Tween 80, polyvinyl alcohol and sorbitol as a stabilizer of $BaSO_4$ turbidity.

The critical level of SO_4–S as extracted by Morgan's solution in alluvial soils of U.P. for rice was found to be 12.5 ppm. On the other hand, in blank cotton soils of Madhya Pradesh, the critical level of SO_4–S as extracted by Morgan's solution for soyabean was found to be 8 ppm. The critical level of sulphur for an extractant varies with the soil and the crop. The available sulphur also decreases with the progressive increase in soil depth.

REFERENCES

Chesnin, L. and C.H. Yien (1950). Turbidimetric determination of available sulphates. Soil. Sci. Soc. Amer. Proc. 15 : 149–151.

Jackson, M.L. (1973). Soil Chemical Analysis, Prentice-Hall of India, Private Ltd., New Delhi.

Takkar, P.N. (1988). Sulphur status of Indian soils pp. S1/2–1 to 31. In TSI–FAI.

Sulphur in Indian Agriculture Sulphur Institute, Washington D.C. and FAI, New Delhi.

2.21.7 0.5 M $NaHCO_3$ (pH 8.5) extractable sulphur in soil determined by the reduction method– Johnson and Nishita (1952)

2.21.7.1 Principle

Sulphur is extracted from soil by 0.5 M $NaHCO_3$ (pH 8.5) using a soil extractant ratio of 1 : 4 with 60 minutes shaking on a mechanical shaker. The extract is filtered through Whatman No. 42 filter paper and 2 ml of the filtrate is transferred to 50 ml of reduction-distillation flask.

The extract in the reduction-distillation flask is treated with the reduction mixture and the distillation flask is quickly attached to condenser of a distillation assembly. The purified N_2 gas is passed through the reduction mixture and the digestion flask is heated by a micro burner to distill H_2S liberated.

The H_2S distilled is made to pass through a gas washing column before its absorption by zinc acetate-sodium acetate buffer solution. The H_2S absorbed is treated with acid solution of p-aminodimethylaniline sulphate and ferric ammonium sulphate solution to form methylene blue colour. The optical density (absorbance) of the methylene blue colour is measured at 670 $m\mu$ wave length using a spectrophotometer or colorimeter.

2.21.7.2 Apparatus/Reagents

Apparatus

Digestion-distillation apparatus (Fig. 2.21.1), Apparatus for preparation of reducing mixture (Fig. 2.21.2) Erlenmeyer flasks, shaking machine, volumetric flasks, pipettes (graduated 2, 5 and 10 ml), cylinder of water pumped N_2, photoelectric colorimeter or a spectrophotometer.

Reagents

(*i*) Ca and S free distilled water. Distill water treated with $KMnO_4$ in a completely glass distill or else use deionized water.

(*ii*) 0.5 M Sodium bicarbonate (pH 8.5). Dissolve 42.0 g of $NaHCO_3$ (AR) in 950 ml of Ca and S free distilled water, adjust its pH to 8.5 by adding few pellets of NaOH and dilute the solution to 1000 ml with distilled water.

(*iii*) Reagents 4 to 12 and sulfur free. Lubricant and rubber tubing as described in total S in plant sample by the reduction method. (See section 3.9.3.2).

2.21.7.3 Procedure

1. Extraction of available sulphur by 0.5 M $NaHCO_3$ (pH 8.5). Transfer 10g of air dry < 2 mm soil sample into a 250 ml Erlenmeyer flask. Add 40 ml of 0.5 M $NaHCO_3$ (pH 8.5) solution. Fit a rubber stopper wrapped in the polythene into the flask mouth and shake the Erlenmeyer flask in a mechanical shaker for 60 minutes. Filter the suspension through a funnel lined with Whatman No. 42 filter paper. Discard about 5 ml of initial filtrate and collect the subsequent filtrate.

2. Reduction of SO_4–S in the soil extract, and distillation of H_2S formed for absorption into zinc acetate-acetic acid buffer solution. Transfer 2.0 ml of the soil extract into the 50 ml digestion distillation flask. Then, add 4 ml of the reduction mixture. Quickly attach the digestion distillation flask to the condenser assembly, and follow steps 2 and 3 of procedure described in total S determination in plant sample by the reduction method (Section 3.9.3.3).

3. Development of the methylene blue colour, measurement of absorbance (optical density), preparation of standard calibration curve, and estimation of sulphur in the aliquot of 0.5 M $NaHCO_3$ (pH 8.5) soil extract. Follow the steps 4 to 9 of procedure described in total S determination in plant sample by the reduction method (See section 3.9.3.3).

2.21.7.4 Calculation

$$\text{Available sulphur in soil (ppm)} = \frac{A \times B}{C \times D} = \frac{A \times 40}{2 \times 10}$$

Where

A is μg of S in aliquot of soil extract as estimated by reference to the standard calibration curve for the SO_4–S by the reduction method of Johnson and Nishita (1952).

B is the total volume of 0.5 M $NaHCO_3$ (pH 8.5) used for the extraction of S from the soil.

C is ml aliquot of the soil extract taken for the reduction and subsequent distillation of H_2S for the development of methylene blue colour.

D is the mass of air dry < 2 mm soil sample taken for the extraction of sulphur from the soil.

2.21.7.5 Comments

The alkaline solution such as 0.5 M $NaHCO_3$ (pH 8.5) extracts soluble plus adsorbed SO_4–S and part of labile organic S from the soil which form the soil available sulphur in soil. Adsorbed SO_4–S in soil in significant amount is particularly encountered in acidic soils. In neutral to alkaline soils coarse in texture under arid to semi-arid conditions are likely to be high in soluble sulphates and low in adsorbed sulphates.

The soil extracts by alkaline solution are coloured, and it is troublesome to determine S extracted from the soil by turbidimetric method of Chesnin and Yien (1950) without the use of decolorization agent such as Darco G–60 an activated charcoal made free from phosphorus.

The sulphur in the 0.5 M $NaHCO_3$ (pH 8.5) soil extract is best determined by the reduction method of Johnson and Nishita (1952) as the coloured extract does not pose any problem in the S determination. Besides, the sulphur estimated by the reduction method also includes labile organic fraction apart from soluble and adsorbed sulfates.

The correlation coefficient of 0.5 M $NaHCO_3$ (pH 8.5) extractable sulphur in the soil-extractant ratio of 1 : 4 with 60 minutes shaking on a mechanical shaker, and sulphur "A" value of 30 surface soils in green house study with isotopic dilution technique employing radio active sulphur was found to be 0.89.

The soils containing < 10 ppm of 0.5 M $NaHCO_3$ (pH 8.5) extractable sulphur by the procedure were generally responsive to sulphur application in the green house test.

REFERENCES

Kilmer, V.J. and D.C. Nearpass (1960). The determination of available sulfur in soils. Soil. Sci. Soc. Amer. Proc. 24 : 337–340.

Johnson, C.H. and H. Nishita (1952). Microestimation of sulfur in plant material, soil and irrigation water. Anal. Chem. 24 : 736–742.

2.21.8 0.5 M $NaHCO_3$ (pH 8.5) extractable sulphur in soil as determined by the turbidimetric method-Kilmer and Nearpass (1960)

2.21.8.1 Principle

The sulphur is extracted from the soil by 0.5 M $NaHCO_3$ (pH 8.5) using a soil extractant ratio of 1 : 4 with 60 minutes shaking. Sulphur free Darco-G-60 or activated charcoal is added to decolourize the coloured extract. The suspension is filtered and an aliquot of the soil extract is treated with $BaCl_2$ crystals in the presence of acid "seed" solution and gum acacia to develop turbidity. The absorbance of the $BaSO_4$ turbidity is measured at 420 mμ wave length using a spectrophotometer or else using a photoelectric colorimeter with 420 mμ filter.

2.21.8.2 Apparatus/Reagents

Apparatus

Volumetric flasks (50 to 1000 ml), Erlenmeyer flask (50 to 250 ml), Pipettes (1 to 10 ml), Spectrophotometer or else a photoelectric colorimeter and shaking machine, pH meter.

Reagents

(*i*) 0.5 M Sodium bicarbonate (pH 8.5). Dissolve 42 g of $NaHCO_3$ (AR) in about 950 ml of sulphur free distilled water. Adjust the pH to 8.5 by dissolving few pellets of NaOH. Dilute the solution to 1 l.

(*ii*) Darco-G-60 or activated charcoal made sulphur free by initial washing with HCl and then washing with the extracting solution after removal of excess acid.

(*iii*) Acid "seed" solution. Transfer 25 ml of conc. HCl (AR) into 50 ml volumetric flask containing 10 ml of S-free distilled water. Add 5.0 ml of 100 ppm SO_4–S solution and dilute the solution to 50 ml with distilled water. Stopper and mix the contents.

(*iv*) 0.5% Gum acacia in 50% glacial acetic acid. Transfer 0.25 g of finely powdered gum acacia in 50 ml volumetric flask. Add 25 ml glacial acetic acid and mix the contents. Then add about 20 ml of distilled water. Shake the flask to dissolve gum acacia. Thereafter, dilute the solution to 50 ml mark.

(*v*) Barium chloride crystals (AR). Grind the $BaCl_2$ crystals in a pestle mortar to pass through a 20 mesh sieve but coarse enough to be retained on a 60 mesh sieve.

(*vi*) 100 ppm SO_4–S solution in 0.5 M $NaHCO_3$ (pH 8.5). Dissolve 0.7703 g of $MgSO_4.7H_2O$ in 1 l of 0.5 M $NaHCO_3$ (pH 8.5)

(*vii*) 0 to 40 ppm SO_4–S in 0.5 M $NaHCO_3$ (pH 8.5). Transfer 0, 10, 15 and 20 ml of 100 ppm SO_4–S solution into separate 50 ml volumetric flask and dilute the solution to 50 ml with 0.5 M $NaHCO_3$ (pH 8.5) to obtain 0, 10, 20, 30 and 40 ppm SO_4–S solution.

2.21.8.3 Procedure

1. Extraction of sulphur from soil by 0.5 M $NaHCO_3$ (pH 8.5). Transfer 10.0 g of air dry < 2 mm soil sample into a 250 ml Erlenmeyer flask. Add 40.0 ml of 0.5 M $NaHCO_3$ (pH 8.5) solution. Fit a rubber stopper wrapped with thin polythene into the flask mouth and shake the Erlenmeyer flask on a mechanical shaker for 60 minutes. Thereafter, remove the rubber stopper, and add 1 g of Darco G–60. Fit the stopper again and continue shaking on the mechanical shaker for 5 minutes.

Filter the suspension through a funnel lined with Whatman No. 42 filter paper. If the filter paper is not free from S, select filter paper of the same mass. Discard 3–4 ml of the initial filtrate and collect the subsequent filtrate (extract) for S determination. Similarly run a blank.

2. Development of the $BaSO_4$ turbidity in the sample and blank extract. Transfer 10.0 ml of sample extract in duplicate into separate 50 ml of Erlenmeyer flask. Add 2 ml of acid "seed" solution and 1 ml of 0.5% gum acacia solution. Mix the contents, and add 0.5 g of $BaCl_2$ crystals to one of the Erlenmeyer flask. Allow it to stand for 1 minute. Then, swirl the contents of the flasks frequently until the crystals are dissolved. Similarly treat 10.0 ml of blank extract with $BaCl_2$crystals and do not run a duplicate without $BaCl_2$ crystals.

3. Measurement of the absorbance of $BaSO_4$ turbidity developed. Within the time interval of 2 to 12 minutes after $BaCl_2$ crystals are dissolved, measure the absorbance (optical density) of the $BaCl_2$ turbidity developed with aliquot of sample extract with added $BaCl_2$ crystals in a photoelectric colorimeter with 420 mμ filter or else in a spectrophotometer at 420 mμ wave length adjusting 0 absorbance with the $BaSO_4$ turbidity developed with the aliquot of the blank extract. Thereafter, measure the absorbance of second aliquot of sample extract treated with only 2 ml of acid "seed" solution, and 1.0 g of 0.5 g gum acacia, and substract its absorbance

from the absorbance of $BaSO_4$ turbidity developed with the soil sample extract on adding $BaCl_2$ crystals. This is necessary to eliminate the effect of any colour in the soil extract. Prior to taking absorbance, the instrument should be switched on after placing cuvette or colorimeter tube filled with distilled water to warm for 10 minutes.

4. Development of $BaSO_4$ turbidity with standard SO_4–S solution. Transfer 30 ml of each concentration in the range of 0 to 40 ppm SO_4–S solution into separate 50 ml Erlenmeyer flask, and add 0.75 g of Darco G–60 into each flask. Stopper and shake the contents on a mechanical shaker for 5 minutes. Thereafter, filter the suspension using separate funnel lined with Whatman No. 42 filter paper, collecting the filtrate of each concentration in a separate clean dry Erlenmeyer flask. Label each flask according to the concentration of the SO_4–S standards.

Thereafter, pipette 10 ml of filtrate from each standard obtained after treatment with Darco G -60 into separate 50 ml Erlenmeyer flask. To each add 2 ml of acid "seed" solution, and 1 ml of 0.5% gum acacia. Mix the contents. Then add 0.5 g of $BaCl_2$ crystals. Allow it to stand for 1 minute. Then, swirl the contents of the Erlenmeyer flask frequently until the crystals are dissolved.

5. Measurement of the absorbance of $BaSO_4$ turbidity with standard SO_4–S solution. Within the time interval of 2–12 minutes after $BaCl_2$ crystals are dissolved, measure the absorbance (optical density) of $BaSO_4$ turbidity developed with aliquot of SO_4–S standards in a photoelectric colorimeter fitted with 420 mμ filter or else in a spectrophotometer at 420 mμ wave length adjusting 0 absorbance with the $BaSO_4$turbidity developed with 0 μg SO_4–S. Prior to taking the absorbance, the instrument should be switched on after placing cuvette or colorimeter tube filled with distilled water to warm for 10 minutes. Record the value of absorbance corresponding to aliquot of each concentration of the SO_4–S standards.

6. Preparation of standard calibration curve. On a linear scale graph paper, plot the value of absorbance corresponding to the μg SO_4–S taken for developing $BaSO_4$ turbidity. Let ordinate be the scale for absorbance and abscissa be the scale for μg SO_4–S taken for developing the turbidity. Connect the point of 0 absorbance with the point of maximum absorbance by a straight line to obtain the standard calibration curve.

7. Estimation of μg SO_4–S in aliquot of soil sample extract. Locate the value of absorbance recorded for the aliquot of soil extract after substracting the absorbance accrueing from the colour of soil extract on the ordinate of the standard calibration curve and draw an imaginary line parallel to the abscissa intersecting the straight line calibration curve. From the point of intersection, draw a line perpendicular to the abscissa. The point at which the perpendicular line intersect the abscissa, read μg SO_4–S in the turbid solution.

2.21.8.4 Calculation

$$\text{Available sulphur in soil (ppm)} = \frac{A \times B}{C \times D} = \frac{A \times 40}{10 \times 10}$$

Where

A is the μg SO_4–S recorded in the turbid solution prepared with aliquot of soil extract by reference to the standard calibration curve.

B is the total volume of 0.5 M $NaHCO_3$ (pH 8.5) used for extracting sulphur from the soil.

C is ml aliquot of the soil extract taken for developing $BaSO_4$ turbidity.

D is mass of air dry < 2 mm soil sample taken for the extraction of sulphur.

2.21.8.5 Comments

The determination of S as SO_4 by the turbidimetric method is more common than the reduction method of Johnson and Nishita (1952).

The alkaline solution such as 0.5 M $NaHCO_3$ (pH 8.5) which yield coloured extract on account of solution of a portion of organic fraction require use of activated charcoal to adsorb coloured constituent for obtaining relatively clear extract.

The reagent extracts soluble and adsorbed sulphates. In addition it extracts also a portion of labile organic S. The adsorbed sulphates are relatively much less in neutral and alkaline soil in comparison with soils having pH below 5.0.

There is need to eliminate the effect of faint colour in the soil extract on measurement of turbidity. It is, therefore, desirable to measure the absorbance of the aliquot of the sample test solution with and without adding $BaCl_2$ crystals.

Normally 1 g of activated charcoal free from sulphur is used for the adsorption of coloured constituents in the extract. If some colour is detected in the filtrate, the use of more amount of activated charcoal would make the filtrate clearer and free from colour.

The soils containing < 10 ppm of 0.5 M $NaHCO_3$ (pH 8.5) extractable SO_4–S are generally rated to be deficient in sulphur. Takkar (1988) reported a critical level of 0.5 M $NaHCO_3$(8.5) extractable sulphur for rice in alluvial soils of U.P. to be 20 ppm and for oats in alluvial brown soils of Punjab to be 22 ppm.

Glasswares must be carefully cleaned to remove any potential nuclei for the formation of irregular sized barium sulphate crystals, and the measuring cell must be cleaned regularly with alkaline EDTA to remove adhering barium sulphate, and maintain sensitivity.

A small amount of sulphate may be added to the aliquot of the sample test solution before measurement of the absorbance of $BaSO_4$turbidity to improve the sensitivity of the method and linearity of the standard calibration curve.

Apart from gum arabic, one can also use galatine, Tween 80, polyvinyl alcohol and sorbitol as a stabilizer of $BaSO_4$ turbidity.

REFERENCES

Bardsley, C.E. and J.D. Lancaster (1960). Determination of reserve sulphur and soluble sulphates in soils. Soil. Sci. Soc. Amer. Proc. 24 : 265–268.

Johnson, C.M. and H. Nishita (1952). Microestimation of sulphur in plant material, soil and irrigation water. Anal. Chem. 24 : 736–742.

Kilmer, V.J. and D.C. Nearpass (1960). The determination of available sulphur in soil. Soil. Sci. Soc. Amer. Proc. 24 : 337–340.

Randall, P.J. (1988). Evaluation of the sulphur status of soils and plants-techniques and interpretation. S1/3–1 to 15. In. TSI–FAI Symposium Sulphur in Indian Agriculture. Sulphur Institute, Washington D.C. and FAI, New Delhi.

Takkar, P.N. (1988). Sulphur states of Indian soils pp. S1/2–1 to 31. In TSI–FAI Symposium. Sulphur in Indian Agriculture. Sulphur Institute, Washington, D.C. and FAI, New Delhi.

2.21.9 Heat soluble sulphur in soil by the method of William and Steinberg (1959)

2.21.9.1 Principle

Soil sulphur occur in organic as well as in inorganic form. A fraction of the organic form is also soluble in water. The further release of organic fraction may be achieved by gentle hydrolysis with the addition of distilled water to the soil and then evaporating it to dryness on a gently boiling water bath, and thereafter drying the soil in oven at 102°C for 1 hour before extraction by suitable reagent. The sulphur extracted was determined by turbidimetric method, and the amount so determined was referred by Williams and Steinbergs (1959) as heat soluble sulphur.

2.21.9.2 Apparatus/Reagents

Apparatus

Erlenmeyer flasks, Burette, Volumetric flasks, pipettes, mechanical shaker, silica basin, Water bath, Hot air oven, centrifuge, spectrophotometer or else a photoelectric colorimeter with 420 $m\mu$ filter.

Reagents

(*i*) 0.15% $CaCl_2$

(*ii*) 0.25% gum acacia solution

(*iii*) Sodium acetate-acetic acid buffer (pH 4.8)

(*iv*) Barium chloride crystals–30 to 60 mesh (AR)

(*v*) 100 ppm SO_4–S

(*vi*) 0–20 ppm SO_4–S

(*vii*) Sulphur free distilled water. The preparation of the above mentioned reagents are described in calcium chloride extractable sulphates in soil by the turbidimetric method (See section 2.21.5.1.2).

2.21.9.3 Procedure

1. Heat treatment of the soil. Transfer 5.0 g of air dry < 2 mm soil into a silica basin, add 20 ml of distilled water. Place the silica basin containing soil and distilled water on a gently boiling water bath; and evaporate the contents to dryness. Thereafter, transfer the silica basin with soil into hot air oven at 102°C, and continue drying of the soil sample in an hot air oven at 102°C for 60 minutes. Remove the silica basin from the hot air oven, and let it cool in a desiccator. Similarly run a blank.

2. Extraction of soluble sulphates from the soil by 0.15% $CaCl_2$. Transfer the soil from silica basin into a 50 ml centrifuge tube. Add 33 ml of 0.15% $CaCl_2$. Triturate the contents in centrifuge tube with a glass rod for 10 minutes. Then place the centrifuge tube in the centrifuge. At least two centrifuge tubes with similar treatments be placed opposite each other for proper balance. Switch on the centrifuge. Gradually increase the speed of the centrifuge drum to 2400 rpm, and let it run for 10 minutes. Then, gradually reduce the speed of the centrifuge drum until it stops. Switch off the centrifuge, and remove the centrifuge tube.

From the centrifuge tube, decant off the supernant clear solution into a funnel lined with Whatman No. 42 filter paper for filtration. Collect the filtrate free from any sediment in a clean Erlenmeyer flask. Place stopper wrapped in thin polythene on mouth of the Erlenmeyer flask. Similarly run a blank.

3. Estimation of the μg SO_4–S in aliquot of the soil sample extract. Transfer 10.0 ml of sample and blank extract into separate 25 ml volumetric flask and follow the steps 2 to 7 of procedure described in soluble SO_4–S determination in 0.15% $CaCl_2$ soil extract by the turbidimetric method (section 2.21.5.1.3).

2.21.9.4 Calculation

$$\text{Heat soluble S in soil (ppm)} = \frac{A \times B}{C \times D} = \frac{A \times 33}{10 \times 5}$$

Where

A is the μg SO_4–S recorded in the $BaSO_4$ turbid solution prepared with aliquot of soil extract by reference to the standard calibration curve.

B is the total volume of 0.15% $CaCl_2$ used for extracting SO_4–S from the soil after heat treatment.

C is ml aliquot of the soil extract taken for developing $BaSO_4$ turbidity.

D is the mass of air dry < 2 mm soil sample for heat treatment of subsequent extraction by 0.15% $CaCl_2$ solution.

2.21.9.5 Comments

The fraction of soil sulphur extracted after heating the soil-distilled water suspension 1 : 4 (w/v) on boiling water bath to dryness followed by heat in hot air oven at 102°C for 1 hour (heat soluble sulphur) was found to be highly correlated with both sulphur uptake and yield of oats grown in pots culture (William and Steinbergs, 1959). It seems that the heat soluble fraction of S in soil may provide a satisfactory index of available sulphur. Heating the soil prior to sulphate extraction substantially increase the amount of water soluble sulphate.

The critical level of heat soluble S on alluvial loam soil of Punjab was found to be 12.5 ppm for wheat. On medium black soils of Gujarat, the critical level of heat soluble S was 10 ppm for groundnut and wheat, and 20 ppm for alfalfa. In U.P., the critical level of heat soluble S was found to be 16 ppm for rice. Thus, the critical level of heat soluble S vary with the crop.

REFERENCES

Chesnin, L, and C.H. Yien (1950). Turbidimetric determination of available sulphate. Soil Sci. Soc. Amer. Proc. 15 : 149-151.

Takker, P.N. (1988). Sulphur status of Indian soils. pp S1/2 - 1 to 31. In TSI-FAI Symposium. Sulphur in Indian Agriculture. Sulphur Institute, Washington D.C. and FAI, New Delhi.

Williams, C.H. and A. Steinbergs (1959). Soil sulphur fraction as chemical indices of available sulphur in soils. Australian J. Agric. Res. 10 : 340-352.

2.21.10 SO_4–S in soil extracts stabilizing $BaSO_4$ turbidity with sorbitol–Swaminathan and Sud (1977)

2.21.10.1 Principle

The aliquot of soil extract is acidified, and treated with 70% sorbitol before adding $BaCl_2$ crystals. The $BaCl_2$ crystals are dissolved by shaking, and the $BaSO_4$turbidity developed is measured after 5 minutes at 470 mμ on a spectrophotometer or a klett photoelectric colorimeter.

2.21.10.2 Apparatus/Reagents

Apparatus

Volumetric flasks, graduated pipettes, spectrophotometer or klett photolectric colorimeter.

Reagents

(*i*) 10.0 N Hydrochloric acid. Dilute 218.2 ml of conc. HCl (AR) to 250 ml with distilled water in a 250 ml volumetric flask.

(*ii*) 6.0 N Hydrochloric acid. Dilute 60 ml of 10 N HCl to 100 ml in a volumetric flask with distilled water.

(*iii*) 70% Sorbitol solution. Dissolve 70 g of Sorbitol in distilled water and make the volume to 100 ml or else obtain 70% aqueous solution of sorbitol.

(*iv*) Barium chloride dihydrate crystals (AR)

(*v*) 0.01M $Ca(H_2PO_4)_2$ solution.

2.21.10.3 Procedure

1. Development of $BaSO_4$ turbidity with the aliquot of soil extract. Extract SO_4–S by equilibrating 5g soil with 50 ml of .01M $Ca(H_2PO_4)_2$ for 30 minutes on a mechanical Shaker. Filter the suspension through whatman 42 paper.

Transfer 10.0 ml of soil extract into a 50 ml volumetric flasks. Then, add 1.0 ml of 6.0 N HCl. Thereafter, add 5 ml of 70% aqueous solution of sorbitol through a pipette having an enlarge jet. Mix the contents and add 1 g of $BaCl_2.2H_2O$ crystals (AR) and shake well until the $BaCl_2$crystals are dissolved. Allow it to stand for atleast 5 minutes. Similarly run a blank taking 10 ml of distilled water.

2. Measurement of the $BaSO_4$ turbidity developed with the aliquot of soil extract. Prior to measurement of the absorbance of $BaSO_4$ turbidity, switch on the spectrophotometer at 470 mμ wave length or else Klett photoelectric colorimeter with 470 mμ filter after placing cuvette or test tube filled with distilled water in the socket provided. Allow the instrument to warm for 10 minutes.

Thereafter adjust 0 absorbance with $BaSO_4$ turbidity developed with the blank extract and measure the absorbance of $BaSO_4$ turbidity developed with the aliquot of the soil extract.

3. Development of $BaSO_4$ turbidity with aliquot of 0–50 ppm SO_4–S standards. Pipette 10.0 ml aliquot of 0–50 ppm SO_4–S standards into separate 50 ml volumetric flasks, and proceed as described in step 1 of procedure beginning with the addition of 1.0 ml of 6 N HCl.

4. Measurement of absorbance of the $BaSO_4$ turbidity developed with aliquot of 0–50 ppm SO_4–S standards. Set the 0 absorbance with $BaSO_4$ turbidity developed with 10 ml aliquot of 0 ppm SO_4–S standard on a spectrophotometer or else on a Klett photoelectric colorimeter at 470 mμ wavelength or filter. Thereafter, measure the absorbance of $BaSO_4$ turbidity developed with 10.0 ml aliquot of the subsequent SO_4–S standards *i.e.* 10, 20, 30, 40 and 50 ppm.

5. Preparation of standard calibration curve. On a linear scale graph paper, use ordinate as scale for the absorbance and abscissa as the scale for SO_4–S taken in μg for developing $BaSO_4$ turbidity. Then plot the absorbance corresponding to the μg SO_4–S taken for developiong $BaSO_4$ turbidity. Connect the point of 0 absorbance with the point of maximum absorbance by a straight line to obtain the standard calibration curve.

6. Estimation of SO_4–S in μg in $BaSO_4$ turbidity developed with the aliquot of soil extract. Divide the absorbance value of $BaSO_4$ turbidity developed with the aliquot of soil extract by the slope of the standard calibration curve expressed in absorbance/μg SO_4–S taken for developing $BaSO_4$ turbidity to obtain content of μg SO_4–S in aliquot of the soil extract.

2.21.10.4 Calculation

$$SO_4\text{–S in soil (ppm)} = \frac{A \times B}{C \times D}$$

Where

A is μg SO_4–S in aliquot of the soil extract by reference to the standard calibration curve.

B is the volume of total soil extractant used for preparing soil extract.

C is ml aliquot of the soil extract taken for developing $BaSO_4$ turbidity.

D is the mass of soil in gram used for the preparation of soil extract.

2.21.10.5 Comments

The aqueous solution of 70% sorbitol possess a high viscosity even after further dilution. Besides, it is not attacked by dilute acid and alkali normally present in soil extract or plant digest. Therefore, sorbital can be also used for stabiliizing $BaSO_4$ turbidity in aliquot of plant digest prepared by wet digestion with conc. HNO_3+70% $HClO_4$ in the ratio of 9:4.

At sorbitol concentration of 20% in the final volume, it regulates the formation of fine sized $BaSO_4$ particles, and preserve the suspension against sedimentation. It hinders also the growth and aggregation of particles for a period of 6 days irrespective of the condition of precipitation such as temperature of the solution, size of $BaCl_2$. $2H_2O$ crystal. The absorbance is also not influenced by the temperature up to 30°C.

REFERENCES

Swaminathan, K and K.C. Sud (1977). An improved turbidimetric procedure for the determination of sulphate in plant and soil. Talanta 24 : 49–50.

2.22 1.0 N CH_3COONH_4 (PH 4.8) EXTRACTABLE Fe IN SOIL BY THE METHOD OF OLSON (1965)

2.22.1 Principle

Fe from soil is extracted by 1.0 N CH_3COONH_4 (pH 4.8). The Fe in soil extract is reduced to Fe^{2+} with $NH_2OH.NH_4Cl$ and then Fe^{2+} is reacted to form a red coloured complex [tri–(1–10)–phenanthroline ferrous ion, $Fe(C_{12}H_{10}N_2)_3{}^{++}$)]. The absorbance of the red coloured complex is measured at 490 mμ wavelength or filter using a spectrophotometer or a photoelectric colorimeter.

2.22.2 Equipments/Reagents

Equipments

Spectrophotometer or else Klett–Summerson photoelectric colorimeter with 490 mμ filter

Reagents

(*i*) 1.0 NCH_3COONH_4(pH 4.8). Dissolve 77.0 g of CH_3COONH_4 (BDH–AR) in about 800 ml of redistilled water using 1000 ml beaker. Then adjust the pH of the solution to 4.8 by adding CH_3COOH (BDH–AR). Add about 150 ml distilled water. Mix the contents and check the pH again. Transfer the solution to 1000 ml vol. flask. Dilute the solution to 1000 ml mark with rinsings from the beaker using redistilled water. Mix the contents and store in a polythene bottle.

(*ii*) 10.0% NH_2OH–NH_4Cl. Dissolve 10 g of NH_2OH–NH_4Cl (hydroxylamine hydrochloride BOH–AR) in redistilled water, and dilute it with redistilled water to 100 ml mark in a 100 ml volumetric flask.

(*iii*) 0.3% Orthophenanthroline solution. Transfer 0.30 g of orthophenanthroline (Merck- Sarabhai) into 250 ml beaker. Add about 75 ml of redistilled water, and place the beaker on a hot plate. Continue heating below boiling temperature until the reagent is dissolved. Cool the solution and transfer it into 100 ml vol. flask.

Dilute the solution to 100 ml mark with the rinsings from the beaker using redistilled water.

(*iv*) 100 ppm Fe solution. Transfer 0.7022 g of fresh $Fe(NH_4)_2 (SO_4)_2.6H_2O$ (BDH–AR) into 1000 ml vol. flask. Add about 500 ml redistilled water. Dissolve the reagent by shaking the flask. Then add 10 ml of 36 N H_2SO_4 (BDH–AR) slowly along the inner wall of the flask with swirling of the contents. Cool the solution and then dilute it to 1000 ml mark with redistilled water. Mix the contents and store in a polythene bottle.

2.22.3 Procedure

1. Extraction of Fe from the soil. Transfer 12.5 g of air dry 2 mm soil, and 50.0 ml 1N CH_3COONH_4 (pH 4.8) into 150 ml conical flask. After fitting a rubber stopper, place the conical flask firmly on a mechanical shaker. Switch on the mechanical shaker and run it at moderate speed for 30 minutes. Remove the flask from the mechanical shaker, and filter the suspension through a funnel lined with Whatman No. 42 filter paper. Discard the initial 5 ml filtrate after rinsing the receiving flask. Collect the filtrate and preserve it for the estimation of Fe concentration. Similarly prepare a blank filtrate using the same volume of extractant but without the soil sample.

2. Development of Fe–orthophenanthroline red coloured complex. Take two 25 ml cap. clean and completely dry volumetric flasks. Pipette in each 10 ml aliquots of the soil extract. Then, add to each 2.0 ml of 10% $NH_2OH.NH_4Cl$ solution. Thereafter, add to one flask 2.0 ml of redistilled water, and to the other 2.0 ml of 0.3% orthophenanthroline solution. Mix the contents and allow it to stand for 30 minutes without making the volume to 25 ml mark.

3. Measurement of the absorbance of soil extracts with and without added orthophenanthroline reagent. Adjust zero absorbance with solution containing 10 ml extractant, and 2 ml each of 10% $NH_2OH.NH_4Cl$ and 0.3% orthophenanthroline at 490 mμ wavelength or filter using a spectrophotometer or colorimeter. Measure the absorbance of soil extract in which orthophenanthroline has not been added to serve as a check for slightly coloured soil extract. Then measure the absorbance of soil extract with added orthophenanthroline after the development of Fe-orthophenanthroline complex. Operate the instruments as described in the estimation of available P in soil (See section 2.11.1.3).

4. Preparation of standard graph for Fe. Transfer 0, 0.1, 0.2 and 0.3 ml of 100 ppm Fe solution into separate clean and completely dry 25 ml volumetric flask, and then add 10.0, 9.9, 9.8 and 9.7 ml of 1.0 N CH_3COONH_4 (pH 4.8) in the 25 ml volumetric flasks respectively to make the total volume to 10.0 ml in each flask. Then, add to each flask 2 ml each of 10.0% NH_2OH-NH_4Cl and 0.3% orthophenanthroline solution. Mix the contents and allow it to stand for 30 minutes without making the volume to 25 ml.

Adjust zero absorbance with 0 μg Fe/14 ml coloured complex at 490 mμ wave length or filter using a spectrophotometer or colorimeter. Then, measure the absorbance of 14 ml coloured complex developed by adding 0, 0.1, 0.2 and 0.3 ml of 100 ppm Fe solution respectively. Note the absorbance corresponding to μg Fe added.

Plot the absorbance or Klett reading on the ordinate of a graph paper and the corresponding Fe concentration in μg Fe/14 ml coloured complex *i.e.* 0 to 30 μg Fe on the abscissa. Connect the absorbance value recorded with 0 and 30 μg Fe/14 ml coloured complex to obtain a straight line calibration curve. Other points should fall nearly on the straight line.

5. Estimation of Fe concentration in soil and blank digest. Substract the absorbance value recorded with the sample extract without phenanthroline from the value of absorbance recorded with the sample extract with orthophenanthroline to obtain the corrected absorbance for the sample extract. Similarly estimate the corrected absorbance for the blank extract.

Estimate Fe in soil and blank extract by reference to the standard graph for Fe using the corrected absorbance of the sample and blank extract. Alternatively estimate the slope *i.e.* absorbance/μg added Fe and divide the corrected absorbance by the slope to obtain μg Fe present in 10 ml extract.

2.22.4 Calculation

$$\text{ppm Fe in soil} = \frac{\mu\text{g Fe in (Sample - Blank) extract} \times \text{Volume of extractant equilibrated with the soil}}{\text{ml of extract used for colour development} \times \text{mass of soil in g}}$$

2.22.5 Comments

Since 1.0 N CH_3COONH_4 (pH 4.8) extract in most soils are slightly coloured because of extraction of organic matter, a correction is made for the colour with each soil. This is done by running a check for each soil, to which all the reagents except orthophenanthroline is added and its absorbance is deducted from the absorbance of soil extract with all the reagents including orthophenanthroline.

It is possible to estimate Fe^{++} and Fe^{+++} in the same solution. Estimation of Fe after the addition of hydroxylamine hydrochloride which reduces Fe^{+++} to Fe^{++} includes both the forms of Fe. Estimation of Fe in the absence of hydroxylamine hydrochloride gives only the Fe^{++} value provided that there is no reducing agent present in the solution.

During the development of ferrous complex of orthophenanthroline (a chelate ring compound of intense red colour), a slightly acidic reaction must be used as some cations will interfere in the alkaline range because of precipitation of their hydroxides. No common anions interfere except orthophosphate through precipitation of Ca and Fe.

On the basis of Fe extracted by 1N CH_3COONH_4 (pH 4.8), Olson and Carlson (1950) attempted to correlate soil test value with the deficiency symptoms as given below :

Extractable Fe in soil (ppm)	Degree of Fe chlorosis in plants
0.01 to 0.30	moderate to severe
0.30 to 2.00	slight to moderate
2.0 to 32.0	nil

It appears from the above observations that the critical level would be 2.0 ppm Fe as extracted by the above extractant for plants sensitive to Fe deficiency. Other extractants EDDHA (Ethylene diamine di–g–hydroxyphenyl acetic acid), DTPA (Diethylene triamine penta acetic acid) have been shown to be useful in predicting Fe deficiency).

Fe chlorosis is caused apart from low extractable Fe by high Mn/Fe ratio, high pH, presence of free lime, excess phosphates, carbonates and high moisture. Excess of Cu, Mn, Ni and Zn are also known to induce Fe chlorosis. Thus, Fe soil test should be assessed keeping in view other factors. It would be preferable to select a plant to fit a soil than change a soil to fit the plant.

REFERENCES

Jackson, M.L. (1973). Soil Chemical Analysis. Prentice-Hall of India Private Limited, New Delhi.

Mortvedt, J.J., P.M. Giordano, and W.L. Lindsay (1972). Micronutrients in Agriculture. Soil Sci. Soc. Amer. Inc., Madison, USA.

Olson, R.V., and C.W. Carlson (1950). Iron chlorosis of sorghums and trees as related to extractable soil iron and manganese. Soil Sci. Soc. Amer. Proc. 14, 109–112.

Olson, R.V. (1965). Iron. In.C.A. Black *et. al* Method of Soil Analysis, Part 2, Agronomy 9, 963–973.

2.23 1.0 N CH_3COONH_4(PH 7.0) EXTRACTABLE Mn IN SOIL-WILLARD AND GREATHOUSE (1917)

2.23.1 Principle

Mn^{++} extracted by 1N CH_3COONH_4 (pH 7.0) is oxidized to MnO_4^- by KIO_4in a strongly acid solution. The reaction is represented by the following equations :

$$(i)\ 15\ KIO_4 + 2M_3\ (PO_4)_2 + 9H_2O \longrightarrow 15\ KIO_3 + 6HMnO_4 + 4H_3PO_4$$

$$(ii)\ 2\ Mn^{++} + 5\ KIO_4 + 3H_2O \longrightarrow 2\ MnO_4^- + 5\ KIO_3 + 6H^+$$

The presence of H_3PO_4 in the medium eliminates interference by Fe^{2+}whereas the presence of excess periodate eliminates interference from small amounts of Cl. The concentration of Mn in the extract is determined by measurement of the absorbance of $HMnO_4$colour at 540 mμ wave length or filter using a spectrophotometer or a photoelectric colorimeter.

2.23.2 Reagents

(*i*) 1.0 N CH_3COONH_4 (pH 7.0) : Dissolve 77.0 g of CH_3COONH_4 (BDH–AR) in 500 ml redistilled water and dilute the solution to about 975 ml. Then adjust the solution pH to 7.0 with ammonia solution (BDH–AR) or with CH_3COOH (BDH–AR). Then dilute the solution to 1000 ml with redistilled water.

(*ii*) Conc. HNO_3 : Use distilled conc. HNO_3 (BDH–AR).

(*iii*) 30% H_2O_2 (BDH–LR–100 volume).

(*iv*) KIO_4 crystals (BDH–AR).

(*v*) 85% H_3PO_4–Mn free (Merek–GR).

(*vi*) 1000 ppm Mn solution. Transfer 2.8767 g of pure dry $KMnO_4$ (BDH–AR) into 1000 ml vol. flask. Then add 250 ml of distilled water and dissolve the reagent by shaking the flask. Thereafter, add 20 ml of 18 N H_2SO_4 and heat the solution to boil on a hot plate. At this stage, add Na_2SO_3 in small portions until the permanganate colour disappear. Take care not to add Na_2SO_3 in large excess. Heat the solution for 30 minutes to expell SO_2. Let the solution cool to room temperature and then dilute it to 1000 ml mark with redistilled water. Mix the contents by shaking and then transfer the solution into a polythene bottle.

(*vii*) 100 ppm Mn solution. Dilute 25.0 ml of 1000 ppm Mn solution to 250 ml with redistilled water.

(*viii*) 10 ppm Mn solution. Dilute 25.0 ml of 100 ppm Mn solution to 250 ml with redistilled water.

2.23.3 Procedure

1. Extraction of Mn from soil. Transfer 10.0 g of soil and 100 ml of 1N CH_3COONH_4 (pH 7.0) into 250 ml Erlenmeyer. flask. Stopper the flask and place it firmly on a mechanical shaker. Switch on the mechanical shaker and run it for one hour at moderate speed. Thereafter, run the mechanical shaker intermittently for 6 hours.

Remove the flask from the shaker. Filter the suspension through a funnel lined with Whatman No. 44 filter paper . Discard the initial 5 ml filtrate after rinsing the receiving flask with it. Collect the subsequent filtrate.

2. Evaporation of an aliquot of soil extract to dryness. Transfer 40 ml aliquot from the soil extract into 250 ml beaker and place it on a hot plate at low heat to evaporate the contents to dryness. Continue heating until CH_3COONH_4 ceases to fume. Remove the beaker from the hot plate and let it cool to room temperature.

3. Treatment of the residue with distilled conc. HNO_3 and 30% H_2O_2. Then add 5 ml of distilled conc. HNO_3 and 2 ml of 30% H_2O_2 and place the beaker on a hot plate at low heat. Evaporate the contents to dryness. Remove the beaker from hot plate, and let it cool to room temperature. Again treat the residue with 5 ml of distilled conc. HNO_3 and 2 ml of 30% H_2O_2. Place the beaker on hot plate at low heat. Evaporate the contents to dryness. Remove the beaker from the hot plate and let it cool to room temperature.

4. Subsequent treatment of the residue for the oxidation of Mn^{++} into $HMnO_4$. Add 2 ml of distilled conc. HNO_3 and 5 ml of 85% H_3PO_4–Mn free. Heat the contents to boil gently. Remove the beaker from the hot plate and let it cool. After cooling, add 10 ml of redistilled water and 0.3 g of KIO_4 crystals, and heat the contents to boil gently. Remove the beaker from the hot plate, and let it cool.

5. Dilution of the permanganate solution in a 50 ml vol. flask. Transfer the permanganate coloured solution from the 250 ml beaker into a 50 ml vol. flask. Rinse the beaker twice with 5 ml portions of redistilled water and transfer the rinsings from the beaker into the same 50 ml volumetric flask. Dilute the contents to 50 ml mark with redistilled water. Stopper the flask and turn it upside down twice so as to mix the contents.

Similarly run a blank following the entire procedure from step 1 to 5 after transfering 40 ml aliquot of the extractant *i.e.* 1N CH_3COONH_4 (pH 7.0).

6. Preparation of standard graph for Mn. Transfer, 0, 5, 10 and 15 ml of 10 ppm Mn solution into separate 250 ml beaker. Then, add to each 40 ml of 1N CH_3COONH_4 (pH 7.0) solution and proceed as described for the sample extract following steps 1 to 5.

After 15 minutes of development of $KMnO_4$ colour diluted to 50 ml with redistilled water, adjust zero absorbance with 0 μg Mn/50 ml coloured solution at 540 mμ wavelength or filter using a spectrophotometer or a photoelectric colorimeter. Then, measure the absorbance of 50, 100 and 150 μg Mn/50 ml of $HMnO_4$ coloured solution.

Record the absorbance corresponding to μg Mn added per 50 ml of coloured complex. Plot the value of absorbance on the ordinate of a linear scale graph paper corresponding to the concentration of Mn in μg/50 ml of $HMnO_4$coloured complex plotted on the abscissa. Connect absorbance corresponding to 0 and 150 μg Mn/ 50 ml of $HMnO_4$ coloured complex by a straight line to draw a calibration curve. All the points should nearly fall on the straight line.

7. Estimation of Mn in aliquot of sample and blank extract. Note the absorbance of the sample and blank extract. Mark the absorbance on the ordinate of standard graph for Mn and from that point draw an imaginary line parallel to the abscissa. The point at which the imaginary line intersect the straight line calibration curve, draw a vertical downward line perpendicular to the abscissa and note the concentration of Mn at the point of intersection. Alternatively estimate the slope of the straight line calibration curve *i.e.* absorbance/μg Mn and then divide the absorbance of sample and blank extract by the slope to obtain μg Mn in aliquot of sample and blank extract

2.23.4 Calculation

$$\text{ppm Mn in soil} = \frac{\mu\text{g Mn in aliquot of (Sample - Blank)} \times \text{total volume of extractant in ml}}{\text{Volume of aliquot in ml} \times \text{mass of soil sample in g}}$$

2.23.5 Comments

In soils, the sum of water soluble, exchangeable and reducible Mn constitute the active Mn. The water soluble and exchangeable Mn comprises of Mn^{++} and it is referred as readily available Mn. The reducible Mn represent those higher oxides which can be reduced by easily oxidizable organic substances *.e.g.* hydroquinone. Generally in soils, the concentration of water soluble Mn < exchangeable Mn <

reducible Mn. Reducible Mn in soils increases with the decrease in oxygen tension on submergence.

Several factors influence the available Mn in soil. With increasing pH, Mn^{++} is converted into manganic oxides (Mn^{3+} and above) which is water insoluble and is also not extracted by 1N CH_3COONH_4(pH 7.0). Consequently negative correlation exist between exchangeable Mn or active Mn and soil pH.

There exist also negative correlation between exchangeable Mn and calcium carbonate content in soil. Surface soils have higher exchangeable Mn than sub-surface soils because of the presence of high content of organic fraction, but organic matter greater than 3% may result in formation of insoluble complexes with Mn. Exchangeable Mn tend to increase with the increase in clay content not exceeding 40%. The clay of ferruginous nature have higher exchangeable Mn.

There exist negative correlation between reducible Mn and ferrous iron in soils. The soils high in ferrous iron are low in Mn and vice versa. Thus Fe/Mn ratio is likely to influence available forms of Mn in soil.

Air drying of soils result in increase in the extractable bivalent Mn in slightly acidic as well as slightly alkaline soils presumably due to non-biological reduction represented by the equation :

$$MnO_2 + 4H + 2e = 2H_2O + Mn^{++}$$

In view of the above, the critical level of extractable Mn needs to be assessed considering other factors. The water soluble and exchangeable Mn is extracted from the soil by 1N CH_3COONH_4 (pH 7.0). Many extractants other than 1.0 N CH_3COONH_4 have been employed for the extraction of Mn from soil. However, some workers have found 1N CH_3COONH_4 (pH 7.0) as the best extractant when pH was considered. Other extractants which appear to be promising under some conditions are H_2O, 0.2% hydroaquinone in 1N CH_3COONH_4, 0.1 N H_3PO_4, 3.0N $NH_4\ H_2PO_4$ and 0.05 N HCl + 0.25 N H_2SO_4.

In India coarse textured soils belonging to the zonal soil Sierozems in certain parts of Rajasthan and Gujarat as well as strongly alkaline soils of Punjab and Haryana have been reported to be deficient in available Mn. Mn deficiency also occur in calcareous soils. Mn deficiency occur in acidic soils on liming in excess.

REFERENCES

Adams, F. (1965) Manganese pp. 1011–1018. In. C.A. Black (ed). Methods of Soil Analysis. part 2, Amer. Soc. Agron. Inc. Publisher. Madison, Wisconsin USA.

Jackson, M.L. (1973). Soil Chemical Analysis, Prentice-Hall of India Private Limited, New Delhi.

Kanwar, J.S. and N.S. Randhawa (1974). Micronutrient Research in Soils and Plants in India (a review), Indian Council of Agricultural Research, New Delhi.

Mortvedt, J.J., P.M. Giordano, and W.L. Lindsay (1972). Micronutrients in Agriculture. Soil Sci. Soc. Amer. Inc. Madison, Wisconsin, USA.

Sherman, C.D. J.S. Melargue, and W.S. Hodgkins (1942). Determination of active manganese in soils. Soil Sci. 253–257.

Willard, H.H., and L.H. Greathouse (1917). The colorimetric determination of manganese by oxidation with periodate. J. Amer. Chem. Soc. 39 : 2366–2397.

2.24 ZINC DETERMINATION IN SOIL–SHAW AND DEAN (1952) AND HOLMES (1945)

2.24.1. Method 1

2.24.1.1 Principle

Diphenyl thiocarbazone (C_6H_5 NHNHCSN : NC_6H_5) known as dithizone is a complexing agent. It reacts with metals forming metal dithizonates through replacement of either one or both hydrogen atoms. The acid form of dithizone is soluble in CCl_4 and the ammonium salt is soluble in water containing a slight excess of NH_4OH.

When the solution of dithizone in CCl_4 is shaken with an aqueous metal solution at an appropriate pH, the metal dithizonate is formed and dissolves in the CCl_4, to which it imparts its characterstic colour usually some shade of red, violet or orange. If the reaction takes place in acidic solution and dithizone is in excess, the CCl_4 phase will show a mixed colour resulting from the green of the remaining dithizone, and the red or violet of the dithizonate. If the solution is sufficiently alkaline, the excess dithizone distributes mostly into the aqueous phase, and the colour of the CCl_4 phase will be that of the dithizonate.

The procedure of Shaw and Dean (1952) has been adopted for the dithizone extractable Zn from soil, whereas the procedure adopted for estimating Zn in solution is the one suggested by Holmes (1945). There are two methods for estimating Zn in solution. In the first method Cu, Co, Pb and Ni are held in carbamate (sodium diethyl dithiocarbamate) complex form in the aqueous layer whereas the Zn is separated into the CCl_4 at pH 8.8. The interference from Ni is satisfactorily removed. The method permits determination of Zn from 1 to 25 μg. In the second method, Cu present as impurity in Zn solution or else in extract from soil is removed as Cu–dithizonate in acidic medium (pH 2.5). Then the aqueous solution free from Cu is extracted by dithizone in the presence of carbamate in alkaline medium (pH 8.3 to 8.8) to form Zn dithizonate. The carbamate complexes Ni, Co, Pb etc and removes their interference. It permits determination of Zn from 1 to 10 μg.

2.24.1.2 Apparatus/Reagents

Apparatus

Pear shaped separatory funnels –125 ml, 250 ml and 41 capacity, volumetric flasks–25 ml and 1 l capacity, separatory funnel stand.

Mechanical shaker, colorimeter with 535 nm filter or spectrophotometer.

Reagents

(*i*) Zn-free distilled water. Redistill water from a metal still in a pyrex still or pass through an ion exchange column.

(*ii*) Approx. 6N HCl. Distill a mixture of equal volume of Zn-free distilled water and concentrated HCl (BDH–AR grade) in pyrex still and collect the constant boiling distillate.

(*iii*) 1.0 N HCl. Standardize approx. 6N HCl against 0.1 N Na_2CO_3 (0.53 g Na_2CO_3/100 ml distilled water) using 0.1% methyl orange indicator and then dilute the distilled HCl to the desired extent with Zn-free distilled water to obtain 1.0 N HCl.

(*iv*) 0.02N HCl. Dilute 20 ml of 1.0 N HCl to 1000 ml with Zn-free distilled water.

(*v*) 1.0 N NH_4OH. Distill conc. NH_4OH (BDH–AR) in pyrex glass still into Zn-free distilled water in a pyrex container packed in ice. Standardize a portion of NH_4OH solution after dilution by a standard acid using phenolphthalein indicator. Thereafter, dilute the distilled ammonium hydroxide solution to the desired extent to obtain 1.0 N NH_4OH solution. Store in a refrigerator in a reagent bottle with a screw cap.

(*vi*) 0.1% Phenolphthalein indicator. Dissolve 0.1 g phenolphthalein indicator in 95% ethanol

(*vii*) CCl_4 (BDH–AR). It be distilled before use.

(*viii*) Thymol blue indicator.

(*ix*) 0.2% carbamate solution. Dissolve 0.2 g of sodium diethyldithiocarbamate (Eastman Kodak Co.) in 100 ml of Zn-free distilled water. Keep the solution in a brown bottle, and store it in cool and dark place. Prepare the solution every week.

(*x*) 0.01% Dithizone in CCl_4. Transfer 0.2g of diphenylthiocarbazone (Eastman Kodak Co.) and 1000 ml of CCl_4 (BDH–AR) into a 41 separatory funnel. Shake the contents frequently for 15 minutes to dissolve the solid dithizone. Thereafter, add 2000 ml of Zn–free 0.02 N NH_4OH (40 ml of 1.0 N NH_4OH diluted to 2000 mi with Zn–free distilled water). Shake the mixture to transfer dithizone to the aqueous phase. Discard the CCl_4 phase (light green in colour). Then rinse the aqueous phase with several 100 m' portions of CCl_4 and discard the CCl_4 phase each time. Thereafter, add 500 ml of CCl_4 and 50 ml of Zn-free 1N HCl. Shake the mixture to transfer dithizone to the CCl_4 layer. Collect the CCl_4 dithizone phase and dilute it to 2 l with CCl_4. Store the solution in a glass stoppered bottle covered wholly with a black paper and keep it in a refrigerator.

(*xi*) 9.0% Ammonium citrate solution. Dissolve 90 g of ammonium citrate (BDH–AR) in Zn-free distilled water and dilute the solution to 11 with Zn–free distilled water. Transfer the entire solution into a 41 separatory funnel, and add enough Zn–free concentrated NH_4OH to bring the pH to 8.5. Remove Zn impurities in the ammonium citrate solution by extraction with portions of dithizone in CCl_4 until the latter reagent no longer changes colour. Then extract the solution with portions of CCl_4 until the citrate solution is free from dithizone colour.

(*xii*) 100 ppm Zn solution. Dissolve exactly 0.100 g of pure Zn in 50 ml of 0.02N H_2SO_4 and dilute it to 1000 ml with Zn-free distilled water.

(*xiii*) 2 ppm Zn solution. Dilute 10 ml of 100 ppm Zn solution to 500 ml with Zn-free distilled water.

(*xiv*) 1.0 N $NH_4C_2H_3O_2$ (pH 7.0). Dissolve 77.0 g of ammonium acetate (BDH–AR) in Zn-free distilled water and dilute it to 1000 ml. Adjust the pH to 7.0 with Zn-free NH_4OH. Purify the buffer in a large separatory funnel with 0.01% dithizone in CCl_4. Discard the organic phase until it no longer changes its colour. Remove dithizone dissolved in aqueous phase by repeated extraction with pure CCl_4.

2.24.1.3 Procedure

1. Formation of Zn-dithizonate complex in CCl_4 phase in the presence of carbamate in an alkaline medium. Transfer 40 ml of 0.02N HCl into a 125 ml separatory funnel and then add solution containing 5 to 25 μg Zn. Thereafter, add 50 ml of 0.4 M ammonium citrate buffer and 3 ml of 0.2% carbamate solution. Adjust pH of the solution in separatory funnel between 8.5 to 9.0 with redistilled NH_4OH or HCl (use thymol blue indicator-its colour become purple at pH 8.9). Thereafter, add 10.0 ml of 0.01% dithizone in CCl_4. Then shake the mixture for 5 minutes and place the separatory funnel on a stand and allow the phases to separate.

2. Removal of free dithizone from the Zn–dithizonate in CCl_4 phase. Transfer the CCl_4 phase containing Zn-dithizonate into another 125 ml separatory funnel. Then add 25 ml of 0.01N NH_4OH to the CCl_4 phase and shake the mixture again for 3 minutes to extract the excess dithizone in the aqueous phase.

On completion of the 2nd step, if the CCl_4 phase containing the metal dithizonate has some off colour, then transfer the CCl_4 phase into a separatory funnel containing fresh 40 ml of 0.02N HCl and repeat the procedure described in the 1st and 2nd step using fresh dithizone reagent in CCl_4.

3. Dilution of the Zn-dithizonate in CCl_4 and measurement of its absorbance (optical density) by use of a colorimeter or spectrophotometer. Transfer 5 ml of CCl_4 phase containing Zn-dithizonate into a 25 ml volumetric flask and dilute it to the 25 ml mark with pure CCl_4. Then, read the absorbance of the diluted Zn–dithizonate at 535 to 540 mμ (nm) wave length or filter adjusting zero absorbance with pure CCl_4.

4. Preparation of standard graph for Zn. Note the absorbance (optical density) of the Zn dithizonate in CCl_4 phase obtained by taking 5 to 25 μg Zn. Then plot the absorbance on a "Y" axis against the corresponding amount of Zn in μg on "x" axis of a linear scale graph paper. Draw a straight line by connecting the absorbance value of pure CCl_4 *i.e.* 0 μg Zn with the absorbance value obtained with the maximum amount of Zn taken (*i.e.* 25 μg Zn).

5. Collection and preparation of soil sample. Soil sample be collected using a plastic shovel. It should be air dried and then crushed with a wooden pestle and be passed through a Zn-free 1 mm sieve (18 mesh per inch). The sieve may be made of iron wire or Zn-free wire commercially available. The sample be then stored in a cardboard ice cream container.

6. Extraction of Zn from soil by 1N ammonium acetate (pH 7.0) + 0.01% dithizone and its subsequent treatment following the procedure described by Shaw and Dean (1952). Weigh 2.5 g soil into a plastic bottle. Add 25 ml of Zn-free 1.0 N ammonium acetate (pH 7.0) and 25 ml of 0.01% dithizone in CCl_4 (Zn-free). Stopper and shake the contents for one hour on a mechanical shaker.

Transfer the soil-extractant equilibrium suspension into a 250 ml separatory funnel and place it on a stand to allow the layers to separate. Remove the bottom organic phase into a dry 60 ml centrifuge tube. Centrifuge the suspension for 10 minutes at 2400 rpm to develop a continuous CCl_4 phase.

Remove 10 ml aliquot from the CCl_4 phase by lowering a 10 ml pipette into the CCl_4 phase. Keep on blowing air through the pipette tip as it is lowered through the aqueous and soil layer. Take care also that the pipette tip does not touch the walls of the centrifuge tube. The water and soil adhering to the pipette tip on withdrawing pipette from the CCl_4 phase be washed with a jet of Zn-free distilled water from a wash bottle. Adjust the volume to the mark and wash the pipette tip again.

Transfer the 10 ml aliquot from the CCl_4 phase into a clean separatory funnel containing 50 ml of 0.02N Zn-free HCl. Agitate the mixture for 3 minutes to effect extraction of Zn from CCl_4 phase into the HCl phase. Discard the CCl_4 phase as it contains interfering elements and some oxidized dithizone. Rinse the aqueous phase twice each time with 10 ml portions of CCl_4. Agitate the contents by hand and discard the CCl_4 each time.

7. Determination of Zn extracted into the aqueous phase. After extraction of Zn from CCl_4 phase into aqueous phase, when rinsing of aqueous phase with CCl_4 is over, proceed with the addition of 5 ml of 0.4 M ammonium citrate as described in step 1 of procedure. Thereafter follow step 2 and 3 of procedure 2.24.1.3.

2.24.1 Calculation

$$\text{Available Zn in soil} = \frac{A \times C \times E}{B \times D \times F}$$

Where A = μg Zn in aliquot of Zn dithizonate by reference to the standard graph for Zn.

B = ml aliquot taken from CCl_4 phase containing Zn-dithizonate for determining absorbance.

C = ml of dithizone reagent used for extracting Zn from aqueous phase.

D = Volume of dithizone extract (CCl_4 phase) taken for extracting Zn into aqueous phase.

E = Volume of dithizone reagent in CCl_4 equilibrated with the soil.

F = Mass of soil in g.

2.24.2 Method 2

2.24.2.1 Principle

As described in section 2.24.1.1

2.24.2.2 Apparatus/Reagents

Apparatus

Same as described in Method 1 (section 2.24.1.2). However, the separatory funnel used in Method 2 is of 250 ml capacity.

Reagents

Almost same reagents as listed in method 1 (section 2.24.1.2). 9% solution of ammonium citrate is used in Method 1. However, in method 2, ammonium citrate used is of higher concentration *i.e.* 11.25%.

2.24.2.3 Procedure

1. Removal of Cu present as impurity in standard Zn solution and plant digest. Transfer 10 ml of 1N HCl (Zn–free) into a 250 ml separatory funnel and add 1 to 10 μg Zn using 2 ppm Zn solution. Then add 5 ml of 11.25% ammonium citrate solution and few drops of thymol blue indicator. Thereafter, add 1N NH_4OH (Zn- free) to adjust the pH to 2.5 which is indicated by change in the colour of solution from red to yellow. Now, add 5 ml of 0.01% dithizone reagent in CCl_4. Shake the separatory funnel for 5 minutes and then place it on a stand for few minutes to allow the layers to separate. Discard the CCl_4 phase containing Cu.

In case of plant sample, prepare 250 ml of triacid digest using 5 g finely ground plant sample. Transfer 5 ml of the tri acid digest.

2. Extraction of Zn from aqueous phase (free from Cu) by 0.01% dithizone reagent in CCl_4. Add 5 ml of 11.25% ammonium citrate to the aqueous phase and adjust its pH to 8.3 by adding NH_4OH solution to the phenolphthalein end point. Add 10 ml of 0.01% dithizone in CCl_4. Agitate the separatory funnel. Place it on a stand and allow the layers to separate.

Remove the CCl_4 phase containing Zn into a clean 250 ml capacity separatory funnel. Repeat the extraction of Zn with 10 ml portions of 0.01% dithizone in CCl_4 twice and collect the organic phase into the same separatory funnel. When the aqueous phase turns orange in colour, it becomes free from Zn. Then add 10 ml of pure CCl_4 to extract droplets of dithizone in aqueous phase. Collect the Zn-dithizonate into the same 250 ml separatory funnel.

3. Removal of other impurities in the metal dithizonates. Add 50 ml of 0.02N HCl into the separatory funnel containing metal dithizonates, and shake the contents for 2 minutes. Discard the CCl_4 phase. Add pure CCl_4 to extract adhering droplets of dithizone and discard the CCl_4 phase.

4. Extraction of Zn in the presence of carbamate. Add 5 ml of 11.25% ammonium citrate solution (Zn free) into 50 ml of 0.02N HCl containing Zn obtained on completion of the 3rd step. Then adjust the solution pH to 8.3 by adding 1N NH_4OH (Zn free) to the phenolphthalein end point. Thereafter, add 10 ml of 0.2% carbamate solution and 10 ml of 0.01% dithizone reagent in CCl_4. Shake the contents for 2 minutes. Place the separatory funnel on a stand and allow the layers to separate.

Transfer Zn-dithizonate in CCl_4 into a clean separatory funnel. Then add 25 ml of 0.01N NH_4OH solution, and shake well. Then place the separatory funnel on a stand and allow the layers to separate. Excess dithizone will be in the aqueous phase and Zn dithizonate in CCl_4 will be in the lower phase.

5. Removal of Zn–dithizonates and measurement of its optical density after dilution with pure CCl_4. Lower a dry pipette blowing air through the pipette while introducing the pipette through the aqueous layer into the bottom organic phase containing the Zn–dithizonates. Pipette 5 ml of the Zn dithizonate into 25 ml of volumetric flask. Dilute the contents to 25 ml mark with pure CCl_4. Measure the optical density of the diluted Zn dithizonates at 535 mμ (nm) wave length adjusting 0 optical density with pure CCl_4.

6. Standard graph for Zn. Note the absorbance (optical density) of the Zn dithizonate in CCl_4 phase obtained by taking 2 to 10 μg Zn. Then plot the absorbance on a "Y" axis against the corresponding amount of Zn in μg on "x" axis of a linear scale graph paper. Draw a straight line by connecting the absorbance value of pure CCl_4 *i.e.* 0 μg Zn with the absorbance value obtained with the maximum amount of Zn taken *i.e.*, 10 μg Zn.

7. Collection and preparation of the soil sample. Follow the procedure described in step 5 of Method 1 (Section 2.24.1.3).

8. Extraction of Zn from soil by 1N ammonium acetate (pH 7.0) + 0.01% dithizone in CCl_4 and its subsequent treatment following the procedure described by Shaw and Dean (1952). Follow the procedure described in step 6 of method 1 (Section 2.24.1.3).

9. Determination of Zn extracted into the aqueous phase. After extraction of Zn from CCl_4 phase into aqueous phase, when rinsing of aqueous phase is over, proceed with the addition of 11.25% ammonium citrate as described in step 2 of method 2 (section 2.24.2.3). Thereafter, follow steps from 3 to 5 of method 2.

2.24.2.4 Calculation

$$\text{Available Zn in soil (ppm)} = \frac{A \times C \times E}{B \times D \times F}$$

Where, A = μg Zn in aliquot of Zn dithizonate by reference to the standard graph for Zn.

B = ml aliquot taken from CCl_4 phase containing Zn-dithizonates for determining absorbance.

C = ml of dithizone reagent used for extracting Zn in the presence of carbamate.

D = Volume of dithizone extract (CCl_4 phase) taken for extracting Zn into aqueous phase.

E = Volume of dithizone reagent in CCl_4 equilibrated with the soil.

F = mass of soil in g.

2.24.2.5 Comments

The dithizone extracble Zn from soil was first determined by Shaw and Dean (1952). They reported that the critical level of Zn in soil at pH 5.0 would be 0.3 ppm and it would increase linearity to 2.3 ppm at pH 8.0.

In neutral and alkaline soil, dithizone extractable Zn was determined by Viets *et. al.* (1953). They observed that sweet corn showed symptoms of Zn deficiency when dithizone extractable Zn in soil was 0.4 ppm or less. They reported also that on 84% of the soils, the crops responded to Zn application when dithizone extractable Zn was less than 0.55 ppm. On 74% of soils, the crop did not respond to Zn application when dithizone extractable Zn was more than 0.55 ppm. They proposed 0.55 ppm of dithizone extractable Zn in soil as the critical level.

Massey (1957) observed negative correlation between Zn uptake and soil pH. Wear (1959) found no response to Zn application when the soil pH was less than 5.9 or when the dithizone extractable Zn was more than 0.9 ppm.

Apart from pH, the critical level of Zn in soil is also influenced by the level of P and free lime. Lindsay and Norvell (1969) reported that Zn sensitive crops would require Zn application when grown on soils low in Zn or else marginal in Zn but high in P. Nelson, Boawn and Viets (1959) reported that in calcareous soils, the presence of free lime is of more importance than the soil pH in the interpretation of soil tests for Zn. They reported the critical level of Zn for calcareous soils to be near 1.0 ppm with no titrable alkalinity. The critical level of Zn increased to 7.5 ppm when the titrable alkalinity was 400 meq/100 g soil.

The other reagents which have been used for extracting Zn from soils are 0.1 N HCl, 2N $MgCl_2$, EDTA + $(NH_4)_2CO_3$ and DTPA + $CaCl_2$(pH 7.3). The use of chelating agent-DTPA has been found to be satisfactory for the assessment of soil Zn. DTPA extractable Zn is, however, conveniently determined by atomic absorption spectrophotometer.

REFERENCES

Black, C.A. 1965. Methods of Soil Analysis. Part–2. Chemical and Microbiological Properties–Amer. Soc. Agron. Madison, Wisconsin, U.S.A.

Holmes, R.S. 1945. Determination of total copper, zinc, cobalt, lead in soils and soil solution. Soil Sci. 59 : 77–84.

Jackson, M.L. 1958. Soil Chemical Analysis. Prentice Hall of India Pvt. Ltd., New Delhi, India.

Lindsay, W.L., and W.A. Norvell 1969. Equilibrium relationship of Zn^{2+}, Fe^{2+}, Ca^{2+} and H^{+} with EDTA and DTPA in soils. Soil Sci. Soc. Amer. Proc. 35 : 62–68.

Massey. H.F. 1957. Relationship between dithizone extractable zinc in the soil and zinc uptake by corn plants. Soil Sci. 33 : 123–129.

Nelson, J.L., L.C. Boawn, F.G. Viets, Jr. 1959. A method for assessing zinc status of soils using acid extractable zinc and "titratable alkalinity" values. Soil Sci. 88 : 275–283.

Shaw, E., and L.A. Dean. 1952. The use of dithizone as an extractant to estimate the zinc nutrient status of soils. Soil Sci. 73 : 341–344.

Viets, F.G. Jr., L.C. Boawn, C.L. Crawford, and C.E. Nelson, 1953. Zinc deficiency in corn in Central Washington. Agron. J. 45 : 559–565.

2.25 COPPER IN SOIL EXTRACTS BY CARBAMATE METHOD-CHENG AND BRAY (1953) AND HOLMES (1945)

2.25.1 Method 1

2.25.1.1 Principle

Sodium diethyl dithiocarbamate reacts with copper to give copper diethyl dithiocarbamate–a golden yellow complex having the following formula.

$$(C_2H_5)_2 = N - C \begin{smallmatrix} S \\ S \end{smallmatrix} Cu_2$$

Copper carbamate is soluble in organic solvents such as isoamyl alcohol, isoamyl acetate and carbon tetrachloride. In this Cu carbamate is extracted by CCl_4 to decrease the effect of interfering ions such as Ca, Mg, Fe, Mn and Ni. Incorporation of mixture of versenate (ethylene diamine tetra acetic acid) and citrate solution in the reaction mixture eliminates all interferences except that from bismuth. However, bismuth is ordinarily present in negligible amounts in both soils and plants. Versenate forms a soluble chelate complex with bi and trivalent metals in the following way :

$Na_2H_2Y_4 + Me^{++} \rightarrow Na_2MeY_4 + 2H^+$ where Me represents a metal.

2.25.1.2 Reagents

(*i*) Cu-free distilled water. Redistill distilled water in a pyrex glass distill.

(*ii*) 1.0% carbamate solution. Dissolve 1.0 g of sodium diethyl dithio carbamate in 100 ml of Cu–free distilled water and filter the solution.

(*iii*) Versenate-Citrate mixture. Dissolve 20g of ammonium citrate and 5g of disodium salt of ethylene diamine tetra acetic acid in 100 ml of redistilled water (Zn free).

(*iv*) Ammonium Hydroxide (BDH–AR Sp. Gr. 0.91).

(*v*) Carbon tetrachloride (BDH–AR). Distill pure CCl_4 before using.

(*vi*) 100 ppm Cu solution. Dissolve 0.3930 g of $CuSO_4.5H_2O$ (BDH–AR) in redistilled water containing 5 ml of conc. H_2SO_4 (BDH–AR). Dilute the solution to 1 l mark with redistilled water.

(*vii*) 1 ppm. Cu solution. Dilute 10 ml of 100 ppm Cu solution to 1 l mark with redistilled water after adding 5 ml of conc. H_2SO_4 (BDH–AR).

2.25.1.3 Procedure

1. Formation of Cu–carbamate complex. Transfer 0, 2.5, 5.0, 7.5, 10.0, 12.5 and 15.0 ml of 1 ppm Cu solution into separate 250 ml capacity separatory funnels. Add 10 ml of versenate–citrate mixture into each separatory funnel. Then add 1–2 ml of NH_4OH (BOH–AR Sp.Gr. 0.91) to adjust pH of the solution to the phenol phthalein end point. Shake the separatory funnel to mix the solution well. Thereafter add 2 ml of 1.0% carbamate solution and mix the contents.

2. Extraction of Cu–carbamate complex from the reaction mixture. Add 10 ml of distilled pure CCl_4 into the separatory funnel and shake the contents vigorously for 3–4 minutes. Place the separatory funnel on a stand and allow the layers to separate. Collect the CCl_4 phase (lower layer) containing Cu–carbamate in a dry 25 ml graduated cylinder.

Add 5 ml of CCl_4 and shake the separatory funnel for 1 minute. Allow the layers to separate. Collect the lower CCl_4 phase into the same 25 ml graduated cylinder. Make the volume of CCl_4 phase (containing Cu-carbamate) collected in 25 ml graduated cylinder to 15 ml by adding distilled (pure) CCl_4.

3. Measurement of absorbance (optical density) of Cu–carbamate complex in CCl_4. Measure the absorbance (optical density) of the Cu–carbamate complex after 15 minutes in a photoelectric colorimeter using green filter or else in a spectrophotometer at 500 nm (mμ) wave length adjusting 0 optical density with pure CCl_4.

However, Jackson (1958) recommended 440 mμ filter for Cu-carbamate complex (golden brown colour).

4. Standard graph for Cu. Record the optical density of Cu–carbamate complex in CCl_4 obtained with 0–15 μg Cu deducting the blank reading *i.e.* optical density of Cu carbamate complex with 0 μg Cu. Thereafter, plot the optical density in "Y" axis of a linear graph paper against Cu in μg/15 ml of Cu–carbamate complex in CCl_4 plotted on the "x" axis. Connect 0 optical density adjusted with pure CCl_4 and the optical density of Cu–carbamate complex corresponding to 15 μg Cu to obtain a straight line curve.

5. Extraction of Cu from the soil by 1.0 N $NH_4C_2H_3O_2$ (pH 7.0) and digestion of the extract following the procedure of Cheng and Bray (1963). Leach 50g air dry 2 mm soil sample with the 100 ml portions of 1.0N $NH_4C_2H_3O_2$ (pH 7.0). Collect the leachate in a 300ml tall beaker and evaporate the contents to dryness on a water bath. Cool the beaker and then add 10 ml of distilled conc. HNO_3 for initial digestion. After half an hour, heat the contents to boil. Remove the beaker from hot plate and let it cool. Then add 5 ml of distilled conc. HNO_3 and 60% $HClO_4$. Place the beaker on a hot plate, cover it with a watch glass leaving the beaker partially open. Heat the contents until digestion is complete *i.e.* when the contents is reduced to a very small volume without any charring. Cool and add 10 ml of approx. 6N HCl and evaporate it to dryness. Cool and add 20 ml of dilute (1 + 4) distilled HNO_3 and heat to dissolve the contents.

6. Formation and extraction of Cu-carbamate complex. Pipette 10 ml of Cu dissolved in HNO_3 into a 250 ml separatory funnel and follow the procedure from step 1 to 3 beginning with the addition of 10 ml of versenate-citrate mixture.

For the blank determination take 200 ml of 1N $NH_4C_2H_3O_2$ (pH 7.0) in 300 ml tall beaker and follow step 5 of the procedure beginning with the evaporation of contents to dryness and then complete step 6. Thereafter follow the step 1 to 3 of procedure in Method 1.

7. Determination of Cu in Cu-carbamate complex. Note the optical density of Cu-carbamate complex recorded and refer to the standard graph for Cu. Locate the value of optical density on the "Y"axis of standard graph for Cu and draw an imaginary line from that point parallel to the base which will intersect the calibration curve. From the point of intersection, draw an imaginary line perpendicular to the base. Note the content of Cu in μg/15 ml of Cu-carbamate complex given on the "x" axis.

2.25.1.4 Calculation

$$\text{Available Cu in soil (ppm)} = \frac{A \times B}{C \times D}$$

Where, A = μg Cu/15 ml of Cu-carbamate complex in CCl_4 (Sample-blank)

B = Volume of dilute distilled HNO_3 used to solubilize Cu after complete digestion of 1N NH_4 $C_2H_3O_2$ (pH 7..0) extract from soil.

C = ml aliquot of dilute HNO_3 solution containing Cu taken for formation of Cu-carbamate complex

D = mass of soil in g taken for extraction of Cu by 200 ml of 1N NH_4-$C_2H_3O_2$ (pH 7.0)

2.25.2 Method 2-Holmes (1945)

2.25.2.1 Principle

The Cu and Zn extracted from soil as dithizonates in CCl_4 following the procedure of Shaw and Dean (1952) is transferred into an acid solution 0.02N with respect to HCl. Then, Cu is extracted at pH 2.5 in the presence of ammonium citrate by dithizone in CCl_4. The Zn dithizonate is transferred into the aqueous phase which is retained for Zn determination.

The CCl_4 phase containing Cu-dithizonate is treated with acidic potassium iodide to remove bismuth and other metals that form complex with iodine.

The Cu-dithizonate in CCl_4 is removed and washed with 0.1N NH_4OH to make it free from excess dithizone. The Cu-dithizonate is then transferred into a tall beaker and it is first digested with conc. HNO_3 and then with conc. $HClO_4$. The contents are heated on a hot plate until almost all of $HClO_4$ is expelled off. The traces of HNO_3in the residue is diluted by adding redistilled water, and then it is removed by heating the solution to boil.

The Cu in the residue is dissolved in hot 10 N HCl and then, its pH is adjusted to 8.5 by adding NH_4OH. The carbamate reagent is added to form Cu-carbamate complex. The optical density of Cu-carbamate complex is measured at 440 nm wave length or filter. The Cu content is determined by reference to the standard graph obtained on plotting optical density of Cu- carbamate complex similarly prepared with 0 to 10 microgram Cu. The procedure is adopted from Holmes (1945).

2.25.2.2 Apparatus/Reagents

Apparatus

Pear shaped separatory funnel-125 ml, 250 ml, 1 l and 2 l capacity, separatory funnel stand, volumetric flasks 100 ml, 1l, and 2l capacity, complete glass distill with 2 l capacity boiling flasks, and heating bowl with temperature regulator : tall beaker 100 ml capacity, mechanical shaker and photoelectric colorimeter with 440 nm filter or a spectrophotometer.

Reagents

(*i*) Redistilled water or distilled water from a metal still passed through an ion exchange column.

(*ii*) Approx. 6N HCl.

(*iii*) 1.0N HCl.

(*iv*) 0.02 N HCl.

(*v*) 1.0N NH_4OH.

(*vi*) 0.1% Phenolphthalein.

(*vii*) CCl_4 (BDH–AR).

(*viii*) 0.0125% Dithizone in CCl_4.

(*ix*) Thymol blue indicator.

(*x*) 0.2% carbamate solution.

(*xi*) 22.5% Ammonium citrate solution.

(*xii*) 1.0 N $NH_4C_2H_3O_2$ (pH 7.0). The above reagents be prepared following the procedure described in Zn determination in soil–method 1 (See section 2.24.1.2).

(*xiii*) 0.1 N Sodium sulphite. Dissolve 0.63 g of Na_2SO_3 in redistilled water and dilute the solution to 100 ml.

(*xiv*) 2.0% Potassium iodide solution. Dissolve 10 g of potassium iodide in 190 ml of Cu free distilled water. Add 5 ml of 1.0 N HCl and 5 ml of 0.1 N sodium sulfite to reduce the free iodine. Transfer the solution into a 1l capacity separatory funnel. Then add 10 ml of 0.125% dithizone in CCl_4. Shake the separatory funnel and discard the dithizone extract in CCl_4 until no discolouration of dithizone occurs. Discard the dithizone extract and remove excess of dithizone by repeated extraction with CCl_4. If upon standing, free iodine, occur in the solution, it should be reduced with sodium sulfite or removed with CCl_4. Thereafter, dilute the contents to 500 ml with Cu-free distilled water.

(*xv*) 100 ppm Cu solution. Dissolve 0.3930 g of $CuSO_4.5H_2O$ (BDH–AR) in redistilled water containing 5 ml of conc. H_2SO_4 (BDH–AR). Dilute the solution to 1l mark with redistilled water.

(*xvi*) 10 ppm Cu solution. Transfer 10.0 ml of 100 ppm Cu solution into a 100 ml volumetric flask. Then add 0.5 ml of conc. H_2SO_4 (BDH–AR) and dilute the solution to 100 ml with redistilled water.

2.25.2.3 Procedure

1. Extraction of Cu by 0.0125% dithizone in CCl_4. Take an aliquot of Cu solution containing 0 to 10 microgram Cu and transfer it to a clean 125 ml separatory funnel containing 50 ml of 0.02 N HCl prepared from redistilled 6N HCl. Add 5 ml

of the 22.5% ammonium citrate solution. Titrate the acidity with NH_4OH until the pH is 2.5 using thymol blue as the indicator (when red colour of the indicator begins to turn yellow). Then, add exactly 5.0 ml of 0.0125% dithizone in CCl_4 and shake for 5 minutes.

Transfer the CCl_4 phase to a clean 125 ml separatory funnel. Repeat the extraction with a less concentrated dithizone in CCl_4. Combine the two dithizonate extraction for Cu determination, and retain the aqueous phase for the zinc determination.

2. Removal of bismuth and other metals that form complex ion with iodine. To the Cu–dithizonate in CCl_4 in the 125 ml of separatory funnel, add 10 ml of acidic potassium iodide solution and shake for 2 minutes. Transfer Cu dithizonate in CCl_4 phase into a clean separatory funnel.

3. Removal of excess dithizone from Cu-dithizonate in CCl_4 phase. Add 50 ml of 0.1N ammonium hydroxide into 125 ml separatory funnel containing Cu-dithizonate in CCl_4. Shake for 1 minute and allow the layers to separate. Transfer ammonia washed Cu-dithizonate quantitatively to a tall 100 ml beaker with spout.

4. Decomposition of Cu dithizonate in CCl_4 by digestion first with conc. HNO_3 (distilled BDH–AR) and then with conc. $HClO_4$ (BDH–AR). Add 0.5 ml of conc. HNO_3 (distilled) into 100 ml tall beaker containing Cu dithizonate. Cover the beaker with a watch glass and evaporate the contents on a water bath. Thereafter, cool the beaker and add 0.5 ml of conc. $HClO_4$ into the beaker and evaporate the contents on a hot plate until almost all of the perchloric acid is driven off. Then cool the beaker and add 10 ml of redistilled water and heat to boil the contents to remove all traces of nitric acid.

5. Formation of Cu-carbamate complex and its extraction by CCl_4. When nitric acid is expelled off, cool the beaker and add 10 ml of hot 10.0 N HCl to dissolve the residue and transfer the solution into a clean 125 ml separatory funnel. Again rinse the beaker with 10 ml of hot 1N HCl and transfer the rinsing into the same 125 ml separatory funnel. Thereafter, add 2-3 drops of 0.1% phenolphthalein indicator, and then add ammonia solution to raise the pH to 8.5.

Now, add 5 ml of 0.2% carbamate reagent and exactly 10.0 ml of pure CCl_4. Shake the contents for 3 minutes and allow the layers to separate. Draw off the CCl_4 phase containing Cu–carbamate complex in a 25 ml dry graduated cylinder. Then add 5.0 ml of pure CCl_4 and shake the separatory funnel for 1 minute. Allow the layers to separate. Collect the lower CCl_4 phase into the same 25 ml graduated cylinder. Make the volume of CCl_4 phase containing Cu-carbamate to 15 ml with pure CCl_4.

6. Measurement of the optical density of Cu carbamate in CCl_4. Set a photoelectric colorimeter using a 440 millimicron blue filter or else set a spectrophotometer at 440 nm wavelength. Adjust 0 optical density with pure CCl_4 and measure the optical density of the Cu-carbamate complex obtained with 0 to 10 microgram Cu.

7. Standard graph for Cu. Record the optical density of Cu-carbamate complex in CCl_4 obtained with 0–10 microgram Cu deductiung the blank reading *i.e.* optical

density of Cu-carbamate complex with 0 microgram Cu. Thereafter, plot the optical density on "y"–axis of a linear scale graph paper against Cu in microgram/15 ml of Cu–carbamate complex in CCl_4 plotted on the "x"axis. Connect 0 optical density adjusted with pure CCl_4 and the optical density of Cu-carbamate complex corresponding to 10 microgram Cu to obtain a straight line curve.

8. Collection and preparation of soil sample. Follow the procedure described in step 5, Method 1 of Zn determination in soil (See section 2.24.1.3).

9. Extraction of Cu from soil by 1N $NH_4C_2H_3O_2$ (pH 7.0) + 0.0125% dithizone in CCl_4. Follow the procedure described in step 6 Method 1 of Zn determination in soil (See section 2.24.1.3).

10. Determination of Cu in the soil extract. Transfer 10 ml aliquot from 25 ml of Cu + Zn dithizonate in CCl_4 extracted from 2.5g soil into a 125 ml separatory funnel containing 50 ml of 0.02N HCl prepared from distilled 6N HCl. Thereafter proceed with the addition of 5 ml of ammonium citrate in step 1 and follow the subsequent procedure outlined in step 2–6 of Cu determination in soil by Method 2 (Section 2.25.2.3).

11. Determination of Cu in Cu-carbamate complex. Note the optical density of Cu carbamate complex recorded, and refer to the standard graph for Cu. Locate the value of optical density on the "y" axis of standard graph for Cu, and draw an imaginary line from that point parallel to the base which will intersect the calibration curve. From the point of intersection, draw an imaginary line perpendicular to the base. Note the content of Cu in microgram per 15 ml of Cu-carbamate complex given on the "x" axis. Alternatively divide the optical density of Cu-carbamate complex from the sample and blank aliquot by the slope of the straight line graph *i.e.* optical density/microgram Cu in Cu-carbamate complex to obtain the content of μg Cu/15 ml of Cu-carbamate complex.

2.25.2.4 Calculation

$$\text{Available Cu in soil (ppm)} = \frac{(S - B) \times E}{A \times M}$$

Where

S and B are microgram Cu/15 ml of Cu–carbamate complex in CCl_4 determined by reference to the standard graph for Cu after measurement of the optical density in sample and blank respectively.

E is the volume of 0.0125% dithizone in CCl_4 used in ml for complexing Zn and Cu extracted from soil by 1N $NH_4C_2H_3O_2$ (pH 7.0).

A is ml aliquot of Zn + Cu dithizonates in CCl_4 taken for Cu determination.

M is mass of soil in gram taken for the extraction of Zn and Cu.

2.25.2.5 Comments

The copper from soil has been extracted by hot H_2O, NH_4NO_3, $NH_4C_2H_3O_2$, dilute HNO_3, dilute HCl, 0.02 to 0.05M EDTA, citrate-EDTA and dilute dithizone solution. Extractable Cu as well as Cu uptake may be influenced by factors such as pH, organic matter level and the level of Al and Fe in soil.

Van Luit and Henkins (1967) reported that the Cu concentration in grasses was found to be highest when soil pH was in the range of 4.5 to 5.5. They also noted that the Cu concentration in foliage decreased by 1 ppm for each 1% increase in humus up to a content of 5%. The Cu deficiency is usually associated with organic soils. The critical level of Cu should be high in organic soils.

Blevins and Massey (1959) reported a negative correlation between Cu uptake and Al extracted from soil by 0.01N $CaCl_2$. Low amount of Fe and Al tend to increase Cu uptake. Thomas (1968) reported that 80% of total Cu in soil was found to be bound to hydrated Fe oxide present, and so the extracting agent for Cu should release a portion of Fe bound Cu. Excess P reduce Cu uptake whereas excess Cu reduce Mo availability.

Cheng and Bray (1953) developed citrate-EDTA method for Cu determination in soil. They found that the Cu extracted by the reagent was closely associated with amount of Cu extracted by 0.1N HCl. Blevins and Massey (1959) reported that Cu uptake by millet plant gave significant correlation with the Cu extracted by citrate EDTA and dithizone reagent.

Grewal, Lal and Randhawa (1969) found 1N $NH_4C_2H_3O_2$ to be a better extractant for Cu in soils than 1N HNO_3, 1N HCl and 0.02 M EDTA as judged by response of maize and wheat to Cu in pot experiment.

The critical level of Cu extracted by $NH_4C_2H_3O_2$ (pH 4.8) has been reported to be 0.2 ppm.

REFERENCES

Black, C.A. (1965). Methods of Soil Analysis. Part 2. Amer. Soc. Agron. Madison, Wisconsin, USA.

Blevins, R.L., and H.F. Massey (1959). Evaluation of two methods of measuring available soil copper, and the effects of soil pH and extractable aluminium on copper uptake by plants. Soil. Sci. Soc. Amer. Proc. 23 : 296–298.

Cheng, K.L., and R.H. Bray (1953). Two specific methods of determining Cu in soil and plant material, Anal. Chem. 25 : 655–659.

Grewal, J.S., C. Lal and N.S. Randhawa (1969). Evaluation of different methods for the determination of available copper in soils of Ludhiana. Indian J. Agr. Sci. 39 : 877–885.

Holmes R.S. (1945). Determination of total copper, zinc, cobalt, and lead in soils and soil solutions. Soil Sci. 59 : 77–84.

Jackson, M.L. (1958). Soil Chemical Analysis. Prentice Hall of India Pvt. Ltd. New Delhi.

Thomas, F. (1968). Bound copper in six soils of different geological origin Chem. Erde 27 : 369–380.

Van Luit, B., and C.H. Henkins (1967). Effect of the copper status of the soil on the copper content of grass and clover. Versl. Landbouwk, Onder 2. No. 695, 33p.

2.26 DETERMINATION OF BORON IN SOIL

2.26.1 (Method 1) : Determination of boron in soil extract by reaction with carmine–Hatcher and Wilcox (1950)

2.26.1.1 Principle

Micro amounts of boron in soil extracts is reacted with quinalizarin or carmine in sulphuric acid medium and with curcumin in hydrochloric acid medium to form colour complex for colorimetric determination.

When curcumin is used to form colour complex with boron, the colour intensity is measured at 540 mμ wavelength up to 1.5 μg B in aliquot of test solution, and at 580–600 mμ wavelength if boron content is greater than 1.5 μg B in aliquot of test solution. The colour intensity is measured at 620 and 655 mμ wave length, when colour complex with B is formed by quinalizarin and carmine respectively.

Boron from soil is extracted by the procedure of Berger and Truog (1939) employing hot distilled water as an extractant.

2.26.1.2 Apparatus/Reagents

Apparatus

Erlenmeyer flasks, polythene bottles, volumetric flasks, beaker, pipettes, water bath, hot plate, muffle furnace, photoelectric colorimeter or a spectrophotometer.

Use glasswares made from B–free glass. The glass wares be weathered using conc. H_2SO_4 and then it be rinsed with conc. HCl. Thereafter, it be thoroughly washed by tap water and rinsed with single distilled water from a metal distill.

Reagents

(*i*) Distilled water. Use water distilled from a metal distill.

(*ii*) Saturated calcium hydroxide solution. Add 10g of $Ca(OH)_2$ to 500 ml of distilled water. Shake well and allow to settle.

(*iii*) Conc. Sulphuric acid (AR–grade).

(*iv*) Conc. Hydrochloric acid (AR grade).

(*v*) 0.092% carmine solution. Dissolve 0.092 g of carmine in 100 ml of conc. H_2SO_4

(*vi*) Approx. 6N HCl. Mix equal volume of conc. HCl (AR grade) and distilled water

(*vii*) 0.36 N sulphuric acid. Dilute 10 ml of conc. H_2SO_4 (AR grade) to 1000 ml with distilled water from a metal distill.

(*viii*) 40% Potassium carbonate solution. Dissolve 40g of K_2CO_3 (AR grade) in distilled water and make the volume to the 100 ml mark with distilled water.

(*ix*) 50% Calcium chloride solution. Dissolve 50g of $CaCl_2$ (AR grade) in distilled water and make the volume to 100 ml.

(*x*) 100 ppm B solution. Dissolve 0.5718 g of H_3BO_3(AR grade) in distilled water and make the volume to 100 ml.

(*xi*) 0 to 7 ppm B solution. Transfer 0, 1, 2, 3, 4, 5, 6 and 7 ml of 100 ppm B solution into separate 100 ml volumetric flasks, and dilute the solution to 100 ml with distilled water.

2.26.1.3 Procedure

1. Formation of boron-carmine coloured complex with dilute standard B solution. Transfer 2 ml each of dilute standard B solution (0 to 7 ppm) into separate Erlenmeyer flasks or polythene plastic bottles. Then add 2 drops of conc. HCl and 10 ml of conc. H_2SO_4 to each container. Shake the containers and cool to room temperature. Then, add 10 ml of 0.092% carmine solution to each container and again shake the contents. Allow the solution to stand for 45 minutes for full development of the boron-carmine coloured complex.

2. Measurement of the absorbance of boron-carmine coloured complex. Place colorimeter tube or cuvette filled with coloured complex formed with 0 μg of B (*i.e.* 2 ml distilled water) into the socket provided in photoelectric colorimeter or spectrophotometer. Thereafter, fitt 655 mμ filter into the colorimeter or regulate the wave length knob to 655 mμ in the spectrophotometer and switch on the instrument. Let it run for 10 minutes to warm.

Then set the absorbance to zero and measure the absorbance of the subsequent standards in sequence recording the absorbance against the corresponding μg B taken for the formation of boron-carmine coloured complex.

3. Standard graph for boron. On a linear scale graph paper, plot the absorbance value on the ordinate and the corresponding μg of B taken for the formation of boron-carmine complex on the abscissa. After plotting the absorbance recorded, connect zero absorbance with the maximum absorbance recorded by a straight line to obtain a straight line standard curve for B.

4. Extraction of hot water soluble boron from soil by the procedure of Burger and Truog (1939). Transfer 60 g of air dry 2 mm soil sample into a B–free Erlenmeyer flask or polythene bottle. Then add 120 ml of metal distilled or demineralized water. Attach a reflex condenser to the Erlenmeyer flask–a thin glass tubing about 1 m long may be inserted into a cork, and placed tightly into the mouth of the erlenmeyer flask. Heat the flask until the first sign of boiling and then reflux the contents for 5 minutes. Allow the flasks to cool without removing the condenser.

Alternatively, stopper the polythene bottle containing 60g soil and 120 ml distilled water and keep the bottle in a hot water bath for 30 minutes with frequent stirring. Thereafter, remove the polythene bottle from the water bath and let it cool.

5. Filtration of the soil extracts. When the polythene bottle has cooled to the room temperature, open the bottle and add 3 drops of 10% $CaCl_2$, and filter the suspension with suction on a Buchner funnel lined with B-free filter paper, or else decant the suspension into centrifuge tubes, and run the centrifuge for 15 minutes at 2000–2450 rpm to obtain a clear solution. Collect the clear solution in a B-free glass container.

6. Evaporation of the filtrate, ignition of the residue, and solubilization of the ignited residue. Transfer 80 ml of the filtered extract into a large silica dish. Then

add 2 ml of saturated $Ca(OH)_2$ solution, and 5 drops of K_2CO_3. Evaporate the extract on a water bath to dryness. Then keep the silica dish in a muffle furnace, and ignite it gently at 450°C for one hour. Thereafter, switch off the muffle furnace, and let it cool to room temperature. Remove the silica dish, and add measured volume of 0.36 N H_2SO_4 to dissolve the residue stirring with a glass rod and transfer the solution into centrifuge tubes, and run the centrifuge at 2000–2440 rpm for 10 minutes. Decant off the clear solution. Similarly run a blank.

7. Formation of boron-carmine coloured complex using 2 ml aliquot of clear (test) solution obtained from step 6 and measurement of its absorbance by a photoelectric colorimeter or a spectrophotometer. Transfer 2 ml aliquot of the clear (test) solution into a polythene bottle and follow step 1 and 2 of procedure given in Method 1 (section 2.26.1.3) beginning with the addition of 2 drops of conc. HCl and 10 ml of conc. H_2SO_4.

8. Estimation of B content in boron-carmine coloured complex. Record the absorbance of the boron-carmine coloured complex obtained with 2 ml of sample test solution setting 0 absorbance with a coloured complex obtained with 2 ml of blank test solution. Then, locate the absorbance of sample test solution on the ordinate of standard curve for B, and draw an imaginary line parallel to the abscissa intersecting the straight line standard curve. From the point of intersection, draw a line perpendicular to the abscissa and locate the point of intersection at the abscissa and read the content of boron in μg. Alternatively, divide the absorbance of the sample test solution aliquot by the slope of the standard curve to obtain the content of boron in 2 ml aliquot of the sample test solution in μg.

2.26.1.4 Calculation

$$\text{ppm B in soil} = \frac{A \times B \times C}{D \times E \times F}$$

Where

A is μg B in aliquot of sample test solution as determined by reference to the standard graph for boron.

B is total volume (ml) of 0.36N H_2SO_4 used for solubilizing B after ignition of residue obtained on evaporation of 80 ml of soil extract.

C is total volume of distilled water used for extraction of B from the soil.

D is ml aliquot of sample test solution taken for developing boron-carmine complex.

E is ml of water extract from soil taken for evaporation in order to prepare the sample test solution.

F is mass of air dry 2 mm soil sample taken for the extraction of B by hot distilled water.

2.26.2 (Method 2). Determination of boron in test solution prepared from soil extracts by reaction with quinalizarin-sulphuric acid solution-Berger and Truog (1939)

2.26.2.1 Principle. As described in section 2.26.1.1

2.26.2.2 Apparatus/Reagents

Apparatus

Glasswares and instruments etc are listed in Method 1 (section 2.26.1.2). In addition a dispenser for quinalizarin-H_2SO_4 solution as described in Jackson (1973) is required. The glasswares must be boron-free. Ordinary soft glass is also satisfactory.

Reagents

Reagents other than described in Method 1 (section 2.26.1.2) are given below.

(*i*) 98% Sulphuric acid. Mix fuming and conc. H_2SO_4 in such a way so as to make 98% H_2SO_4. Determine their strengths volumetrically by titration against sodium hydroxide of known normality.

(*ii*) Quinalizarin-sulphuric acid solution. Dissolve 25 mg of quinalizarin in 1000 ml of 98% H_2SO_4 and store the reagent in boron-free glass bottle under a dry atmosphere. Incoming air be passed through an anhydrous $CaCl_2$ solution.

2.26.2.3 Procedure

1. Formation of boron–quinalizarin blue coloured complex with graded amount of boron. Transfer 1.0 ml aliquot of the standard B solution containing 0–8 μg B into separate test tubes. Then add exactly 10.0 ml of quinalizarin-sulphuric acid solution to each test tube. Stopper the test tubes, mix the contents thoroughly, and cool the solution to 25°C or to the room temperature. The reaction mixture should not touch the rubber stopper while mixing the contents.

2. Measurement of absorbance of boron-quinalizarin complex and preparation of standard curve for boron. In a photoelectric colorimeter or a spectrophotometer, set 0 absorbance with coloured complex obtained with 0 μg B at 620 mμ wave length and determine the absorbance of the subsequent standards. Record the absorbance of the coloured complex corresponding to the μg B taken for the formation of boron-quinalizarin complex.

Thereafter, plot the absorbance on the ordinate of a linear scale graph paper corresponding to the μg B taken for colour development plotted on the abscissa. Connect zero absorbance with the maximum absorbance by a straight line to obtain standard curve for boron.

3. Formation of boron-quinalizarin complex with sample and blank test solution prepared in Method 1 (section 2.26.1.3) and measurement of absorbance of the coloured complex. Pipette 1.0 ml aliquot of sample and blank test solution prepared in Method 1 (section 2.26.1.3) following steps 4 to 6 of the procedure into separate test tubes. Then, add exactly 10.0 ml of quinalizarin-sulphuric acid solution. Stopper the test tubes, mix thoroughly taking care that the solution does not touch the rubber stopper. Cool the coloured complex to 25°C or else to the room temperature.

In a photoelectric colorimeter or a spectrophotometer, set 0 absorbance with coloured complex obtained with 1 ml aliquot of the blank test solution at 620 mμ wave length and then determine the absorbance of the coloured complex obtained with 1.0 ml of sample test solution.

It is desirable that the optical properties of the test tubes of colorimeter or spectrophotometer be tested by developing coloured complex with the same amount of boron in the test tubes and comparing the absorbance recorded. As the absorbance of the same standard vary with the temperature fresh standard curve be prepared if the room temperature differ a great deal. In order to use the standard curve for boron made previously, the colour complex with one of the previous boron standard be prepared in the test tube, and the spectrophotometer reading of the coloured complex be adjusted to the absorbance value recorded previously.

4. Estimation of boron in the sample test solution. Locate the observed absorbance of coloured complex with aliquot of the sample test solution on the ordinate of standard curve for boron, and draw a line parallel to the abscissa intersecting the straight line curve. From the point of intersection, draw a line perpendicular to the abscissa. Note the point of intersection and read the μg boron given on the abscissa.

2.26.2.4 Calculation

$$\text{ppm B in soil} = \frac{A \times B \times C}{D \times E \times F}$$

Where

A is μg B in aliquot of sample test solution as determined by reference to the standard graph for boron

B is the total volume (ml) of 0.36 N H_2SO_4 used for solubilizing B after ignition of residue obtained on evaporation of 80 ml of soil extract.

C is total volume of distilled water used for extraction of B from soil.

D is ml of aliquot of sample test solution taken for developing boron quinalizarin complex.

E is ml of water extract from soil taken for evaporation in order to prepare the sample test solution.

F is mass of air dry 2 mm soil sample taken for the extraction of B by hot distilled water.

2.26.3 (Method 3). Boron determination in the sample and blank test solution by formation of boron-curcumin complex– Dible *et al.* (1954)

2.26.3.1 Principle. As described in section 2.26.1.1

2.26.3.2 Apparatus/Reagents

Apparatus

Vol. flasks 1000 ml, 500 ml, 250 ml and 50 ml, Pipettes 1, 5, 25, and 50 ml capacity, Beaker 250 ml capacity, Oven or water bath regulated at 55 ± 3°C, Photoelectric colorimeter with 540 mμ and 580–600 mμ filter.

Reagents

(*i*) 95% Ethanol (AR). It may be distilled to make it free from B :

(*ii*) Curcumin-oxalic acid solution. Dissolve 0.04g of finely ground curcumin (Eastman Kodak No. 1179) and 5.0 g of $H_2C_2O_4.2H_2O$ in 100 ml of boron free 95% ethanol. Store the solution in amber coloured bottle in a cool dark place or else it may be stored in a refrigerator for not more than a week as the reagent decomposes slowly.

(*iii*) 100 ppm B solution. Prepare the reagent as described in Method 1.

(*iv*) 10 ppm B solution. Dilute 50 ml of 100 ppm B solution to 500 ml with redistilled water

(*v*) 0–5 ppm B solution. Pipette 0, 1, 2, 3, 4 and 5 ml of 10 ppm B solution into separate 50 ml volumetric flask. Then add 5 ml of 0.5 N HCl and dilute the solution to 50 ml with distilled water.

2.26.3.3 Procedure

1. Formation of boron-curcumin coloured complex using 0-5 ppm boron solution. Transfer 1.0 ml of slightly acidified 0 to 5 ppm B solution into separate 250 ml beaker. Use the same 1 ml pipette for transferring both standard as well as the test solution. Then add 4.0 ml of the curcumin-oxalic acid solution. Mix the contents by rotating the beaker. Then evaporate the solution to dryness in an oven (or in a bath) regulated at 55 ± 3°C. Thereafter, bake the residue for 15 minutes at 55 ± 3°C to ensure dryness. The coloured substance rosecyanine is developed during the evaporation and drying.

2. Extraction of rosecyanin in 95% ethanol. Cool the beaker containing the dried residue to room temperature. Then, add 25 ml of 95% ethanol. Triturate the residue to extract the colour. Thereafter, filter the solution through Whatman No. 2 filter paper directly into the colorimeter tube. The slight contamination of the filter paper with boron is of no significance as the colour forming step is past.

3. Measurement of the absorbance of the boron-curcumin coloured complex. The absorbance of the coloured complex be taken within 2 hours of the development of boron-curcumin complex since rosecyanine gradually hydrolyzes to curcumin.

Set the colorimeter with 540 mμ filter and place the colorimeter tube containing coloured complex from 0 μg B into the socket provided and let the instrument warm for 10 minutes. Then adjust the absorbance to zero with coloured complex containing 0 μg of added B.

Thereafter, take the absorbance of the subsequent standards and record the absorbance corresponding to the amount of added B in μg. As the absorbance reaches 0.60 *i.e.* transmittance 25% indicate over 1.5 μg B, replace 540 mμ filter by 580–600 mμ filter and reread the absorbance. Similarly measure the absorbance of coloured complex containing higher amount of B at 580–600 mμ wave length.

4. Standard graph for boron by the curcumin method. Record the absorbance (optical density) of boron-curcumin coloured complex prepared with 0 to < 1.5 mμ B, and with 1.5 μg B and more, and plot the absorbance values of the two ranges separately on a linear scale graph paper. Plot the absorbance on the ordinate and the corresponding μg of B taken for formation of boron-curcumin complex on the

abscissa. Connect the lowest and the maximum absorbance of the each range by a straight line to obtain the standard curve for boron.

5. Extraction of hot water soluble B from soil by the procedure of Burger and Truog (1939) and preparation of the sample and blank test solutions. Follow steps 4 to 6 of procedure given in Method 1 (See section 2.26.1.3) of boron determination in soil. After igniting the extract residue in silica dish placed in a muffle furnace, cool and add measured volume of 0.05 N HCl to dissolve the residue stirring with a glass rod. Then transfer the solution into centrifuge tubes. Run the centrifuge at 2000–2440 rpm for 10 minutes. Decant off the supernant clear solution into a glass stoppered flask.

6. Formation of boron-curcumin coloured complex with sample and blank test solution and measurement of the absorbance of the coloured complex. Use the same 1.0 ml pipette for taking aliquot of the blank and sample test solution. Transfer 1.0 ml aliquot of the clear test solution into separate 250 ml beakers, and follow steps 1 to 3 of procedure (Section 2.26.3.3) omiting taking of 1 ml aliquot of 0–5 ppm B solution. Record the absorbance of the coloured complex obtained with sample test solution setting 0 absorbance in a colorimeter or spectrophotometer at 540 or else 580–600 mμ wavelength with coloured complex obtained with blank test solution.

7. Estimation of B in boron-curcumin complex prepared from sample test solution. Refer to the appropriate standard curve for boron on the basis of the wave length of the filter used in determining the absorbance of the coloured complex obtained with sample test solution.

Locate the absorbance value on the ordinate, and draw an imaginary line parallel to the base *i.e.* abscissa intersecting the standard curve. From the point of intersection draw a line perpendicular to the abscissa. Note the boron content in μg given at the point of intersection.

2.26.3.4 Calculation

$$\text{ppm B in soil} = \frac{A \times B \times C}{D \times E \times F}$$

Where

A is μg B in aliquot of sample test solution as determined by reference to the standard graph for boron.

B is the total volume (ml) of 0.05 N HCl used for solubilizing B after ignition of residue obtained on evaporation of 80 ml of soil extract

C is the total volume of distilled water used for the extraction of boron from soil.

D is ml aliquot of sample test solution taken for developing boron-curcumin complex.

E is ml of water extract from soil taken for evaporation in order to prepare the sample test solution.

F is the mass of air dry 2 mm soil sample taken for the extraction of B by hot distilled water.

2.26.3.5 Comments

The glasswares must be boron free such as Kavalier or Corning 728.

Ordinary soft glass may be used for most of the reagents. All glasswares be weathered in conc. H_2SO_4 and it should be washed with conc. HCl before use. Pyrex and borosilicate glasswares must not be used. Polythene plastic bottles are free from boron and it can be used for storage of reagents as well as for soil extraction. Use analytical reagents for the chemical analysis.

Berger and Truog (1939) extracted boron by boiling a 1 : 2 soil-water suspension (w/v) for 5 minutes. The boron added to the soil could be completly recovered by the method. Berger and Truog (1940) used amount and severity of black spot on roots of beet root as an index of degree of deficiency of boron. They reported that for minimization of black spots in beets, the hot H_2O soluble boron in soil must be atleast 0.75 ppm. Others reported critical level of hot H_2O extractable boron in different soils from 0.15 to 0.50 ppm.

Later Berger proposed desirable level of hot H_2O soluble boron for high, medium and low B crops. The desirable level of boron in soil appear to be greater than 0.50 ppm for high B crops such as apples, alfalfa, cloves, beets, cabbage, cauliflower, sunflower and radish, 0.1 to 0.5 ppm boron for medium B crops such as tobacco, tomatoes, lettuce, peach, cherry, cotton, peanut, carrots, onions and pears, 0.1 ppm boron or less for low B crops such as wheat, oats, barley, corn, peas, beans, citrus, straw berry and soyabean.

Wear and Patterson (1962) found that alfalfa grown on a course textured soil had a higher uptake of boron per unit of hot H_2O soluble B in soil than plants grown on a fine textured soil. Martins (1968) reported greater uptake of B as the soil pH increased. Cox and Reed (1965) observed that a change in pH altered the level of hot H_2O extractable B in soil. Acid soils formed under moderate to high rainfall are low in boron because of removal of soluble boron by leaching. The boron tend to accumulate in soils of low rainfall regions. Some cultivated soils become high in soluble boron because of use of water high in boron for irrigation. The saline soils are generally much higher in soluble boron than non-saline soils as sodium salt of boron are highly soluble.

It is apparent that hot H_2O extractable boron is well accepted for evaluating B status of soil. However, consideration of soil pH and texture is likely to improve predictions.

REFERENCES

Berger, K.C. and E. Truog (1939). Boron determination in soils and plants Ind. Eng. Chem. Anal. Ed. 11 : 540–554.

Berger, K.C. and E. Truog (1940). Boron deficiencies as revealed by plant and soil tests. J. Amer. Soc. Agron. 32 : 297–301.

Berger, K.C. and E. Truog (1945). Boron availability in relation to soil reaction and organic matter content. Soil Sci. Soc. Amer. 10 : 113–116.

Cox, F.R., and P.H. Reid (1964). Calcium-boron nutrition as related to concealed damage in peanuts. Agron. J. 56 : 175–176.

Cox, F.R., and R.J. Kamprath (1972). Micronutrient Soil Tests. In. Mortevedt *et. al.* (ed). Micronutrients in Agriculutre. Soil Sci. Soc. Amer. Inc., Madison, Wisconsin USA.

Dible *et al* 1954. Anal. Chem. 26 : 418.

Hatcher, J.T. and L.V. Wilcox (1950). Colorimetric determination of boron using carmine. Anal. Chem. 22 : 250–252.

Jackson, M.L. (1973). Soil Chemical Analysis, Prentice-Hall of India Pvt. Ltd. New Delhi.

Martens, D.C. (1968). Plant availability of extractable boron, copper and zinc as related to selected soil properties, Soil. Sci. 106 : 23–28.

Wear, J.L. and R.M. Patterson (1962). Effect of soil pH and texture on the availability oı water soluble boron in the soil. Soil Sci. Soc. Amer. Proc. 26 : 344–346.

2.27 MOLYBDENUM IN SOIL–GREGG (1953)

2.27.1 Principle

Molybdenum determination is based on formation of an amber to orange red coloured compound on addition of a reducing agent such as $SnCi_2$ to an acid solution of molybdenum salt in the presence of thiocyanate. The coloured complex is extracted from the solution by an organic solvent.

The absorbance of the coloured complex is measured by a spectrophotometer at 470 nm wavelength or else by a photoelectric colorimeter using an appropriate filter. Mo from the soil is extracted by $(NH_4)_2C_2O_4$(pH 3.3), following the procedure of Gregg (1953).

2.27.2 Apparatus/Reagents

Apparatus

Volumetric flasks, burettes, pipettes, separatory funnels, spectrophotometer or else a photoelectric colorimeter, polythene bottles, and mechanical shaker.

Reagents

(*i*) Redistill water. Redistill water from a metal distill in a completely glass distillation apparatus

(*ii*) 5.0% Potassium thiocyanate solution. Dissolve 50 g of KCNS (AR grade) in 1000 ml redistilled water

(*iii*) 5.0 N Sodium nitrate solution. Dissolve 106.25 g $NaNO_3$ in 250 ml of redistilled water.

(*iv*) 5% Sodium carbonate. Dissolve 5g Na_2CO_3 in 100 ml redistilled water.

(*v*) 10% Stannous chloride. Dissolve 25 g $SnCl_2$ in 25 ml of conc. HCl and dilute the solution to 250 ml with redistilled water.

(*vi*) Carbon tetrachloride-isoamyl alcohol. Mix equal volume of carbon tetrachloride and isoamyl alcohol.

(*vii*) Conc. Hydrochloric acid (BDH–AR).

(*viii*) 100 ppm Fe solution. Dissolve 0.3509 g of $(NH_4)_2$ SO_4 $FeSO_4.6H_2O$ in 500 ml of redistilled water.

(*ix*) 100 ppm Molybdenum solution. Dissolve 0.1840 g of $(NH_4)_6$ Mo_7 O_{24}. $4H_2O$ in 900 ml of redistilled water and dilute the solution to 1000 ml.

(*x*) 5 ppm Molybdenum solution. Dilute 50 ml of 100 ppm Mo solution to 1000 ml with redistilled water.

(*xi*) 0.4N Ammonium oxalate (pH 3.3). Dissolve 24.8067 g of $(NH_4)_2$ C_2O_4 reagent in distilled water and make the volume to about 950 ml. Then adjust the pH of solution to 3.3 and dilute the solution to 1000 ml with redistilled water.

2.27.3 Procedure

1. Formation of Mo-thiocyanate complex. Pipette aliquot of standard Mo solution containing 2.5 to 50 μg Mo into separate separatory funnel. Then add 50 ml of 5% Na_2CO_3 and 8 ml of conc. HCl. Cool the funnel to room temperature. Then, add 1ml of 5N $NaNO_3$ and shake the separatory funnel well. Thereafter, add 5 drops of 100 ppm Fe solution, 5 ml of 5% KCNS solution, and 5 ml of 10% $SnCl_2$ in approx. 1.2 N HCl.

2. Extraction of Mo-thiocyanate (amber to orange red coloured complex) by organic solvent. Add 10 ml of carbon tetra chloride-isoamyl (1 : 1) mixture into the separatory funnel and shake for 2 minutes and allow the phases to separate. Collect the lower organic phase containing Mo-thiocyanate complex.

Thereafter, add 5 ml of carbon tetra chloride-isoamyl alcohol mixture into the separatory funnel and shake the contents for 2 minutes. Place the separatory funnel on a stand. Allow the phases to separate. Collect the lower organic phase and combine it with the previously collected organic phase.

3. Measurement of the absorbance (optical density) of the Mo-thiocyanate coloured complex. Measure the optical density of the Mo-thiocyanate coloured complex obtained with 2.5 to 50 μg Mo at 470 mμ wave length setting 0 optical density with carbon tetrachloride-isoamylalcohol (1 : 1) mixture using a photoelectric colorimeter or else a spectrophotometer.

4. Preparation of standard graph for Mo. Record the optical density of the Mo–thiocyanate coloured complex corresponding to the μg Mo taken for the formation of the coloured complex. Plot the optical density of the coloured complex on "Y" axis of linear scale graph paper against the corresponding μg Mo taken plotted on the "x" axis. Connect 0 optical density with the optical density of coloured complex with 50 μg Mo by a straight line to obtain the standard graph for Mo.

5. Extraction of Mo from soil with 1N ammonium oxalate (pH 3.3) following the procedure of Greg (1953). Transfer 25 g of air dry 2 mm soil sample into a polythene bottle. Then, add 100 ml of 1N ammonium oxalate (pH 3.3) to the polythene

bottle. Stopper and shake the contents for 4 to 6 hours on a mechanical shaker. Thereafter, allow the suspension to stand overnight.

Then filter the suspension through Whatman no. 42 paper discarding the initial filtrate. Collect the subsequent filtrate and transfer 50 ml of the extract *i.e* filtrate into a silica dish and evaporate it to dryness on a water bath or hot plate.

Transfer the silica dish containing the residue into a muffle furnace and ignite the residue at 450°C for 4 hours to destroy the organic matter.

After cooling the silica dish, add 25 ml of 1 + 1 HCl (approx. 6N) to dissolve the residue. Transfer the solution into a 250 ml separatory funnel. Rinse the silica dish with redistilled water to make silica dish free from acid, and transfer the rinsings to the same separatory funnel. Similarly run a blank.

6. Formation of Mo-thiocyanate complex, its extraction by organic solvent and measurement of its optical density. Follow the steps 1 to 3 given in the procedure (section 2.27.3) beginning with the addition of 50 ml of 5% Na_2CO_3 in step 1.

7. Determination of the content of Mo in the Mo-thiocyanate complex. Record the optical density of the Mo-thiocyanate complex obtained with the sample and blank extract following steps 5 and 6, and determine the content of Mo in Mo-thiocyanate complex extracted with 15 ml of carbon tetra chloride-isoamyl alcohol (1 : 1) mixture by reference to the standard graph. Alternatively, estimate the slope of the standard graph for Mo i.e. optical density per μg Mo taken for the formation of Mo-thiocyanate complex. Then, divide the optical density of the Mo-thiocyanate complex with the sample and blank extract by the slope to estimate the content of Mo in μg present in the coloured complex.

2.27.4 Calculations

$$\text{Mo in soil (ppm)} = \frac{(S - B) \times E}{A \times M}$$

Where

S and B denote μg Mo in sample and blank extract taken for the formation of Mo-thiocyanate complex.

E is the total volume of 1.0 N ammonium oxalate (pH 3.3) used for the extraction of Mo from the soil.

A is the ml aliquot of ammonium oxalate (pH 3.3) extract taken for the formation of Mo-thiocyanate complex.

M is the mass of soil sample in g taken for the extraction of Mo from the soil.

2.27.5 Comments

The extraction of Mo by $(NH_4I_2C_2O_4$ (pH 3.3) has been widely used and checked against field trials. The response from Mo application has been obtained in soils containing less than 0.2 ppm upto 0.4 ppm Mo. Mo from soil was extracted by Gregg (1953) by several extractants. He found oxalate to be a satisfactory extractant for Mo. The critical level of Mo was found to be 0.20 ppm at pH 5.0.

The other reagents which have been employed for the extraction of Mo from soil are distilled water, soxhlet extraction with distilled water, $NH_4C_2H_3O_2$(pH 7.0),

0.1 N NaOH, 0.3N NH_4F, 0.1 N HCl + 0.03 N NH_4F. Haley and Melsted (1957) reported that alkali extracted the most Mo and the fluoride solution the least. Mo extracted by alkaline and oxalate correlated well with the plant composition.

The availability of Mo is reduced in acid soils. Mo deficiency is usually associated with soils below pH 5.0. Liming acid soils result in an increase in available Mo. A positive correlation between available Mo and pH has been also reported. In usar (sodic) soils the extractable Mo increases with the increase in ESP probably due to increase in the solubility of organic matter. Some workers have observed negative correlation between organic matter and available Mo. In some soils negative correlation between available Mo and clay content has been reported. Soils high in free Fe_2O_3 are often deficient in available Mo. Mo disorders is also induced by excess of Mn, Cu and SO_4

REFERENCES

Cox, F.R. and E.J. Kamprath (1972). Micronutrient soil tests In. Mortvedt, J.J. *et. al.* (ed). Micronutrients in Agriculture. Soil. Sci. Amer. Inc., Madison, Wisconsin, USA.

Grigg, J.L. (1953). Determination of available molybdenum of soils New Zeal. J. Sci. Technical 54 : 405–414.

Haley, L.R. and S.W. Melsted (1957). Preliminary studies of molybdenum in Illinois soils. Soil Sci. Soc. Amer. Proc. 21 : 316–319.

Johnson, C.M. and T.H. Arkley (1954). Anal. Chem. 26 : 572–573.

Sandell, E.B. (1944). Colorimetric Determination of Traces of Metals. Interscience Publishers, Inc. N.Y.

2.28 NUTRIENT ELEMENTS DETERMINATION IN SOIL EXTRACTS BY ATOMIC ABSORPTION SPECTROPHOTOMETER

2.28.1 Calcium, Magnesium determination in Soil Extract by Atomic absorption Spectrophotometer

2.28.1.1 Principle

Same as described in section 2.28.2.1

2.28.1.2 Equipments/Reagents

Equipment

Same as listed in 2.28.2.2

Reagents

(*i*) Deionized distilled water.

(*ii*) Morgan's extractant (sodium acetate 10% in 3% glacial acetic acid). Dissolve 100 g of sodium acetate (AR–grade) in deionized distilled water. Transfer the sodium acetate solution into a 1l volumetric flask. Then add 30 ml of glacial acetic acid and make up the volume to 1000 ml mark with deionized-distilled water.

(*iii*) Strontium Chloride, 4.56%. Dissolve 45.58 g of pure St Cl_2. $6H_2O$ in deionized distilled water, and make up the volume to 1000 ml mark with deionized distilled water.

(*iv*) Phosphorus–1000 ppm in 5000 ppm K solution. Dissolve 4.3932 g of KH_2PO_4 dried at 40°C and 9.5334 g of KCl dried at 105°C for 2 hours in deionized distilled water and make the volume to 1000 ml mark with deionized distilled water.

(*v*) Calcium and Magnesium Stock Solutions calcium–1000 ppm solution. Transfer 2.4972 g $CaCO_3$ (AR–grade, dried at 105°C for 2 hours) into a 250 ml Erlenmeyer flask. Place a small funnel and slowly add 50 ml of 1N HCl prepared from distilled approx. 6N HCl (AR–grade). Boil the solution to expell CO_2. Allow it to cool. Then transfer the solution into 1 l volumetric flask. Rinse the Erlenmeyer flask with portions of deionized distilled water, and transfer the rinsings to the same volumetric flask. Then make up the volume to 1000 ml mark with deionized distilled water.

Calcium–50 ppm solution. Transfer 50 ml of 1000 ppm Ca solution into 1 l volumetric flask and make up the volume to 1000 ml mark with deionized distilled water.

Magnesium-1000 ppm solution. Transfer 10.1339 g of pure $MgSO_4$. $7H_2O$ (AR–grade) into 1 l volumetric flask. Add deionized distilled water and shake to dissolve magnesium sulphate. Then make up the volume to 1000 ml mark with deionized distilled water.

Magnesium–100 ppm solution. Dilute 50.0 ml of 1000 ppm Mg solution to 500 ml mark with deionized distilled water.

Magnesium -5 ppm solution. Dilute 25 ml of 100 ppm Mg solution to 500 ml mark with deionized distilled water.

(*vi*) Calcium and Magnesium Working Standards

Calcium 0 to 10 ppm solution. Transfer 0, 10, 20, 30, 40 and 50 ml of 50 ppm Ca solution into separate 250 ml volumetric flask. Then, add to each flask 25.0 ml of Morgan's extractant, 12.5 ml each of strontium chloride (4.56%) and calcium compensating solution. Make up the volume of each volumetric flask to 250 ml mark by adding deionized-distilled water. It will give 0, 2, 4, 6, 8 and 10 ppm Ca working standards.

Magnesium 0 to 1.0 ppm solution. Transfer 0, 10, 20, 30, 40 and 50 ml of 5 ppm Mg solution into separate 250 ml volumetric flask. Add to each volumetric flask 25 ml of Morgan's extractant, and 12.5 ml each of 4.56% strontium chloride and magnesium compensating solution. Then make up the volume to 250 ml mark with deionized distilled water. It will give 0, 0.2, 0.4, 0.6, 0.8 and 1.0 ppm Mg working standards.

(*vii*) Compensating solution for Ca determination by a.a.s. Pipette 5.0 ml of Phosphorus-1000 ppm in 5000 ppm K solution and 6.0 ml of 1000 ppm Mg solution into a 250 ml volumetric flask. Then make the volume to 250 ml mark with deionized distilled water.

(*viii*) Compensating solution for Mg determination by a.a.s. Pipette 1.0 ml of phosphorus 1000 ppm in 5000 ppm K solution into a 250 ml volumetric flask. Then add 12.5 ml of 1000 ppm Ca solution, and make the volume to 250 ml mark with deionized distilled water.

2.28.1.3 Procedure

1. Preparation of soil extract. Transfer 5.0 g of air dry < 2 mm soil sample into a 100 ml Erlenmeyer flask. Then, add 25.0 ml of Morgan's extractant. Stopper and shake the contents on a mechanical shaker at medium speed for 5 minutes. Thereafter, filter the suspension through a double folded Whatman No. 1 filter paper. Discard the initial filtrate and collect the subsequent filtrate in a 100 ml Polystyrene bottle.

2. Preparation of sample and blank test solution for Ca determination by a.a.s.

Sample test solution. Transfer 5.0 ml soil extract into a 50 ml volumetric flask, and then add 2.5 ml of 4.56% strontium chloride and make up the volume to 50 ml mark with deionized distilled water.

Blank test Solution. Transfer 5 ml of Morgan's extractant into a 50 ml volumetric flask. Then add 2.5 ml of 4.56% strontium chloride and 2.5 ml of compensating solution for Ca. Then, make up the volume to 50 ml with deionized distilled water.

3. Preparation of sample and blank test solution for Mg determination.

Sample test Solution. Transfer 5.0 ml of soil extract into a 50 ml volumetric flask. Then, add 2.5 ml of 4.56% strontium chloride and make up the volume to 50 ml mark with deionized distilled water.

Blank test solution. Transfer 5.0 ml of Morgan's extractant, 2.5 ml of 4.56% strontium chloride, and 2.5 ml of compensating solution for Mg. Then make up the volume to 50 ml mark with deionized distilled water.

4. Determination of absorbance of the working standards and the sample and blank test solution. Let the operator set the atomic absorption spectrophotometer as per specification for Ca determination given in the instruction manual. Also set zero knob position as per operation manual. Run the instrument for a period of 20 minutes to allow it to warm.

Then atomize 0, 2, 4, 6, 8 and 10 ppm Ca solution in sequence noting down the corresponding absorbance recorded by the spectrophotometer. Thereafter, atomize the sample and blank test solution prepared from soil extract and Morgan's extractant and note their respective absorbance.

After Ca determination by a.a.s., let the operator set the instrument as per specifications for Mg determination given in the instruction manual. Also set zero knob position as per operation manual.

Then atomize 0, 0.2, 0.4, 0.6, 0.8 and 1.0 ppm Mg working standards in sequence. Note down the absorbance recorded by the instrument corresponding to each Mg working standard. Thereafter atomize the sample and blank test solution prepared from soil extract and Morgan's extractant and note the absorbance recorded by the spectrophotometer.

5. Preparation of standard Ca and Mg calibration curve and estimation of concentration of element in the sample and blank test solution. On a linear scale graph paper with absorbance scale on the ordinate and the concentration scale on the abscissa, plot the value of absorbance corresponding to each concentration, and then connect the points plotted by a straight line to obtain the calibration curve.

The concentration of element in the sample and blank test solution is determined by reference to the calibration curve for each element. The values of concentrations of the element in the sample and blank test solution is substituted in the formula given under the heading "Calculation" for estimating the concentration of extractable elements in soil. With the attachment of a computerized reading system, the data recording, processing and print out of the results is simultaneously done without any loss of time.

2.28.1.4 Calculation

$$\text{Extractable Element in Soil (g / 100g sample)} = \frac{(S - B) \times VST \times VE \times 100}{A \times M \times 10^6}$$

Where

S is ppm element in the sample test solution as determined by reference to the elements calibration Curve.

B is ppm element in the blank test solution as determined by reference to the elements calibration Curve.

VST is the volume of the sample test solution in ml.

VE is the volume of Morgan's extractant equilibrated with the soil.

A is the ml aliquot of soil extract taken for the preparation of the sample test solution.

M is the mass of soil sample in g taken for equilibration with the Morgan's extractant

10^6 is the number of microgram in one gram. When placed in the denominator, it converts microgram element into gram.

2.28.2 Micronutrient determination in soil extracts by atomic absorption spectrophotometer

2.28.2.1 Principle

A sample is converted into an atomic vapour, usually by a flame, and irradiated by the light from a source whose emission lines are those of the metal being sought. The absorbance of the light by the vaporized samples is related to the concentration of the desired metal in it.

The narrow emission lines which are to be absorbed by the sample are generally provided by a hallow cathode lamp a source filled with neon on argon at a low pressure, which has a cathode made of element being sought. Such a lamp emits only the spectra of the desired element, together with that of the filter gas.

In the atomiser burner system the solution to be analysed is drawn by a capillary and converted by means of a stream of compressed air to a fine spray which after condensation of larger droplets is mixed with acetyline and burned in a long flame at a stainless steel burner. The light from the lamp, after traversing the flame enters a monocromator which has been set at a wave length of the resonance line of the metal to be determined and then falls on a photomultiplier tube.

The light from the lamp is modulated and the signal from the photomultiplier is amplified by an A.C. amplifier, whose output is read on a meter.

The atomic absorption spectrophotometer is mainly employed for the determination of metals whose resonance lines lie in the ultraviolet and visible region of the spectrum. Methods exist for the indirect determination of phosphorus, sulphur, and chloride by combining them with a metal, and then analyzing for the metal.

The atomic absorption spectrophotometer be used according to the instruction manual by an operator–a trained technician made familiar with the instrument and its operation. The specifications for the individual element analysis in respect of concentration of working standards, Range, Wavelength, Lamp and operating current, slit width and height, Filter, Meter response, Zero knob position, Burner height, Air flow, Fuel and Fuel flows be noted and strictly employed for the satisfactory results.

2.28.2.2 Equipments/Reagents

Equipments

Atomic Absorption Spectrophotometer installed in an air conditioned room. A large size smoke hood connected to an outlet is placed above the gas burner of the spectrophotometer.

Precision Physical and Chemical Balances, Mechanical Shaker, Water Treatment Plant to prepare deionized water. Completely glass distillation unit with B_{24} joints and with 2 l capacity boiling flask and heating mantle, Laboratory glasswares (corning or pyrex), Polythene containers of different sizes.

Reagents

(*i*) Zn, Cu, Mn and Fe Solutions–100 ppm. Transfer 0.4398 g of $ZnSO_4.7H_2O$, 0.3929 g of $CuSO_4.5H_2O$, 0.3602 g of $MnCl_2.4H_2O$ and 0.4977 g of $FeSO_4.7H_2O$ into separate 1l volumetric flasks. Add deionized or double distilled water. Shake to dissolve the contents and make the volume to 1000 ml mark with the deionized water.

(*ii*) Working standards for Zn and Cu. Transfer 0, 0.2, 0.4, 0.6, 0.8, 1.0, 1.4, 1.8, 2.2 and 2.6 ml of 100 ppm Zn or Cu solution into separate 100 ml volumetric flasks and dilute each with the extracting solution upto the 100 ml mark to obtain 0 to 2.6 ppm Zn or Cu solution.

(*iii*) Working standards for Fe and Mn. Transfer 0, 2.5, 5.0, 7.5, 10.0, 15.0 and 20.0 ml of 100 ppm of Fe or Mn solution into separate 100 ml volumetric flasks and dilute each with the extracting solution up to the 100 ml mark to obtain 0 to 20 ppm Fe or Mn solution.

(*iv*) Micronutrient Extracting Solution–0.005 M DTPA, 0.01M $CaCl_2$ and 0.1 M Triethanolamine (pH 7.3). Transfer 1.965 g of DTPA (Diethylene triamine pentacetic acid.) 13.3 ml of Triethanolamine and 1.45 g of $CaCl_2 . 2H_2O$ into 1l volumetric flask. Add about 800 ml deionized water. Shake to dissolve the contents and adjust the pH to 7.30 with HCl and make the volume to 1 l with deionized or double distilled water.

2.28.2.3 Procedure

1. Extraction of micronutrient elements from the soil. Transfer 10g of air dry < 2 mm soil sample into a100 ml Erlenmeyer flask. Then add 20 ml of 0.005 M DTPA, 0.01M $CaCl_2$ and 0.1 M Triethanolamine (pH 7.3). Stopper the Erlenmeyer flask and shake the contents for 2 hours on a mechanical shaker at a medium speed. Thereafter, filter the contents through a funnel lined with Whatman No. 42. filter paper. Discard the initial few ml filtrate after rinsing the receiving flask and preserve the subsequent filtrate for the determination of micronutrient element by the atomic absorption spectrophotometer. Simillarly run a blank.

2. Preparation of the standard graph for each micronutrient element by the atomic absorption spectrophotometer. Let the operator run the atomic absorption spectrophotometer as per specifications for the desired element given in the instrument manual. Run the instrument for a period of 20 minutes to allow it to warm up.

Thereafter, first run a series of standards of the desired element beginning with the concentration of 0 ppm and raising the concentration of the element by increment of 0.2 to 0.4 ppm upto a concentration of 2.6 ppm in case of zinc and copper. However, in the case of Fe and Mn, the element concentration be raised by increment of 2.5 ppm in the range of 0–20 ppm.

The absorbance recorded on atomization of solution of known concentration of an element be noted corresponding to the concentration of the element. Then on a linear scale graph paper with absorbance scale on the ordinate and the concentration scale on the abscissa, the value of absorbance corresponding to each concentration be plotted and then the points be connected by a straight line to obtain the calibration curve. With the attachment of a reading system, the same is achieved by a computer.

3. Determination of the concentration of an element in the soil extracts and the blank. When the atomic absorption spectrophotometer is in operation, after running the working standards, the soil extract and the blank be atomized, and the value of absorbance recorded by the atomic absorbance spectrophotometer be noted. The concentration of element in the soil extract and blank is estmated by reference to the calibration curve for each element. The values of concentrations of an element in sample and blank extract is substituted in the formula given in section 2.28.2.4 for estimating the content of extractable elements in soil. With the attachment of a computerized reading system, the data recording, processing and print out of the results is simultaneously done without any loss of time.

2.28.2.4 Calculation

$$\text{Extractable micronutrient element in soil (ppm)} = \frac{(S - B) \times VE}{M}$$

Where

S is ppm concentration of the element in the soil extract as estimated from the recorded absorbance by reference to the element calibration curve.

B is ppm concentration of the element in the blank test solution estimated from the recorded absorbance by reference to the element calibration curve

VE is the total volume of the extractant equilibrated with the soil sample.

M is the mass of soil sample in gram taken for equilibration with the extractant.

2.28.2.5 Comments

The extraction of micronutrient elements in soil by 0.005M DTPA, 0.01M $CaCl_2$ and 0.1M Triethanolamine (pH 7.3) was developed by Lindsay and Norvell (1969). They found the extractant satisfactory for the extraction of soil Zn and Fe. Brown et al (1971) also reported that Zn extraction by DTPA is a better method of assessment of soil Zn. The DTPA extractable soil Zn may be interpreted as low, 0 to 0.5 ppm ; Marginal, 0.5 to 1.0 ppm ; and Adequate, greater than 1.0 ppm. Zn fertilization is needed when Zn level in soil is low. When soil P is high Zn fertilization is needed also on soils found marginal in soil Zn when crops sensitive to low Zn level are grown.

Lindsay and Norvell (1969) also found soil Fe extracted by 0.005 M DTPA, 0.01 M $CaCl_2$ and 0.1 M Triethanolamine to be useful in the diagnosis of Fe deficiency. The critical level may range from 2.5 ppm to 4.5 ppm of soil Fe. However, soil Fe extracted by 1N $NH_4C_2H_3O_2$ (pH 4.8) gave satisfactory correlation with the degree of chlorosis (Olson and Carlson, 1950). Fe chlorosis was found to be moderate to severe with Fe level between .01 to 0.3 ppm. Between 0.3 to 2.2 ppm Fe, the chlorosis was slight to moderate. However, Fe chlorosis is enhanced apart from low extractable Fe, by high Mn/Fe ratio, high pH, and presence of free lime.

$NH_4C_2H_3O_2$ has been also found to be the best extractant for soil Mn when the pH factor was included–Browman *et. al.* 1969).

The acid $NH_4C_2H_3O_2$ is generally used for the extraction of Cu (Fiskell, 1965). Soils with less than 0.2 ppm Cu are suspected of being deficient. Grewal, Lal and Randhawa (1969) also found 1N $NH_4C_2H_3O_2$ a satisfactory soil extractant for copper in soils of Ludhiana.

The micronutrient elements in soil extracted by either 0.005 M DTPA, + 0.01M CaCl2 + 0.1 M Triethanolamine or else by $NH_4C_2H_3O_2$ can be determined by the atomic absorption spectrophotometer. It is also convenient to determine $NH_4C_2H_3O_2$ extractable exchangeable Ca and Mg in soil by the atomic absorption spectrophoto meter.

The rating of micro nutrients extracted by 0.005 M DTPA + .01 M $CaCl_2$ + 0.1 M TEA (pH 7.3) and determined by AAS in soil healthy lab are given in Table 2.28.1.

Table 2.28.1

Micronutrients	Nutrient Low	content in Medium	ppm High
Zinc	upto 0.60	61 to 1.20	71.20
Iron	upto 4.0	4.1 to 8.0	78.0
Copper	upto 0.2	0.21 to 0.4	70.4
Manganise	upto 2.0	2.1 to 4.0	74.0

The micronutrient carrier dose kg/hg recommended for soils tested as low and medium are given in Table 2.28.2.

Table 2.28.2

Micronutrients Status	Micronutrient Carriers			
	Zinc sulphate	Ferrous Sulphate	Copper Sulphate	Manganese Sulphate
Low	50	50	5	20
Medium	25	25	2.5	10

REFERENCES

Browman, M.G., G. Chesters, and H.B. Pionke, 1969. Evaluation of tests for predicting the availability of soil manganese to plants. J. Agr. Sci. 72 : 335–340.

Brown. A.L., J. Quick and J.L. Eddings (1971). A comparison of analytical methods for soil Zn. Soil Sci. Soc. Amer. Proc. 35 : 105–107.

Fiskell, J.G.A. 1965. Copper In. C.A. Black *et. al.* (ed.) Method of Soil Analysis, Part 2. Amer. Soc. Agron, Madison, Wisconsin USA.

Grewal, J.S., C. Lal, and N.S. Randhawa 1969. Evaluation of different methods for the determination of available copper in soils of Ludhiana. Indian J. Agr. Sci. 39 : 877–885.

Olson, R.V. and C.W. Carlson. 1950. Iron chlorosis of sorghum and trees as related to extractable soil iron and manganese to plants. Soil Sci. Soc. Amer, Proc. 14 : 109–112.

Robinson, J.W. 1966. Atomic Absorption Spectroscopy. Marcel Dekker Inc., New York, USA.

Slavin, W. 1968. Atomic Absorption Spectroscopy. Inter-Science Publishers–John Wiley and Sons, New York, USA.

Chemical Analysis of Plant Samples

3.1 SAMPLING AND PREPARATION OF SAMPLE–JONES Jr (1972), AMMA (1989)

3.1.1 Introduction

The plant analysis is undertaken to investigate nutritional disorders in plants as well as for advisory work on the basis of nutrient concentrations in most sensitive plant parts at certain stage of plant growth and development. It helps also in better interpretation of the results of field trials. When plant samples for analysis is taken at crop maturity, it enables to estimate amount of nutrients removed by the crop from the soil.

In the present day agriculture, the plant analysis technique are applied to various trees *e.g.* rubber, coconut, coffee, fruit crops *e.g.* apples, citrus, plantation crops *e.g.* sugarcane, pineapples, sugarbeets and cereals *e.g.* rice and wheat. There has been also significant development in methods of analysis. As the nutrient element content of the plant is a reflection of the available nutrient status of the soil, the plants ability to absorb nutrient in the established environment is reflected in the nutrient concentration in the plant at any one time. The plant analysis technique involves several steps such as sampling, sample preparation, Ashing (sample digestion), elemental analysis and interpretation.

3.1.2 Sampling

For proper sampling, guide line may be obtained from the concerned research laboratory regarding plant part to select, number of parts/plant to collect, and the number of plants to sample. For micronutrient studies in general, leaves are preferred, usually mature leaves just below the growing tip on main branches and stems. Sampling is normally recommended just to or at the time the plant begins its reproductive stage of growth. In case of mobile nutrient elements such as N, P & K lower part of the plant is sampled for analysis. At a certain stage of plant development, plant-parts show maximum difference in nutrient concentrations resulting from inadequate supply of nutrient elements from the soil. In cereals like wheat, nutrient concentration at 3.0 g plant dry weight stage is reported to be a good index of nutrient status of the plant.

When plants begin setting and developing fruit or seed, the nutrient concentration of the vegetative portions of the plant changes substantially. Consequently, sampling of plant parts be preferably done prior to flowering or else before pollination. Seed is usually not useful for analysis in order to determine the nutrient status of the plant. The sampling time also depend on the season. In rubber plantation, rainy season is not considered suitable for leaf sampling because of removal of nutrients by rain water.

Normally sampling is not recommended when plant part is covered with soil or dust, damaged by insects, mechanically injured or diseased. Dead plant tissues should not be included. Sampling is also not recommended when plants are under moisture or temperature stress. Sampling be preferably done by a trained personnel well aware of desirable sampling techniques. The sampling technique for rubber plantation employed by the Rubber Research Institute is described below.

(*i*) For young unbranched trees. Select 10 plants from each block having no new flushes at the top. The upper most whorl should be fully mature. Remove 4 basal leaves from the terminal whorl.

(*ii*) For branched immature trees and trees under tapping. Trees to be selected for sampling depend on the plantation area. For an area up to 5 ha, select 10 trees. Select 16, 23 and 30 trees from an area of 10, 15 and 20 ha respectively. For the selection of trees, follow a diagonal pattern or V pattern, *i.e.*, going from one corner to the middle of the opposite side and back to the other corner.

Collect leaf samples in the month of August, September, October when the leaves are 6-8 months old. From each tree selected, collect 4 basal leaves from lower branches in shade, four basal leaves from spur leaves (small offshoot) with only one whorl from the trunk or main branches are also suitable for sampling. The leaflets to be taken from each leaf depend on number of trees selected. When number of selected trees is 30, collect middle leaflet. When trees selected is 15, collect two leaflets on either side and when trees selected is 10, collect all the leaflets so that 120 leaflets get selected to form a composite sample. The composite sample be brought to the laboratory on the same day or else it should be kept in a refrigerator.

3.1.3 Sample preparation and storage

3.1.3.1 Washing

The plant parts contaminated with soil and dust be preferably washed or rinsed with water. For micronutrient studies, distilled or deionized water be used for rinsing the plant parts. The removal of surface contamination is more effective when 0.1% or 0.2% detergent such as teepol solution is used. By incorporating EDTA in the detergent solution, it is possible to remove contamination from spray of ferrous sulphate solution.

At early stage when the whole plant is uprooted, the soil adhering to the roots, and the basal part of the plant be washed under tap water. Thereafter, the plant sample be immersed in the detergent solution for 20 seconds. Thereafter, it should be rinsed first with tap water and then with distilled water before spreading it on a large clean paper for drying quickly. The air dry plant samples be placed preferably in paper bags for transport to laboratory.

Some of the more soluble inorganic constituents may be lost by solution on washing. As such, some scientists prefer wiping leaf surfaces with cheese cloth moistened in a detergent solution. Leaf analysis without removal of surface contamination is not recommended for assessing Fe and Mn concentration.

3.1.3.2 Drying

Drying of plant sample at 60 to 80°C prevent dry weight losses due to continuing respiration. There is also transformation of proteins into simpler nitrogenous compounds during plant respiration. In some plant species, drying is done at 100°C which also result in loss in dry weight due to volatalization of volatile organic compounds.

Normally plant samples are dried at 70°C in a well ventilated oven or plant sample drier till constant dry weight is attained. The plant sample dried at 70°C be placed in an oven at 100°C for a short period for further drying without any charring to obtain the dry weight of the sample. The drying of plant sample at 100°C for a long period is likely to cause charring of samples. The plant samples may be spread in trays for rapid drying in an oven.

3.1.3.3 Grinding

The dried plant tissues are usually grinded in a Willey Mill. Finely grinded plant tissues passed through 100 mesh sieve contained 3 to 4 times more Fe and Zn than the whole sample. Coarse fraction show lower concentration. Finer fractions usually contain more soil and dust contaminants than the coarse fraction. The metals contained in the grinding equipment also contaminate the grinded sample. Grinding in a Willey Mill may contaminate the sample with Fe and Cu. The contamination with Cu is significantly reduced on use of stainless steel screens, and tool steel cutting faces. For reducing contamination from Cu and Zn, brass fittings grinding mills should be replaced by tool or stainless steel. For accurate analysis of Fe, the plant sample may be grinded in agate mortars.

3.1.3.4 Storage

The grinded samples be kept in air tight polythene containers such as plastic bottles with a screw cap to prevent adsorption of water from the humid environment. The bottle containing plant sample be kept in a dark and cool place. Storage at 0°C in the dark is preferred for long period of storage. It is desirable to dry the sample at 90°C in an oven for 20 hours prior to elemental analysis.

3.1.4 Comments

After the collection of plant samples, it should be washed, rinsed with detergent, air dried on new paper for 30 minutes, weighed, cut into pieces with stainless steel knife etc, dried in oven at 70°C and then grinded in a Willey Mill. It should be preferably digested for elemental analysis. The samples should not be kept too long in store as with time chemical changes occur, and it could be also spoiled by pathogen, insect and pest.

REFERENCES

Amma, M.K. 1990. Plant and Soil Analysis. Rubber Research Institute, Rubber Board, Kottayam-686009. Keral.

Jones (Jr) J. Benton. 1972. Plant Tissue Analysis for Micronutrients pp. 319-348. In J.J. Mortvedt *et. al.* (ed.) Micronutrients in Agriculture. Soil Sci. Soc. Am., Madison, Wisconsin, U.S.A.

Yoshida, S., D.A. Forno, J.H. Cock, and K.A. Gomez. 1972. Laboratory Manual for Physiological Studies of Rice. Intern Rice Res. Inst. P.O. Box 583, Manila.

3.2 METHODS OF ASHING PLANT TISSUES

Wet ashing is accomplished by digestion with HNO_3, $HClO_4$ and H_2SO_4or else with HNO_3 and $HClO_4$ especially for S determination. The method is unsuited for B determination as the element is volatalized during wet ashing.

Dry ashing is done in muffle furnace at temperature varying from 400 to 500°C for 2–8 hours. The plant sample taken in platinum or vitreous silica crucible is treated with $Mg(NO_3)_2$ or Na_2CO_3 prior to placing in a muffle furnace. Dry ashing at 500°C is considered satisfactory for the determination of Fe, Mn, Cu in biological materials. The use of stainless steel lined muffle furnaces eliminate contamination from Al and Zn during ashing.

3.2.1 Wet ashing of plant sample with ternary acid mixture– Jackson (1958)

3.2.1.1 Principle

The plant sample is treated with mineral acid (conc. HNO_3, H_2SO_4 & $HClO_4$) and heated for more rapid decomposition. The volatile constituents disappear and non-volatile mineral elements enter into solution.

The heating is continued until digest is reduced to few ml of clear white residue. The residue is dissolved in HCl. The test solution is prepared by dilution and filtration.

3.2.1.2 Reagents

(*i*) Ternary acid mixture. Mix 100 ml of conc. HNO_3, 10 ml of conc. H_2SO_4 and 40 ml of 60% $HClO_4$ or else 30 ml of 72% $HClO_4$.

(*ii*) Conc. HNO_3. AR grade. The reagent be distilled before use if test solution to be prepared is for micronutrient analysis.

3.2.1.3 Procedure

1. Predigestion of plant sample with conc. HNO_3. Transfer 1.0 g plant sample into a 300 ml tall pyrex or borosilicate beaker and wet the sample with 8 ml of conc. HNO_3. Add few glass beads, and cover the beaker with a watch glass. Let it stand for at least 2 hours. Then heat gently on a hot plate until the solution of the plant constituents. Thereafter, heat briskly without causing serious bumping keeping a portion of the beaker mouth uncovered with the watch glass. Continue heating until the volume of the contents is reduced to about 4 ml. Remove the beaker from the hot plate and allow it to cool.

2. Digestion with ternary acid mixture. Add 5 ml of ternary acid mixture into the beaker in which 1.0 g plant sample has been predigested with conc. HNO_3 and transfer the beaker on a hot plate for brisk heating. Keep a portion of the beaker mouth open to permit the gases to escape. Continue heating until evolution of copious dense white fumes subsides leaving about 3 ml of colourless solution in the beaker which on cooling gives a whitish residue. If the residue is not white and signs of begining of charring is seen, remove the beaker from the hot plate and let it cool. Then add 2 ml of ternary acid mixture and place the beaker on hot plate. Continue heating until the contents is reduced to about 2-3 ml of colourless clear solution. In order to avoid excessive heating, a layer of sand may be spread over the hot plate. If white residue is not obtained, the residue be treated again with 2ml of the ternary acid mixture and the heating be continued till the contents is reduced to about 2-3 ml. Thereafter, the beaker be removed from the hot plate or sand bath. Allow the beaker to cool.

3. Preparation of plant test solution. Add 5 ml of glass distilled water into the beaker containing digested residue. Heat the contents gently on a hot plate until white fumes appear. Remove the beaker and let it cool. Then, add 10 ml of glass distilled water and heat the contents to dissolve the residue.

Take a 100 ml clean volumetric flask fitted with a funnel lined with Whatman No. 42 paper. Transfer the contents of the beaker into the funnel using a glass rod collecting the filtrate in the volumetric flask. Rinse the beaker with about 15 ml portions of glass distilled water and transfer each rinsing into the funnel in order to transfer the digested residue quantitatively. Collect the filtrate from each washing into the same volumetric flask. Then wash the residue on filter paper with small portions of glass distilled water and collect the washing until the volume of filtrate reaches 100 ml mark. Stopper the flask, label it and then preserve it for the elemental analysis other than Ca, Mg and S. Similarly run a blank digestion and subsequently prepare the blank test solution following the above mentioned procedure.

If Ca and Mg is to be determined, add 5ml of conc. HCl instead of glass distilled water into the beaker containing the plant digested residue. Swirl the contents and heat it to warm. Take a clean 100 ml volumetric flask fitted with a funnel lined with Whatman No. 42 paper. Transfer the contents of the beaker into the funnel and collect the filtrate in the volumetric flask. Then, add another 5 ml of conc. HCl into the beaker to dissolve the remaining residue in the beaker. Transfer the solution into the same funnel, collecting the filtrate in the same volumetric flask. Repeat subsequent washing of the contents of the beaker with small portions of 6N HCl and transfer the washing into the same funnel until the filtrate reaches 100 ml mark. Similarly run a blank digestion and then prepare test solution using conc. and 6 N HCl instead of glass distilled water as described above.

3.2.2 Wet ashing of plant sample with conc. HNO_3 & $HClO_4$ – Richards ed. (1954)

3.2.2.1 Principle

The plant sample is digested only with conc. HNO_3 and $HClO_4$ until the digest is reduced to few ml of clear white residue. The residue is dissolved in HCl, diluted and filtered to prepare test solution. The test solution can be analysed for S also in addition to other nutrient elements.

3.2.2.2 Reagents

(*i*) Conc. HNO_3. A.R. grade. The reagent be distilled using completely glass distillation apparatus if test solution to be prepared is for micronutrient analysis.

(*ii*) Conc. $HClO_4$: A.R. grade 72%

3.2.2.3 Procedure

1. Wet digestion of plant sample. Transfer 1.000 g of oven dry finely ground plant sample and few glass beads into 300 ml tall beaker (pyrex or borosilicate), and 20 ml of conc. HNO_3 wetting the entire sample. Cover the beaker with a watch glass and allow it to stand for 2 hours until the initial reaction subsides.

Thereafter, place the beaker on a hot plate kept inside a fume chamber. Continue heating gently until solid particles nearly disappear. Remove the beaker from the hot plate and allow it to cool. Then add 10 ml of 72% $HClO_4$ and place the beaker on hot plate. Continue heating gently first and then vigorously until dense white fumes given off during heating subsides, and the contents become a colourless clear solution. Discontinue heating when the solution is reduced to a volume of about 2-3 ml. Do not heat the contents to dryness which may lead to charring and loss of nutrients.

2. Preparation of sample test solution. Allow the contents of the beaker to cool. Thereafter add 3ml of distilled (1 : 1) HCl and heat the contents on a hot plate until the solution is reduced to about 2 ml. Cool and add glass distilled water. Stir the contents with a glass rod and transfer it to a 100 ml volumetric flask. Continue transfer of the digested residue with portions of glass distilled water into the same 100 ml volumetric flask until the entire digest has been quantitatively transferred into the same flask. Make the volume to 100 ml with glass distilled water. Stopper the flask and mix the contents by turning the flask upside down 4-5 times and let it stand over night. Thereafter, filter the diluted digest through a dry Whatman no. 42 filter paper. Discard the initial 2-3 ml filtrate after rinsing the receiving flask with it. Collect the subsequent filtrate and preserve it as sample test solution for the determination of nutrient elements including sulphur. Similarly prepare a blank test solution without taking plant sample.

3.2.3 Dry Ashing of plant sample with $Mg(NO_3)_2$ –Richards (1954)

3.2.3.1 Principle

The finely ground plant sample is wetted with $Mg(NO_3)$ in silica crucible and after initial heating at 180°C, it is transferred to a muffle furnace at low heat. Heating is continued until the charge is completely oxidized. The ash after cooling is treated with HCl (1+1) and evaporated to dryness. The residue is dissolved in HCl (1+1) filtered and diluted to prepare test solution.

3.2.3.2 Reagents

(*i*) Magnesium nitrate solution. Add 113 g of MgO (AR) to 300 ml distilled water, and stir to a paste. Add HNO_3(1+1) until MgO is in solution. Add a little excess MgO and boil. Filter and dilute to 1l.

(*ii*) Distilled (1+1) hydrochloric acid (analar). Add 500 ml of distilled water into boiling flask of a glass distillation apparatus. Then add 500 ml of conc. HCl (Analar) into the boiling flask through the feeding funnel. Continue heating and collect the distillate in a receiving flask. Continue distillation until 700 ml of distillate is collected.

3.2.3.3 Procedure

1. Dry ashing of the plant sample. Transfer 2.000 g of finely ground plant sample into large vitreous silica crucible or casserole. Add 10 ml of $Mg(NO_3)_2$ solution and make the entire plant sample in contact with the $Mg(NO_3)_2$solution. Then, heat the contents gently on a hot plate raising the temperature to 180°C. Continue heating until the reaction is complete.

Thereafter, transfer the hot large vitreous silica crucible with plant residue into an electric muffle furnace at low heat (muffle furnace inside must not show any redness). Let the silica crucible remain in the muffle furnace until the charge is thoroughly oxidized which is indicated by the absence of any black particles. During the period of heating in the muffle furnace, remove the vitreous silica crucible, and let it cool for a while. Then, break up the charge, and return the crucible in the muffle furnace. When the charge is thoroughly oxidized, remove the silica crucible from the muffle furnace, and let it cool. The muffle furnace be switched off if not required.

2. Preparation of the sample test solution after dry ashing of plant sample. When the vitreous silica crucible containing oxidized plant residue has become sufficiently cool, moisten the ash with glass distilled water, and add 10 ml of hydrochloric acid (1+1) which should provide an excess of acid. Heat the crucible on a water bath to evaporate the contents to dryness. Thereafter, add HCl (1+1) to moisten the residue. Again evaporate the contents to dryness to dehydrate silica

Then, add 5 ml of HCl (1+1) and sufficient glass distilled water to bring the salts into solution. Place the silica crucible on a steam bath until the salts are completely dissolved. Filter the solution through a funnel lined with Whatman no. 42 filter paper fitted into mouth of a 100 ml volumetric flask. Wash the residue on filter paper with glass distilled water. Collect the filtrate and washings into the same 100 ml volumetric flask. Continue washing the residue with glass distilled water until the volume of filtrate and washings collected reaches 100 ml mark. Preserve the sample test solution for the elemental analysis including S. Similarly run a blank dry ashing without plant sample and also prepare the blank test solution following the procedure described above.

3.2.4 Dry ashing of plant sample after treatment with dilute sulphuric acid alcoholic solution–Prasad (1970)

3.2.4.1 Principle

The finely ground plant sample is wetted in platinum or silica crucible with dilute alcoholic sulphuric acid solution and ignited. When the flame subsides, the crucible is transferred to a muffle furnace. The inside temperature is raised to 525°C and heating is continued until grey white ash is obtained. The residue on cooling is treated with 2N HCl, filtered and washed with hot distilled water. The filtrate and washing is collected to prepare test solution for the analysis of particularly Ca and Mg.

3.2.4.2 Reagents

(*i*) Dilute sulphuric acid alcoholic solution. Transfer 80 ml of ethyl alcohol into a 100 ml volumetric flask. Then, add 5 ml of conc. H_2SO_4 slowly along the inner wall of the flask. Let the contents cool. Then, make the volume to 100 ml with ethyl alcohol.

(*ii*) Hydrochloric acid approx 2N. Dilute 16.7 ml of conc. HCl (approx. 12N) to 100 ml in a 100 ml volumetric flask.

3.2.4.3 Procedure

1. Dry ashing of plant sample. Place 1.0 g sample of dried ground plant material into a platinum, silica or porcelain crucible. Then moisten the plant sample with the dilute sulphuric acid alcoholic solution through a pipette.

Thereafter, ignite the sample using the pipette used for adding dilute H_2SO_4 alcoholic solution. Torch to remove excess alcohol. When the flame subsides, transfer the crucible into muffle furnace, and gradually raise the temperature to 525°C. Maintain a temperature of 525°C for 90 minutes or longer until grey-white ash free of carbon is obtained. Similarly run a blank without plant sample.

2. Preparation of plant simple test solution. Remove the crucible with ash from the muffle furnace and allow it to cool.

Add 5 ml of 2N HCl to the crucible containing plant ash, and let it stand for 15 minutes. Then place a funnel lined with Whatman No. 1 filter paper into the mouth of a 100 ml volumetric flask. Transfer the residue dissolved in 2N HCl into the funnel lined with filter paper using portions of hot distilled water. Collect the filtrate in the volumetric flask.

Rinse the crucible several times with 10 ml portions of hot distilled water and transfer each rinsing into the funnel.

Thereafter, leach the residue remaining on filter paper with hot distilled water. Collect the leachate in the volumetric flask containing the filtrate. Continue leaching the residue with hot distilled water until the leachate plus filtrate reach 100-ml mark. Cool the contents and make the volume to 100 ml with distilled water. Similarly treat the blank residue in crucible for preparing the blank test solution.

The above procedure of dry ashing of plant sample has been suggested by Prasad (1970) for the determination of Ca and Mg in plant sample.

3.2.5 Dry ashing of plant sample with CaO by the method of Chapman and Pratt (1961)

3.2.5.1 Principle

Finely ground plant sample mixed with CaO one fourth of its mass is made into a paste by adding some distilled water in a large size silica or porcelain crucible.

The crucible is transferred to muffle furnace and then inside temperature is raised to 550°C for atleast 90 minutes. The ash is wetted with distilled water and after warming it is broken with glass rod. The contents is filtered and washed with hot distilled water. The filtrate and washings are collected and diluted to obtain test solution particularly for the determination of chlorine.

3.2.5.2 Reagents

Calcium Oxide. Take AR grade CaO from a freshly opened reagent bottle.

3.2.5.3 Procedure

1. Dry ashing of plant sample. Place a 2.0 g sample of dried ground plant material in a silica or porcelain crucible of large size. Mix the plant sample with 0.5g of CaO, and sufficient distilled water to make a thin paste.

Transfer the crucible with treated plant sample in a muffle furnace. Gradually raise the temperature of muffle furnace to 550°C. Maintain this temperature for at least 90 minutes. Remove the crucible with ash from the muffle furnace and allow it to cool. Simillarly run a blank taking 0.5g CaO only in a cruciible and follow the same ashing procedure as for sample to obtain crucible with ash for preparing blank test solution.

2. Preparation of plant sample test solution. Place the crucible with ash on a hot plate. Add in small portion hot distilled water up to a total of 15 ml while warming the crucible. Then break up the ash with a large diameter blunt glass rod.

Fit a funnel lined with Whatman No. 44 filter paper 4 cm in dia into the mouth of a 100 ml volumetric flask. Transfer the plant ash-distilled water suspension from the crucible into the funnel lined with filter paper and collect the filtrate in the volumetric flask. Rinse the crucible with several 10 ml portions of hot distilled water transferring each rinsing into the funnel and collect the filtrate in the same volumetric flask. Now leach the residue on filter paper with hot distilled water until the volume of filtrate plus leachate reaches 100 ml mark. Allow the contents to cool, and then make the volume to 100 ml with distilled water. Stopper and mix the contents. Similarly prepare the blank test solution from residue obtained on dry ashing of 0.5 g CaO in a crucible.

The dry ashing with CaO after making sample– CaO paste have been suggested by Chapman and Pratt (1966) for the determination of chlorine in plant sample.

3.2.6 Dry ashing for K determination in plant sample as proposed by M-Peech–Jackson (1958)

3.2.6.1 Principle

Plant sample is ashed for a longer period, but at lower temperature. The ash obtained is treated with 1 N HNO_3 and again heated in a muffle furnace for a brief period. The ash after cooling may be dissolved in HCl for preparation of the sample test solution. Ashing at low temperature prevent loss of volatile mineral constituents.

3.2.6.2 Reagents

1. Nitric acid 1.0N

3.2.6.3 Procedure

Transfer 1.0g of dried and grounded plant sample into a porcelain crucible. Then place the crucible in a muffle furnace at 400 to 450°C for several hours preferably over night.

Remove the porcelain crucible from the muffle furnace and let it cool. On cooling, treat the ash with 1N HNO_3 wetting the ash completely. Then evaporate the wetted ash to dryness on a hot plate.

Place the ash treated with 1N HNO_3 after drying into a muffle furnace at 400°C ± 10°C just for 10 minutes. Thereafter, remove the porcelain crucible from the muffle furnace.

Now add 5 ml of 1N HCl into the porcelain crucible and heat to warm the contents and filter through a funnel lined with Whatman No. 1 filter paper.

Collect the filtrate into a 100 ml volumetric flask. Transfer the residue in porcelain crucible quantitatively into the funnel using portions of distilled water. Collect the filtrate in the same 100 ml volumetric flask.

Wash the residue on filter paper with distilled water. Combine the leachate with the filtrate and continue washing the residue until the combined leachate and filtrate reaches 100 ml mark. Stopper, and mix the contents.

3.2.7 Dry ashing of plant sample for P determination by the Vanadomolybdate method–Jackson (1958)

3.2.7.1 Principle

No P is lost from plant sample during dry ashing, if ashing is carried in the presence of $Mg(NO_3)_2$ or $Mg(OAC)_2$ or else in their alcoholic solution at temperature not exceeding 600°C.

3.2.7.2 Reagents

(*i*) Magnesium nitrate-0.5 N

(*ii*) Sulphuric acid approx. 2.0 N : Dilute 2.6 ml of conc. H_2SO_4 (36N) to 100 ml with distilled water.

3.2.7.3 Procedure

Transfer 0.5 g of dried and finely ground plant sample into a 100 ml capacity evaporating dish. Add 5ml of 0.5N Mg $(NO_3)_2$ or $Mg(OAC)_2$, and 10 ml distilled water. Evaporate the mixture to dryness on a steam bath. Remove the evaporating dish from water bath and leave it for air drying.

Place the dry evaporating dish in a muffle furnace and ignite at 600°C, until the residue becomes uniformly grey in colour. Usually 30 minutes time is sufficient for ashing.

Remove the evaporating dish from the muffle furnace, and allow it to cool. Then, add 10 ml of approx. 2NH_2SO_4. Rotate the evaporating dish to bring the acid in contact with the entire ash. There after add 15 ml of distilled water and evaporate the contents to less than 5 ml on a steam bath.

Remove the evaporating dish from steam bath, and add 20 ml of distilled water. Allow the evaporating dish to cool. Rub down the inner sides of evaporating dish with a glass rod provided with rubber policeman.

Transfer the contents into a funnel lined with Whatman 42 filter paper. Collect the filtrate in a 100 ml volumetric flask. Transfer the contents quantitatively using portions of warm distilled water. Collect the filtrate in the same 100 ml volumetric flask.

Wash the residue on filter paper and combine the leachate with the filtrate. Continue washing the residue on filter paper until the volume of filtrate cum leachate reaches 100 ml mark. Stopper and mix the contents by turning the volumetric flask upside down holding the stopper firmly.

3.2.8 Dry ashing of plant sample for B determination

3.2.8.1 Principle

The finely ground plant sample is wetted with saturated $Ca(OH)_2$ solution, and is then evaporated to dryness. Then, it is ignited on a burner to drive off volatile material. The residue is ignited in a muffle furnace. The B in residue is dissolved in measured volume of either 0.1N HCl or 0.36N H_2SO_4. The suspension is filtered or centrifuged to obtain a clear test solution. The B dissolved in HCl is reacted with curcumine and B dissloved in H_2SO_4 is reacted with quinalizarin or carmine to form a coloured complex for the colorimetric determinate of boron.

3.2.8.2 Apparatus and Reagents

Apparatus and Equipments. Platinum, silica or porcelain evaporating dish, Hot air oven, gas burner, muffle furnace and centrifuge.

Reagents

(*i*) Distilled water : Use water distilled from a metal distill

(*ii*) Saturated calcium hydroxide solution 2% : Add 10g of $Ca(OH)_2$to 500 ml distilled water. Shake well and allow it to settle.

(*iii*) 0.10 N Hydrochloric acid : Dilute 8.4 ml of conc. HCl to 100 ml with distilled water.

(*iv)* 0.36 N sulphuric acid : Dilute 10.0 ml of conc. H_2SO_4 (AR) to 100 ml with distilled water

3.2.8.3 Procedure

Transfer 1.0-2.0 g of finely ground oven dry (at 70°C) plant sample into a platinum, silica, or porcelain evaporating dish. Wet the plant sample with 5 ml of saturated (2%) $Ca(OH)_2$ solution, and evaporate the contents to dryness in an oven at 105°C. Remove the dish and ignite on a burner carefully to drive off volatile material. Thereafter, place the evaporating dish into a muffle furnace and ignite at 550°C for an hour to obtain a white or grey ash.

After dry ashing remove the evaporating dish containing the residue from the muffle furnace and cool to room tempertaure. Add exactly 10.0 or 15.0 ml of 0.1NHCl, break up the ash with glass rod, and triturate with the police man to dissolve boron. Centrifuge the suspension for 10 minutes at 2400 rpm to obtain a clear test solution. Similarly prepare a blank test solution.

When quinalizarin or carmine is to be used to form coloured complex with B, after igniting the plant residue in muffle furnace at 550°C, dissolve B present in ignited residue in 10.0 or 15.0 ml of 0.36 N H_2SO_4 instead of 0.1N HCl, and then centrifuge the suspension at 2400 rpm for 10 minutes to obtain a clear test solution.

3.2.8.4 Calculation : Nil

3.2.5 Comments

The wet oxidation with $HClO_4$ avoids the loss of K through volatalization. Besides, it gives a clear solution of all constituents except Si. However, Si is quantitatively dehydrated, and precipitated and so, it can be removed by filtration or centrifugation.

During $HClO_4$ digestion, there is danger of loss of H_3PO_4, if the temperature of digestion mixture exceeds 230°C. There is also danger of explosion by contact of plant sample with $HClO_4$, but this danger is completely overcome by predigestion of plant sample with conc. HNO_3.

In ternary acid digestion, the temperature is raised rapidly to 180°-200°C. With the increase in the temperature of the digestion mixture, the power of oxidation of $HClO_4$ increases more rapidly than the rate of evaporation. On heating, $HClO_4$ yields anhydrous $HClO_4$ which further dissociates into nascent Cl and O. At a temperture of 203°C, the decomposition of anhydrous $HClO_4$ is complete.

The condensation of $HClO_4$ fumes on fume hood and duct can be hazardous. Only fume pipe made of borosilicate glass is safe. For safe outlet of $HClO_4$ fume, concrete pipe fume duct may be connected to an exhaust fan installed at the roof. The digestion can be also carried in the open on a large size hot plate or else in outside verandah.

The digestion process be continued until a clear solution remains after acids are largely volatalized. When the residue in the flask is clear, white and only slightly moistened with H_2SO_4, the digestion is complete. At this stage $HClO_4$ is largely removed.

The digestion need to be prolonged enough to decrease $HClO_4$ sufficiently to prevent precipitation of K as $KClO_4$. The residue be dissolved in HCl as chloride salts are soluble and sufficient concentration of HCl prevents rehydration of Si.

The triacid digest is suitable for the analysis of major and micronutrient elements. When digest residue is taken with HCl, an aliquot from the sample test solution may be evaporated to dryness before P determination by Vanado- molybdate method in order to control the concentration of HCl.

In wet oxidation of plant sample with only conc. HNO_3 and $HClO_4$, H_2SO_4 is omitted in the digestion mixture. The omission of H_2SO_4, enable determination of sulphur in the residue. In addition, precipitation of Ca and Mg as sulphates in significant amounts is unlikely in the digest. Consequently, the diacid digest of plant sample can be also used for Ca and Mg determination with greater precision. Other elements such as Na and K can also be determined. However, its use for other elemental analysis should be under taken only after making adequate comparison of recovery against triacid digest.

In dry ashing of plant sample, ignition at temperature exceeding 600°C cause loss of both P and K. However, treatment of plant sample with $Mg(NO_3)/Mg(OAC)_2$or with their alcoholic solution prevent nutrient losses during ashing, but metallic cation added cannot be determined on the same ash. The treatment of plant tissues with Na_2CO_3 before ashing serve to retain S and Cl which are otherwise volatalized at high temperature.

Dry ashing is considered useful for plant samples high in Ca as Ca become precipitated as $CaSO_4$ in wet oxidation with ternary acids including H_2SO_4. When plant tissues is treated with $Mg(NO_3)_2$ nutrient losses during ashing is prevented which makes possible to determine P, S and Cl also. If ashiing of plant tissue is under taken at temperature of about 400°C loss of K is prevented. In dry ashing care be taken to completely avoid charring. When charring occurs ashing remain incomplete and a portion of P and trace elements become occluded or insoluble. Ignition of sample inside Schöniger flask in excess oxygen may be found convenient for oxidation of S to enable S determination by developing $BaSO_4$ turbidity. The dry ashing of plant sample after wetting with saturated $Ca(OH)_2$ has been suggested for the determination of boron in the plant test solution.

REFERENCES

Amma, M.K. 1990. Plant and Soil Analysis. Rubber Research Institute Rubber Board. Kottayam 686009, Keral.

Chapman, H.D and P.F. Pratt. 1961. Chlorine p. 97–100. In H.D. chapman and P.F. Pratt (ed) : Methods of Analysis for Soil Plants, and waters University of Calfornia, River side, USA.

Jaekson, M.L. 1958. Soil Chemical Analysis. Practice Hall of India Pvt. Ltd., New Delhi (1973)

Jones (Jr) J. Benton 1972. Plant Tissue Analysis for micro nutrients p 319–348. In J.J Mortvedt et. al. (ed.) Miconutrients in Agriculture. Soil Science Society of America. Madison, USA.

Richards, L.A. (ed.) 1954. Diagnosis and Improvement of Saline and Alkali soils. Oxford and IBH Publishing Co., New Delhi (1968)

3.3 NITROGEN DETERMINATION IN PLANT SAMPLE

3.3.1 Nitrogen determination by micro Kjeldahl digestion and distillation method–Amma (1989)

3.3.1.1 Principle

200 mg of plant sample and an equal amount of catalyst mixture is transferred to a micro Kjeldahl digestion flask. On adding 3 ml of conc. H_2SO_4 the digestion

flask is transferred to digestion assembly with a fume hood.The contents are first heated slowly. When the initial vigorous reaction is over, the contents is heated briskly until it becomes clear. Organic N is transformed into ammonium sulphate during the digestion process.

The digested contents are transferred to distillation flask of a micro Kjeldahl distillation unit which has a water filled large flask with round bottom for generating steam to boil the contents placed in distillation flask.The distillation flask is connected to a condenser, and the condenser outlet is dipped in 25 ml of 2% boric acid containing mixed indicator. On adding 10 ml of 40% NaOH into distillation flask, the feeding funnel is closed and water in large round flask is heated to boil for generating steam for boiling the contents in distillation flask. In 9 minutes of distillation about 55 ml of distillate is collected in boric acid forming ammonium borate which makes the weak boric acid alkaline.

The ammonium borate is back titrated by adding 0.02 N HCl until the reaction mixture changes its colour from green to pink. The standard acid used in back titration of ammonium borate is a measure of N content of the plant sample.

3.1.2.2 Reagents

(*i*) Conc. H_2SO_4 (BDH-Analar).

(*ii*) Salt Mixture. Mix 25.0g of K_2SO_4, 5.0g of $CuSO_4 . 5H_2O$ and 0.5 g of metallic selenium by grinding in a mortar.

(*iii*) Boric acid 2% containing brom cresol green-methyl red indicator. Dissolve 20.0g of H_3BO_3 (BDH -Analar) in redistilled water and dilute the contents to 1000 ml after adding 20.0 ml of brom cresol green-methyl red indicator.

(*iv*) Brom Cresol Green-Methyl Red indicator. Dissolve 0.1 g of brom cresol green and 0.07 g of methyl red in 100 ml of ethyl alcohol.

(*v*) Sodium hydroxide 40% solution. Transfer 400 g of NaOH (BDH-LR) in 1000 ml beaker. Then add 600 ml of redistilled water under a fume hood. Stir the contents with a glass rod. Cool the contents and transfer in to a screw top bottle.

(*vi*) Sodium Carbonate 0.1N. Dissolve 0.530 g of anhydrous Na_2CO_3 (BDH-AR) in about 20 ml of redistilled water and dilute the contents to 100 ml mark in a vol. flask.

(*vii*) Methyl orange indicator, 0.1%. Dissolve 0.1 g of methyl orange (BDH) in 40 ml of redistilled water and dilute the contents to 100 ml mark in a vol. flask.

(*viii*) Hydrochloric acid 0.02N. Dilute 4.0 ml of 5.0N HCl to 1000 ml with redistilled water. Check the normality of the acid against. 0.1N Na_2CO_3 using methyl orange indicator.

3.3.1.3 Procedure

1. Digestion of the plant sample. Transfer 200mg of dried, finely ground plant sample into a 100 ml Kjeldahl flask. Then add 200 mg of salt mixture and 3ml of conc. H_2SO_4 . Let the contents react for an hour. Then transfer the Kjeldahl flask to the heating mantle of a digestion unit with fume hood. Switch on the heating element

and digest at low heat until initial vigorous reaction is over. Take care that frothing remains below the neck of Kjeldahl flask.

Thereafter heat briskly for subsequent digestion and continue heating until the digest become clear. Switch off the heating mantle and let the Kjeldahl flask cool. Then remove Kjeldahl flask from the digestion assembly.

2. Distillation of the digested mixture. Add 10 ml of redistilled water into the plant digest and allow the contents to cool. Meanwhile set the micro Kjeldahl steam distillation assembly provided with large flask filled with distilled water for generating steam, distillation flask in which digested material is transferred, condenser flask with outlet tube which remains dipped in 25 ml of 2% H_3BO_3. There is also one collection flask for siphoning residue after distillation is over.

Then transfer plant digest quickly into the distillation assembly through the inlet funnel. Rinse the Kjeldahl flask thrice with 5 ml portions of redistilled water and transfer each rinsing into the distillation flask. Then transfer 10 ml of 40% NaOH and 10 ml of redistilled water into the distillation flask through the inlet-funnel. Close the inlet opening after adding redistilled water by means of stop cock provided.

Heat the flask filled with redistilled water to boil. On boiling the steam will pass into the distillation flask. Continue heating the steam generating flask and let the contents in distillation flask boil for 9 minutes. During the period collect about 55 ml distillate into 150 ml conical flask containing 25ml of 2.0 % H_3BO_3. Lift the condenser outlet from the receiving conical flask and let the distillate fall into the conical flask for 1 minute. Rinse the dipped portion of the condenser outlet with 3 ml of redistilled water and collect the rinsings in the same conical flask. Remove the conical flask containing the distillate, and discontinue heating of the steam generating flask.

Immediately siphon the contents of the distillation flask into the adjoining collecting flask automatically by cooling the steam generating flask. Place a wet thick cloth just beneath the neck of the steam generating flask and remove it the moment contents of the distillation flask get siphoned into the collecting flask.

Rinse the distillation flask thrice with 25 ml portions of redistilled water and siphon the contents automatically into the collecting flask by cooling the portion beneath the neck of steam generating flask. Then drain off the contents of collecting flask into a large beaker by opening the outlet. Close the outlet after emptying the collecting flask.

3. Titration of the Distillate. Titrate NH_3 distilled into 2% H_3BO_3 solution containing brom cresol green-methyl red indicator with the standard 0.02 N HCl using a microburette. Titrate the distillate with the standard acid until the colour of the reaction mixture changes from green to distinct pink. Run a blank digestion without plant sample, distill the digested mixture collecting the distillate and titrate with the standard acid to the same end point as that of sample titration.

3.3.1.4 Calculation

$$\% \text{ N in plant sample} = \frac{(S-B) \times N \times ME \times 100}{W}$$

Where

S is ml of standard acid used in the titration of the sample distillate

B, is the ml of standard acid used in the titration of the blank distillate

N is the normality of the acid used in the titration of the distillate

ME, is the milligram equivalent weight of nitrogen *i.e.* .014 g

W is the mass of plant sample taken in gram for the preparation of the sample digest.

3.3.1.5 Comments

N content of plant samples may vary from < 1.0% to even more than 4.0% depending upon the plant species, and the plant parts analysed. The concentration of N in vegetative parts of the plant may vary with the stage of plant development. The stage at which maximum difference in N concentration occur due to supply of N at different rates needs to be identified for more realistic diagnosis of N deficiency. In diagnosis of nutrient deficiency, the concentration of other nutrient elements are of significance, when there is optimum concentration of N in the plant.

The plant sample is dried in an oven at 70°C and then it is grinded in a Willey Mill to a desired fineness. Plant sample from 50 to 250 mg is used for total N determination depending upon the content of N in the plant sample. The micro Kjeldahl method of N determination in plant sample is rapid, less expensive, satisfactory and suitable for estimation of N in large number of samples.

The use of redistilled water is desirable, if distilled water is found to be distinctly alkaline.

REFERENCES

Amma, M.K. 1989. Plant and Soil Analysis. Rubber Res. Inst. Rubber Board, Kottayam, Keral, India.

Horwitz, W. 1970. Methods of Analysis. Association of Official Analytical Chemist P.O. Box. No. 540, Benjamin Franklin Sta. Washington, D.C. 20044.

Yoshida, S., D.A. Forno, J.H. Cock, and K.A. Gomez, 1972. Laboratory Manual For Physiological Studies of Rice. Intern. Rice Res. Inst. P.O. Box 583, Manila.

3.3.2 Total nitrogen in plant sample by the modified macro-Kjeldahl method–Jackson (1958)

3.3.2.1 Principle

The oven dry plant sample grinded to pass 40 mesh screen is digested in a 800 ml Kjeldahl flask with conc. H_2SO_4, anhydrous $Na_2SO_4$and $CuSO_4.5H_2O$ by heating under a fume hood until a clear digest containing $(NH_4)_2SO_4$ is obtained.

The digest on cooling is diluted with distilled water. After adding paraffin wax, glass beads and zinc granules, the Kjeldahl flask is fitted into a distillation assembly. The digest is made strongly alkaline by adding NaOH solution through a feeding funnel. Then, the digest is distilled by boiling until 120 ml distillate is collected in 2% H_3BO_3 containing brom cresol green-methyl red indicator.

The ammonium borate formed is titrated with a standard acid. The total N in sample is estimated from the standard acid titre used in the neutralization of ammonium borate.

The reaction that occur as exemplified by carbohydrate and an amino acid is presented below :

Digestion

I. $(C_6H_{10}O_5)_n + nH_2SO_4 = 6nC + 5nH_2O$

$C + 2H_2SO_4 = CO_2 + 2H_2O + 2SO_2$

II. $R-COONH_2 + 2H_2SO_4 + 5H_2 = 2(NH_4)_2SO_4 + 2R-COOH$

Distillation

$2(NH_4)_2SO_4 + 4NaOH = 2Na_2SO_4 + 4H_2O + 2NH_3$

$4NH_3 + H_2O = 4NH_4OH$

$NH_4OH + H_3BO_3 = NH_4\ (B(OH_4))$

Titration

$2NH_4\ (B(OH_4)) + H_2SO_4 = (NH_4)_2SO_4 + 2H_3BO_3 + 2H_2O$

3.3.2.2 Reagents

(*i*) Catalyst mixture. Anhydrous Na_2SO_4-10g and pure dry $CuSO_4.5H_2O$ - 1.0g per sample.

(*ii*) Zinc granules. (*iii*) Paraffin wax.

(*iv*) Conc. sulphuric acid (AR., sp. gr. 1.84).

(*v*) Sodium hydroxide solution, 40%.

(*vi*) Boric acid solution, 2% containing mixed indicator.

(*vii*) Hydrochloric acid, 10N. (*viii*) Hydrochloric acid, 2N.

(*ix*) Hydrochloric acid, .05N. (*x*) Sodium carbonate, 0.1N.

(*xi*) Methyl orange, 0.1%. (*xii*) Methyl red-bromcresol green indicator.

See section 2.8.2. for the preparation of reagents (*v*) to (*xii*).

3.3.2.3 Procedure

1 Digestion of plant sample. Grind oven dry plant sample to pass through a 40 mesh screen. Weigh 0.25 g to 0.50 g plant sample on a square cut butter paper and fold the paper containing plant sample and drop it in the centre of 800 ml macro Kjeldahl flask provided with B-24 joint on its mouth.

Add 25 ml of conc. H_2SO_4 (AR. sp.gr. 1.84) wetting the folded plant sample, and let it stand for 30 minutes. Similarly run a blank in another 800 ml Kjeldahl flask taking a square cut butter paper of the same size and mass as taken for weighing the plant sample.

After half an hour of adding conc. H_2SO_4, add 10g of anhydrous Na_2SO_4 and 1g of pure dry $CuSO_4.5H_2O$ into both the Kjeldahl flasks. Place the Kjeldahl flasks on the digestion unit inside the fume hood. Light the burner and adjust the gas flame below the level of the acid in the Kjeldahl flask.

Heat the contents of the flasks slowly till frothing continues, and thereafter heat briskly until the digest become clear. Continue heating the contents briskly further for just one hour. Put off the gas burner, and allow the Kjeldahl flask to cool.

2. Distillation of the sample and blank digest. Remove the Kjeldahl flask after adding 1g of paraffin wax, few glass beads and few pieces of zinc granules and place it on the heating unit of the distillation assembly.

Now place the glass adapter with side arm on the mouth of Kjeldahl flask. Fit the inlet grounded joint of the distillation unit into the top mouth of the adapter. Place the feeding funnel into the mouth of the side arm. Now add 150 ml of distilled water into the Kjeldahl flask through the feeding funnel.

Now take a 250 ml–300 ml Erlenmeyer flask and make a mark with glass pencil at 160 ml volume. Fill it with 40 ml of 2% H_3BO_3 solution containing mixed indicator and place it below the condenser dipping the condenser outlet tube extension inside the H_3BO_3 solution. Now add 10 ml of distilled water into the feeding funnel. Lift the glass rod stopper slightly to permit distilled water flow into Kjeldahl flask slowly. If air begins to bubble through the H_3BO_3 solution, the entire distillation assembly is air tight.

Now fill the feeding funnel with 120 ml of 40% NaOH solution in portions, and lift the glass rod stopper to let NaOH solution flow into the Kjeldahl flask. Thereafter, fill feeding funnel with 20 ml distilled water and then let it flow into the Kjeldahl flask to wash down the conc. alkali adhering to the funnel outlet. Place the glass stopper back and then fill the feeding funnel again with 20 ml distilled water and let it remain there during the period of distillation of ammonia.

Switch on the electric heater and heat the contents to boil, and let it boil continuously until 120 ml distillate is collected in the receiving flask filled with 40 ml of 2% H_3BO_3 containing mixed indicator.

Lift the condenser outlet extension tube dipped in the distillate-H_3BO_3 solution, rinse the portion that was dipped in distillate with a jet of distilled water allowing the rinsings to collect in the same receiving flask. Test the fresh distillate with a red litmus paper. If there is no change in the colour of red litmus paper, the distillation of ammonia is complete. Remove the Erlenmeyer flask containing the distillate, and switch off the heating unit. Allow hot Kjeldahl flask to cool before its separation from the distillation apparatus.

Similarly distill the blank digest as described above.

3. Titration of the distillate with standard acid. Fill 50 ml burette with .05 N HCl and check whether there is any leakage. Do not use burette which leaks.

Add standard acid through burette in small portions into the Erlenmeyer flask. Swirl the contents after each addition of standard acid. Continue titration with standard acid until the colour of the distillate changes from green to faint pink. With one or two drop more of standard acid, the colour of the reaction mixture will change from faint pink to distinct pink. Note down the standard acid titre for sample distillate. Then, titrate the blank distillate after refilling the burette, similarly and to the same end point as that of the sample distillate. Note down the standard acid titre for the blank distillate.

3.3.2.4 Calculation

$$\% \text{ Total N in plant sample} = \frac{(S-B) \times N \times ME \times 100}{W}$$

where,

S is standard acid titre in ml for the sample distillate.

B is standard acid titre in ml for the blank distillate.

N is the normality of acid determined against 0.1N Na_2CO_3 using 0.1% methyl orange as indicator.

ME is the milligram equivalent weight of N *i.e.*, 0.014 g

W is the mass of plant sample taken in g for total N determination.

3.3.2.5 Comments

Na_2SO_4 is incorporated into the digestion mixture to raise the boiling point of the conc. H_2SO_4.

$CuSO_4.5H_2O$ is considered satisfactory as a catalyst for plant sample.

However, in soil samples selenium is used with Na_2SO_4 with or without Hg or HgO or $CuSO_4$.

The use of Se hastens the time of digestion. However, the digest becomes clear rather too soon, and there is a danger of allowing less time for digestion. The use of Se may also cause loss of N with large amount of Na_2SO_4 and if the digestion is prolonged, Hg is a much better catalyst as compared to Cu and it compares well with the Se.

The laboratory should be free from NH_3 gas and HNO_3 fumes during the determination of nitrogen by the kjeldahl method.

Proper supervision is needed during digestion and distillation. At the initial stage of digestion, if heating is not slow, the reaction mixture may rise and flow out of the kjeldahl digestion flask. Brisk heating during initial period of distillation may cause excessive frothing to such an extent that the reaction mixture rises and enters into the condenser.

At later stage of distillation brisk heating may cause violet bumping as to cause damage to the kjeldahl distillation flask, and to the condenser unit. When electric heating unit is employed, as electricity goes off, boric acid and distillate from receiving flask get sucked into kjeldahl distillation flask if condenser outlet extension tube is not immediately taken out of the receiving flask.

Nitrogen concentration in plant samples

Bould (1965) suggested tentative standards for assessing nutrient status of strawberries at fruiting stage based on concentration of nitrogen in laminae of recently matured leaves of one year plant. The percent nitrogen on dry matter basis may be interpreted as follows :

	Deficient	Marginal	Sufficient
N-concentration	< 2.0%	2.0 to 2.5%	>2.6 to 3.0%

Lepodevin and Robinson (1964) suggested a tentative standard for assessing nitrogen status of 18 weeks old sugarcane leaves in British Guiana. The percent nitrogen on dry matter basis may be interpreted as follows :

	Deficient	Marginal	Sufficient
N-concentration	< 1.8%	1.8 to 2.0%	>2.0%

The nitrogen recommended for top dressing worked out to 50, 25 and 0 kg N/ha for nitrogen status rated as Deficient, Marginal and Sufficient respectively.

Watson (1962, 1963) reported level of nitrogen in rubber leaves of accessible branches below which response is likely, and level above which response is unlikely. The results are summarized below :

Particulars	Nitrogen concentration in leaves	
	Response likely	**Response unlikely**
Rubber leaves from accessible branches	< 3.4%	>3.6%
Rubber leaves exposed to sunlight	< 3.2%	>3.6%
Rubber leaves under shade	< 3.3%	>3.7%

Immes (1960) reported optimum concentration of nitrogen in sugarcane leaves of different age in months. The optimum concentration of nitrogen in sugarcane leaves decreased progressively as the leaves advanced in age.

	Age of sugarcane leaves in months					
	3	4	5	6	7.5	9
Optimum N%	2.25	2.02	1.93	1.85	1.20	1.15

In a study at the Agricultural Institute, Allahabad, it was observed that the sample of wheat plant taken 38th day after sowing showed significant difference in nitrogen concentration resulting from different rates of N application. A researcher has proposed that plant samples in cereals be taken at 3g dry weight stage for the assessment of nutrient element concentration.

The intermediate range of N concentration in selected crops are presented in the table given below :

Crop	Plant tissue	Particulars	%N on oven dry mass basis
Apples	leaves	in July	>1.80 – 2.00
Apricot	leaves	mid summer	3.00
Orange	leaves	4–7 months of age	2.00 – 3.16
Coffee	leaves	Recently matured on lateral branches	>2.60 – < 2.90
Grapes	conchord petiole	Youngest mature in late June.	.20–.30 NO_3 – N
Potato	leaves	40–45 days old, 4th node from top	6.00 – 7.50
Sugar beet	Petiole	Youngest mature leaf	.10–3.5 ppm NO_3N
Banana	leaf		>2.60
Barley	straw	at harvest	>0.20
Corn	leaves	below uppermost ear	2.90
Peanut	leaves		3.73 - 3.81
Rubber	leaf lamineae		3.07 - 3.81
Sugarcane	leaf blade	3rd to 6th leaf from apex - 6 to 8 months	1.80 - 2.00
Tea	leaves	3rd from terminal of tender shoots	
		May toMid July	3.50

Source. Jones (1975)

Nitrogen deficiency visual symptoms

A plant is deficient in nitrogen as long as an increase in the nitrogen level will improve the economic return of the plant, plant parts or plant product for which the crop is grown.

When nitrogen deficiency is slight, there is loss of normal green colour, and it is associated with slight retardation of shoot elongation. Under moderate to acute deficiency of nitrogen, there is uniform loss of green colour of the foliage which is also associated with the reduction in vegetative growth. Leaves drop prematurely. Foliage is sparse, and light green to yellow in colour. Fruit production is greatly reduced.

The plants may show deficiency symptoms of nitrogen inspite of having adequate concentration of nitrogen in plant tissues under deficiency of molybdenum. Mo reduces nitrates for being utilized by the plant. Consequently when Mo is deficient, nitrates accumulate in plant tissues in toxic concentration.

Cobalt is essential for the symbiotic fixation of nitrogen by legumes. Severe nitrogen deficiency may occur in legumes such as alfalfa in absence of cobalt or inoculum.

The visual symptoms of nitrogen deficiency in selected crops are presented in the table given below :

Crop	Visual symptoms of nitrogen deficiency
Apple	Leaves-small, pale green, premature shedding of older leaves, which become coloured, twigs short and thin, sparse blossom, fruit small and highly coloured.
Barley	Leaves yellowish green, older leaves dry up, stalks-slender, erect, purplish green, tillers few, heads small.
Coffee	Leaves pale green, primary veins yellow, entire leaf coloured yellow, older leaves shed, new leaves small, later die and shed, yield much reduced.
Corn	Leaves yellow, older leaves dry up at tips and drying proceed downward along mid rib, stalks slender.
Cotton	Plants light green, lower leaves yellow, dry to brownish colour, stalks less branched and slender.
Grape	Uniform light green leaves, weak vine growth, early cessation of shoot elongation.
Mango	Shoot growth retarded, leaves small and yellow.
Orange	Leaves-yellow, small ; tree stunted, limited shoot growth, twigs die back, low yield under prolong deficiency.
Papaya	Growth retarded, leaves small, stiff and light yellow ; petiole-short, and tend to grow downward from the trunk.
Potato	Leaves-light green, older or lower leaves turn yellow and shed ; stems–few and slender ; growth - upright, tubers small.
Sugarbeet	Leaves-pale green ; growth habit-thin, erect ; storage roots - small and high in sucrose.
Sugarcane	Leaves-yellowish green to yellow, dry prematurely at tip and margin older leaves-reddish purple, leaves-narrow ; stem-thin ; tillering retarded.
Tomato	Leaves-light green, lower leaves turn yellow, and dry up, veins become deep purple ; stems-hard and purple ; flower buds-turn yellow and drop.

The excess supply of nitrogen encourages vegetative growth, succulence, crop lodging, frost injury, virus diseases, delays flowering, fruiting and maturity, impairs quality of fruit, fibre, tuber, tobacco, decreases the content of oil and sugar.

REFERENCES

Bould, C. 1965. The nutrient requirement of fruit crops. Span 8 : 81-83.

Horwitz, W. 1970. Methods of Analysis, A.O.A.C.,.P.O. Box No. 540, Benjamin Franklin, Sta. Washington, D.C. 20044.

Innes, R.F. 1960. The nitrogen, phosphorus and potassium requirement of sugarcane. J. Sci. Fd. Agric. 11 : 299 - 300.

Jackson, M.L. 1958. Soil Chemical Analysis. Prentice-Hall of India, Private. Ltd., New Delhi.

Jones, W.W. 1975. Nitrogen pp. 310-323. In H.D. Chapman (ed.). Diagnostic Criteria for Plants and Soils. Eurasia Publishing House (P) Ltd., Ram Nagar, New Delhi - 110055.

Le Pordevin, N.A. and L.A. Robinson, 1964. Foliar analysis procedure as employed in the Booker group of sugar estates in British Guiana. Part. I. Sampling and Analytical technique Part. II. The interpretation of results. Fertilite No. 2 mp. 3-11 and 12-17.

Prasad, R. 1970. A Practical Manual for Soil Fertility. Division of Agronomy, Indian Agric.` Res. Inst., New Delhi.

Watson, G.A. 1962. Plant Investigations. Relations between nutrient status and growth and yield. Rep. Rubb. Res. Inst. Malaya for 1961 pp. 33-34.

Watson;, G.A. 1963. Nutrition of Heven. Relationship between tree nutrient status, and growth & yield. Rep. Rubb. Res. Inst. Malaya for 1962 pp. 20-21.

3.4 DETERMINATION OF TOTAL P IN TRIACID DIGEST OF PLANT SAMPLE BY DEVELOPING VANADOMOLYBDO PHOSPHORIC YELLOW COLOUR—KITSON AND MILTON, 1944

3.4.1 Principle

A portion of P extracted from plant sample by digestion with triacid is made to react in nitric acid medium with 5 ml of 5.0% ammonium molybdate and 5 ml of 0.25% ammonium metavanadate solution to form a yellow coloured complex of vanadomolybdophosphoric. The intensity of the yellow colour complex is proportional to the concentration of P in the medium. The reaction mixture is diluted to 50 ml mark with redistilled water. Then after 10 minutes, the absorbance of the vanadomolybdophosphoric yellow colour is measured at 470 mμ either by Klelt Summerson colorimeter or else by a spectrophotometer.

3.4.2 Reagents

(*i*) Nitric acid, 5.0N. Dilute 79.22 ml of conc. HNO_3 (BDH sp. gr. 1.42 and, Assay 70.0%) to 250 ml with redistilled water.

(*ii*) Ammonium metavanadate, 0.25%. Dissolve 1.25 g of NH_4VO_3 (made in Austria) in about 250 ml of boiling redistilled water. Cool the solution somewhat and then add 10 ml of conc. HNO_3 (BDH-AR). Cool the solution by letting tap water fall over the 500 ml volumetric flask for some time. Then dilute the contents to 500 ml mark with the redistilled water. Mix the contents and store in amber glass bottle.

(*iii*) Ammonium molybdate, 5.0%. Dissolve 25.0 g $(NH_4)_6Mo_7O_{24}.4H_2O$ (made in Hungary) in 400 ml of redistilled water warmed to about 50°C. Cool the solution and filter through Whatman No. 44 filter paper. Then dilute the contents with redistilled water up to 500 ml mark. Mix the contents and store in amber glass bottle.

(*iv*) Phosphorus solution, 100.0 ppm : Weigh 5g of KH_2PO_4 (BDH-AR) and dry it at 40°C for 4 hours in an oven. Then dissolve 0.4393 g of dried KH_2PO_4 in about 400 ml of redistilled water. Add 25.0 ml of $7NH_2SO_4$. Mix and cool the solution. Then dilute it to 1000 ml mark with redistilled water. Mix the contents and store in polythene bottle.

3.4.3 Procedure

1. Development of the Vanadomolybdophosphoric yellow colour with triacid plant digest.

Transfer 5.0 ml of triacid plant digest into a 50 ml volumetric flask. Then add 5.0 ml of 5.0 N HNO_3, 5.0 ml of 0.25% NH_4VO_3, and 5.0 ml of 5.0% $(NH_4)_6Mo_7O_{24}\cdot 4H_2O$ in sequence and dilute the contents to 50 ml mark. Mix the contents after placing the stopper by turning the flask upside down. Allow 10 minutes time for the development of full colour.

2. Measurement of the absorbance /% transmittance of the Vanadomolybd-ophosphoric yellow colour with a Klett Summerson colorimeter or else with a spectrophotometer at 470 mμ wave length.

The filter of appropriate wavelength is fitted into the filter frame and placed inside the colorimeter opposite the tungsten lamp, or else wave length knob is rotated to obtain the desired wave length in a spectrophotometer.

There is need for two clean cuvettes or colorimeter tubes with identical optical properties. The second test tube is filled with redistilled water upto the 10 ml mark and it is placed inside the socket provided in the instrument. The instrument leads are connected to the mains through a voltage stabilizer. The instrument is switched on to run for 10 minutes for warming the instrument. The first test tube is filled with coloured complex developed on adding 0 ml of 100 ppm P solution. The second test tube is removed from the socket and the first test tube is fitted into the socket. Rotate fine galvanometer adjustment knob gently to set the galvanometer needle to zero absorbance or 100% transmittance. Empty the second test tube and fill it with coloured complex developed on adding 5 ml portion of triacid plant digest after rinsing the test tube with the coloured complex. Remove the first test tube from the socket and place the second test tube in the socket provided. As the galvanometer reading deflect to a point on the Absorbance / % Transmittance scale of the spectrophotometer, read and note the reading against the sample reference.

The operation of Klett Summerson colorimeter is some what different. It is provided with a switch close to the tungsten lamp by which the tungsten lamp is switched on and off. It has another ON/OFF switch beneath the coarse galvanometer. knob on the right hand side of the instrument body. This switch is for opening and closing the slit through which a particular wave length transmitted through the test tube containing coloured complex is allowed to strike the photocell connected to a galvanometer.

Connect the instrument leads to a voltage stabilizer connected to the mains. Fill the second colorimeter tube with redistilled water and place it in the socket provided. Remove filter frame and fit 470 mμ filter and place the filter frame back into the colorimeter. Set the dial reading at zero mark. If galvanometer needle has deflected away from zero, set the galvanometer needle to zero by carefully adjusting the coarse galvanometer knob right on top of the front portion of the instrument. Then switch on the first and the second switch for 10 minutes for warming up the instrument. If the galvanometer reading deflects, adjust it to zero by gentle rotating the fine galvanometer adjustment knob. After the lapse of 10 minutes, remove the second colorimeter tube filled with redistilled water from the instrument and place the first test tube filled with coloured complex developed with 0 μg of added P into the socket provided. Thereafter pour out the redistilled water from the second test tube and fill it with coloured complex developed on adding 5 ml portion of triacid digest prepared from plant sample. Check the galvanometer needle and set it at zero. Then, remove the first test tube and place the second test tube in the socket. The galvanometer needle will deflect from zero, now rotate the reading dial knob until the galvanometer needle rest at the zero mark. Note down the dial reading and note it as Klett reading against the sample reference.

Rotate the reading dial in the reverse direction until its zero coincides with the zero mark. Remove the second test tube from the socket and place the first tube filled with colour complex developed with 0 ml of 100 ppm P solution. Set the galvanometer needle to zero mark by adjusting the fine galvanometer needle knob. The second test tube may be emptied, and washed with redistilled water for filling the colour complex developed with 5 ml portions of triacid digest from other plant samples. The Klett reading or absorbance (optical density) be determined as described for the sample digest. Record the Klett reading or absorbance corresponding to the plant sample in the cuvette or colorimeter test tube.

3. Preparation of standard graph for P by developing vanadomolybdophosphoric yellow colour. Transfer 0, 2, 4 and 6 ml of 100 ppm P solution into separate 50 ml volumetric flasks and then add 5.0 ml each of 5N HNO_3, 0.25% NH_4VO_3, and 5% $(NH_4)_6Mo_7O_{24}.4H_2O$ into each 50 ml volumetric flasks. Dilute the contents to 50 ml mark with distilled water. Mix the contents and let it stand for 10 minutes for the full development of yellow colour.

Measure the absorbance (optical density) of colour complex with graded amount of P adjusting 0 absorbance with the colour complex containing 0μg of added P. Plot the absorbance (optical density) of the colour complex developed on 'y' axis of a linear scale graph corresponding to the μg P added for the development of colour complex on 'x' axis. If % transmittance has been noted instead of absorbance (optical density), then use a semilog graph paper. Plot the % transmittance on a log scale and μg P added for development of the colour complex on "x" axis. All the points should fall more or less on the straight line. By reference to the standard graph for P, estimate the concentration of P in the sample and the blank triacid digest.

3.4.4 Calculation

$$\text{ppm P in plant sample} = \frac{(S - B) \times V}{A \times W}$$

$$\text{\% P in plant sample} = \frac{\text{ppm P in plant sample} \times 100}{10^6}$$

where,

S is microgram P per 50 ml of coloured complex made from an aliquot of sample digest as estimated by reference to the standard P graph.

B is microgram P per 50 ml coloured complex made from aliquot of blank digest as estimated by reference to the standard P graph.

V is total volume of the plant sample triacid digest in ml.

A is an aliquot of sample triacid digest in ml taken for the colour development.

W is the mass of plant sample in gram taken for the preparation of triacid digest.

3.4.5 Comments

With the passage of time the potency of the reagents may deteriorate. It is therefore desirable to check the standard graph for P from time to time. If the absorbance is found to deviate considerably from the previous record of the same standards, the old graph be discarded and new graph be prepared.

There is effect of temperature on the intensity of colour developed. The intensity of colour is less during the winter as compared to the intensity of colour in other seasons.

The use of the distilled water and good glasswares are desirable in colorimetric determinations. The reagents should be preferably of the analytical grade.

The glasswares must be clean and free from phosphorus. The glasswares may be rinsed with chromic acid to destroy greasy material. Soaking glasswares in 0.002% teepol solution for some time is desirable for proper cleaning. The glasswares begin to shine also on being soaked in teepol solution. The rinsing with conc. HCl before final washing with tap water would certainly make the glasswares free from P.

The concentration of P in the plant sample is usually about one tenth of the concentration of N. Chapman (1975) has summarized the concentration of P in different plant species in the range of low-, medium and high. The concentration of P varies with the plant species, plant part analysed and the stage of plant development. The medium level of P in most crops is less than 0.30% on plant dry weight basis. The concentration of P in wheat grain and straw was found to be 0.40% and 0.08% respectively. P in sugarbeet and sugarcane leaves ranged from 0.20% to 0.25%. Turnips and tomato leaves contained more than 0.30% P. The medium concentration of P in mustard leaves was 0.71%.

The symptoms of moderate to acute deficiency of phosphorus manifest in off colour green foliage with purple veination especially on the underside of leaves. The petioles have a purplish cast. The root development is poor. The plants appear stunted and produce less fruit.

REFERENCES

Chapman, H.D. (1975). Diagnostic Criteria For Plants and Soils. Eurasia Publishing House (P) Ltd., Ram Nagar, New Delhi, 110055.

Jackson, M.L. (1958). Soil Chemical Analysis, Prentice Hall of India Private Limited, New Delhi (1973).

Kitson and Milton, 1944. Ind. Eng. Chem., A.E. 16 : 379.

3.5 K DETERMINATION IN PLANT DIGEST BY FLAME PHOTOMETER-TOTH AND PRINCE (1949)

3.5.1 Principle

Oven dry plant sample grinded to pass 40 mesh sieve is digested by wet ashing procedure using ternary acid mixture after predigestion with conc. HNO_3as described in section 3.2.1 on sample preparation and digestion. Normally 100 ml plant test solution is prepared on digestion of 1g plant sample. The plant sample test solution is diluted and K in the diluted plant sample test solution is determined conveniently by a flame photometer.

Diluted plant sample test solution is atomized to spray in air-propane blue flame of a flame photometer. K ions on gaining energy from the flame begin to dissipate energy in the form of 767 millimicron wave length which is directly proportional to the concentration of K. The emission from K is made to pass through K filter before striking a photocell connected to a galvanometer. The photocell converts light energy into electrical energy which is measured by the galvanometer. The instrument is first calibrated by atomizing K standards, and then diluted plant sample and blank test solution is atomized, and K in the test solution is determined by reference to a calibration curve.

3.5.2 Reagents

(*i*) Standard potassium solution, 1.0 g K/l. Dissolve 1.9066 g pure dry KCl (AR) in distilled water, and dilute it to 1000-ml mark with distilled water. The concentration of K in this solution is 1000 ppm. Store the solution in polythene bottle.

(*ii*) Standard potassium solution, 0 to 20 ppm K. To obtain 0, 5, 10, 15 and 20 ppm K solution, transfer 0, 2.5, 5.0, 7.5 and 10.0 ml of 1000 ppm K solution into separate 500 ml volumetric flask. Add distilled water to make the volume to 500 ml.

If internal lithium standard is specified by the manufacture the estimated amount of lithium/500 ml K standard be added to each flask before making the volume to 500 ml mark with distilled water.

If interference from metallic cations is anticipated, the interfering cations be also added to the K standards in amounts equal to their concentration in the diluted sample test solution prior to making the volume of K standard to 500 ml with distilled water. Store each standard in 500 ml polythene bottles.

3.5.3 Procedure

1. Operation of the flame photometer. Place K filter against the burner by rotating the filter disc, and then operate the flame photometer as described in Na determination in plant sample (section 3.8.3).

2. Calibration of the instrument. Atomize 0 and 20 ppm K solution, and adjust the flame photometer reading to 0 and 100 respectively. Then atomize the intermediate K standards, and note down the corresponding flame photometer reading.

3. Preparation of calibration curve for K. On a linear scale graph paper, set flame photometer reading scale on the ordinate, and K concentration scale on the abscissa. Then, plot flame photometer reading on the ordinate corresponding to K concentration in ppm on the abscissa. Connect flame photometer reading for 0 and 20 ppm K solution by a straight line to obtain the calibration curve for K.

4. K determination in diluted plant sample and blank test solution. Dilute plant sample test solution by transferring 2.0 ml and 1.0 ml of sample test solution into separate 50 ml volumetric flask, and dilute each to 50 ml mark with distilled water. Similarly, dilute the blank test solution.

Now, atomize to spray 0 and 20 ppm K solution into the flame of flame photometer, and set the flame photometer reading to 0 and 100. Then atomize diluted plant sample and blank test solution and observe the flame photometer reading. If diluted plant sample test solution give flame photometer reading within the range of 0 to 100, note down the flame photometer reading, and the corresponding times of dilution.

3.5.4 Calculation

$$\text{ppm K in plant sample} = \frac{(S - B) \times dt \times T}{W}$$

$$\text{meq of K/100 g plant sample} = \frac{\text{ppm K in plant sample}}{EW \times 10}$$

$$\text{\% K in plant sample} = \text{meq K/100 g plant sample} \times ME$$

Where, S is ppm K in diluted plant sample test solution

B is ppm K in diluted blank test solution

dt is times of dilution of the plant sample test solution.

T is total volume of plant sample test solution prepared after wet ashing.

W is the mass of oven dry grinded plant sample taken for wet ashing in order to prepare plant sample test solution.

EW is the equivalent weight of K i.e. 39.1 g.

ME is the milligram equivalent weight of K i.e. 0.0391 g.

3.5.5 Comments

Direct reading without lithium as internal standard is satisfactory when air-propane blue flame is used.

In flame photometric determination of K, the interference by Na occurs if Na : K ratio is 5 or greater, the interference by Ca occur if Ca : K ratio is 10 or greater, and the interference by Mg occur if Mg : K ratio is in excess of 100. The interference in K determination is very rarely encountered.

The plant tissue analysis are undertaken to ascertain nutrient deficiency before the appearance of deficiency symptoms. The familiarity with visual symptoms of nutrient deficiecy is also important in identifying healthy and nutrient deficient plant. Their elemental analysis ascertain nutrient concentration in deficient and healthy plants. It also serve as a confirmatory test.

The intermediate range of K concentration in selected crops are summarized in Table 3.5.1, whereas the critical level of K in selected crops is summarized in Table 3.5.2. As K concentration in plant tissue decreases, when it declines to a value at which growth first becomes retarded is termed as the critical concentration.

The K deficiency symptoms appear first on recently matured leaves not on young, immature leaves at the growing point. The deficiency symptoms become more pronounced on older leaves. K is mobile and it is readily translocated to the growing point. The characterstic symptoms of K deficiency is scorching of recently matured leaves. Once the leaf get scorched, senescence is irreversible. Consequently in trees it results in die back. Chlorosis appear near the margin. The chlorosis develops from light yellow to tan, becoming brown and finally dry to a scorch.

Table 3.5.1 Intermediate range of K concentration in different crops.

Crops	**Plant tissue**	**% K concentration on dry weight basis**
Alfalfa	Leaves	0.83
Apples	Leaves	.42–1.65
Grape fruit	Leaves from fruity terminals	1.10–2.16
Lemon	Leaves in October	1.55-2.62
Orange	Leaves	0.70-1.50
Corn	Leaves	0.74-1.49
Cotton	petiole	>0.60
Grapes	Leaves	0.45–0.78
Oats	Leaves	1.15
Peas	Leaves	0.50
Potato	upper leaves	2.10–3.80
Tea	old leaves	0.95–1.25
Tobacco	Leaves in September	4.37–5.29

Table 3.5.2 Critical level of K in selected crops

Crop	**Plant tissue**	**K concentration % Dry weight basis**
Alfalfa	Tops at 1/10 th bloom	1.00
Apple	Leaves-mature spur	1.00
Barley	Old leaves	0.92
White clover	Top	0.80
Corn	6th leaf at silking	1.00
Cotton	Petiole youngest mature	1.00
Grape	Petiole mid season	0.44
Orange	Leaves	1.21
Peanut	Leaves at bloom stage on fresh weight basis	0.50
Sugarbeet	mature blade and petiole (Na >1.50%)	1.00
Sugarcane	leaf laminae-3rd fully opened leaf	1.08

In legumes first sign of K deficiency is white speckling or freckling of the leaf blade. In presence of Na in leaves, dominant symptoms of K deficiency are tanning, browning and drying of the leaf margin and these progress inward until the entire leaf blade is scorched or necrotic. The deficiency symptoms of K deficiency in selected crops are summarized in Table 3.

Table 3.5.3 K deficiency symptoms in selected crops.

Crops	K-deficiency symptoms
Alfalfa	Leaflets chlorotic with necrotic small white dots, margins brown and has ragged appearance.
Apple	bluish green leaves showing interveinal chlorosis, marginal necrosis and scorching, later leaves dry up. Fruit is small and poorly developed.
Barley	Leaves bluish green, later turn yellow, then brown, die at tips and margins, scorching, weak and short stacks, profuse tillering but few ears, shrivelled grains, purplish brown spots on narrow leaves.
Beans	Leaflets curl or cup downward, necrotic brown areas (Phaseolus spp) at margins.
Cabbage	Yellowing in leaves, Heads-small, soft and puffy.
Cauli-flower	dark green leaves, yellowing in older leaves, margins become brown.
Grapefruit	Top most leaves die back, pale and bronze coloured leaves, fruit small and less.
Orange	Leaves curl, and twist, fruit is small with less acidity
Coffee	Young leaves crowded, dark coloured between veins
Corn	Yellowish green streaks, become corrugated in older leaves scorching at tip and margin, poorly filled cob at tips, short internodes.
Cotton	Leaves light yellowish green, tips and margins curl downward, finally leaves become reddish brown.
Grape	Leaves yellow, brown spots, develop holes, fruit ripening un-even.
Peas	Leaves become dark green with yellowing and fading at mar-gins, leaflets cupped down ward, growth retarded, pods poorly filled.
Potato	Young leaves bluish green, old leaves yellowish, browning at tips and margins, cupped leaflets, tuber flesh is bluish.
Rape	Leaves become bluish green, leaf suface wrinkled with some chlorosis, older leaves show drooping and premature withering.

Rice	Leaves become bluish green, chlorosis and necrosis in older leaves, tips become brown, poor grain filling.
Sugarcane	Young leaves dark green, later turn pale yellow, reddish colouration on upper surface, lower leaves die back at tips and margins. Some chlorosis and necrosis also seen.
Tobacco	Chlorosis and necrosis at leaf tips and margins, leaves bluish green, cupped at tips and margins, stalks are slender.
Wheat	Leaves bluish green, later become yellow, then brown and finally die at tips and margins, scorching, short and weak, profuse tillering but few ears, grains shrivelled.

REFERENCES

Jackson, M.L. 1973, Soil Chemical Analysis. Prentice Hall of India Private Ltd., New Delhi.

Richards, L.A., (ed). 1954. Diagnosis And Improvement of Saline And Alkali Soils, USDA Hand Book 60 : Oxford and IBH Publishing Co., New Delhi.

Toth, S.J., A.I. Prince, A. Wallace, and D.S. Mikkelsen, 1948. Rapid quantitative determination of eight mineral elements in plant tissues by a systematic procedure involving use of a flame photometer. Soil Sci. 66 : 459-466.

Ulrich, A. and K. Okhi. 1975. Potassium pp. 362-393. In H.D. Chapman (ed). Diagnostic Criteria. For Plants And Soils. Eurasia Publishing House (P) Ltd., Ram Nagar, New Delhi.

3.6 CALCIUM IN PLANT SAMPLES

3.6.1 Ca determination in plant test solution by EDTA titration using Ca-indicator-Cheng and Bray (1951) and Barrow and Simpson (1968)

3.6.1.1 Principle

Ca in plant test solution is determined by complexometric titration with. 01N EDTA solution at pH 12 in the presence of Ca– indicator. Mg is precipitated at high pH and the interference from heavy metals is prevented by adding 1.0% KCN and disodium diethyl dithio carbamate crystals which form stable complexes with heavy metals.

3.6.1.2 Reagents

(*i*) Calcium chloride solution- .01N. Dissolve 0.5000 g of pure dry $CaCO_3$ (AR) in 10 ml of (1+3) HCl and dilute it to 1000 ml mark with distilled water.

(*ii*) EDTA solution -0.01N. Transfer 2.0 g of disodium dihydrogen ethylene diamine tetra acetic acid and .05g of $MgCl_2.6H_2O$ into 1 l volumeter flask. Add distilled water to dissolve the reagents, and dilute it to 1000 ml mark with distilled water. Standardize the solution against 0.01N $CaCl_2$ using Ca-indicator (ammonium purpurate).

(*iii*) Sodium hydroxide solution–4.0 N. Dissolve 16 g of NaOH pellets (AR) in distilled water and make the volume to 100 ml.

(*iv*) Potassium cyanide solution-1.0%. Dissolve 1.0 g KCN in 100 ml distilled water and label it as deadly poison.

(*v*) Disodium diethyl dithio carbamate crystals (reagent grade).

(*vi*) Ca-indicator. Mix 0.5 g of ammonium purpurate and 100 g of powdered K_2SO_4(AR) in a pestle mortor.

3.6.1.3 Procedure

Transfer 10 ml aliquot of plant test solution into a 150 ml Erlenmeyer flask. Evaporate the contents to dryness on a hot plate at low heat or else on water bath. Remove the flask from hot plate and allow it to cool. Add 25 ml distilled water and warm to dissolve the residue. Alternatively transfer 10 ml aliquot of plant test solution and 15 ml of distilled water into a 150 ml Erlenmeyer flask.

Then, add a pinch of disodium diethyl dithio carbamate crystals and 5 drops of 1.0% KCN solution through indicator bottle. Mix the contents by swirling the flask, Now, add 4N NaOH drop by drop through a pipette in order to raise the solution pH to 12.0 to effect precipitation of Mg.

Thereafter, add a pinch of Ca-indicator and mix the contents. The solution colour on adding Ca- indicator be neither faint nor deep pink. Fill a microburette with EDTA solution and place it on a burette stand. Place a white porcelain plate at the base of burette stand. Transfer the Erlenmeyer flask containing the sample solution on the porcelain plate and titrate with EDTA in sufficient day light.

Add EDTA in small portions and swirl the contents after each addition. Near the end point add few drops at a time and swirl the contents. Continue titration until the colour of the reaction mixture changes from pink to lavender *i.e.* purple. Note down the EDTA titre. Repeat the titration until the difference between two sample titre does not exceed by more than .05 ml. Similarly run a blank titration taking an aliquot of the blank test solution.

3.6.1.4 Calculation

$$\text{meq Ca/100 g plant sample} = \frac{(S - B) \times T \times N \times 100}{A \times W}$$

$$\text{\% Ca in plant sample} = \text{meq Ca/100 g plant sample} \times ME$$

where

S is EDTA titre for the aliquot of plant sample test solution

B is EDTA titre for the aliquot of blank test solution

T is total volume of plant test solution prepared

N is normality of EDTA solution as against standard $CaCl_2$ using Ca indicator

ME is milligram equivalent weight of Ca *i.e.* 0.02.

A is the aliquot of plant sample test solution in ml taken for EDTA titration

W is the mass of plant sample in gram used for the preparation of plant test solution.

3.6.2 Ca determination in plant test solution by flame photometer using air-propane blue flame–Toth et. al. (1948)

3.6.2.1 Principle

Ca in solution are atomized into air-propane blue flame of a flame-photometer. Ca ions gain energy from the flame aquiring a higher energy level and begin to dissipate energy by emission of a certain wave length 6220 Å which is allowed to pass through a Ca-filter before striking a photocell connected to a galvanometer. The photocell generates current in microamperes which is directly proportional to the concentration of Ca ions in the solution. The current generated is measured by galvanometer in digital values such as 0–100 or more.

The instrument is first calibrated by atomizing Ca-standards into the air-propane blue flame. Thereafter, the plant test solution is atomized and the flame photometer reading is recorded to determine Ca concentration in the test solution by reference to the standard graph. Direct reading with air-propane blue flame is satisfactory even without incorporation of certain concentration of $LiCl_2$ as an internal standard.

3.6.2.2 Reagents

(*i*) Calcium standard-50.0 meq Ca/l. Dissolve 2.503 g of pure $CaCO_3$ (calcite crystals) in 50 ml of 3N HCl and dilute to 1000 ml mark in 1l volumetric flask.

(*ii*) Calcium standards-0 to 4 meq 1l. Transfer 0,2, 4, 6 and 8 ml of 50.0 meq Ca/l reagent into separate 100 ml volumetric flasks and dilute the contents to 100 ml mark with distilled water.

3.6.2.3 Procedure

1. Operation of the instrument. Switch on the flame photometer, and regulate the filter disk to set Ca-filter i.e. 6220 Å maximum emission in between the flame and photocell of the instrument. Then switch on the compressor and regulate the air inlet knob until desired air pressure is attained.

Now connect the gas tubing from the gas cylinder into the gas inlet pipe. Open the gas inlet knob slightly. Place a lighted match stick above the built-in gas burner, and let the lab. asistant put the gas regulator knob to the "ON" position immediately when the gas burner gets lighted. Regulate the gas inlet knob to obtain a blue flame of sufficient length.

2. Calibration of the instrument. Thereafter, atomize first distilled water for some time to effect preliminary washing of the gas burner. Then, atomize Ca-standard 4 meq Ca/l and set the flame photometer reading to 100. Atomize Ca standard 0 meq Ca/l and adjust the corresponding flame photometer reading to 0. Again atomize Ca-standards-0 and 4 meq Ca/l and adjust the corresponding flame-photometer reading to 0 and 100. Then atomize Ca-standards containing 1, 2, & 3 meq Ca/*l* and record the corresponding flame photometer reading.

3. Testing sample and blank digest by the instrument. Again check the flame photometer reading on atomizing Ca-standards. 0 and 4 meq Ca/l. Set the reading again if considered necessary, and then atomize sample and blank test solution noting

down the reading recorded by the instrument. Dilute the sample and blank test solution to the desired extent.

4. Switching off the instrument. When the instrument is not in use, atomize distilled water into the flame. If the work is complete, first bring the knob of gas regulator placed on the gas cylinder to the off position. This will stop supply of gas to the burner, and the flame will get extinguished. Then, switch off the compressor, and finally switch off the flame photometer.

5. Preparation of standard graph for calcium. On a linear graph paper, set flame photometer reading 0-100 on the ordinate scale, and set 0-4 meq Ca/*l* on the abscissa scale. Then, plot the flame photometer reading corresponding to Ca concentration in meq/l. Connect flame photometer reading for 0 to 4 meq Ca/*l* by a straight line.

6. Estimation of Ca concentration in the plant and blank test solution. Find out the concentration of Ca in the sample and blank test solution by reference to the standard graph for Ca. Check the flame photometer reading on the ordinate and then draw an imaginary line parallel to the base intersecting the straight line curve. From the point of intersection, draw a line perpendicular to the base noting the Ca concentration on the abscissa scale. Alternatively, divide the flame photometer reading of the sample and blank test solution by 25 to obtain the concentration of Ca in meq/1

3.6.2.4 Calculation

$$\text{meq Ca/100g plant sample} = \frac{(S - B) \times D \times T}{1000 \times W} \times 100$$

% Ca in plant sample = meq Ca/100 g plant sample × ME

Where

S is Ca concentration in meq/l in sample test solution

B is Ca concentration in meq/l in blank test solution

D is times of dilution of the sample and blank test solution prior to atomization into the flame photometer.

T is total volume of sample test solution prepared

W is the mass of plant sample taken in gram for wet or dry ashing.

ME is the milligram equivalent of Ca *i.e.* 0.02.

3.6.3 Ca in plant sample test solution by the oxalate (titrimetic) method-Blasdale (1909)

3.6.3.1 Principle

Ca is precipitated as Ca oxalate from an acid solution of calcium salt. The precipitate is filtered and washed. Thereafter, the precipitate is dissolved in H_2SO_4. Ca react with sulphate to form calcium sulphate and oxalate ($C_2O_4^-$) combine with $2H^+$ to form oxalic acid ($H_2C_2O_4$) which is titrated by standard $KMnO_4$ solution. The reactions that occur are as follows :

$$CaCl_2 + H_2C_2O_4 \xrightarrow{\text{at pH 4.0}} CaC_2O_4 + 2HCl \quad ...(i)$$

$$CaC_2O_4 + H_2SO_4 \longrightarrow CaSO_4 + H_2C_2O_4 \quad ...(ii)$$

$$5H_2C_2O_4 + 2KMnO_4 + 3H_2SO_4 \longrightarrow 2MnSO_4 + K_2SO_4 + 8H_2O + 10CO_2 \quad ...(iii)$$

If Ca-oxalate is precipitated in neutral or alkaline solution, Ca and Mg phosphates will also be precipitated. The Ca-oxalate must, therefore, be precipitated under conditions which preclude the Ca-precipitates of the alkaline earth metal phosphates. The requirements are fulfilled by carrying out precipitation in a weekly acid medium. Ca-oxalate is completely precipitated at pH 4.0. At pH 4.5, Mg, Fe, Al phosphate (and oxalate) are not precipitated. Hence, whole of Ca separates as oxalates.

3.6.3.2 Reagents

(*i*) Brom cresol green (sodium salt)-0.1% solution. Dissolve 0.1g of brom cresol green in distilled water and dilute the solution to 100 ml.

(*ii*) Hydrochloric acid-approx. 6N. Transfer 40 ml distilled water and 50 ml of conc. HCl into 100 ml volumetric flask and dilute the contents to 100 ml with distilled water.

(*iii*) Oxalic acid solution –1.0N. Dissolve 63.02 g of $H_2C_2O_4 \cdot 2H_2O$ in distilled water and dilute the solution to 1000 ml mark in 1 l volumetric flask with distilled water.

(*iv*) Ammonium hydroxide (1+1). Transfer 40 ml distilled water and 50 ml conc. NH_4OH solution in 100 ml volumetric flask and dilute the contents to 100 ml mark with distilled water.

(*v*) Sulphuric acid solution 1 : 9 (*v/v*). Transfer 45 ml distilled water and 5 ml conc. H_2SO_4 into 100 ml volumetric flask and mix the contents.

(*vi*) Standard potassium permanganate solution –.05N. Transfer 1.5804 g of $KMnO_4$ into a 1 l flask. Add distilled water to dissolve the reagent. Then dilute the solution to 1000 ml mark with distilled water. Prepare the fresh solution when required.

3.6.3.3 Procedure

1. Transfer an aliquot (10 ml) of plant sample test solution into a 300-400 ml tall beaker, and dilute the contents to 100 ml. Then, add 2-3 drops of brom cresol green indicator, and acidify the contents by adding 6N HCl in drops until the colour changes to yellow, and then add 0.5 ml of 6N HCl in excess.

2. Thereafter, add 0.5 ml of oxalic acid for each 100 ml of solution. Heat the contents to boiling. Then add ammonium hydroxide (1 : 1) to neuralize the solution. Now add oxalic acid 1.0N in excess gradually @ 5ml for each 100 ml of solution with constant stirring. Make the hot solution slightly alkaline with ammonium hydroxide (1 : 1). Boil the contents gently for several minutes. Cool and let it stand until the precipitate of calcium oxalate settles. During the cooling, if necessary add ammonium hydroxide (1 : 1) in order to keep the solution faintly alkaline. The brom cresol green is green at pH 4.6. It turns yellow on decrease in pH and it turns blue on increase in pH.

3. On a large size tripod, place a funnel lined with Whatman No. 40 filter paper and keep 400 ml beaker below the funnel. Transfer the contents of the beaker

including precipitate of Ca-oxalate into the funnel receiving the filtrate in 400 ml beaker. Transfer precipitate remaining in the beaker quantitatively by repeated washing of the beaker with small portions of distilled water. Thereafter, wash the precipitate transferred on filter paper with distilled water until the filtrate is free from soluble oxalates. Preserve the beaker (400 ml Cap.) containing the filtrate for the determination of magnesium.

4. Keep the original 300-400 ml beaker below the funnel, and puncture the tip of the filter paper lined on the funnel with a glass rod, and then wash down the precipitate into the breaker with distilled water. Now pour 50 ml of sulphuric acid (1 : 9 v/v) through the funnel lined with the used filter paper collecting the washings into the 300-400 ml beaker placed below the funnel. Rinse the funnel with distilled water and collect the rinsings in the same beaker. Heat the beaker nearly to boiling and then titrate the liberated oxalic acid with 0.05 N $KMnO_4$ until the colour of the solution changes to faint pink. Now, remove the used filter paper from the funnel and place it into the same beaker in which titration of liberated oxalic acid is being done. Continue titration with 0.05 N $KMnO_4$ until appearance of persistent pink colour at least for 30 seconds. Note down the titre reading. Similarly run a blank with 10 ml aliquot of the blank test solution.

3.6.3.4 Calculation

$$\text{meq Ca/100 g plant sample} = \frac{(S - B) \times N \times T \times 100}{A \times W}$$

% Ca in plant sample = meq Ca/100 g plant sample × ME

Where

S is $KMnO_4$ titre in ml for aliquot of the sample test solution

B is $KMnO_4$ titre in ml for aliquot of the blank test solution

N is normality of the $KMnO_4$ solution

T is total volume of the plant sample test solution prepared in ml

A is ml aliquot of plant sample test solution taken for Ca determination

W is the mass of plant sample taken in g for preparing the test solution.

ME is the milligram equivalent weight of Ca i.e. 0.02

3.6.4 Calcium determination in plant test solution by the oxalate-(titrimetric)-method 2 Blasdale (1909)

3.6.4.1 Principle

As given in section 3.6.3.1.

3.6.4.2 Reagents

(*i*) Conc. Hydrochloric Acid

(*ii*) Conc. Acetic acid

(*iii*) Brom cresol green - .04%. Transfer 40 mg of brom cresol green in a 100 ml vol. flask. Dissolve in distilled water and dilute the solution to 100 ml mark.

(*iv*) Conc. Ammonium hydroxide

(*v*) Ammonium hydroxide (1 : 1). Transfer 50 ml conc. NH_4OH into 100 ml volumetric flask containing 40 ml distilled water. Dilute the contents to 100 ml mark with distilled water

(*vi*) Oxalic acid-10% solution. Transfer 10 g of $H_2C_2O_4$ $2H_2O$ into 100 ml volumetric flask. Add distilled water and shake to dissolve the reagent. Dilute the solution to 100 ml mark.

(*vii*) Conc. Sulphuric acid.

(*viii*) Potassium permanganate solution -.05 N. Dissolve 1.5804 g $KMnO_4$ in distilled water and dilute the reagent to 1000 ml mark with distilled water.

3.6.4.3 Procedure

1. Transfer 10 ml of plant sample test solution into 300 ml tall beaker. Then add 5 ml of conc. HCl and 5 ml of conc. CH_3COOH and 0.5 ml of brom cresol green indicator. Yellow colour indicate pH below 4.6.

2. Now add drop by drop conc. NH_4OH with constant stirring till the appearance of green colour indicative of pH 4.6. Heat the contents to boiling and then add 10 ml of 10% $H_2C_2O_4.2H_2O$. The solution colour turns yellow indicating solution pH below 4.6. Add drop by drop NH_4OH (1 : 1) from a burette with rapid stirring until the appearance of strong green colour. This is to bring about precipitation of all Ca as pH reaches about 4.6. In slightly acidic medium, Ca oxalates precipitates slowly in hot solution.

3. Transfer the beaker containing precipitates of Ca oxalates on a water bath, and continue digestion of the contents for an hour. Remove the beaker, and allow it to cool. Transfer the contents quantitatively with repeated use of portions of distilled water on a funnel lined with Whatman No. 40 filter paper, and collect the filtrate in a clear 300 ml beaker. Preserve the filtrate for the determination of Mg.

4. Place a beaker below the funnel lined with folded filter paper containing the precipitate. Wash the precipitate with warm distilled water until the filtrate is free from chloride. Discard the washings.

5. Remove the filter paper containing CaC_2O_4 precipitate and transfer into the original beaker. Add 75 ml of 1.0 N H_2SO_4 and sufficient distilled water to immerse the filter paper. Place the beaker on a hot plate, and heat the contents to a temperature of 90°C. Promptly titrate with standard 0.05 $KMnO_4$ solution until pink colour appears and persist at least for 30 seconds. Note down the titre. Similarly run a blank taking 10 ml aliquot of the blank test solution.

3.6.4.4 Calculation

$$\text{meq Ca/100 g of plant sample} = \frac{(S - B) \times N \times T \times 100}{A \times W}$$

% Ca in plant sample = meq Ca/100 g plant sample × ME

Where

S is $KMnO_4$ titre in ml for aliquot of the sample test solution.

B is $KMnO_4$ titre in ml for aliquot of the blank test solution.

N is the normality of the $KMnO_4$ solution.

T is the total volume of the plant sample test solution prepared in ml.

A is ml aliquot of the plant sample test solution taken for Ca determination.

W is the mass of plant sample taken in g for preparing the test solution.

ME is the milligram equivalent weight of Ca *i.e.* 0.02.

3.6.5 Comments

Plant samples be preferably digested by wet ashing procedure using conc. HNO_3 and conc. $HClO_4$. In wet ashing procedure with ternary acid mixture, there is risk of incomplete recovery of Ca due to its precipitation as $CaSO_4$. Dry ashing after wetting the sample with 5% conc. H_2SO_4 in 95% ethanol (v/v) has been suggested by Prasad (1970).

In flame photometric determination, direct readings without an internal standard give satisfactory results if an air-propane blue flame is used.

If certain concentration of $LiCl_2$ is specified by the instrument manufacturer as an internal standard, then the Ca standard as well as the plant sample test solution should have the specified concentration of $LiCl_2$.

For compensating the interference of elements such as K, Na, & Mg, these elements may be added to the Ca standards in amounts depending on their concentration in the test solution. The interference by K, Na and Mg is not an appreciable factor unless the concentration of these elements is at least 5 times the concentration of calcium. The interferences from K, Na and Mg is seldom encountered in determination of Ca in plant sample by flame photometer.

If K concentration in sample test solution is 150 ppm which exceeds the concentration of Ca by more than 5 times. Besides, $LiCl_2$concentration of 5 meq/l is required as an internal standard. Then to prepare Ca standard of 4 meq/l, transfer 8 ml of $CaCl_2$-50 meq/l, to 100 ml volumetric flask containing 10 ml of $LiCl_2$-50 meq/l and 15 ml of 1000 ppm K solution and dilute the contents to 100 ml with distilled water. The resulting Ca standard will have 4 meq Ca/l, 5 meq Li/l & 150 ppm K.

If the Ca concentration in the test solution is just 2 meq/l and the desired level of $LiCl_2$ is 5 meq/l. Then, transfer 2.5 ml of $LiCl_2$-50 meq/l into 25 ml volumetric flask containing 12.5 ml of sample test solution, and dilute the contents to 25 ml with distilled water. The resulting test solution will be diluted 2 times and it will have $LiCl_2$ concentration of 5 meq/l.

The Ca concentration in different crop tissues rated to be low is given below :

Crops	—	Plant parts	—	Ca concentration
Alfalfa	—	tops	—	< 1.75%
Cauliflower	—	tops	—	< 0.93%
Coffee	—	leaves-3 month old	—	< 0.80%
Peanut	—	lower blade in early stage	—	< 0.32% water soluble on fresh weight basis
Potato	—	leaves	—	< 0.49%
Rice	—	leaves	—	0.24-0.26%
Tobacco	—	leaves	—	0.94 -1.30%
Tomato	—	leaves	—	0.58 %
Wheat	—	Tops of young plants	—	0.14%

The symptoms of Ca deficiency in different crops are summarized below:

Crops	—	symptoms of Ca-deficiency
Beans	—	blackening and death of plants
Cabbage	—	Leaves rolled up at margins, ragged and distorted, white in narrow bands
Celery	—	thin growth, pale colour, death of growing point, leaflet tips and margins become distorted and turn black.
Citrus	—	Twigs die back from tops, yellowing at leaf margin.
Coffee	—	death of terminal buds and yellowing of leaf margins which later show necrosis, roots finally rot and plant die.
Grapes	—	Interveinal and marginal chlorosis followed by necrotic pin head spots near margin, vine tips die back.
Potatoes	—	Young leaves are small, and pale green, rolled toward upper surface with marginal necrosis, leaf dies, Tubers are dwarfed and useless with spots in pith region.
Turnips	—	Leaves show marginal and interveinal chlorosis followed by marginal chlorosis, and rolling up.

The visual symptoms of Ca deficiency do not appear in the early stages. When moderate to acute deficiency occur at later stage, root growth is markedly impaired and root rotting frequently occurs. The young leaves of new plants are affected first. The young leaves are often distorted and small. The margins are irregular in form, and the leaves show spotted and/or necrotic areas. There may be dieback of the terminal bud.

The calcium deficiency occur in soils which are distinctly acid (pH 5.0 or less) or else in soils having pH higher than 8.5 in 1 : 2 soil-water suspension. In acid soil crops are benefited by lime whereas in alkaline soils, crops are benefited by gypsum application. Calcium deficiency gradually develop with continuous use of sulphur, and sulphur containing fertilizer. Prolong growing of crops removing considerable amount of calcium such as alfalfa, accumulation of excecssive exchangeable potassium, and sodium result in calcium deficiency.

REFERENCES

Barrow, H.L. and E.C. Simpson. 1962. An EDTA method for the direct routine determination of calcium and magnesium in soils and plant tissues. Soil Sci. Soc. Amer. Proc, 26 : 443-445

Blasdale, W.C. 1909. The quantitative separation of calcium from magnesium. Amer. Chem. Soc. Jour. 311: 917.

Bear, F.E. 1965. Chemistry of the Soil. Oxford and IBH company, New Delhi.

Chapman, H.D. 1975. Calcium pp 65-92. In. Chapman, H. Diagnostic Criteria For Plants and Soils. Eurasia Publishing House (P) Ltd New Delhi.

Cheng, K.L. and Bray, R.H. 1951. Determination of calcium, magnesium in soil and plant material, Soil Sci. 72 : 42

Jackson M.L. 1958. Soil Chemical Analysis. Prenticel-Hall of India Private Ltd New Delhi (1973).

Diehl, H. et. al. 1950. The versenate titration for total, hardness Amer. Water Works Jour. 42 : 42-48.

Prasad, R. 1970. A Practical Manual For Soil Fertility, Division of Agronomy, IARI, New Delhi.

Richards, L.A., et. al. (ed). 1954. Diagnosis and Improvement of Saline. Alkali Soils. USDA Hand Book 60, Oxford IBH Publishing Company, New Delhi.

3.7 Mg IN PLANT DIGEST

3.7.1 Mg determination in plant sample test solution by EDTA titration using Mg indicator-Cheng and Bray (1951)

3.7.1.1 Principle

Ca and Mg are both complexed by EDTA. However, Ca-EDTA complex is more stable than Mg-EDTA complex. When Ca-indicator is used, with the completion of complexing of Ca by EDTA, the Ca-indicator changes its colour from wine red or pink to purple. When Mg indictor is used, on the completion of complexing of both Ca and Mg by EDTA, the Mg-indicator will change its colour from wine red or pink to clear blue or green. Hence for the estimation of Mg, it would be desireable to determine the concentration of Ca and Ca + Mg separately. Thereafter, the concentration of Ca be substracted from the concentration of Ca + Mg for obtaining the concentration of Mg. For routine analysis of Ca and Mg in plant test solution, EDTA titration is satisfactory.

3.7.1.2 Reagents

(*i*) Magnesium chloride - .01N. Dissolve 0.508 g Mg $Cl_2.6H_2O$ in distilled water and make the volume to 500 ml with distilled water.

(*ii*) Ammonium chloride-ammonium hydroxide buffer solution. Dissolve 67.5g of NH_4Cl in 570 ml of conc. NH_4OH and make the volume to 1 l.

(*iii*) Eriochrome Black T indicator. Dissolve 0.5 g of Eriochrome Black T (EBT) and 4.5 g of $NH_2OH.HCl$ (hydroxyl amine hydrochloride) in 100 ml of 95% ethanol. Store in a refrigerator. Prepare fresh reagent every three weeks.

(*iv*) Hydroxylamine hydrochloride solution–5.0%. Dissolve 5.0 g of NH_2OH . HCl in 100 ml distilled water

(*v*) Potassium ferrocyanide solution –4.0 %. Dissolve 4.0 g of K_4 Fe $(CN)_6$. $3H_2O$ (reagent grade) in 100 ml of distilled water.

(*vi*) Triethanolamine -50%. Transfer 50 ml of triethanolamine (reagent grade) into 100 ml volumetric flask, and dilute it to 100 ml mark with distilled water.

(*vii*) Calcium chloride solution- .01N. (as given in section 3.6.1.2).

(*viii*) EDTA solution - .01N. (as given in section 3.6.1.2).

(*ix*) Potassium cyanide solution -1% (as given in section 3.6.1.2).

3.7.1.3 Procedure

Transfer 10 ml aliquot of plant sample test solution into a 150 ml Erlenmeyer flask. Evaporate the contents to dryness at low heat on a hot plate. Remove the flask from hot plate and allow it to cool. Add 25 ml distilled water, and warm to dissolve the residue. Alternatively, transfer 10 ml aliquot of plant test solution and 15 ml of distillated water into a 150-ml Erlenmeyer flask.

Then add 1 drop of .01N $MgCl_2$, and 5 drops each of 1% KCN, 4% potassium ferrocyanide, 5% hydroxylamine hydiochloride and (1+1) triethanolamine in the Erlenmeyer flask containing sample test solution. Now, add NH_4Cl. NH_4OH buffer to raise the solution pH to 10.0. Thereafter, add 2 ml NH_4Cl. NH_4OH buffer in excess.

Immediately add 2 drops of EBT indicator through a 2 ml pipette and titrate with a micro burette filled with .01N EDTA solution. Add EDTA solution in small portions and swirl the flask contents after each addition of EDTA until the solution colour changes from wine red or pink to purple.

Then add EDTA solution drop by drop and swirl the contents after addition of each drop of EDTA. Continue titration with EDTA until the solution colour changes from purple to clear blue or green. Note down the EDTA titre. Repeat the titration until EDTA titre for sample in duplicate do not exceed by more than 0.05 ml. Similarly run a blank titration taking 10 ml of the blank test solution following the entire procedure omitting taking aliquot of sample test solution. Also take an aliquot of 5 ml of 0.01N $CaCl_2$solution into 150 ml Erlenmeyer flask and perform EDTA titration using Mg indicator following the above mentioned procedure. Run a blank titration taking 5ml + 20 ml of distilled water.

3.7.1.4 Calculations

Normality of EDTA

$$\text{solution} = \frac{5 \times .01}{\text{EDTA titre for 5 ml of } CaCl_2 - \text{EDTA titre for Blank}}$$

$$\text{meq Ca + Mg /100 g plant sample} = \frac{(S - B) \times N \times T \times 100}{A \times W}$$

meq Ca/100 g plant sample = As estimated by EDTA titration of sample test solution using Ca indicator

meq Mg/100 g plant sample = meq Ca + Mg /100 g plant sample – meq Ca/ 100 g plant sample

% Mg in plant sample = meq Mg/100 g plant sample × ME

Where

S is EDTA titre for the aliquot of plant sample test soluution using Mg indicator

B is the EDTA titre for the aliquot of blank test solution using Mg indicator

T is the total volume of the plant test solution prepared.

N is normality of EDTA solution as against standard $CaCl_2$ using Mg indicator

A is the aliquot of plant sample test solution in ml taken for EDTA titration.

W is the mass of plant sample in g used for the preparation of plant test solution.

ME is the milligram equivalent weight of Mg i.e. .01216.

3.7.2 Mg determination in plant test solution by gravimetric method after separation of Ca by precipitation as CaC_2O_4 and retaining the ppt on a filter paper and allowing Mg to pass with the filtrate — A.O.A.C. (1950).

3.7.2.1 Principle

Mg in the filtrate is precipitated as magnesium ammonium phosphate in alkaline medium. The precipitate is filtered, washed, dried and ignited in a muffle furnace. Thereafter, the residue is cooled and weighed as magnesium pyrophosphate ($Mg_2P_2O_7$)

3.7.2.2 Reagents

(*i*) Hydrochloric acid- approx. 6N. Transfer 50 ml conc. HCl into a 100 ml volumetric flask containing 50 ml distilled water.

(*ii*) Diammonium hydrogen phosphate solution-20%. Transfer 20 g of $(NH_4)_2HPO_4$ into a 100 ml volumetric flask. Add 80 ml distilled water and shake the flask intermittently for an hour. Then dilute the solution to 100 ml mark. Shake to mix the contents again. Thereafter, filter the reagent through Whatman No. 44 filter paper and collect the filtrate. Prepare a fresh lot of this reagent for each set of samples.

(*iii*) Ammonium hydroxide solution (1+1). Transfer 50 ml of conc. NH_4OH into a 100 ml volumetric flask containing 40 ml distilled water. Dilute the reagent to 100 ml mark with distilled water.

(*iv*) Concentrated Ammonium hydroxide AR.

(*v*) Dilute NH_4OH solution. Add 5 ml of conc. NH_4OH into 100 ml vol. flask containing 95 ml distilled water.

3.7.2.3 Procedure

1. Collect the 400 ml capacity beaker filled with filtrate obtained by filtration and washing of precipitated Ca as CaC_2O_4 from an aliquot of plant sample test solution for the determination of Mg. (See step 3 of section 3.6.3.3 or 3.6.4.3).

2. Acidify the filtrate by adding approx. 6N HCl solution. Then add 2 ml of 6N HCl in excess, and evaporate the filtrate on a hot plate until the volume of filtrate is reduced either to 100 or 50 ml as desired. If the mass of magnesium pyrophosphate is expected to be 0.0500 g or more, reduce the volume of the filtrate to 100 ml. Otherwise, evaporate the filtrate until the volume is reduced to about 50 ml. Then, allow the beaker to cool.

3. Add 5 ml of 20% $(NH_4)_2$ HPO_4 solution for each 50 ml filtrate remaining in the beaker after evaporation. Then, add (1+1) NH_4OH drop by drop with stirring until the solution is made strongly alkaline. After a few minutes, add 10 ml of conc. NH_4OH for each 100 ml of filtrate remaining in the beaker. Stir well at intervals until precipitate is formed. Stirring is very essential. Otherwise, the precipitate may not appear especially when small quantity of Mg is present.

4. Cover the beaker with a watch glass and allow it to stand over night in a cool place. On the following day, filter the contents through Whatman No. 44 ashless filter paper retaining the precipitate of magnesium phosphate on filter paper. Wash the precipitate with dilute NH_4OH.

5. Transfer the filter paper with the precipitate on to a weighed silica or porcelain crucible. Allow the filter paper and the precipitate to dry. Transfer the silica crucible into a muffle furnace. Ignite the precipitate by raising the temperature to 500°C for few hours until white residue is obtained. Remove the residue with the silica crucible from muffle furnace and place it in a desiccator. Allow it to cool and then weigh the silica crucible + residue as Mg_2 P_2O_7 on a chemical balance up to an accuracy of fourth place after decimal.

3.7.2.4 Calculation

$$\text{meq Mg/100 g plant sample} = \frac{(m_2 - m_1) \times R \times T \times 100}{A \times W \times ME}$$

$$\text{\% Mg in plant sample} = \text{meq Mg/100 g plant sample} \times ME$$

where

m_1 is the mass of crucible in g

m_2 is the mass of crucible + $Mg_2P_2O_7$ in g

R is the ratio of mass of Mg_2/Mg_2 P_2O_7

T is the total volume of plant sample test solution prepared

A is the aliquot of plant sample test solution taken for the determination of Ca and Mg

W is the mass of plant sample in g taken for the preparation of the sample test solution.

ME is the milligram equivalent of Mg i.e. .01216

3.7.5 Comments

The plant samples be preferably digested by wet ashing using conc. $HNO_3$and conc. $HClO_4$. Dry ashing of plant samples after treatment with alcoholic solution containing 5% conc. H_2SO_4 has been also suggested. The deficient and intermediate range of Mg in plant tissues of different crops are summarized below.

Table 3.7.1 Deficient and intermediate range of Mg in different plant parts of different crops.

Crops	Plant Parts	Concentration of Mg %	
		Deficient level	**Intermediate range**
Apple	Middle leaf of current terminal growth in August	< .20	>0.24
Cabbage	Leaf base	.04	.09
Lemon	Leaf mature	.08	.23
Orange	Leaf mature	.026–.128	.17–.52
Corn	Leaf	.13	.23–.25
Grapes	recently mature leaf on fruiting spur	.06–.17	.12–.22
Potato	Leaves	.22—.24	.40—.78
Sugarbeet	Leaves	.08–.20	.21–1.70
Tomato	leaf laminae	.06	.25
Tobacco	Whole leaves	.08–.20	.18–.65
Wheat	Leaf	.10–.13	.16–.17

In oil palm Mg chlorosis has been highly correlated with the ratio of CaO/MgO in the ash of leaves.

The visual symptoms of Mg deficiency appear first on older leaves. There is a loss of healthy green colour between the veins followed by chlorosis and/or development of brillant green colour. The chlorosis may start at the leaf margins or tips, and progress inward interveinally.

In more severe cases, the whole leaf may turn yellow or other brilliant colours. Necrosis may develop interveinally or along margins or tips of leaves. In some plants the leaves curl. Premature defoliation usually occurs. This may result in reduced growth and yield.

The visual symptoms of Mg deficiency in selected crops are described below :

Table 3.7.2 Mg deficiency symptoms in different crops

Crops	Mg deficiency symptoms
Alfalfa	Chlorosis along margins and tips, leaves may wither and dry up
Barley	Leaf yellowing along margins
Bean	Chlorosis in older leaves, interveinal mottling also occur
Cabbage	Older leaves become chlorotic between veins.
Cauliflower	Older leaves show chlorotic spots between veins.
Celery	Leaves become chlorotic at tips and margins advancing interveinally
Grapefruit	Loss of green colour in small patches, mid rib fading to yellow and spreading outward until entire leaf become yellow.
Lettuce	Older Leaves show chlorotic, mottling and eventually complete yellowing.
Onions	Leaves also break at tips, foliage dies prematurely, growth is slow.
Pepper	leaves turn pale green followed by interveinal yellowing, lower leaves may drop, margins curl upward. Plants are small,fruit is sparse and undersized.
Potato	Older leaves show central, interveinal or marginal chlorosis followed by necrosis, leaves are brittle in final stages and turn up at margins.
Rice	Leaves are chlorotic with white tips.
Rubber	Olive or yellowish green chlorotic interveinal patches appear which turn bright yellow orange. Later, large dark brown necrotic patches occur at regular spacing between lateral veins.
Sorghum	Leaves have a light green colour between veins which develops with purplish striping.
Spinach	Leaves develop necrotic areas between veins and later become papery and white in appearance.
Tea	Old leaves appear dull and brittle. Leaves are of pale colour, and lack the dark bluish green colour of healthy plant. A diffuse yellowish brown to coffee coloured mottling develop.
Tobacco	Loss of green colour in lower leaves especially at tips and margins and in between veins. Stalks are slender, roots are long, little branched and slimy in appearance.
Tomatoes	Older leaves show interveinal chlorosis and necrosis. petioles tend to hang down the stalk, leaf margins turn up, stalks are slender, Roots are long with few branches.
Wheat	Leaves show distinct mottling (yellowish green patches in between veins) followed by yellowing and striping, growth is stunted.

Mg deficiency in crops can be partly corrected by foliar spray of Mg salt. Chloride and nitrate salts are more useful for foliar spray of Mg than sulphate salt as Mg is more readily absorbed. Soil application of Mg salt may show beneficial effects after 2-3 years. Mg deficiency is likely to occur when exchangeable Mg is only 6% of the exchange complex. A lower exchangeable Mg/exchangable K ratio in soil also aggravate Mg deficiency.

Mg deficiency occur most commonly in acid sandy soils with moderate to heavy rainfall. Mg deficiency has been also reported from alkaline soil, imperfectly drained soil, as well as from muck soil. Irrigation water having high Ca/Mg ratio give rise to Mg deficiency.

In several crops liberal use of potassic fertilizers tend to aggravate Mg deficiency. However, nitrogen application tend to increase Mg in plant, and thereby reduce Mg deficiency. Some have reported antagonistic effect of ammonium salt on Mg concentration in plants. Application of calcite @ 12.5 t/ha reduced Mg concentration in soyabean. Mg deficiency increase with decrease in pH. Mg deficiency in potatoes was more difficult to correct in presence of large amount of sodium.

REFERENCES

A.O. A.C. 1950. Official And Tentative Methods of Analysis Of The Association of Official Agricultural Chemists Ed. 7, 910 pp. Washington

Barrow, H.L. and E.C. Simpson. 1968. An EDTA method for the direct routine determination of calcium and magnesium in soils and plant tissues. soil Sci. Soc. Amer. Proc. 26 : 443-445.

Cheng, K.L. and Bray, R.H. 1951. Determination of calcium and magnesium in soil and plant material. Soil Sci. 72 : 449-452.

Jackson, M.L. 1973. Soil Chemical Analysis. Prentice- Hall of India Private Limited, New Delhi.

Tom, W.E., 1975. Magnesium pp. 225-263 in. Chapman, H.D. (ed) Diagnostic Criteria for Plants and Soils. Eurasia Publishing House (P) Ltd., New Delhi.

3.8 SODIUM IN PLANT DIGEST–TOTH *et al* (1948)

3.8.1 Principle

The flame photometric method is very convenient for routine analysis of Na in sample test solution. Na on being heated in air–propane blue flame gets heated, and begin to dissipate energy by emission of 5890 Å wave length. The emission is directly proportional to the Na concentration, and it is measured by a galvanometer connected to a photocell set behind Na-filter facing the blue flame.

After installing Na filter, and after lighting the burner flame, the instrument is switched on, and the Na standard solutions are atomized into the flame. The instrument readings are calibrated prior to atomizing sample and blank test solution with appropriate dilution. Na concentration in the diluted sample and blank test solution is estimated from the corresponding flame photometer reading by reference to the standard graph for sodium.

3.8.2 Reagents

(*i*) Sodium Standard - 5 meq Na/l. Dissolve 0.2922 g of dry NaCl in distilled water, and dilute the solution to 1000 ml mark with distilled water in 1 l volumetric flask.

(*ii*) Sodium standards. For preparing 0, 0.1, 0.2, 0.3 and 0.4 meq Na/l solution. Transfer 0, 2, 4, 6 and 8 ml of 5 meq Na/l into separate 100 ml volumetric flasks. Then, add interfering elements such as K and Ca in amounts equal to their concentration in diluted test solution using KCl-12.5 meq/l and $CaCl_2$ 50 meq/l. Also add estimated aliquot of $LiCl_2$ .05 N to serve as internal standard, and then make the volume to 100 ml with distilled water. Normally direct instrument readings without $LiCl_2$ as internal standard is satisfactory when air-propane blue flame is used.

3.8.3 Procedure

1. Operation of the flame photometer. Transfer 5 ml aliquot of sample and blank test solution into 100 ml volumetric flask and dilute the contents to 100 ml with distilled water. Then, switch on the compressor unit and attain the desired air pressure by regulating the air inlet knob. Switch on the instrument and place a lighted match stick above the built in gas burner. Immediately get the gas valve in the gas cylinder regulator opened to supply gas. When the burner gets lighted, regulate the gas inlet regulator until blue flame is obtained.

2. Calibration of the flame photometer and testing of the plant sample test solution. Now atomize 0.4 meq Na/l solution and set the instrument reading to 100. Then atomize 0 meq Na/l solution, and adjust the instrument reading to zero. Again atomize the above two Na standards, and adjust the reading desired if considered necessary. Now atomize the intermediate Na standards, and note the instrument reading corresponding to the concentration of Na in the standard solution.

Recheck the instrument reading by atomizing the primary Na standards. Then, atomize diluted sample and blank test solution noting the corresponding instrument reading. If the sample reading exceeds 100, further dilution of the sample test solution be done. The instrument reading be taken again after the necessary dilution.

3. Preparation of standard graph for Na. On a linear scale graph paper, mark a suitable scale on the ordinate for 0–100 range of flame photometer reading. Mark on the abscissa a suitable scale for the Na concentration from 0 to 0.4 meq Na/l. Thereafter plot flame photometer readings corresponding to 0 and 0.4 meq Na/l and connect the two points by a straight line. Thereafter plot the flame photometer readings for the intermediate standards.

4. Na concentration in the sample and blank test solution. Now, locate the point for the flame photometer reading of diluted sample test solution on the ordinate, and draw an imaginary line parallel to the abscissa intersecting the straight line curve. From the point of intersection, draw a perpendicular to the abscissa and note the value of Na concentration on the abscissa scale. Alternatively divide the flame photometer reading by 250 to obtain Na concentration in meq/l. Similarly estimate the concentration of Na in the blank test solution.

3.8.4 Calculation

$$\text{meq of Na/100 g plant sample} = \frac{(S - B)\, dt \times T \times 100}{W \times 1000}$$

% Na in plant sample = meq Na./100 g plant sample × ME

where S is meq Na/l in diluted sample test solution estimated by reference to the standard graph for Na.

B is meq Na/l in diluted blank test solution estimated by reference to the standard graph.

dt is the dilution times of the sample test solution

T is the total volume of the sample test solution in ml.

W is the mass of plant sample taken in g for the preparation of the sample test solution

ME is the milligram equivalent weight of Na i.e. 0.023 g

3.8.5 Comments

The plant samples be digested by wet ashing procedure employing either ternary acid mixture or else HNO_3 and $HClO_4$.

The concentration of $LiCl_2$ as an internal standard is specified by the instrument manufacturer. However, direct reading is considered satisfactory when air-propane blue flame is used.

For compensating the interference from other elements in flame photomteric determination the interfering elements are added to Na standards.

When the ratio of K/Na is 5 or else Ca/Na is 10, the interference from K and Ca be expected in Na determination. For compensating the interference from expected elements, the test solution be analysed to ascertain the concentration of Na and interfering elements in the test solution. Thereafter estimated amount of interfering elements be added to aliquot of each Na standard and then the solution be diluted to the desired extent.

The critical levels of Na in plants has not been established. However, Na is considered to be essential element for *Atriplex vericaria*. Heward and blue green algae *Anabaena cylindrica*.

The crops such as table beet, sugar beet, and Mangel respond to Na even in the presence of ample K. Some crops such as turnip, radish, table beet, cucumber, sugarbeet, oats, carrots, cabbage, rape, barley, rye, wheat, millet and lettuce respond to Na on K deficient soil. Beets, celery, turnip and spinach respond to NaCl.

The symptoms of Na deficiency in some plants manifest in cessation of growth, yellowing, white necrotic areas along the tips and margins of cotyledons, and older leaves. In beet, Na deficiency symptoms being leaves smaller in size, increase in purple pigments on the under side, leaves also become thin and dark indicating early senescence.

Na in some plants can substitute a small fraction of K requirement. In other crops, Na can substitute a smaller fraction of K requirement. Some plants do not show better growth from Na supply in presence of ample K.

When Na is in excess, leaf injury occur in the form of necrotic areas on the tips, margins, or in the interveinal areas. There is lack of vigour in roots. In alfalfa at high ESP there is reddish colouration of the leaves, and collapse of petiole.

The control of excess Na require correction of excess soluble salts or high exchangeable Na by deep leaching with good quality water and by lowering of water table.

REFERENCES

Lunt, O.R. 1975. Sodium pp 409-432, in Chapman H.D.; Diagnostic Criteria for Plants and Soils, Eurasia Publishing House (P) Ltd. Ram Nagar, New Delhi

Richards, L.A,. (ed). 1954. Diagnosis And Improvement of Saline And Alkali Soils. USDA Hand Book 60, Oxford and IBH Publishing House Co. New Delhi.

Toth, S.J. Prince, A.L. Wallace, A. and Mikkelsen, D.S. 1948. Rapid quantitative determination of eight mineral elements in plant tissues by a systematic procedure involving use of a flame photometer. Soil Sci. 66 : 459-466.

3.9 SULPHUR DETERMINATION IN PLANT SAMPLE

3.9.1 Total S in plant digest by gravimetric method as $BaSO_4$, A.O.A.C. (1950)

3.9.1.1 Principle

The sample test solution is prepared by the dry ashing procedure described in section 3.2.3. Then 50 ml of the test solution is transferred to 250 ml beaker. The test solution is acidified and diluted to 100 ml with distilled water. The diluted test solution is heated to boil and $BaCl_2$ solution is added in excess drop by drop and digested on a water bath for several hours.

The digest is then filtered through ignited and weighed gooch crucible. The precipitate is quantitatively transferred into gooch crucible and is made free from chloride by washing with distilled water. Then the gooch crucible is placed again in muffle furnace at low red heat. The crucible is weighed after cooling to estimate SO_4-S as $BaSO_4$.

3.9.1.2 Equipments etc. and reagents

Equipments and apparatus. Muffle furance, water bath, hot plate, volumetric flask, 100 ml capacity, 50 ml burette, beaker-250 ml capacity, watch glass, gooch crucible, desiccator.

Reagents

(*i*) Conc. hydrochloric acid (Analar)

(*ii*) Barium chloride solution 10.0% : Dissolve 10g of $BaCl_2$ crystals (Analar) in distilled water and dilute it to 100 ml.

(*iii*) 0.05% methyl orange indicator in ethanol : Transfer 50 mg of methyl orange indicator into 100 ml volumetric flask. Add about 50 ml of ethyl alcohol, and shake the contents to dissolve the indicator and make up the volume to 100 ml with ethyl alcohol.

3.9.1.3 Procedure

1. Preparation of plant sample and the blank test solution. Following the procedure described in section 3.2.3, prepare plant sample test solution by dry ashing of 2g oven dry finely ground plant material. Similarly prepare the blank test solution.

2. Precipitation of sulphate in the test solution as barium sulphate. Transfer 50 ml of the sample test solution containing in 0.1 to 2.0 meq of sulphate into a 250 ml beaker. Add 2 drops of 0.05% methyl orange indicator in ethanol. Acidify with conc. HCl if required, and then add 1 ml in excess. Add distilled water to dilute the solution to approximately 100 ml.

Then, heat the contents to boil, and add 10% $BaCl_2$ solution in excess drop by drop with constant stirring. Cover the beaker with a watch glass and let it stand on a steam bath for several hours until the volume is reduced to approximately 50 ml. Remove the beaker from the steam plate and let it cool.

3. Washing, ignition and weighment of the $BaSO_4$ precipitate. Transfer the $BaSO_4$ precipitate in beaker in small portions into an ignited and preweighed Gooch crucible. Transfer the precipitate quantitatively into the Gooch crucible using distilled water. Thereafter, wash the precipitate in Gooch crucible with distilled water until the precipitate become free from chloride. Test filtrate for the presence of chloride with few drops of dilute $AgNO_3$ solution.

When the precipitate become free from chloride, dry the contents of the crucible, and then ignite the Gooch crucible with $BaSO_4$ precipitate inside an electric muffle furnace at low red heat for two hours. Switch off the electric muffle furnace and let it cool. Transfer the Gooch crucible from the muffle furnace into a desiccator, and let it cool furthur. Thereafter, weigh the cooch crucible plus $BaSO_4$ precipitate on a chemical balance upto fourth place after decimal and substract the mass of empty Gooch crucible to find out the mass of $BaSO_4$ precipitate from the 50 ml of the plant sample test solution.

3.9.1.4 Calculation

$$\text{ppm S in plant sample} = \left[\frac{A \times B \times C}{D \times E}\right] \times 10^6 = \left[\frac{A \times 0.137376 \times 100}{50 \times 2}\right] \times 10^6$$

$$\text{\% S in plant sample} = \left[\frac{A \times B \times C}{D \times E}\right] \times 100 = \left[\frac{A \times 0.137376 \times 100}{50 \times 2}\right] \times 100$$

$$\text{SO}_4\text{–S in meq/ 100 g plant sample} = \left[\frac{A \times B^1 \times C}{D \times E \times F}\right] \times 100 = \left[\frac{A \times 0.4116 \times 100}{50 \times 2 \times .048}\right] \times 100$$

where

A is mass of ignited $BaSO_4$ obtained from aliquot of the sample text solution

B is mass of sulphur per g of Barium sulphate

B^1 is mass of sulphate per g of barium sulphate

C is the total volume of the plant sample test solution

D is aliquot of sample test solution taken in ml for the precipitation of sulphate as barium sulphate.

E is the mass of oven dry finely ground plant sample taken for the preparation of sample test solution after dry ashing.

F is milligram equivalent weight of sulphate sulphur.

3.9.1.5 Comments

The gravimetric method of total S determination in plant form the basis for assessing the indirect methods of total S estimation in soil and plant.

The total S in plant parts show much less variation in concentration than the concentration of SO_4-S in plant tissues with the change in S status or else with the supply of sulphur.

The critical level of total S in % in plant parts analysed are summarized below:

Table 3.9.1.

Crop	Optimum S concentration for high yield %	Plant part analysed	Stage of crop growth growth
Wheat	0.253	Leaves	Pre fllowering
Maize	0.20	Leaves	Tasseling
Paddy	0.197	Whole plant	35 days after transplanting
Sorghum*	0.163	Leaves	50 percent flowering
Groundnut	0.67	Leaves	50 percent flowering
Mustard	0.321	Leaves	50 per cent flowering
Sunflower*	0.258	Leaves	50 percent flowering
Soyabean*	0.36	Leaves	Pre flowering
Green gram	0.390	Leaves	Pre flowering
Black gram	0.361	Leaves	30 days after sowing
Cowpea	0.184	Leaves	Flowering
Peas	0.53	Leaves	45-50 days after sowing
Gram*	0.24	Leaves	75 days after sowing
Sugarcane	0.16	Leaves	90 days after sowing
Cotton*	0.412	Leaves	Pre flowering
Lucerene*	0.17	Leaves	50 days after sowing

*Unpublished data Source ; Singh (1988)

REFERENCES

AOAC (1950) 2.8 (e) p. 8, and 6.35, p.104.

Singh, H.G. (1988) Sulphur management in fine textured calcareous soils 1988 Sulphur in Indian Agriculture. Sulphur Institute, Washington, D.C. and FAI, New Delhi.

3.9.2 S in plant sample by turbidimetric method after wet digestion with conc.HNO_3 and $HClO_4$- adapted from Chesnin and Yien (1950)

3.9.2.1 Principle

The sample and blank test solution is prepared by wet digestion with conc. HNO_3 and $HClO_4$ as described in section 3.2.2.

In an aliquot of the sample test solution, $BaSO_4$ turbidity is developed in the presence of sodium acetate-acetic acid buffer and gum acacia by dissolving $BaCl_2$ crystals. The absorbance of the $BaSO_4$ turbidity is measured at 420 mμ wave length (blue filter) on a spectrophotometer or else on a Klett photoelectric colorimeter. The content of sulphur is estimated by reference to the standard calibration curve.

3.9.2.2 Equipments Apparatus/Reagents

Equipments-Apparatus. Completely glass distill, tall beakers 300 ml capacity (pyrex/borosil), hot plate, fume hood, volumetric flasks 100 ml capacity, burette, acid bulb pipette-20 and 10ml capacity, graduated pipette-20, 10, 5, 2 & 1 ml capacity, spectrophotometer or klett photoelectric colorimeter

Reagents

(*i*) Conc. nitric acid (AR). Preferably distill conc. HNO_3 through a completely glass distill.

(*ii*) Perchloric acid -72% (AR).

(*iii*) Sulphur free distilled water. Distill distilled water treated with 25 mg K MnO_4/l using a completely glass distill or use deionized water.

(*iv*) Barium chloride crystals (AR). Grind $BaCl_2$ crystals to pass through 0.50 mm sieve, but which is retained by 0.25 mm sieve.

(*v*) Sodium acetate-acetic acid buffer (pH 4.8). Dissolve 100g of CH_3COONa (AR) in about 500 ml of distilled water. Then, add 31 ml of 99.5% CH_3COOH_4 (AR). Mix the contents and add 450 ml of distilled water. Adjust the pH of solution to 4.8 and make the volume to 1 l with distilled water.

(*vi*) 0.25% gum acacia solution. Transfer 0.25g of finely ground gum acacia into a 100 ml vol. flask. Add distilled water and shake to dissolve the contents. Dilute the solution to 100 ml mark, and leave it overnight. There after, filter the solution and collect the filtrate.

(*vii*) 100 ppm SO_4-S solution. Dissolve 0.5370g of $CaSO_4.2H_2O$ in sodium acetate-acetic acid buffer and dilute the solution to 1000 ml with the buffer solution.

(*viii*) 1.0 ppm SO_4-S solution. Dilute 1.0 ml of 100 ppm SO_4–S solution to 100 ml with sodium acetate-acetic acid buffer.

3.9.2.3 Procedure

1. Preparation of sample and blank test solution. Prepare the plant sample test solution from 1.0 g finely ground sample following the wet digestion procedure described in section 3.2.2. Similarly prepare the blank test solution without taking the plant sample.

2. Development of $BaSO_4$ turbidity in aliquot of the sample and blank test solution. Transfer 5 ml of sample and blank test solution into separate 25 ml volumetric flasks. Add to each 10 ml of sodium acetate-acetic acid buffer and 1 g of $BaCl_2$ crystals (0.25 to 0.50 mm size). Allow it to stand for a minute and then swirl the contents until the crystals are dissolved. Thereafter, add to each 1 ml of 0.25% gum acacia solution, if the concentration of SO_4-S is <20 ppm. If the concentration of SO_4-S is in between 20 to 40 ppm, add 2 ml of 0.25% gum acacia solution. Mix the contents and dilute it to 25 ml with distilled water.

3. Measurement of the absorbance of $BaSO_4$ turbidity developed with aliquot of the sample and blank test solution. Use Klett photoelectric colorimeter fitted with 420 mμ (blue) filter or else a spectrophotometer on adjusting its wave length at 420 mμ. Fill the cuvette or colorimeter test tube with distilled water, and place it in the socket provided. Thereafter, switch on the instrument to light the tungsten lamp and let it warm for 10 minutes. Then, fill the cuvette with the turbid solution obtained by developing turbidity in aliquot of the blank test solution. Place the cuvette in the socket provided and adjust the meter reading to 100% transmittance i.e. 0 absorbance or optical density. Fill the second cuvette with turbid solution obtained by developing turbidity in aliquot of the sample test solution, and place it in the socket provided after removing the first cuvette from the socket. Note the change in the % transmittance or its absorbance i.e. optical density. The optical density of the turbid solution be determined within 5 to 30 minutes of developing $BaSO_4$ turbidity.

4. Measurement of the absorbance of $BaSO_4$ turbidity developed with SO_4-S standards. Transfer 0, 2, 4, 6, 8 and 10 ml of 1.0 ppm SO_4-S solution into separate 25 ml volumetric flasks, and then develop $BaSO_4$ turbidity in each as described in step 2 of the procedure 3.9.2.3 omitting the addition of the sample or blank test solution.

Measure the absorbance of the turbidity developed with the SO_4–S standards adjusting 0 absorbance each time with the $BaSO_4$ turbidity developed with 0 μg SO_4-S. Record the absorbance of the turbidity developed corresponding to each SO_4-S standard.

5. Preparation of the SO_4-S standard calibration curve. On a linear scale graph paper, set the scale of absorbance on the ordinate and the corresponding scale for μg SO_4–S taken for developing turbidity on the abscissa. Plot the values of absorbance recorded corresponding to μg SO_4-S taken for developing $BaSO_4$ turbidity. Connect the point of 0 absorbance with the point of maximum absorbance by a straight line to obtain the standard calibration curve.

6. Estimation of the content of SO_4-S in aliquot of the plant sample test solution. Locate the point of absorbance of $BaSO_4$ turbidity developed with the aliquot of plant sample test solution on the ordinate of the standard calibration curve. Draw a line from the point parallel to the abscissa, intersecting the straight line calibration curve. From the abscissa, draw a perpendicular line that would meet the point of intersection. Read the content SO_4-S on the abscissa at the origin of the perpendicular.

3.9.2.4 Calculation

$$\text{Total S in plant sample (ppm)} = \frac{A \times B}{C \times D} = \frac{A \times 100}{5 \times 1.0}$$

$$\%\ \text{S in plant sample} = \frac{\text{Total S in plant sample (ppm)} \times 100}{10^6}$$

where A is μg SO_4-S estimated in $BaSO_4$ turbidity developed with the aliquot of the sample test solution.

B is the total volume of sample test solution prepared by the wet digestion procedure

C is ml aliquot of the sample test solution taken for developing $BaSO_4$ turbidity.

D is mass of oven dry finely ground plant sample taken for wet ashing to prepare sample test solution.

3.9.2.5 Comments

Glass wares must be carefully cleaned to remove any potential nuclei for the formation of irregular sized barium sulphate crystals, and the measuring cell must be cleaned regularly with alkaline EDTA to remove adhering barium sulphate, and maintain sensitivity.

A small amount of sulphate is generally added to the aliquot of the sample test solution before measurement of the absorbance of $BaSO_4$ turbidity to improve the sensitivity of the method and linearity of the standard calibration curve.

Apart from gum acacia, one can also use gelatin, tween 80, polyvinyl alcohol, and sorbitol as a stabilizer of $BaSO_4$ turbidity.

The optimum concentration of total sulphur in plant parts of different crops are summarized under comments in the determination of total sulphur in plant sample after ignition with $Mg(NO_3)_2$ (see section 3.9.1.5).

REFERENCES

Chesnin, L. and C.H. Yien (1950). Turdimetric determination of available sulphates. Soil Sci. Soc. Amer. Proc. 15 : 149-151.

3.9.3 Sulphur in plant sample by the methylene blue method– Johnson and Nishita (1952) after wet digestion with conc. HNO_3 and $HClO_4$

3.9.3.1 Principle

The test solution from oven dry finely ground plant sample is prepared by wet digestion with conc. HNO_3 and 60% $HClO_4$. Care is taken to reduce the concentration of oxidants below the interference level.

Not more than 2 ml of the test solution containing < 300 μg SO_4-S is transferred into 50 ml capacity digestion flask. After adding reduction mixture prepared with hydriodic acid, formic acid and hypophosphorous acid, the digestion flask is quickly attached to the condenser of the distillation apparatus. The nitrogen gas purified by passing through a solution of mercuric chloride in 2% $KMnO_4$solution is bubbled through the reduction mixture in digestion flask. In order to prevent excessive bumping during heating the digestion flask, it is essential that the gas inlet tube terminate below the liquid level in the flask, approximately 3 to 5 mm from the bottom of the flask. The digestion flask is heated by microburner and the contents is boiled gently for one hour. As the digestion flask is heated, cooling water is allowed to flow in the condenser.

The H_2S liberated is distilled to pass through pyrogallol sodium phosphate gas washing column, and then it is absorbed in zinc acetate-sodium acetate buffer solution. The H_2S absorbed is treated with acid solution of p-amino dimethylaniline sulphate solution and ferrous ammonium sulphate solution to form methylene blue colour. At first the solution is red, but it changes to blue as the dye is formed. The colour developed is stable for 24 hours when the solution is stored in a stoppered flask in subdued light. The reactions may be represented by

$$2\,(CH_3)_2\underset{\substack{|\\HA}}{N}-C_6H_4-NH_2 + H_2S \xrightarrow{Fe^{3+}} \left[(CH_3)_2N{=}C_6H_3{\left(\begin{smallmatrix}=N-\\-S-\end{smallmatrix}\right)}C_6H_3-\underset{\substack{|\\HA}}{N}(CH_3)_2\right]^{+}A^{-}$$

where A^- represent an anion

The optical density of the methylene blue colour is measured at 670 mμ wave length using a spectrophotometer or photoelectic colorimeter.

Digestion-Distillation Apparatus. The digestion-distillation apparatus as shown in Figure 3.9.3.1 consists of a boiling flask, water-jacketed condenser, connecting U-tube, gas-washing column, and detachable connecting tube (4 mm. in outside diameter) leading to a receiving vessel. A gas-washing bottle (not shown in Figure 3.9.3.1) is placed between the nitrogen tank and the inlet arm of the boiling flask. The incoming nitrogen is washed by bubbling through a solution of mercuric chloride in 2% potassium permanganate to remove possible traces of sulfur gases.

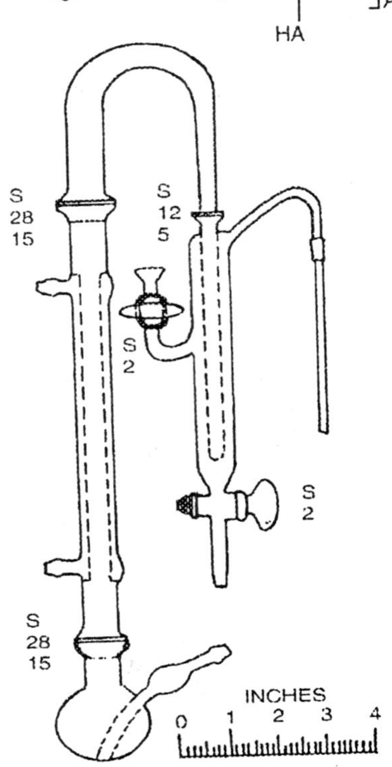

Fig. 3.9.3.1 Digestion Distillation Apparatus.

The sample flask is a 50-ml. round-bottom boiling flask (Pyrex No. 4325) fitted with a 7-mm. gas inlet tube as illustrated in Figure 3.9.3.1. In order to prevent excessive bumping, it is essential that the gas inlet tube terminate below the liquid level in the flask, approximately 3 to 5 mm from the bottom of the flask. Heat is supplied by a microburner protected from drafts by a suitable shield. An asbestos pad must protect the boiling flask from direct flame.

Special attention should be paid to the angle at which the side arm emerges from the gas washing column. This angle is necessary, in order that the side arm may remain filled with solution to prevent trapping of hydrogen sulfide.

The delivery tube leading to the receiving vessel is attached to the connecting arm of the gas washing column by a short piece of rubber tubing.

3.9.3.2 Apparatus & Reagents

Equipments/Apparatus. Digestion-distillation apparatus, Apparatus for preparation of the reducing mixture, hot plate, water bath, oven, muffle furnace, fume hood, microburner, tall beakers, volumetric flasks, pipettes, thermometers, spectrophotometer or photoelectric colorimeter.

Reagents etc.

(*i*) 69% Nitric Acid (AR)

(*ii*) 60% Perchloric acid (AR)

(*iii*) 1.0 N Hydrochloric acid : Add 83 ml of 37% HCl to 800 ml distilled water and dilute the solution to 1l.

(*iv*) Reducing mixture

Mix 400 ml of hydriodic acid (HI, sp. gr. 1.7, methoxyl grade), 1000 ml of hypophosphorus acid (H_1PO_2, 50%) and 200 ml of formic acid (90%) in 1 litre three-necked-round bottom flask (Fig. 3.9.3.2). Place the flask in a heating mantle connected to a rheostat in a well-ventilated hood. Place a fitted thermometer graduated to 360°C in the centre neck after lubricating the joint with a small amount of S free ground joint lubricant. Connect the exhaust fumes outlet to one of the side necks and the N_2 gas inlet to the other neck.

Place a 250 ml Erlenmeyer flask containing about 150 ml of cold water under the fume exhaust outlet. Allow a stream of purified N_2 gas to bubble through the acid mixture. Heat the mixture slowly to 115°C and hold at 115 to 117°C for 10 minutes.

Do not let the temperature of the solution exceed 117°C. Extremely poisonous fumes of phosphine (PH_3) may be liberated from the reagent if it is heated about 120°C or spilled on a hot surface.

Cool the mixture by turning off the power and allowing the mixture to stand in the hood. For more rapid cooling, remove the flask from the heating mantle and place it on a cork ring. Let N_2 gas continue bubbling through the mixture during the cooling period. When the mixture has cooled to the room temperature, transfer it to 1 l bottle, and store it in the dark.

(*v*) Nitrogen purification solution. Add 5-10g of $HgCl_2$/100 ml to 2% $KMnO_4$ solution.

(*vi*) Pyrogallol-sodium phosphate solution. Dissolve 10 g of NaH_2PO_4. H_2O and 10g of $[C_6H_1(OH_1)]$ in 100 ml of S-free distilled water with the aid of a stream of N_2 bubbling through the solution. This reagent be prepared daily.

(*vii*) Sulphur-free distilled water. Distilled and deionized water is normally free of S. If necessary, prepare by distilling water containing alkaline permanganate from all-glass still.

(*viii*) Zinc acetate sodium acetate H_2S absorbing solution. Dissolve 50g of $[Zn (CH_1COO)_2.2H_2O]$ and 12.5g of $(CH_1COONa.3H_2O)$ in about 800 ml of distilled water. Dilute the solution to 1 l. If turbid, filter the solution through funnel lined with Whatman No. 42 filter paper.

(*ix*) Aminodimethylaniline solution. Dissolve 2g of p-aminodimethylaniline sulphate (no. 1333, Eastman Kodak Co., Rochester, N.Y.) in about1500 ml distilled water in a 2l volumetric flask. Place the bottle in a pan of cold water, and slowly add 400 ml of conc. H_2SO_4(AR) and 195 ml of distilled water. Add H_2SO_4in small portions and mix the contents after each addition and then allow some time for the mixture to cool. Dilute the contents to 2 l after cooling.

(*x*) Ferric ammonium sulphate solution. Transfer 25g of $[Fe_2 (SO_4)_3.(NH_4)_2 SO_4. 24H_2 O]$ into 250 ml volumetric flask. Add 195 ml of distilled water and 5 ml of conc. H_2SO_4. Shake to mix the contents and let it stand at room temperature for several days for the reagent to dissolve.

(*xi*) 1000 ppm SO_4-S solution. Dissolve 5.435 g of K_2SO_4 (AR) in about 800 ml of S-free distilled water and dilute the solution to 1000 ml-mark.

(*xii*) 0 to 50 ppm S solution. Transfer 0, 1, 2, 3, 4 and 5 ml of 1000 ppm S solution into separate 100 ml volumetric flask and dilute the solution to 100 ml mark with S-free distilled water to obtain 0, 10, 20, 30, 40 and 50 ppm of SO_4-S solution.

(*xiii*) Sulphur-free ground-joint lubricant. Mix approximately 5g of silicone stopcock lubricant (Dow Corning Corp, Midland Mich.) with 5 ml of hydriodic acid (H_1) and 5 ml of hypophosphorus acid (H_1PO_2) in a 100 ml beaker. Place a 50 ml round bottom boiling flask filled with cold water on the beaker to serve as a condenser. Gently heat the contents of beaker to boiling on a hot plate in a well ventilated hood with frequent stirring for about 45 minutes. At the end of the boiling period, pour off the acid mixture, and wash the lubricant thoroughly with S-free water.

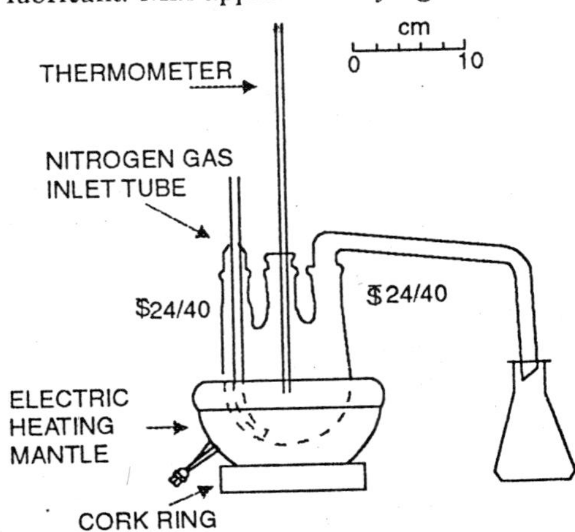

Fig. 3.9.3.2 Apparatus for preparation of the reducing mixture.

(*xiv*) Sulphur-free short pieces of rubber tubing. Place the short pieces of rubber tubing in a mixture containing 10 ml of H1 and 10 ml of H_1PO_2 in a 100 ml beaker. Place a 50 ml boiling flask containing cold water on the beaker to serve as a condenser. Heat the contents of the beaker to boil for about 45 minutes. At the end of the boiling period, pour off the acid mixture, and wash the short pieces of the rubber tubing with S-free water.

3.9.3.3 Procedure

1. Wet digestion of plant sample. Transfer 100 mg of oven dry finely ground plant sample into 30 ml microkjeldahl flask, add 2ml of conc. HNO_3 and digest on the steam bath for 30 minutes. Add 1.0 ml of 60% $HClO_4$, and heat slowly over a microburner. Increase the heat, and continue the digestion for atleast 1 hour after fumes of perchloric acid appear.

Cool and add 3 ml of 6N HCl to facilitate complete removal of nitrate. Heat the digest containing the HCl again until copious fumes of $HClO_4$ again appear. Cool and make the volume of the test solution to 50 ml in a volumetric flask. Take aliquot usually not exceeding 2 ml for the determination of sulphur by the reduction method.

2. Setting of the digestion-distillation apparatus. Lubricate all spherical joints of the reduction-distillation apparatus with minimal amount of the S-free Lubricant.

Pipette 10 ml of pyrogallol-sodium phosphate solution into the gas washing column, and connect it to the apparatus holding it in place with a clamp.

Now condition the apparatus for reason of adsorption and solubility of H_2S in the apparatus and wash solution. For conditioning the apparatus, a preliminary run should be made with a sample or standard containing a small portion of reducible S. The sample need not be taken through the development and measurement of colour step. The digestion-distillation apparatus and wash solutions are equilibrated after the first run with respect to H_2S, and determination of the total S in the sample may be carried on in the usual manner.

After conditioning the apparatus, add 10 ml of the Zn(OAc)-NaOAc solution to 100ml glass stoppered volumetric flask to be used as a receiver, and add approximately 50 ml distilled, deionized water. Connect a delivery tube to the side arm of the gas washing solution, using a piece of S-free rubber tubing. Insert the delivery tube into the liquid almost to the bottom of the receiving flask, and clamp the receiving flask in place.

3. Treatment of test solution containing SO_4-S with the reducing mixture and distillation of H_2S formed. After transfering 2ml aliquot of the sample test solution into the 50 ml digestion flask. Add 4 ml of the reducing mixture by a rapid delivery pipette. Then moisten the condenser's joint with a drop of water, and immediately attach the 50 ml digestion flask to the condenser.

Thereafter, adjust the N_2flow so that about 2 bubbles/sec issue from the receiving flask. Also make sure that cold water is flowing through the condenser. Let N_2 flow for about5 minutes. Then, light the microburner, and boil the contents gently for one hour.

After one hour of boiling the contents gently, remove the receiving flask, leaving the connecting tube in the $Zn(OAc)_2$-NaOAc solution by disconnecting the delivery tube and leaving the small pieces of the rubber tubing attached to the side arm of the distillation apparatus.

4. Development of the methylene blue colour with H_2S absorbed in $Zn(OAc)_2$-NaOAc solution. Now, add 10.0 ml of the p-aminodimethyl aniline solution using a rapid delivery pipette. Quickly stopper the flask and mix the contents thoroughly. Unstopper the flask, and add 2.0 ml of ferric ammonium sulphate solution. Stopper the flask, mix the contents, and dilute the solution to 100 ml with distilled water.

5. Measurement of the absorbance of methylene blue colour. After 20 minutes but before 24 hours, measure the absorbance of methylene blue colour at 670 mμ wave length adjusting 0 absorbance (optical density) with the blank test solution run through the entire procedure as described for the sample test solution.

6. Reduction of 0 to 50 μg SO_4-S and distillation of H_2S formed for absorption into $Zn(OAc)_2$-NaOAc solution. Set the digestion-distillaion apparatus following step 2 of procedure. Then, pipette 1 ml aliquot of 0, 10, 20, 30, 40 and 50 ppm SO_4–S solution in the 50 ml distillation flask one at a time. Thereafter, add 4 ml of the reducing mixture, and proceed as described in step 3 beginning with the moistening of the condenser joints with one drop of water.

7. Development of methylene blue colour with H_2S absorbed in $Zn(OAc)_2$ NaOAc solution, and measurement of its absorbance. Follow steps 4 and 5 of the procedure, and record the value of absorbance of each standard at 670 mμ wave length adjusting 0 absorbance with the methylene blue colour formed with 0 μg of SO_4–S. Allow 10 minutes times after switching on the spectrophotometer or photo-electric colorimeter to warm before measurement of the absorbance.

8. Preparation of calibration curve for sulphur by the methylene blue method. On a linear scale graph paper, plot the value of absorbance on the ordinate, and the corresponding content of SO_4–S taken on the abscissa. Connect 0 absorbance with the value of maximum absorbance by a straight line to obtain the calibration curve.

9. Estimation of SO_4–S in the aliquot of sample test solution on development of methylene blue colour. Note the absorbance of the methylene blue colour with H_2S formed from sample test solution. Locate the absorbance value of the sample test solution on the ordinate of standard calibration curve for sulphur and draw an imaginary line parallel to the abscissa. From the point of intersection of the calibration curve, draw a line perpendicular to the abscissa, and record the content of μg SO_4–S in the 2 ml aliquot of the sample test solution

3.9.3.4 Calculation

$$\text{Total S in plant sample (ppm)} = \frac{A \times B}{C \times D}$$

where A is μg SO_4–S in aliquot of plant sample test solution as determined by reference to the standard calibration curve for S by the methylene blue method.

B is total volume of the plant sample test solution prepared from the plant sample after wet digestion with conc. HNO_3 and 60% $HClO_4$.

C is ml aliquot of plant sample test solution taken for reduction and subsequent distillation of H_2S for the development of methylene blue colour.

D is mass of oven dry finely ground plant sample taken for wet digestion with conc. HNO_3 and 60% $HClO_4$.

3.9.3.5 Comments

In water containing 0.1 ppm Cu, sulfide is quantitatively precipitated in the wash solution as well as in $Zn(OAc)_2$ solution, and is not available to form methylene blue. It is, therefore, desirable to use deionized water or water prepared by distillation from all glass distill in the preparation of reagents.

The volatile compounds arising from the reaction of nitrates with the reducing mixture are incompletely trapped in the sodium phosphate–pyrogallol washing solution, and prevent the synthesis of methylene blue. Consequently, nitrate remaining from digestion of plant sample with conc. HNO_3 and 60% $HClO_4$ can result in low sulphur values if nitrate content is more than 6 mg per sample. The evidence of nitrate interference is a rapid and pronounced darkening of the sodium phosphate–pyrogallol wash solution. In the absence of nitrate in the sample, a freshly prepared sodium phosphate–pyrogallol wash solution will remain almost colourless for as many as 6 to 8 determinations. Preliminary tests such as the diphenylamine test for nitrates, and other oxidizing agents should be made on samples of unknown history. The removal of nitrate by taking the sample to dryness with conc. HCl may not remove nitrate to the desired extent. As an alternative, the extraction of plant sample (after wetting with 95% ethyl alcohol) with 10% $BaCl_2$ solution in a centrifuge tube followed by centrifugation and then decanting off the supernant clear solution will make the solid residue of plant material free from nitrates. The alternative method is reported to be satisfactory for the elimination of nitrate interference with minimum effect in the sulphate content of the sample.

The rate of nitrogen flow through the apparatus should be sufficiently rapid to sweep the system completely free of hydrogen sulphide in a reasonable time. The nitrogen flow rates of 100 to 200 ml per minute are entirely adequate. The slower rates require extremely long times to ensure complete transfer of hydrogen sulphide.

The free flame be not allowed to play on the surface of the flask above the level of the reducing mixture. The best results are obtained by use of an asbestos mat on tip of the flame shield.

All tubing in the N_2 lines to the boiling flask must be tygon or similar inert material. Rubber tubing should not be used.

The methylene blue solution is stable for atleast several days if stored in the dark. The colour fades rapidly on exposure to sun light. When the solution is kept away from sunlight the colour is stable for 24 hours.

The relationship between amount of SO_4–S in μg and the absorbance of the resulting methylene blue solution at 670 mμ wave length is usually linear in the range of 2 to 50 μg SO_4–S. If the aliquot of test solution contains > 50 or < 300 μg SO_4–S, the methylene blue colour developed be diluted by distilled water containing the same concentration of $Zn(OAc)_2$–NaOAc, p–aminodimethylene aniline, and ferric ammonium sulphate reagent to the desired extent prior to the measurement of the absorbance. The μg SO_4–S estimated by reference to the standard calibration curve be multiplied by the dilution factor for estimating total SO_4–S in the aliquot of the sample test solution.

The regeneration of the used reagents should be avoided. The used reducing mixtures should be collected in a wash bottle and disposed of properly.

Dean (1966) proposed that H_2S evolved can be adsorbed in a solution containing NaOH, and treated with Bi reagent to develop colloidal bismuth sulfide, the absorbance of which can be measured at 400 mμ wave length.The standard curve obtained obeys Beer's Law upto 200 μg of SO_4–S. It is, however, equally affected by excess of nitrate. Nitrate in excess of 6 mg per sample brought a reduction of 75% of sulphate determined.

The removal of sulphur by crops differ considerably. Maize, wheat and grasses remove about 9 to 13.5 kg S/ha. Onions and clovers remove 16.8 to 22.5 kg S/ha. High sulphur requiring crops are Lucerne (22.5 to 26.9 Kg S/ha), Turnips (28.0 to 39.3 Kg S/ha) and cabbage (21.3 to 42.7 kg S/ha) as reported by Coleman (1966). In general cruciferous crops and members of the Liliaceae family require considerable more S than most crops. S is required for the synthesis of proteins as well as for other constituents such as mercaptans, and glucosides.

Sulphur is absorbed by plants from the soil in the form of sulphates. Sulphur present as sulphur dioxide in the atmosphere is also absorbed by leaves, and it is rapidly converted into sulphate. Sulphur is essential for the formation of chlorophyll. The chloroplasts fragments contain 10 times more N than S.

When S is deficient, symptoms are similar to nitrogen, but it appears first on newer leaves due to restricted mobility of S from older leaves. There is also change in the concentration of S, N/S ratio as well as in the concentration of SO_4–S in the plant tissues.

Sulphur deficient plants are generally small, weak, retarded, have light green or yellowish coloured leaves, weak stems, and do not ripe in time. Crop produce low yields of inferior quality. Sulphur deficiency can reduce crop yields even when the deficiency symptoms are not visible.

Sulphur applied to crops on sulphur deficient soils can increase crop yields on an averge by 15 to 30 percent. Sulphur application in oil seed crops in India increased the oil content in sunflower, groundnut, soyabean, taramira, rai and mustard by 5 to 9%. Sulphur increases the amino acid content and percent protein. It also provides an ideal N : S ratio in forage. There is increase in the starch content in tuber crops. Besides, HCN toxicity is also reduced in forage.

Harward et. al. (1962) worked out relationship between yield of alfalfa and total S in forage. He reported critical level of total S to be 0.22%. Plants below this level showed progressively lower relative yield, and plants higher than this level showed little response to added S. Alfalfa grown on S–deficient soils contained less than 0.15% total S in plant. Jones (1963) reported that the S concentration and critical levels tend to be higher for leaves than for stems and petioles. In other plant species, the critical level of S vary from 0.15% to 0.30%.

Pumphrey and Moore (1965) reported that in alfalfa a N : S ratio of 11 : 1 indicates adequate supply of S to the plants. In 22 alfalfa fields, N : S ratio below 11 indicated adequate S supply while higher values of 15 to 25 indicated progressively more acute deficiencies of S. For ruminants, the desirable N : S ratio in forage is 10 : 1 to 15 : 1. The N : S ratio in plant remains to be relatively constant at different growth stages when nutrients are expressed on a dry weight basis.

The concentration of total S vary with the plant parts and age. The temperature, insecticidal and fungicidal sprays also effect concentration of S in plant. The critical level of S in plant tissues of different plant species was summarized by Eaton (1975).

REFERENCES

Allaway, W.H. and J.F. Thompson (1966). Sulfur in the nutrition of plants and animals. Soil Sci. 101 : 240–247.

Coleman, R. (1966). The importance of sulfur as a plant nutrient in World Crop Production. Soil Sci. 101 : 230–239.

Dean, G.A. (1966). A simple colorimetric finish for the Johnson–Nishita micro–distillation of Sulphur. Analyst (London) 91 : 530–532.

Eaton, R.M. (1975) sulfur pp 444–475. In Chapman, H.D. (ed.) Diagnostic Criteria for Plants and Soils. Eurasia Publishing House (P) Ltd., Ram Nagar, New Delhi.

Esminger, L.E. and J.R. Freney (1966). Diagnostic techniques for determining sulfur deficiencies in crops and soils. Soil Sci. 101 : 283–290.

Harward, M.E., T.T. Chao and S.C. Fang (1962). The sulfur status and sulfur supplying power of Oregon soils. Agron. J. 54 : 101–105.

Johnson, C.M. and H. Nishita (1952). Microestimation of sulfur in plant materials, and irrigation water, Anal. Chem. 24 : 736–742.

Jones, M.B. (1963). Effect of sulfur applied, and date of harvest on yield, SO_4–S concentration and total S uptake of 5 annual grassland species. Agron. J. 55 : 251–254.

Pumphrey, F.V. and D.P. Moore (1965). Diagnosing sulfur deficiency of alfalfa from plant analysis. Agron. J. 57 : 364–366.

Tabatabai, M.A. (1982) Sulphur pp 501–535. In Methods of Soil Analysis. Part 2. Chemical and Microbiological Properties. Agronomy Monograph No. 9 (second edition). American Society of Agronomy, Madison, Wisconsin, U.S.A.

3.9.4 S in plant sample by method of Chopra and Kanwar (1976)

3.9.4.1 Principle

The finely ground plant material is ignited in excess of oxygen in a Schöniger flask to oxidize sulphur into sulphates. The products of combustion are absorbed in

0.1N NaOH containing 3% H_2O_2. The combustion products of proteins absorbed in NaOH containing dilute H_2O_2 are evaporated to dryness and then extracted with 10 ml of distilled water and transfered to 25 ml volumetric flask. The $BaSO_4$turbidity is developed in the presence of sodium acetate-acetic acid buffer and gum acacia solution. The combustion products of plants material is also absorbed in 0.1 NaOH containing 3% H_2O_2, but the solution is passed through Amberlite 120–H^+ form to reduce the concentration of cations. The solutions and washings are transferred into 50 ml vol flask and it is treated with $BaCl_2$ crystals in the presence of sodium acetate-acetic acid buffer and gum acacia to develop $BaSO_4$ turbidity. The alsorbance of the turbidity is measured at 420 mμ wave length using a spectrophotometer or else in a photoelectric colorimeter fitted with 420 mμ filter.

3.9.4.2 Apparatus & Reagents

Apparatus. Thomas–Schoniger combustion apparatus–available from Arthur H. Thomas Co. Philadelphia, U.S.A.

Reagents

(*i*) Distilled water free from sulphur. Distill water containing $KMnO_4$ in a completely glass distill or use deionized water.

(*ii*) 0.1 M sodium hydroxide solution. Dissolve 4g NaOH pellets (AR) in 1 l of distilled water.

(*iii*) 3.0 % Hydrogen peroxide (reagent grade).

(*iv*) 0.25% gum acacia solution. Dissolve 2.50 g of gum acacia in 1 l of distilled water. Allow it to stand over night, and filter.

(*v*) Sodium acetate–acetic acid buffer (pH 4.8). Dissolve 100 g CH_3COONa (AR) in 500 ml distilled water. Add 31 ml of CH_3COOH (AR) and make the volume to 1 l with distilled water.

(*vi*) Barium chloride crystals (30–60 mesh). Use $BaCl_2$ crystals that pass through 30 mesh screen, but which is retained on 50 or 60 mesh sieve.

(*vii*) 100 ppm SO_4–S solution. Dissolve 0.5434 g of K_2SO_4 (AR) in 1 l of distilled water.

(*viii*) 50 ppm SO_4–S solution. Dilute 50 ml of 100 ppm SO_4–S solution to 100 ml with distilled water in 100 ml vol. flask .

3.9.4.3 Procedure

1. Preparation of the sample test solution. Grind the plant sample dried in oven at 70°C in a Willey mill passing it through 80 or 100 mesh sieve. Transfer 5–20 mg plant sample containing 0.025 to 0.5 mg sulphur in the middle of tared filter paper carrier, fold the filter paper and fit it in the platinum holder.

Transfer 2.0 ml of 0.1N NaOH, 1.5 ml of distilled water, and 2.0 ml of 3% H_2O_2 into the Schoniger flask. Flush the solution with oxygen gas for 30 seconds. Ignite the wick, and insert the stopper quickly. Tilt the flask, and keep it in the inverted position for 23 minutes after the combustion of the sample. Allow the flask to stand for 30 minutes with occasional shaking.

Thereafter, in case of protein sample, heat the flask on a water bath to evaporate the contents to dryness. Then, add 5 ml of distilled water, swirl the flask to dissolve the residue, and transfer the solution into 25 ml volumetric flask. Again repeat the extraction of residue containing SO_4–S with 5 ml of distilled water and transfer the solution into the same volumetric flask. Similarly run a blank.

In case of plant sample combustion, open the flask, and transfer SO_4–S absorbed in NaOH solution and there after washings through Amberlite resin column. Collect the solution passing through the amberlite resin into 50 ml volumetric flask. Similarly run a blank.

2. Development of the $BaSO_4$ turbidity with the sample and blank test solution. In test solution prepared from protein sample, add 10 ml of sodium acetate–acetic acid buffer (pH 4.8) into 25 ml volumetric flask containing 10 ml of the sample test solution. Then, add 1 ml of 0.25% gum acacia solution. Mix the contents and add 1g $BaCl_2$crystals (60 mesh). Swirl the flask frequently until the crystals are dissolved. Then make the volume to 25 ml with distilled water. Similarly treat the blank test solution.

In test solution prepared from plant sample, add 20 ml of sodium acetate–acetic acid buffer (pH 4.8) solution into the 50 ml volumetric flask containing the sample test solution. Then, add 2 ml of 0.25% gum acacia solution. Mix the contents and 1g $BaCl_2$ crystals (50 mesh). Swirl the flask frequently until the crystals are dissolved. Make the volume to 50 ml mark with distilled water. Similarly treat the blank test solution.

3. Measurement of the absorbance of $BaSO_4$ turbidity developed with the sample test solution. Within 5 to 10 minutes of development of $BaSO_4$ turbidity, measure the absorbance of the turbidity developed with the sample test solution on a spectrophotometer or else in photoelectric colorimeter fitted with 420 mμ filter adjusting 0 absorbance with the blank test solution. Prior to measurement of the absorbance, switch on the instrument after placing cuvette or colorimeter tube filled with distilled water to allow it to warm for 10 minutes.

4. Development of the $BaSO_4$ turbidity with the standard SO_4–S solution. In separate 25 ml vol. flask, pipette 0–10 ml of 50 ppm SO_4–S solution, and then add 10 ml of sodium acetate–acetic acid buffer and 1 ml of 0.25% gum acacia solution. Mix the contents and add 1g of $BaCl_2$crystals (60 mesh). Allow it to stand for 1 minute. Then swirl the contents frequently to dissolve the $BaCl_2$ crystals. Thereafter dilute the solution to 25 ml mark.

If the turbidity in the sample test solution was prepared in 50 ml volumetric flask, pipette 0 to 10 ml of 100 ppm SO_4–S solution into separate 50 ml volumetric flask. Then add 20 ml of sodium acetate–acetic acid buffer (pH 4.8) and 2 ml of 0.25% gum acacia solution. Mix the contents and add 1g of $BaCl_2$ crystals (50 mesh). Allow it to stand for 1 minute. Then, swirl the contents frequently until $BaCl_2$crystals are dissolved. There after, dilute the solution to 50 ml.

5. Measurement of the absorbance of $BaSO_4$ turbidity developed with 0 to 10 ml aliquots of standard SO_4–S solution. Within 2–10 minutes of developing $BaSO_4$

turbidity, measure the absorbance on a spectrophotometer or else on a photoelectric colorimeter fitted with 420 mμ filter adjusting 0 absorbance with turbidity developed with 0 μg SO_4–S either in 25 ml or 50 ml volumetric flask. Prior to measurement of the absorbance, the instrument should be switched on after placing cuvette or test tube filled with distilled water in the socket provided . Allow 10 minutes time for the instrument to warm.

6. Preparation of standard calibration curve. Plot the absorbance of SO_4–S standards on a linear scale graph paper. Let the crdinate be the scale for absorbance and abscissa be the scale for μg SO_4–S taken for developing the turbidity. Connect the point of 0 absorbance with the point of maximum absorbance by a straight line to obtain a straight line standard calibration curve. Draw a separate graph for the turbidity developed in 25 ml and in 50 ml vol. flask.

7. Eastimation of SO_4–S in the sample test solution. Refer to the appropriate standard calibration curve. Locate the absorbance value of the sample test solution on the ordinate, and draw an imaginary line parallel to the abscissa intersecting the standard calibration curve. From the point of intersection, draw a line perpendicular to the abscissa and read the value of SO_4–S in μg at the point of intersection on the abscissa.

3.9.4.4 Calculation

S in plant sample (ppm)

$$= \frac{\mu g\ SO_4 - S \text{ in sample test solution by reference to the standard calibration curve for } SO_4 - S}{\text{mass of sample in gram}}$$

$$\%\ S \text{ in plant sample} = \frac{\text{ppm S in plant sample}}{10^6} \times 100$$

3.9.4.5 Comments

The cystine and methonine containing 26.69 and 21.49% S was accurately determined by the method described. The method is satisfactory for the determination of sulphur in sulphur containing organic compounds including plant material.

In turbidimetric determination, glasswares must be carefuly cleaned to remove any potential nuclei for the formation of irregular sized barium sulphate crystals and the measurement cell must be cleaned regularly with alkaline EDTA to remove adhering sulphate and maintain sensitivity.

A small amount of sulphate may be added to the aliquot of the sample test solution before measurement to improve the sensitivity of the method and linearity of the standard line.

REFERENCES

Chopra, S.L.(1964). A turbidimetric method for the micro determination of sulphur in proteins. Indian J. Chem. 2 : 78–79.

Chopra, S.L. and J.S. Kanwar. (1976). Microdetermination of sulphur in proteins, plant–leaves and feeding stuff pp 350–351. In Analytical Agricultural Chemistry. Kalyani Publishers, Ludhiana, New Delhi.

3.9.5 SO_4–S in plant material by the reduction method developed by Johnson and Nishita (1952)

3.9.5.1 Principle

About 25 mg of finely ground plant sample containing < 6 mg NO_3 is transferred to 50 ml of digestion-distillation flask. If the plant sample contains 6 mg or more of NO_3, NO_3 from sample is extracted by 10 ml of 10% $BaCl_2$ solution and the aqueous phase after centrifugation is discarded. The residue is quantitatively transferred to 50 ml digestion-distillation flask and the volume is reduced to 2 ml or less.

Then the sample is treated with a reduction mixture and the flask is attached to a distillation apparatus. The purified N_2 gas is passed through the reduction mixture and the digestion flask is heated by a microburner to distill H_2S liberated. The H_2S liberated is made to pass through a gas washing column before its absorption by zinc acetate–sodium acetate buffer solution. The H_2S absorbed is treated with acid solution of P aminodimethyl aniline sulphate and ferric ammonium sulphate solution to form methylene blue colour. The optical density (absorbance) of the methylene blue colour is measured at 670 mμ wave length using a spectrophotometer or photo electric colorimeter.

3.9.5.2 Apparatus/Reagents

Equipments and apparatus

Digestion distillation apparatus, Apparatus for preparation of reducing mixture, conical flasks, volumetric flasks, pipettes, thermometers, water bath, hot plates, fume hood, micro burner, centrifuge, cylinder of water pumped N_2, spectrophotometer or photoelectric colorimeter.

Reagents

(*i*) 10% Barium chloride solution. Dissolve 10g of $BaCl_2$ crystals in 100 ml of distilled water.

(*ii*) 95% Ethyl Alcohol.

(*iii*) Reagents 3 to 12 as described in the determination of total S in plant by the reduction method (section 3.9.3.2)

3.9.5.3 Procedure

1. Removal of NO_3 from the finely ground plant sample. Transfer 25mg finely ground plant sample into a 15 ml centrifuge tube and moisten it with sufficient 95% ethyl alcohol to ensure that all the sample has been wetted. Add 5 drops of 10% $BaCl_2$ solution. Immerse the tube to a depth of 1.25 cm in a boiling water bath for 10 to 15 minutes. The alcohol and water vapours condensing on the walls of the tube aid in wetting any sample particle adhering to the side of the tube.

Add 5 ml of 10% $BaCl_2$ solution and mix thoroughly with a glass rod. Wash down the sides of the tube with an additional 5 ml portion of 10% $BaCl_2$solution. Triturate the contents for 10 minutes with a glass rod. Then centrifuge the suspension at 2400 rpm for 10 minutes.

Discard the aqueous phase after centrifugation. Transfer the residue quantitatively into 50 ml of digestion–distillation flask. Reduce the volume to 2 ml or less by evaporating the contents on a water bath.

2. Treatment of the plant sample residue in 50 ml capacity digestion–distillation flask with the reduction mixture and distillation of the H_2S formed for absorption into zinc acetate–sodium acetale buffer solution.

Add 4 ml of the reducing mixture into 50 ml digestion–distillation flask containing plant sample residue. Immediately attach the reduction–distillation flask to the condenser assembly. Follow steps 2 and 3 of procedure (section 3.9.3.3) described in total S determination in plant sample by the reduction method of Johnson–Nishita (1952).

3. Development of the methylene blue colour, measurement of its absorbance (optical density), preparation of standard calibration curve and estimation of SO_4–S in 25 mg plant sample taken for S estimation.

Follow the steps 4 to 9 of procedure described in total S determination in plant sample by the reduction method (section 3.9.3.3)

3.9.5.4 Calculation

$$SO_4\text{–S in plant sample (ppm)} = \frac{A}{B} = \frac{A}{0.025}$$

where A is μg SO_4–S in 25 mg plant sample as estimated by reference to the standard calibration curve after developing methylene blue colour.

B is the mass of finely ground plant sample in gram.

3.9.5.5 Comments

Jones and Martin (1964) observed SO_4–S to be a more sensitive indicater of sulphur status of plant than total S concentration. There may exist linear relationship between SO_4–S and total S in plant. With the supply of S, there is relatively more increase in SO_4–S than in total S. There is also greater difference between SO_4–S values for deficient and non–deficient tissues.

The critical level of SO_4–S in legumes is reported to be 200 ppm with adequate nitrogen supply. However, SO_4–S concentration in alfalfa has been reported to be much higher. Under low N supply, the critical level of SO_4–S in the plant increases. Any deficiency affecting protein synthesis such as Mn or N deficiency will cause So_4–S to accumulate in the plant. The concentration of both total S and SO_4–S vary with the plant parts, and age. The temperature, insecticidal and fungicidal sprays also effect S concentration in plants. The critical level of S in plant tissues of different plant species is summarized by Eaton (1975).

REFERENCES

Eaton, F.M. (1975) Sulfur pp. 444–475. In. Chapman, H.D. (ed.) Dignostic criteria for plants and soils. Eurasia Publishing House (P) Ltd., Ram Nagar New Delhi.

Johnson, C.M. and H. Nishita (1952). Microestimation of sulfur in plant–material, soil and irrigation water. Anal. Chem. 24 : 736–742.

Jones, M.B. and W.E. Martin (1964). SO_4–S concentration as an indicater of S status in various California dryland pasture species. Soil Sci. Soc. Amer. Proc. Agro. J. 28 : 539–541.

3.10 IRON DETERMINATION IN PLANT DIGEST – SAYWELL AND CANNIGHAM (1937)

3.10.1 Principle

An aliquot of triacid digest of plant sample is evaporated to dryness. The residue is solubilized in small volume of redistilled water and evaporated to dryness. The treatment is repeated once. Then, the solution is treated with sodium acetate–acetic acid buffer, a reducing agent and finally with orthophenonthroline to form Fe^{++} orthophenanthroline red coloured complex, the intensity of which is proportional to the Fe concentration. The absorbance of the red coloured complex is measured at 490 $m\mu$ wave length or filter using a spectrophotometer or colorimeter. The Fe concentration in the plant digest is estimated by reference to the calibration curve prepared by plotting absorbance corresponding to Fe concentration on a graph paper.

3.10.2 Equipments/Reagents

Equipments

A spectrophotometer or a Klett. Summerson photoelectric colorimeter

Reagents

(*i*) 0.1N HCl. Dilute 10.0 ml of 5.0 N HCl to 500 ml with redistilled water. Use AR redistilled HCl.

(*ii*) Sodium acetate–acetic acid buffer. Dissolve 82.9 g of CH_3COONa (BDH–AR) in 150 ml of redistilled water and then make the volume to 1000 ml mark. with CH_3COOH (BDH–AR).

(*iii*) 10.0% $NH_2OH–NH_4Cl$. Dissolve 10.0 g of $NH_2OH.NH_4Cl$ (BDH–LR) in redistilled water and dilute the contents to 100 ml.

(*iv*) 1.5% orthophenanthroline solution. Transfer 1.5g of orthophenanthroline (Merck–Sarabhai) into 100 ml volumetric flask. Then add 50 ml of ethyl alcohol. Mix the contents by shaking the flask. Then dilute the solution with ethyl alcohol. (Bengal chemicals) to 100 ml mark.

(*v*) 100 ppm Fe solution. Transfer 0.7019 g of Fe $(NH_4)_2(SO_4)_2.6H_2O$ (BDH–AR) into 1000 ml volumetric flask. Add about 500 ml of redistilled water. Shake the flask to dissolve the reagents. Add 10.0 ml of conc. H_2SO_4(BDH–AR). Shake it to mix the contents, and allow it to cool. Then make the volume to 1000 ml mark with redistilled water. Shake to mix the contents and store in a polythene bottle.

(*vi*) 10 ppm Fe solution. Dilute 10.0 ml of 100 ppm Fe solution to 100 ml with redistilled water.

3.10.3 Procedure

1. Evaporation of aliquot of sample and blank digest to dryness. Transfer 2.0 ml of triacid plant and blank digest into separate 100 ml tall beakers. Keep the

beakers on a hot plate at low heat for evaporating the digests to dryness. Remove the beakers from the hot plate and let the beakers cool. Then add 3.0 ml of redistilled water to each beaker and evaporate the contents on a hot plate at low heat to dryness. Again add 3.0 ml of redistilled water in each beaker and evaporate the contents to complete dryness. Remove the beakers from hot plate and let the beakers cool to room temperature.

2. Solubilization of residue in dilute HCl. Add 10.0 ml of 0.1N HCl in each beaker and heat the beakers on a hot plate to warm the contents. Transfer the solution from each beaker into separate 50 ml volumetric flask. Rinse the beakers twice with 2.5 ml portions of 0.1N HCl and transfer both the rinsings into the corresponding 50 ml volumetric flask.

3. Development of Fe–Orthophenanthroline red coloured complex. Add to each 50 ml vol. flask containing digest - residue solubilized in 15.0 ml of 0.1N HCl, 5.0 ml of CH_3COONa–CH_3COOH buffer, 2.0 ml of 10.0% NH_2OH–NH_4Cl and 2.0 ml of 1.5% orthophenanthroline. Dilute the solution to 50 ml mark with redistilled water. Mix the contents and let it stand for 30 minutes.

4. Measurement of absorbance of Fe–orhophenanthroline coloured complex in aliquot of triacid plant sample and blank digest. Adjust zero absorbance with solution containing 15.0 ml of 0.1 N HCl, 5 ml of CH_3COONa-CH_3COOH buffer, 2 ml of 10% NH_2OH NH_4 Cl, 2 ml of 1.5% orthophenanthroline and redistilled water to dilute the contents to 50 ml at 490 mμ wave length or filter using a spectrophotometer or colorimeter. Then, measure the absorbance of Fe–Orthophenanthroline complex in aliquot of plant sample and blank digest. Operate the instruments as described in the estimation of P in triacid plant digest.

5. Preparation of standard graph for Fe. Transfer 0, 2, 4 and 6 ml of 10 ppm Fe solution into separate 50 ml vol. flasks . Then add 15 ml of 0.1 N HCl, 5.0 ml of CH_3 COONa–CH_3COOH buffer, 2.0 ml of 10.0% NH_2OH–NH_4Cl and 2.0 ml of 1.5% ortho phenanthroline to each flask. Dilute the contents to 50 ml with redistilled water. Mix the contents and let it stand for 30 minute. Measure the absorbance of the red coloured complex at 490 mμ wave length or filter adjusting zero absorbance with colour complex developed with 0 μg Fe. Note the absorbance corresponding to μg Fe added.

Plot the absorbance or Klett reading on the ordinate of linear scale graph paper and the corresponding Fe concentration in μg Fe/50 ml coloured complex i.e. 0 to 60 μg Fe on the abscissa. Connect the absorbance value recorded with 0 and 60 μg Fe/50 coloured complex to obtain a straight line calibration curve. All other points should fall nearly on the straight line.

6. Estimation of Fe in aliquot of triacid plant sample and blank digest. Note the absorbance of coloured complex developed from 2 ml aliquot of triacid plant sample and blank digese. Mark the recorded absorbance on 'Y' axis of the calibration graph. Then draw an imaginary line parallel to the abscissa and note the point at which it crosses the straight line calibration curve. From the point of inersection draw a vertical downward line perpendicular to the abscissa. Note the Fe concentration at the point vertical downward line intersect the abscissa.

Alternatively, estimate the slope of the calibration straight line curve i.e. absorbance/μg of added Fe. Then divide the recorded absorbance by the slope to obtain μg Fe present in 2 ml aliquot of triacid digest of sample and blank digest.

3.10.4 Calculation

ppm Fe in plant sample

$$= \frac{\mu g \text{ Fe in digest aliquot (Sample - blank)} \times \text{total volume (ml) of triacid digest}}{\text{aliquot taken for colour development in ml} \times \text{mass of plant sample taken for preparation of triacid digest in g}}$$

3.10.5 Comments

It is desirable to distill conc. HCl (BDH–AR) after adding equal volume of redistilled water and collecting distillate equal to 60% of the total volume of diluted acid :

The glass wares be rinsed first with a detergent and washed with tap water. Thereafter, it be rinsed with conc. HCl prior to final washing with tap water. The washed glasswares be rinsed twice with redistilled water.

Unless the leaf or plant material is washed in dilute acid or detergent, the results of chemical analysis can not be valid because of surface contaminants such as soil dust. Washing with water is not satisfactory.

In general, Fe concentration in recently matured leaves on dry mass basis may be interpreted as given below :

Fe concentration (ppm)	—	Interpretation
50 or less	—	deficiency likely to occur
50 to 250	—	deficiency unlikely to occur

The Fe concentration in the plant vary with the plant parts as well as with the age and position of the plant parts. Besides Fe concentration vary with the plant species. The plants in the early stage have higher concentration of Fe which decreases as the plant advance in age. The concentration of Fe in grains is lower than the Fe concentration in straw in Wheat, Maize and Bajra. The concentration of Fe in leaves of different crops rated to be above the deficient concentration are summarized below:

Fe concentration (ppm) in different crops above deficient level

Rice	Maize	Sunflower	Tobacco	Tomatoes	Peas	Lemon	Orange
>80	56–178	113	68–140	107–150	28–39	77	42–137

The Fe deficiency manifest in interveinal chlorosis particularly in immature to recently matured leaves. Fe deficiency can be confirmed by tissue analysis of chlorotic and healthy tissues of the same age and position or else by observing chlorophyll development in parts showing Fe chlorosis on foliar spray of 0.5% $FeSO_4$ solution with and without chelating agents such as EDTA, HEEDTA, DTPA, EDDHA. The veins remain green but there is lack of chlorophyll development in interveinal areas.

REFERENCES

Jackson, M.L. (1958). Soil Chemical Analysis. Prentice–Hall of India Private Limited, New Delhi. (1973).

Mortvedt, J.J. P.M. Giordano, and W.L. Lindsay (1972). Micronutrients in Agriculture. Soil. Sci. Soc. Amer. Inc., Madison, U.S.A.

Saywell and Canningham 1937. Ind. Eng. Chem. A.E., 9 : 67.

3.11 MANGANESE DETERMINATION IN PLANT DIGEST-WILLARD AND GREATHOUSE (1917)

3.11.1 Principle

An aliquot of triacid plant digest is evaporated to dryness. The residue is treated with conc. HNO_3 and 30% H_2O_2 and it is again evaporated to dryness. The residue is then solubilized in redistilled water and Mn free 85% H_3PO_4. Thereafter $K1O_4$ crystals is added and the solution is heated to develop $HMnO_4$ colour the intensity of which is proportional to the Mn concentration. The reaction in respect of development of the $HMnO_4$ colour is represented by the following equations :

(*i*) $15\ KIO_4 + 2Mn_3\ (PO_4)_2 + 9H_2O \rightarrow 15KIO_3 + 6HMnO_4 + 4H_3PO_4$

(*ii*) $2Mn^{++} + 5KIO_4 + 3H_2O \rightarrow 2MnO_4^{-} + 5KIO_3 + 6H^+$

The residue obtained from evaporation of triacid plant digest containing manganese is heated in the presence of nitric acid and hydrogen peroxide to ensure complete removal of chloride, bromide, sulphite, nitrite, ferrous Iron, oxalate and other oxidizable anions. In the presence of hydrogen peroxide, Manganese does not become insoluble as manganese dioxide.

The presence of H_3PO_4 in the medium eliminates interference by Fe^{2+}whereas the presence of excess periodate eliminates interference from small amounts of Cl. The concentration of Mn in the coloured complex is determined by the measurement of the absorbance of the $HMnO_4$ coloured solution at 540 mμ wavelength or filter using a spectrophotometer or a photoelectric colorimeter.

3.11.2 Equipments/Reagents

Equipments

Spectrophotometer or photoelectric colorimeter

Reagents

(*i*) 1000 ppm Mn solution. As per procedure described in Mn determination in soil extract section 2.23.2.

(*ii*) 50 ppm Mn solution. Dilute 25.0 ml of 1000 ppm Mn solution to 500 ml with redistilled water.

(*iii*) $18NH_2SO_4$. Transfer 49.0 ml of conc. H_2SO_4 (BDH–AR) into 100 ml volumetric flask containing 30 ml redistilled water. Cool the solution and then dilute it to 100 ml with redistilled water.

(*iv*) Na_2SO_3 (BDH–LR).

(*v*) Distilled conc. HNO_3. Distill conc. HNO_3 (BDH–AR) in a glass distillation unit and collect the distillate equal to 2/3rd of the initial volume of conc. HNO_3

(*vi*) 30% H_2O_2 (BDH–LR).

(*vii*) 85% H_3PO_4. Mn Free (Merck–GR).

(*viii*) KIO_4 crystals (BDH–AR).

3.11.3 Procedure

1. Evaporation of triacid plant digest to dryness and its subsequent treatment. Transfer 7.5 ml of triacid plant digest into 250 ml corning or pyrex beaker. Evaporate the digest to dryness on a hot plate at low heat. Cool the beaker and then add 3 ml of distilled conc. HNO_3 and 2 ml of 30% H_2O_2. Evaporate the contents to dryness on a hot plate at low heat. When the residue in the beaker appears to be completely dry, remove the beaker from the hot plate and let it cool.

After cooling the beaker, add 2 ml redistilled water and 5 ml of 85% H_3PO_4. Then heat the contents to boil. Remove the beaker from the hot plate and let it cool. Then add 10 ml redistilled water into the beaker. Swirl the beaker to mix the contents.

2. Development of $HMnO_4$ solution and dilution of the coloured solution in 50 ml vol. flask. Add 0.3 g of KIO_4 crystals into the beaker and place the beaker on a sand bath. Continue heating till the appearance of $HMnO_4$ colour in the entire solution. Remove the beaker from sand bath. After cooling, transfer the solution into 50 ml volumetric flask. Rinse the beaker twice with 5 ml portions of redistilled water and transfer the rinsings into the same 50 ml volumetric flasks. Dilute the contents to 50 ml with redistilled water. Stopper the volumetric flask and mix the solution thoroughly. Similarly run a blank by taking 7.5 ml of triacid blank digest into separate 250 ml beaker and following the entire procedure as described for the triacid plant digest.

3. Preparation of standard graph for Mn. Transfer 0, 2, 4 and 6 ml of 50 ppm Mn solution into separate 250 ml corning or pyrex beakers and evaporate the solution to dryness on a hot plate at low heat. There after proceed as described for the triacid plant digest omitting the initial step of taking 7.5 ml of triacid plant digest and evaporating it to dryness.

Adjust zero absorbance with 0 μg Mn/50 ml of $HMnO_4$ coloured solution at 540 mμ wave length or filter using a spectrophotometer or photoelectric colorimeter. Then, measure the absorbance of 100, 200 and 300 μg Mn/50 ml of $HMnO_4$ coloured solution. Operate the instrument as described in the estimation of P in the triacid plant digest section 3.4.3.

Note the value of absorbance corresponding to 0 to 300 μg Mn/50 ml of $HMnO_4$ coloured complex. Plot the value of absorbance on the ordinate of a linear scale graph paper corresponding to μg Mn/50 ml of coloured solution on the abscissa. Connect the absorbance value with 0 and 300 μg Mn/50 ml of $HMnO_4$ coloured complex to obtain a straight line calibration curve. All other points should fall nearly on the straight line.

4. Measurement of absorbance of $HMnO_4$ coloured complex in aliquot of triacid plant and blank digest. Adjust zero absorbance with 0 μg Mn/50 ml $HMnO_4$ coloured solution at 540 mμ wavelength or filter using a spectrophotometer or photoelectric colorimeter. Then measure the absorbance of HMO_4 colour complex developed in the aliquot of triacid plant and blank digest.

5. Estimation of Mn in aliquot of triacid plant and blank digest. Note the absorbance of $HMnO_4$ coloured complex in aliquot of triacid plant and blank digest. Mark the absorbance on the ordinate of standard graph for Mn, and from that point draw an imaginary line parallel to the abscissa. The point at which the imaginary line intersect the straight line calibration curve, draw a vertical downward line perpendicular to the abscissa, and note the concentration of Mn at the point of intersection. Alternatively estimate the slope of the straight line calibration curve i.e., absorbance/μg Mn, and then divide the absorbance of aliquot of triacid plant and blank digest after development of $HMnO_4$colour by the slope to obtain μg Mn in aliquot of triacid plant and blank digest.

3.11.4 Calculation

ppm Mn in plant

$$= \frac{\mu\text{g Mn in aliquot (Sample - Blank)} \times \text{total volume of triacid plant digest}}{\text{Volume of aliquot in ml} \times \text{mass of plant sample in g}}$$

3.11.5 Comments

Mn deficiency normally occurs when the plant tissue concentration is less than 20 ppm Mn in the dry matter. Mn concentration in the plant tissues in the range of 20 to 500 ppm are considered to be ample, and Mn concentration above 500 ppm are probably toxic.

At high concentration of Fe & Al (800–900 ppm), Mn deficiency in soyabean have been recorded even at Mn concentration greater than 20 ppm. In general Mn deficiency become more pronounced on increase in Fe concentration in plant tissues. However Mn toxicity is reduced with the increase in Fe concentration. The Mn requirement of plants decreases at high light intensity and consequently Mn toxicity may become accentuated at high light intensity whereas shaded plants may show Mn deficiency. Mn deficiency occur most often in small grains and large seeded legumes. Mn deficiency also occur in vegetables and fruit crops.

The availability of Mn in soil is related to soil reaction. The available Mn increase markedly when the pH falls below 5.5. Mn deficiency occur in neutral and calcareous soils. Lime induced Mn deficiency may be corrected by use of acid forming fertilizers or elemental S. Carriers containing water soluble Mn be applied by band placement to minimize conversion of soluble Mn into unavailable forms.

Deficiencies of Mn are often found in organic and mineral soils that are poorly drained. It is because of reduction of Mn to Mn^{2+} and its subsequent leaching. However, flooding temporarily correct Mn deficiency due to reduction of Mn to Mn^{2+}, and soil drainage decrease Mn uptake due to oxidation of Mn^{2+}. Mn deficiency is usually aggravated by drought conditions and at low soil temperature.

An increase in oxidizing microflora cause Mn disorder, whereas decrease in Mn oxidizing microflora resulting from use of fumigants may reduce or eliminate Mn deficiency.

Manganese deficiency in citrus consist of dark green bands along the midrib and main veins with lighter green areas between the bands. Young leaves often show a network of green veins on a lighter green background.

When the deficiency become moderate to acute, the dark green areas become dull green and the light green areas persist for the entire life of the leaf. In severe deficiency, the light green areas become even whitish. The foliage become thin, and the twigs tend to die back. Mn deficiency occur on calcareous soils.

REFERENCES

Willard, H.H. and L.H. Greathouse (1917). The colorimetric determination of manganese by oxidation with periodate. J. Amer. Chem. Soc. 39:2366–2397.

3.12 ZINC DETERMINATION IN THE PLANT SAMPLE DIGEST

3.12.1 Principle

Same as described for Zn determination in soil extract (section 2.24.1.1).

3.12.2 Apparatus & Reagents

Apparatus & Equipments. Pear shaped separatory funnel–250 ml capacity, volumetric flasks, separatory funnel stand, mechanical shaker, colorimeter with 535 nm filter or spectrophotometer.

Reagents

(*i*) Zn free distilled water.

(*ii*) Approx. 6 N HCl.

(*iii*) 1.0 N HCl.

(*iv*) 0.02 N HCl.

(*v*) 1N NH_4OH.

(*vi*) Carbon tetrachloride (AR).

(*vii*) 0.1% Phenolphthalein indicator.

(*viii*) Thymol blue indicator.

(*ix*) 0.2% carbamate solution.

(*x*) 0.01% Dithizone in CCl_4.

(*xi*) 11.25% Ammonium citrate solution.

(*xii*) 100 ppm Zn solution.

(*xiii*) 2 ppm Zn solution.

See section 2.24.1.2 for the preparation of above mentioned reagents.

3.12.3 Procedure

1. Preparation of the plant sample test solution. Transfer 5g of oven dry finely grinded plant sample into a 300 ml tall beaker, and prepare 250 ml of the plant sample test solution by digestion with ternary acid mixture after predigestion with conc. HNO_3. Follow the procedure described in section 3.2.1. Simultaneously prepare the blank test solution following exactly the same procedure without taking the plant sample.

2. Processing of the standard Zn solutions and the plant sample test solutions. Take a suitable aliquot of Zn standard solution (usually 0 to 10 μg Zn) or else the plant sample test solution (usually 5 ml) in separate 250 ml capacity pear shaped separatory funnel and process it further following under given four steps in sequence.

(*i*) Removal of Cu present as impurity in standard Zn solution or else in the plant sample test solution.

(*ii*) Removal of other impurities in the metal dithizonates.

(*iii*) Extraction of Zn in the presence of carbamate.

(*iv*) Removal of Zn dithizonates, and measurement of its optical density.

The detailed procedure of the above four steps are described in section 2.24.2.3

3. Preparation of the standard graph for Zn. Note the absorbance i.e. optical density of the Zn–dithizonates (from 0–10 μg Zn) in CCl_4 phase at 535 mμ (nm) wave length. Then, plot the absorbance on a 'y' axis i.e. ordinate against the corresponding amount of Zn in μg on 'x' axis i.e. abscissa of a linear scale graph paper. Then , draw a straight line by connecting the absorbance value of pure CCl_4 i.e. 0 μg Zn with the absorbance value obtained with the maximum amount of Zn taken i.e. 10 μg Zn.

4. Estimation of Zn concentration in the plant sample test solution. Note the absorbance i.e. optical density of Zn–dithizonates in CCl_4 phase extracted from the plant sample and blank test solution. Then, refer to the standard graph for Zn.

Locate the absorbance recorded on the 'y' axis i.e. ordinate, and draw a imaginary line parallel to the 'x' axis i.e., abscissa intersecting the straight line calibration curve for Zn. Draw a perpendicular from the point of intersection of the calibration curve to the abscissa, and read the Zn concentration given on the abscissa at the point of origin of the perpendicular.

3.12.4 Calculation

$$\text{Zn in plant sample (ppm)} = \frac{A \times C \times E}{B \times D \times F}$$

Where A = μg Zn in aliquot of Zn–dithizonates (sample-blank) by reference to the standard graph for Zn.

B = ml aliquot taken from CCl_4 phase containing Zn–dithizonates for determining the absorbance of the complex.

C= ml of dithizone reagent used for extracting Zn in the presence of carbamate reagent.

D = Volume of triacid plant digest taken for extracting Zn in the aqueous phase.

E = Volume of total triacid plant digest prepared from the plant material.

F = mass of plant material in g used in preparation of 250 ml of triacid plant digest.

3.12.5 Comments

The plant digest for Zn determination is also prepared by dry ashing of finely ground plant material in a crucible (platinum or well glazed porcelain) at 560–550°C in an electric furnace, and subsequently solubilized in 1N HCl by heating on a steam bath.

The deficiency of Zn occurs in many plants when the leaf concentration is less than 20 ppm Zn in the dry matter. Normal concentration of Zn in plant range from 25–150 ppm. Toxicity of Zn occur when Zn concentration in leaf exceed 400 ppm.

Zn deficiency in paddy plant manifest in bronzing of leaf or else in midrib bleaching and bronzing.

Zn defincinecy in crops is not common on acid soils. Zn content is often higher in top soils. Removing or burrying top soil by land levelling can result in Zn deficiency.

Most of the Zn disorders occur in calcareous solis as Zn is fixed on surface of $CaCO_3$. Besides, the solubility product of Zn soil complexes and carbonates is low (10^{-16} to 10^{-18}), Zn deficiency is accentuated by liming an acid soil.

Zn deficiency becomes more pronounced in cool weather due to decrease in solubility of soil Zn. The Zn deficiency is also induced by high P levels. Cool, wet and cloudy weather result in greater Zn disorder.

There is need for Zn –free reagents as well as glass wares. The glasswares be first weathered with HNO_3 and then after thorough washing, it should be rinsed with Zn–free distilled water. The reagents such as HCl, NH_4OH, CCl_4 and distill water be redistilled in a pyrex–glass distill.

REFERENCES

Holmes, R.S. 1945. Determination of total copper, zinc, cobalt, lead in soils and soil solution. Soil Sci. 59 : 77–84.

Mortvedt, J.J., P.M. Giordano, W.L. Lindsay (ed). 1972. Micronutrients in Agriculture. Soil Sci. Soc. Amer., Inc. Madison, Wisconsin, U.S.A.

3.13 DETERMINATION OF COPPER IN PLANT TRIACID DIGEST-CHENG AND BRAY (1953)

3.13.1 Principle

The triacid digest of plant material containing Cu is reacted with sodium diethyl dithiocarbamate in the presence of versenate–citrate mixture in alkaline medium, pH adjusted to phenolphthalein end point. The Cu–carbamate complex formed is extracted with pure CCl_4. The optical density of the Cu–carbamate complex is measured by a photoelectric colorimeter or else a spectrophotometer at 440 nm wave length.

The Cu content is estimated by reference to a standard graph for Cu prepared on determining optical density of Cu carbamate complex formed with 0–15 microgram Cu. The interference from Ca, Mg, Fe, Mn and Ni is removed by inclusion of versenate–citrate mixture. It does not, however, eliminate interference from bismuth which is ordinarily present in negligible amounts in plant digest.

3.13.2 Apparatus/Reagents

Apparatus

Separatory funnel-250 ml capacity, pipette 1, 2, 5 and 10 ml capacity, volumetric flask-100 and 1000 ml capacity, 25 ml graduated cylinder, photoelectric colorimeter with 440 millimicron filter or else a spectrophotometer.

Reagents

(*i*) Redistilled water using glass distill.

(*ii*) 1.0% carbamate solution.

(*iii*) Versenate–citrate mixture.

(*iv*) Conc. ammonium hydroxide solution.

(*v*) Carbon tetrachloride.

(*vi*) 100 ppm Cu solution.

(*vii*) 1 ppm Cu solution.

(*viii*) 0.1% phenolphthalein indicator.

The preparation of reagents are described in Cu determination in soil (see section 2.25.1.2).

3.13.3 Procedure

1. Preparation of the plant sample test solution. Transfer 5g of oven dry finely ground plant sample into a 300 ml tall beaker. Digest the plant sample with ternary acid mixture after predigestion with conc. HNO_3. Follow the procedure of wet ashing as described in section 3.2.1. Also prepare the blank test solution following the entire procedure without the plant sample.

2. Processing of the standard Cu solution, and the plant sample test solution. Transfer 0, 5, 10 & 15 μg of Cu or else 25 ml of plant sample test solution into separate 250 ml capacity separatory funnel (pear shaped). Add 10 ml of versenate-citrate mixture into each separatory funnel, and adjust the pH of the solution to the phenolphthalein end point by adding about 2 ml of NH_4OH (sp. gr. 0.91 AR). Shake to mix the contents. Thereafter, add 2 ml of 1.0% carbamate reagent, and mix the contents well to form Cu–carbamate complex.

3. Extraction of Cu–carbamate complex. Add 10 ml of pure redistilled CCl_4 into each separatory funnel, and shake the contents for 3–4 minutes vigourously. Place the separatory funnel on a stand, and allow the layers to separate. Collect the CCl_4 phase *i.e.* lower layer containing Cu–carbamate in a dry 25 ml graduated cylinder. Now add 5 ml of CCl_4 into separatory funnel. Shake the contents of separatory funnel for one minute. Then, allow the layers to separate. Collect the lower CCl_4 phase containing Cu-carbamate into the same 25 ml graduated cylinder. Make the volume of Cu-carbamate complex in the graduated cylinder to 15 ml by adding distilled CCl_4.

4. Measurement of absorbance. Measure the absorbance (optical density) of the Cu–carbamate complex after 15 minutes in a photoelectric colorimeter using green filter or else in a spectrophotometer at 500 nm (mμ) wave length adjusting 0 optical denstiy with pure CCl_4.

5. Standard graph for Cu. Record the optical density (absorbance) of Cu–carbamate complex in CCl_4obtained with 0–15 μg Cu deducting the blank reading i.e. optical density of Cu–carbamate complex with 0 μg Cu. Thereafter, plot the optical

density on 'y' axis of a linear scale graph paper against Cu in $\mu g/15$ ml of Cu–carbamate complex in CCl_4 plotted on the 'x' axis. Connect zero optical density adjusted with pure $CCl_4$and the optical density of Cu carbamate complex corresponding to 15 μg Cu to obtain a straight line curve.

6. Estimation of Cu in Cu–carbamate complex. Note the absorbance (optical density) of Cu–carbamate complex and refer to the standard graph for Cu. Locate the value of absorbance on the 'y' axis i.e. ordinate of the linear scale graph paper. Draw an emaginary line from that point parallel to the base which will intersect the calibration curve. From the point of intersection, draw an imaginary line perpendicular to the base. Note the content of Cu in $\mu g/15$ ml of Cu–carbamate complex given on the 'x' axis.

3.13.4 Calculation

$$\text{Cu in plant sample (ppm)} = \frac{(S - B) \times E}{A \times M}$$

Where S and B are microgram Cu/15 ml of Cu–carbamate complex in CCl_4 determined by reference to the standard graph for Cu after measurement of the optical density in sample and blank test solution respectively.

E is the total volume of triacid plant digest prepared.

A is the ml aliquot of the triacid plant digest taken for the Cu determination.

M is the mass of ground plant material taken for the preparation of triacid plant digest.

3.13.5 Comments

The normal range of Cu–concentration in plant tissues vary from 5 to 20 ppm. The deficiencies are likely to occur at Cu concentration less than 4 ppm, and toxicities may occur when Cu concentration exceed 20 ppm.

The maize crop grown on peat or muck soil may become Cu deficient at Cu concentration of 4 to 5 ppm, but when grown on mineral soil Cu deficiency may not be observed at the same level of Cu concentration. The Cu concentration in the plant decreases with the initial growth but later the Cu concentration in plant tissues increases and remains fairly constant till maturity. The Cu concentration in leaf is higher than petiole and stem.

The Cu deficiency in Cu responsive crops grown on organic soils occurred at pH below 4.5 and above 7.0. The excessive uptake of other elements under acidic conditions can cause Cu–deficiency. The deficiency of Cu also become accentuated at higher level of N, P and Zn.

The soils inherently low in total Cu respond to Cu application. The sufficiency of Cu is indicated by the total Cu in soil. For the Cu responsive crops, the total Cu should exceed 4 to 6 ppm for mineral soil, and 20–30 ppm for organic soil. Since leaching of Cu is negligible, a single liberal soil application can supply all of the plant requirement for many years.

For the determination of Cu status of the plant, the plant material should be carefully washed with water as well as with dilute detergent solution such as 0.01%

teepol because of common use of Cu bound materials as foliar spray to control insects and diseases. This would also remove the dust particles adhering to the surface of leaves and other plant parts.

In case of moderate to acute deficiency of copper, the terminal growth is affected, the twigs and growing point tend to die back. The rosetting of terminal leaves precedes die back.

REFERENCES

Cheng K.L., and Bray, R.H. (1953). Two specific methods of determining copper in soil and plant material. Anal. Chem, 25 : 655–665.

Jackson, M.L. (1958). Soil Chemical Analysis. Prentice Hall of India. Pvt. Ltd. New Delhi, India.

Mortvedt, J.J., P.M. Giordano, and W.L. Lindsay (ed) (1972). Micronutrients in Agriculture. Soil Sci. Soc. Amer., Inc., Madison, Wisconsin, USA.

3.14 BORON DETERMINATION IN PLANT SAMPLE

3.14.1 Boron determination in plant sample test solution by formation of boron–curcumin complex–Dible *et al.* 1954

3.14.1.1 Principle

The finely ground plant sample is wetted with saturated $Ca(OH)_2$ solution and is then evaporated to dryness. It is initially ignited on a gas burner and then placed in a muffle furnace at 550°C. The residue containing B is treated with measured volume of 0.1N HCl. Thereafter, the suspension is centrifuged to obtain a clear test solution. The test solution is reacted with curcumin to form a coloured complex for the colorimetric determination of boron.

3.14.1.2 Apparatus and reagents

Apparatus and Equipments. Pipettes, volumetric flasks, Beakers, Oven, Water bath with thermostat, photoelectric colorimeter with 540 mμ and 580–600 mμ filter or spectrophotometer.

Use glass wares made from B–free glass. Weather glass wares with conc. H_2SO_4 rinse glass wares with conc. HCl, wash it with tap water and rinse with distilled water from a metal distill.

Reagents

(*i*) Distilled water. Distill water using a metal distill.

(*ii*) Ethanol–95% (AR) Distill to make it free from B.

(*iii*) Curcumine–oxalic acid solution. Dissolve 0.04g of finely ground curcumin (Eastman Kodak No. 1179) and 5.0g of $H_2C_2O_4 . 2H_2O$ in 100 ml of boron free 95% ethanol. Store the solution in amber coloured bottle in a cool dark place or else it may be stored in a refrigerator for not more than a week as the reagent decomposes slowly.

(*iv*) 100 ppm B solution. Dissolve 0.5717 g of H_3BO_3(AR) in distilled water and make the volume to 100 ml with distilled water :

(*v*) 50 ppm B solution. Dilute 50 ml of 100 ppm B solution to 100 ml with distilled water

(*vi*) 0–5 ppm B solution. Transfer 0, 1, 2, 3, 4 and 5 ml of 50 ppm B solution into separate 50 ml volumetric flasks. Then add 5ml of 0.5 N HCl into each flask and make the volume to 50 ml with distilled water.

3.14.1.3 Procedure

1. Preparation of sample and blank test solution. Prepare plant sample and blank test solution for B determination by dry ashing procedure described in section 3.2.8.

2. Formation of boron-curcumin complex with–0–5 μg B or else with plant sample and blank test solution. Transfer 1.0 ml of B standards (0–5 ppm) or else sample and plant test solution into separate 250 ml beaker using the same 1 ml pipette. Then add 4 ml of curcumin–oxalic acid solution. Mix the contents by rotating the beaker. Then, evaporate the solution to dryness in an oven (or on a water bath) regulated at 55 ± 3°C. Thereafter, bake the residue for 15 minutes at 55 ± 30°C to ensure dryness. During evaporation and drying there is development of the coloured substance 'rosecyanine'. Then cool the beaker containing the dried residue and extract the colour with 25 ml of 95% ethyl alcohol. Triturate the contents and filter the solution through Whatman No.2 filter paper directly into the colorimeter tube. The slight contamination of the filter paper with boron is of no significance as the colour forming step is past.

3. Measurement of the absorbance of the boron–curcumin coloured complex. The absorbance of the coloured complex be taken within 2 hours of the development of boron-curcumin complex since rosecyanine gradually hydrolyzes to curcumin.

Set the colorimeter with 540 mμ filter or else set the spectrophotometer at 540 nm wave length. Place the colorimeter tube containing coloured complex from 0 μg B into the socket provided and let the instrument warm for 10 minutes. Then adjust the absorbance to zero.

There after take the absorbance of the subsequent standards and record the absorbance corresponding to the amount of added B in μg. As the absorbance reaches 0.60 (i.e. klett reading 300 or transmittance 25%) B content reaches over 1.5 μg. At this stage replace 540 mμ filter by 580–600 mμ filter, and read the absorbance.

4. Standard graph for boron by the curcumin method. Record the absorbance (optical density) of boron–curcumin coloured complex prepared with 0 to < 1.5 μg B and with 1.5 μg B and more. Then plot the absorbance values of the two ranges separately on a linear scale graph paper. Plot the absorbance on the ordinate and the corresponding μg of B taken for the formation of boron–curcumin complex on the abscissa.

Connect the lowest and the maximum absorbance of each range by a straight line to obtain the standard curve for boron.

5. Estimation of B in boron–curcumin complex prepared from sample test solution. Refer to the appropriate standard curve for boron on the basis of wave

length of the filter used in determining the absorbance of the coloured complex obtained with sample test solution.

Locate the absorbance value on the ordinate, and draw an imaginary line parallel to the base i.e. abscissa intersecting the standard curve. From the point of intersection, draw a line perpendicular to the abscissa. Note the boron content in μg given at the origin on the abscissa.

3.14.1.4 Calculation

$$\text{ppm B in plant sample} = \frac{A \times B}{C \times D}$$

where A is μg B in aliquot of (sample-blank) test solution as determined by reference to the standard curve for boron.

B is the total volume of 0.1 N HCl used for dissolving boron in plant sample ash after ignition in muffle furnace.

C is ml aliquot of sample test solution taken for developing boron-curcumin complex

D is the mass of finely ground oven dry plant sample taken for preparation of sample test solution after dry ashing.

3.14.2 Boron determination in plant sample test solution by formation of boron–quinalizarin coloured complex in H_2SO_4 medium

3.14.2.1 Principle

The finely ground plant sample wetted by saturated $Ca(OH)_2$ solution is dried at 55°C. Then, the alkaline sample is initially ignited on a gas burner before placing in a muffle furnace at 550°C.

The residue containing B is dissolved in measured volume of 0.36N H_2SO_4. The suspension is filtered or centrifuged to obtain a clear test solution. An aliquot of the plant test solution is reacted with quinalzarin in sulphuric acid medium to form a coloured complex for the colorimetric determination of boron.

3.14.2.2 Apparatus and Reagents

Apparatus and equipments

Boron free glasswares such as pipettes, flasks, beakers, test tubes, dispenser for quinalzarin in H_2SO_4 medium, Oven, water bath with thermostat, gas burner, muffle furnace, metal distill, centrifuge, photoelectric colorimeter with 620 mμ filter or a spectrophotometer.

Reagents

(*i*) Boron free distilled water from metal distill.

(*ii*) Saturated $Ca(OH)_2$ solution.

(*iii*) 0.36 N H_2SO_4.

(*iv*) 50 ppm Boron.

(*v*) 0–5 ppm Boron.

(*vi*) Sulphuric acid 98%. Mix fuming and conc. H_2SO_4 in such a way as to make 98% H_2SO_4. Determine its strength volumetrically by titration against sodium hydroxide of known normality.

(*vii*) Quinalizarin–sulphuric acid solution. Dissolve 25 mg of quinalizarin in 1000 ml of 98% H_2SO_4, and store the reagent in boron free glass bottle under a dry atmosphere. Incoming air be passed through an anhydrous $CaCl_2$ solution. See section 3.14.1.2 for the preparation of reagents *i*, *iv* & *v*.

3.14.2.3 Procedure

1. Preparation of the plant sample and the blank test solution. Prepare the plant sample test solution for B determination by dry ashing following the procedure given in section 3.2.8.

2. Formation of boron–quinalizarin complex with 0–5 μg B or else with plant sample and blank test solution. Transfer 1.0 ml of B standards (0–5 ppm solutions) or else sample and blank test solutions into separate test tubes using the same 1 ml pipette. Then add exactly 10.0 ml of quinalzarin-sulphuric acid solution into each test tubes. Stopper the test tubes, mix the contents thoroughly and cool the solution to 25°C or else to the room temperature. Take care that the reaction mixture do not touch the rubber stopper while mixing the contents.

3. Measurement of absorbance of boron–quinalizarin complex. Place 620 nm filter in the filter socket of the photoelectric colorimeter or set 620 nm wave length in a spectrophotometer. Place cuvette filled with distilled water in the socket provided. Switch on the instrument and let it warm for 10 minutes before taking measurement on absorbance of the coloured complex.

Set 0 absorbance with the coloured complex obtained with 0 μg B, and then determine absorbance of coloured complexes obtained with the subsequent boron standards. Test the absorbance of coloured complex obtained with sample test solution after setting 0 absorbance with the coloured complex obtained with the blank test solution.

4. Preparation of standard calibration curve for boron–quinalizarine coloured complexes. Record the values of absorbance of boron–quinalizarin coloured complexes corresponding to 0–5 μg of B taken for the reaction. On a linear scale graph paper, set absorbance scale on the ordinate and μg B taken for formation of the coloured complex on the abscissa. Then plot the absorbance values corresponding to μg B taken for the formation of coloured complex.Connect the point of 0 and maximum absorbance by a straight line to obtain the standard calibration curve for boron–quinalizarin coloured complexes.

5. Estimation of B in boron–coloured complex of sample test solution. Locate the observed absorbance of coloured complex with aliquot of the sample test solution on the ordinate of standard calibration curve for boron. Then, draw an imaginary line parallel to the abscissa intersecting the straight line curve. From the point of intersection, draw a line perpendicular to the abscissa and note the value of μg B at the origin.

3.14.2.4 Calculation

$$\text{ppm B in plant sample} = \frac{A \times B}{C \times D}$$

where A is μg B in aliquot of (sample-blank) test solution as determined by reference to the standard curve for boron by use of quinalizarin in sulphuric acid.

B is the total volume of 0.36 N H_2SO_4 used for dissolving boron in plant sample ash after ignition in muffle furnace.

C is ml aliquot of sample test solution taken for developing boron-quinalizarin complex.

D is the mass of finely ground oven dry plant sample taken for the preparation of sample test solution after dry ashing

3.14.3 Boron determination in plant sample test solution by formation of boron–carmine coloured complex–Hatcher and Wilcox (1950)

3.14.3.1 Principle

The plant sample test solution is prepared by dry ashing procedure described in section 3.2.8. Then a 2 ml aliquot of the sample test solution is treated with 2 drops of HCl, 10 ml of conc. $H_2SO_4$and 10 ml of 0.92% carmine to form boron–carmine coloured complex for the colorimethric determination of boron.

3.14.3.2 Apparatus & Reagents

Apparatus and Equipments

Pipettes, flasks, beakers etc. made from boron–free ordinary soft glass. Hot plate, water bath, gas burner, muffle furnace, photoelectric colorimeter or a spectrophotometer.

Before use, glass wares be weathered with conc. H_2SO_4 and subsequently it should be rinsed with conc. HCl, and thoroughly washed with tap water and rinsed with distilled water from a metal distill.

Reagents

(*i*) Boron free distilled water.

(*ii*) Conc. H_2SO_4 (AR).

(*iii*) Conc. HCl (AR).

(*iv*) 50 ppm B solution.

(*v*) 0–7 ppm B solution.

(*vi*) 0.092% carmine solution. Dissolve 0.092 g of carmine in 100 ml of conc. H_2SO_4

The preparation of reagent (*iv*) and (*v*) are described in section 3.14.1.2.

3.14.3.3 Procedure

1. Preparation of the plant sample and the blank test solution. Prepare the plant sample test solution for B determination by dry ashing following the procedure given in section 3.2.8.

2. Formation of boron–carmine coloured complex with B standards or else with sample & blank test solutions. Transfer 2 ml of dilute B standards (0–7 ppm B solutions) or else 2 ml of sample and blank test solutions into separate Erlenmeyer flasks or plastic bottles. Then add 2 drops of conc. HCl, 10 ml of conc. H_2SO_4 and 10 ml of 0.092% carmine solution to each Erlenmeyer flask or plastic bottle. Shake the flasks or bottle to mix the contents and then let it stand for 45 minutes for full development of the boron–carmine coloured complex.

3. Measurement of absorbance of the boron–carmine coloured complexes. Fit 585 or 655 mμ filter into socket provided and place it against the tungsten lamp in the photoelectric colorimeter. In spectrophotometer, adjust the wave length to 585–655 nm. Fill the cuvette i.e. colorimeter test tubes with the coloured complex formed with 0 μg B and then place it in the appropriate socket. Now switch on the instrument, and allow it to warm for 10 minutes. Adjust the absorbance to zero. Thereafter, measure the absorbance of subsequent standards. Record the absorbance against the corresponding μg B taken for the formation of boron-carmine coloured complex. Test the absorbance of coloured complex obtained with the sample test solution after setting 0 absorbance with the coloured complex obtained with the blank test solution.

4. Standard calibration curve for boron-carmine coloured complexes. Record the values of absorbance corresponding to μg B reacted with carmine to form coloured complexes for the colorimetric determination. Set the absorbance scale on the ordinate of a linear scale graph paper and set the scale for the amount of boron taken in μg on the abscissa. Plot the values of absorbance recorded corresponding to the amount of boron taken in μg and then connect point of 0 absorbance with the point of maximum absorbance by a straight line to obtain the standard calibration curve.

5. Estimation of boron content in boron–carmine coloured complex of sample test solution. Locate the observed absorbance of coloured complex with aliquot of the sample test solution on the ordinate of standard graph for boron. From the point draw a line parallel to the abscissa intersecting the straight line curve. From the point of intersection, draw a line perpendicular to the abscissa and read the value of μg B on the abscissa at the base of perpendicular.

3.14.3.4 Calculation

$$\text{ppm B in plant sample} = \frac{A \times B}{C \times D}$$

where A is μg B in aliquot of (sample-blank) test solution as determined by reference to the standard calibration curve for boron by use of carmine.

B is the total volume of 0.36N H_2SO_4 used for dissolving boron in plant sample ash after ignition in muffle furnace.

C is ml aliquot of sample test solution taken for developing boron– carmine complex.

D is the mass of finely ground oven dry plant sample taken for the preparation of sample test solution.

3.14.3.5 Comments

There is loss of boron during acid digestion of plant sample. Dry ashing of plant sample with $Ca(OH)_2$ at 550°C in a muffle furnace for one hour is found to be satisfactory for the extraction of total boron.

Boron free glasswares such as kavalier, corning 728, and glass wares made from soft glass are required in the determination of boron in soil and plant. Pyrex & borosilicates glass wares contain significant amount of boron. Such glass wares must not be used in boron determination. The glasswares must be weathered in conc. H_2SO_4 and washed with conc. HCl before use.

Water distilled by a metal distill and demineralized water by an ion exchange resins are statisfactory. The use of analytical reagents is also necessary for boron determination. The use of same pipette for taking aliquot of boron standards and sample test solution is considered necessary.

The deficiency of boron in a variety of plants occurs when boron concentration in the dry matter is less than 15 ppm. Adequate level of boron in plant vary from 20 to 100 ppm. Boron toxicity occurs normally at concentration greater than 200 ppm. In case of boron sensitive plants, toxicity may occur at boron concentration less than 200 ppm. The plants with low calcium supply have low toterance for boron whereas plants with excessive supply of calcium have a high requirement of boron. The availability of boron in soil to sunflower plants is better indicated by analysis of stems and leaves of sunflower rather than by the analysis of soil in which it was grown.

The boron concentration in plants vary with the plant species. Monocots contain less boron than dicots. The members of the Papilionaceae and Cruciferae are high in boron concentration. The boron concentration vary with the plant parts as well as with the age. The boron concentration in the stalks is lower than the leaf. Upper leaves have higher concentration of boron than lower leaves. The boron tend to accumulate in leaf margins. Petioles of sugar beet contain less boron than leaf blade.

The indicator plants for boron deficiency are table beet, turnip, celery, cauliflower, cabbage amongst vegetables, alfalfa, red and white clover among fodder crops, apples and pears amongst fruit trees. The external symptoms of boron deficiency are :

1. Terminal growth roseting, dieback, discolouration, failure to grow or elongate.

2. Leaves show abnormalities such as thickening, brittleness, curling, wrinkling, wilting and chlorotic spotting.

3. Petioles or stems may be thickened, corky, cracked, or may show water soaked dead areas.

4. Fruits, tubers or roots, the flesh part may show brown flecks, necrosis, cracks or dry rot.

Boron deficiency is most common in soils naturally low in boron such as those derived from acidic igneous rocks, naturally acid soils from which much of the original boron content has been removed by leaching, light textured soils, acid, peat and muck soils, alkaline soils especially those contaning free lime, soils irrigated with water low in boron, soils having salt or carbonates deposits and soils low in organic matter.

Boron toxicities is likely to occur on arid soils, and on soils derived either from parent material rich in boron minerals or from marine deposits high in boron.

REFERENCES

Berger, K.C., and E. Truog (1939). Boron determination in soil and plants. Ind. Eng. Chem. Anal. Ed. 11 : 540–545.

Bradford, G.R. (1975) Boron. In. Chapman, H.D. (ed.) Diagnostic Criteria.

For Plants and Soils. Eurasia Publishing House (p) Ltd. Ram Nager, New Delhi.

Dible et. al. 1954.Anal. Chem. 26 : 418.

Hatcher, J.J. and L.V. Wilcox (1950) Colorimetric determination of boron using

Carmine. Anal. Chem. 22 : 250–252.

Jackson, M.L. (1973). Soil Chemical Analysis. Prentice Hall of India Pvt. Ltd. New Delhi.

Jones, Jr., J.D. (1972). Plant Tissue Analysis for micronutrients. In. Mortvedt, J.J. et. al. Micronutrients in Agriculture. Soil Sci. Soc. Amer. Inc. Madison, Wisconsin, U.S.A.

3.15 DETERMINATION OF MOLYBDENUM IN PLANT SAMPLE BY THE METHOD OF PURVIS AND PETERSON (1956)

3.15.1 Principle

The molybdenum react with thiocyanate in hydrochloric acid solution in the presence of a reducing agent to form amber or orange red colour. The colour compound is a thiocyanate complex of quinquivalent molybdenum which can be extracted in ether or other suitable solvents such as isoamyl alcohol or isoamyl alcohol–carbon tetrachloride mixture.

The reducing agent $SnCl_2$ reduces the Mo to the quinquivalent form, the valency required in an molybdenum–thiocyanate complex. The reducing agent reduces also ferric iron and eliminates colour interference by the red ferric thiocyanate complex. The presence of ferric and nitrate ions result in oxidation of Mo present in the acid solution in reduced form. The presence of these ions may also prevent reduction of Mo below the quinquivalent form after addition of a reducing agent.

The test solution is prepared by digestion of the plant sample initially with conc. HNO_3, 70% $HClO_4$ and subsequently with 30% H_2O_2 and dissolved in HCl solution. Then, the solution is treated with $FeCl_3$, $NaNO_3$, NH_4SCN, and $SnCl_2$ in a separatory funnel to form Mo–thiocyanate complex which is extracted by isopropyl ether and collected in a glass stoppered container.

The absorbance of the coloured complex is measured at 475 mμ in a standard photoelectric colorimeter. The absorbance of the coloured complex for the blank determination is substracted. The Mo content of the test solution is determined by reference to the standard curve prepared by taking absorbance of the coloured complex similarly prepared with 0–20 μg of Mo.

3.15.2 Apparatus/Reagents

Apparatus

200 ml tall beaker, hot plate, 100 ml volumetric flask, pipette 1, 2, 5 ml & 10 ml capacity, separatory funnel 150 ml capacity, separatory funnel stand, glass stoppered container, spectrophotometer or a standard photoelectric colorimeter.

Reagents

(*i*) Redistilled water. Distill water from a metal distill by completely glass distill.

(*ii*) Conc. Nitric acid. Distill conc. HNO_3 (BDH–AR) by completely glass distillation apparatus

(*iii*) 70% Perchloric acid (BDH–AR).

(*iv*) 30% hydrogen peroxide (AR or LR).

(*v*) 10% stannous chloride. Dissolve 10g. Sn Cl_2 in 10 ml of conc. HCl, heat below boiling point and make the volume to 100 ml with redistilled water. Prepare fresh reagent every day.

(*vi*) 10% Ammonium thiocyanate. Dissolve 10g of ammonium thiocyanate in 100 ml of redistilled water.

(*vii*) 4.9% Ferric chloride solution. Dissove 4.9 g. of $FeCl_3$. $6H_2O$ in 100 ml of redistilled water.

(*viii*) 42.5% sodium nitrate solution. Dissolve 42.5g of $NaNO_3$ in redistilled water and make the volume to 100 ml.

(*ix*) Approx. 6N Hydrochloric acid. Mix conc. HCl (BDH AR) with equal volume of distilled water and then distill the diluted HCl by completely glass distillation apparatus .

(*x*) Isopropyl Ether. Place reagent grade isopropyl ether in a glass stoppered separatory funnel, and add one tenth its volume of a mixture containing one third each of 10% $SnCl_2$ and 10% NH_4SCN solutions and one third distilled water. Shake thoroughly and separate. Prepare fresh daily.

(*xi*) 100 ppm molybdenum solution. Dissolve 0.1840 g. of $(NH_4)_6$ Mo_7O_{24}. $4H_2O$ in 900 ml of redistilled water, and dilute the solution to 1000 ml.

(*xii*) 2 ppm Molybdenum solution. Dilute 2.0 ml of 100 ppm Mo solution to 100 ml with redistilled water.

3.15.3 Procedure

1. Formation of Mo–thiocyanate amber to orange red coloured complex with 0–20 μg Mo. Transfer 20 ml of redistilled approx. 6N HCl into a 100 ml volumetric

flask and dilute the acid to 100 ml with redistilled water. Then, transfer 10 ml of the diluted acid into 150 ml glass stoppered separatory funnel. Add 0 to 10 ml of 2 ppm Mo solution, 1 ml of 4.9% $FeCl_3$ solution, 1 ml of 42.5% $NaNO_3$ solution, 5 ml of 10% NH_4SCN solution and 5 ml of 10% $SnCl_2$ solution. Shake thoroughly after each addition.

2. Extraction of Mo–thiocyanate complex by isopropyl ether. Add 10 ml of isopropyl ether into the separatory funnel containing Mo–thiocyanate complex and shake by hand 100 times. Place the separatory funnel on separatory funnel stand and allow ether to separate. Draw off aqueous phase, and deliver ether into a glass stoppered container.

3. Measurement of absorbance (optical density) of Mo–thiocyanate complex extracted by isopropyl ether. Switch on the photoelectric colorimeter fitted with 475 mu filter or else a spectrophotometer set at 475 mμ wave length after placing water filled colorimeter tube or cuvette in the socket provided. After allowing 10 minutes for warming of the instrument, measure the absorbance (optical density) of the coloured complex setting 0 absorbance with redistilled water. Record the absorbance of the Mo–thiocyanate complex corresponding to μg of Mo taken in the range of 0–20 μg for the formation of the coloured complex.

4. Standard curve for molybdenum. On a linear scale graph paper, plot the absorbance value on the 'y' axis (ordinate) and the corresponding μg of Mo taken for colour development on the ``x" axis (abscissa). Connect 0 absorbance (set with redistiled water) with the value of absorbance of coloured complex formed with 20 μg. Mo after substraction of the absorbance obtained with 0 μg. Mo by a straight line to obtain standard curve for Mo.

5. Preparation of test solution from plant sample following the procedure of Evans *et. al.* (1950). Weigh 2g sample of oven dried, finely ground plant material and place it in a tall 200 ml beaker. Add 15 ml of conc. distilled HNO_3(AR grade) cover with a watch glass, and heat on a hot plate at low heat until solid material disappears. Cool the beaker, and add 2ml of 70% $HClO_4$, and digest to dryness at low heat. This should require a minimum of three hours.

Remove the watch glass and add 5 ml of conc. distilled HNO_3 (AR grade) and 1 ml of 70% $HClO_4$ and evaporate to dryness, reducing heat to avoid spattering as residues approach to dryness.

Thereafter, add 2 ml of 30% H_2O_2 and again bring the residue to dryness. Cool and add approximately 50 ml of redistilled water and heat to boil for 1 minutes Add 20 ml of distilled 6N HCl, filter, allow the filtrate to cool, and then make up the volume to 100 ml with redistilled water. Similarly prepare a blank test solution following the entire procedure without the plant sample.

6. Formation of the Mo–thiocyanate complex in test solutions, its extraction by isopropyl ether, and measurement of the absorbance of the coloured complex. Transfer 100 ml of the test solution prepared following the procedure described in step 5 into 150–ml glass stoppered separatory funnel and follow the entire procedure given in step 1–3 beginning with the addition of 1 ml of 4.9% Fe Cl_3 solution. Similarly process the blank test solution

7. Determination of the content of Mo in the sample and blank test solution. Record the absorbance (optical density) of the Mo–thiocyanate complex extracted from the sample and blank test solution following step 1, 2 & 3. Then refer to the standard curve for Mo for determining the Mo content in the test solutions.

Locate the absorbance of the Mo–thiocyanate complex for the sample on the ordinate, and draw an imaginary line parallel to the abscissa intersecting standard curve. From the point of intersection draw a line perpendicular to the abscissa and note the content of Mo in μg given at the point of intersection on the abscissa. Similarly, determine the content of Mo in the coloured complex from the blank test solution.

Alternatively, estimate the slope of the standard curve for Mo i.e. absorbance per μg Mo taken for formation of the Mo–thiocyanate complex. Then divide the absorbance of the Mo thiocyanate complex of sample and blank test solution by the slope to estimate the content of Mo in μg present in the coloured complex.

3.15.4 Calculation

$$\text{Mo in plant sample (ppm)} = \frac{S - B}{M}$$

Where S is μg. Mo in test solution prepared from plant sample

B is μg. Mo in test solution for the blank

M is the mass of plant sample in g taken for the preparation of test solution.

3.15.5 Comments

For the formation and extraction of Mo thiocyanate complex in the test solution from the plant, follow steps 1–4 and 7 of procedure described in section 2.27.3 on molybdenum determination in soil, use only 80 ml of the test solution and multiply Mo estimated in plant sample by 1.25 for not using the entire test solution for the formation of Mo thiocyanate complex.

The test solution from plant sample can be prepared also by dry ashing in a muffle furnace at 550°C for several hours. The ash be then dissolved in 1.2 N HCl, filtered and volume made to 100 ml with 1.2 NHCl.

Molybdenum deficiency usually occurs in most plants when the plant concentration is less than 0.10 ppm Mo in the dry matter (Johnson 1966). However, the sufficiency level of Mo in alfalfa appear to be near 0.5 ppm. In case of pasture plants, Mo concentration greater than 15 ppm Mo can be toxic to cattles. Leaf blade is usually higher in Mo concentration than midrib and stem.

There exist Mo and Mn antagonism in plant nutrition. Consequently Mo and Mn concentration in plant show inverse relationship. Mo and Cu concentration in plant also exhibit inverse relationship. SO_4^- ions in soil exert a depressing effect on Mo uptake.

Mo is required in all crops for the reduction of nitrates to ammonia as the first step in protein synthesis. In legumes, it is also required for nitrogen fixation. Mo deficient plants show characteristic symptoms of nitrogen deficiency.

In legumes nitrogen starvation signs are pale yellow foliage, stunted growth, patchiness in the field. Other crops have poor irregular growth, and an over all pale yellowish cast. Chlorotic areas develop between veins. Marginal chlorosis is seen in broad leaved crop.

Leaf distortions are common–margins often curl or cup upwards around the rim. In cauliflower, and some other crucifers leaf tissues develop irregularly from the mid rib to produce long narrow leaves. This disease, whip tail is the most specific indication of molybdenum deficiency. The death of growing tip is also symptoms of Mo deficiency. In grape fruits Mo deficient leaves have yellow spots. Mo deficient beans have chlorotic areas on leaves in spite of nitrogen supply. On spraying with sodium molybdate solution, the leaves turn green within few days.

Soils high in Fe_2O_3 are often deficient in available Mo. On acidic soils high in free Fe, crop respond to Mo application. Mo deficiency is usually associated with soils below pH 5.5, because of decrease in the solubility of Mo due to reaction with Fe and Al. Plants are able to utilize more Mo at high pH because of increase in the solubility of Mo. Liming increases Mo availability. Well limed or naturally neutral soils may also be depleted of available Mo by cropping or leaching. Phosphate applications enhance available Mo in soil. However, when superphosphate containing 50% calcium sulphate is applied, the effect of P and SO_4 on the availability of Mo would tend to offset each other.

REFERENCES

Hewitt, E.J. (1956) Symptoms of molybdenum deficiencies in plants. Soil Sci. 81 : 159–171.

Jackson, M.L. (1973) Soil Chemical Analysis. Prentice Hall of India, Pvt. Ltd., New Delhi.

Johnson C.M. and Arkley, J.H. (1954) Determination of molybdenum in plant tissue. Anal. Chem. 26 : 572–574.

Johnson, C.M., Pearson, G.A., and Stout, P.R. (1952) Molybdenum nutrition of crop plants. 11. Plant and Soil 4 : 178–196.

Jones, J. Benten. (1972) Plant Tissue Analysis for Micronutrients pp 319–346.

In Mortvedt, *et. al.* (ed.) Micronutrients in Agriculture. Soil Sci. Soc. Amer., Inc., Madison, Wisconsin, USA.

Purvis, E.R. and Peterson, N.K. (1956) Methods of soil and plant analysis for molybdenum. Soil Sci. 81 : 223–228.

Rubins, E.J. (1956) Molybdenum deficiencies in the United States. Soil Sci. 81 : 191–197.

Sandell, E.B. (1950) Colorimetric Determination of Traces of Metals. Interscience Publishers, Inc., New York.

3.16 CHLORIDE DETERMINATION IN PLANT SAMPLE–CHAPMAN AND PRATT (1961)

3.16.1 Principle

The plant sample is digested by dry ashing procedure of Chapman and Pratt (1961). The sample test solution containing Cl is titrated by standard $AgNO_3$ in the presence of CrO_4 in an alkaline medium (pH 8.2). As long as some Cl persist in the solution, formation of red $Ag_2\ CrO_4$ is momentary. When Cl in solution is exhausted through precipitation as AgCl, the red precipitate of Ag_2CrO_4 sharply signals the end point.

3.16.2 Reagents

(*i*) Sulphuric acid approx. 0.1 N. Dilute 2.78 ml of conc. H_2SO_4 (36N) to 100 ml for obtaining approx. 1N H_2SO_4. Dilute 10 ml of approx. 1N H_2SO_4 to 100 ml with distilled water to prepare 0.1 N H_2SO_4.

(*ii*) Sodium carbonate 0.1 N. Dissolve 0.530 g of dry Na_2CO_3 in distilled water, and dilute the solution to 100 ml.

(*iii*) Phenolphthalein indicator, 0.5%. Transfer 0.500 g of phenolphthalein into 100 ml volumetric flask. Then, add 50 ml of ethyl alcohol and 50 ml distilled water. Shake to mix the contents, and filter the solution through Whatman No. 1 filter paper.

(*iv*) Sodium chloride 0.05 N. Dissolve 2.922 g of NaCl (AR–dry) in distilled water, and dilute the solution to 1000 ml in 1 l volumetric flask.

(*v*) Silver nitrate solution–0.05 N. Dissolve 8.4937 g of $AgNO_3$ in distilled water and dilute the solution to 1000 ml in 1l volumetric flask. Check the normality of $AgNO_3$by titration against standard NaCl using K_2CrO_4 as end point indicator.

(*vi*) Potassium chromate indicator. Dissolve 5.0g of K_2CrO_4 in 80 ml distilled water. Then, add saturated solution of $AgNO_3$drop wise with stirring until a permanent red precipitate is produced. Filter the solution and dilute the filtrate to 100 ml.

3.16.3 Procedure

Take an aliquot of 25 ml of plant test solution in a large casserol or else in a 125 ml Erlenmeyer flask and place it on a white porecelain plate. The size of the sample aliquot be so adjusted as to contain 5 to 25 mg of Cl. Adjust the solution pH to 8.2 just colourless to the phenolphthalein end point by adding first 0.1 N Na_2CO_3. When the solution colour becomes purple add few drops of 0.1N H_2SO_4 to change the solution colour from purple to colourless pH remaining just close to 8.0.

Now add 1 ml of K_2CrO_4 indicator, and titrate the solution with standard $AgNO_3$solution till the appearance of first permanent red colouration due to precipitation of Ag_2CrO_4. Similarly run a blank with 25 ml aliquot of the blank test solution following the above procedure beginning with the adjustment of soluton pH to 8.2.

3.16.4 Calculation

$$\text{meq Cl/100 g plant sample} = \frac{(S - B) \times T \times N \times 100}{A \times W}$$

$$\text{\% Cl in plant sample} = \text{meq Cl/100 g plant sample} \times \text{ME}$$

Where

S is ml of $AgNO_3$ titre used for precipitating Cl in 25 ml of sample test solution

B is ml of $AgNO_3$ titre used for precipitating Cl in 25 ml of blank test solution.

T is the total volume of plant sample test solution prepared in ml.

N is the normality of $AgNO_3$ solution as determined by titration against standard NaCl solution.

A is ml aliquot of sample test solution taken for titration.

W is the mass of plant sample in g taken for digestion with acids and for preparing the sample test solution.

ME is the milligram equivalent weight of Cl *i.e.* 0.0354 g

3.16.5 Comments

The CrO_4^{--} react with Ag^+ to form red coloured Ag_2CrO_4 only in alkaline medium (pH 8.0) on the exhaustion of Cl due to precipitation as AgCl.

The low range of Cl concentration in plant tissues of selected crops is given in Table 3.16.1.

Table 3.16.1.

Crops	Plant parts	Low Range Cl-concentration%
Beans	Leaf blade-8 days after flower bud appearance	0.63
	Leaf blade 48 days -do- -do-	0.88
Grapefruit	Leaves 4 to 6 months old	0.07
Corn	Leaves at silking stage	0.20-0.32
	Leaves second from base	0.97
Grapes	Leaves	0.04
Lettuce	Leaves from plants grown on salt plots	2.51
Potato	mature leaves	0.71
Sugar beet	Petiole of recently mature fully expanded leaf	0.40
Tobacco	Leaves	0.07-1.13

Chlorides accumulate in plant tissues with age. Lower leaves are 50 to 90% higher in chlorides than upper leaves. The Cl mobility in plant tissue is also low. The Cl concentration in leaves > stems > roots.

Sulphates and nitrates depress the concentration of chloride. As the exchange capacity of roots increases, the Cl–uptake progressively decreases. Osmotic pressure of the solution increases with the increase in the content of chloride.

Chloride is an essential element. It is necessary for the transportation of carbohydrate in plants. The symptoms of chloride deficiency in selected crops are described in Table 3.16.2.

Table 3.16.2.

Crops	Symptoms of Chloride deficiency
Lettuce	Wilting, leaf mottling, stubby roots with many branches.
Tomato	Leaflet tips wilt, leaf chlorosis, bronzing and necrosis.
Cabbage	Cupping of leaves and absence of cabbage odour.
Carrot	roots with many branches, leaflets turn brown and die.
Sugarbeet	more branching of lateral roots.
Barley	Chlorosis in newly emerging leaves, remain wrapped in tubular form longer and are necrotic.
Alfalfa	Leaflets become smaller and thickened
Corn	Wilting, no other symptoms
Beets	Interveinal chlorosis of younger leaf blade, stubby roots
Potato	bronzing of older chlorotic leaves.

The symptoms of chloride excesses are as follows :-

Burning and firng of leaf tips or margins, bronzing, premature yellowing abscisson of leaves, less frequently chlorosis, reduction in leaf size, and slow rate of growth.

REFERENCES

Chapman, H.D. and P.F. Pratt. 1961. Chlorine pp. 97-100. In H.D. Chapman and P.F. Pratt (ed); Methods of Analysis of Soils, Plants And Water, University of California, Riverside, USA.

Eaton, F.M. 1975. Chlorine PP. 98-135. In Chapman, H.D. (ed) Diagnostic Criteria for Plants and Soils. Eurasia Publishing House (P) Ltd., New Delhi.

Jackson, M.L. 1973. Soil Chemical Analysis, Prentice Hall of India, (Pvt). Ltd., New Delhi.

Yoshida, S., D. A. Forno, T.A. Cook, K.A. Gomez 1972. Laboratory Manual For Physiological Studies of Rice, IRRI, Los Banos, Laguna, Philipines.

3.17 SILICA IN PLANT SAMPLE

3.17.1 Crude silica in plant sample-Yoshida *et. al.* (1972)

3.17.1.1 Principle

The residue obtained on filtration of plant sample extract resulting from wet or dry ashing contains Si as insoluble constituent. The HCl treatment given to the sample extract dehydrate Si and render it insoluble. The filter paper with the residue after drying in an oven at 90°C is transferred to a platinum crucible. It is then charred with a naked flame under a fumehood, and then it is ignited in a muffle furnace at 550°C for 2 hours. The sample ash is dried in a desiccator for 2 hours and there after it is weighed to obtain the mass of crude silica.

3.17.1.2 Equipments and Reagents

No specific reagents or equipments

3.17.1.3 Procedure

Treat the residue obtained from wet or dry ashing with 5 ml distilled water and 10 ml of approx. 6N HCl, and evaporate the contents to dryness on a sand bath. Thereafter, repeat HCl treatment thrice to dehydrate silica completely so as to render Si insoluble.

Place a funnel lined with Whatman No. 44 ashless filter paper into a 100 ml volumetric flask. Transfer the residue into the funnel. Leach the soluble constituents by use of 10 ml portions of hot distilled water several times.

Transfer the silica together with the ashless filter paper after oven drying at 90°C into a clean and dry platinum crucible. Ignite the platinum crucible with a burner flame until smoke is not given off. Then, transfer the platinum crucible into a muffle furnace at 550°C for a period of 2 hours. Thereafter, remove the platinum crucible. Allow it to cool in a desiccator before weighing on a chemical balance with an accuracy of atleast three place after decimal.

3.17.1.4 Calculation

$$\text{\% Crude silica in plant sample} = \frac{(m_2 - m_1) \times 100}{W}$$

where

m_1 is mass of empty platinum crucible in g

m_2 is mass of platinum crucible plus raw silica in g

W is mass of dried ground plant sample taken for wet or dry ashing in g.

3.17.2 Pure silica (SiO_2) in plant sample-Jackson (1958)

3.17.2.1 Principle

The residue obtained on filtration of plant sample extract resulting from wet ashing by ternary acid mixture (HNO_3 + H_2SO_4+ $HClO_4$) contains Si as insoluble constituent. The residue is first dried, and the dry residue on filter is transferred to a platinum crucible and weighed before giving HF treatment. The loss in weight of residue on HF treatment gives an estimate of Si content as Si is volatalized by HF

$$SiO_2 + 4HF \longrightarrow SiF_4 + 2H_2O$$

3.17.2.2 Reagents

(*i*) Sulphuric acid-2N. Dilute 5.6 ml of conc. H_2SO_4 (36N) to 100 ml with distilled water

(*ii*) Hydrogen Fluoride-48% (AR).

3.17.2.3 Procedure

Dry the filter paper containing residue from filtration and leaching of ternary acid digest of plant sample in an oven at 100°C for 2 hours.

Thereafter, transfer the entire silica *i.e.* residue from the filter paper into a preweighed clean platinum crucible and weigh the crucible containing silica. Moisten the silica in platinum crucible with 3 ml of 2 N H_2SO_4 and then add 5 ml of 48% HF slowly.

Place the crucible on a sand bath or on a electric hot plate and heat until the contents become almost dry and SO_3 begins to evolve.

Remove the crucible from hot plate or sand bath and let it cool. Add 1 to 2 ml more HF, and rotate the crucible in an enclined position so as to dissolve any SiO_2 that may be stuck to the sides. Then, evaporate the acid to complete dryness on the sand bath.

Thereafter, heat the uncovered crucible in an enclined position (45°) gradually with a flame. Then rotate the crucible in an enclined position and heat with a flame the remaining part of the crucible. Heat the contents for few minutes on a strong flame. Cool the crucible in a desiccator atleast for two hours, and weigh to find out the loss in weight of pure silica.

3.17.2.4 Calculation

$$\% \text{ pure silica } (SiO_2) \text{ in plant sample} = \frac{\{(m_2 - m_1) - (m_3 - m_1)\}}{w} \times 100$$

where m_1 is the mass of a clean empty and dry platinum crucible in g.

m_2 is the mass of platinum crucible + silica remaining on filter paper after filtration of triacid digest which was transferred to platinum crucible after drying in an oven at 90°C.

m_3 is the mass of platinum crucible plus residue remaining after HF treatment.

W is the mass of dried grounded plant sample in g taken for wet ashing with $HNO_3 + H_2SO_4 + HClO_4$.

3.17.5 Comments

The silica is dehydrated in boiling HNO_3 in few minutes where as HCl dehydration require time consuming baking of silica.

REFERENCES

Jackson, M.L. 1958. Soil Chemical Analysis. Prentice-Hall of India (Pvt.) Ltd., New Delhi. (1973).

Yoshida, S., D.A. Forno, J.H. Cock, and K.A., Gomez. 1972. Laboratory Manual For Physiological Studies of Rice. International Rice Research Institute, Los Banos, Philippines.

3.18 NUTRIENT ELEMENTS DETERMINATION IN PLANT SAMPLE BY ATOMIC ABSORPTION SPECTROPHOTOMETER

3.18.1 Micronutrient Elements Determination in Plant Sample by Atomic Absorption Spectrophotometer

3.18.1.1 Principle

The dried and ground plant sample is digested by concentrated HNO_3+ $HClO_4$, and the residue is treated with few ml of HCl and it is subsequently dissolved in distilled water or else dried, ground plant sample is thoroughly wetted with 1N HCl and is allowed to stand for 24 hours. Thereafter, the equilibrium solution is filtered after a brief shaking to obtain the sample test solution. The micro nutrient elements in the sample test solution are determined by the use of atomic absorption spectrophotometer following specifications for the individual element in respect of range of working standards, Range, Wave Length, Lamp current, Slit, Scale, Filter, Meter Response, Zero Knob Position, Burner Height, Air Flow, Fuel and Fuel Flow.

When the atomic absorption spectrophotometer is in operation, the sample test solution is delivered into a spray chamber as a mist where it is mixed with fuel (acetylene/propane), and oxidant (air/Nitrous oxide), and is atomized and burnt in a broad flame at a stainless steel burner. Pulsed light (400 Hz) from a hollow cathode lamp emitting the spectrum of the element to be determined is passed through the flame, and monocrometer.

The change in the intensity of light at selected wave length which occurs when atoms are introduced in the flame, is detected by a photomultiplier. The signal is isolated, amplified electronically and displaced on a meter as absorbance. The change in the intensity of light is caused by the presence of free atoms, and the absorbance indicated is proportional to the number of atoms present in the flame.

3.18.1.2 Equipments/Reagents

Equipments

Atomic Absorption Spectrophotometer is installed in an air conditioned room. A large size smoke hood connected to an outlet is placed above the gas burner of the spectrophotometer.

Precision Physical and Chemical Balances, Mechanical Shaker, Water Treatment Plant to prepare deionized water, completely glass distillation unit with B_{24}joints and with 2 l capacity boiling flask and heating mantle, Laboratory glasswares (Corning or Pyrex), Polyethylene containers of different sizes.

Reagents

(*i*) Deionized–Distilled Water or Double Distilled Water. Pass distilled water through an ion exchanger or fill boiling flask of a completely glass distillation unit with distilled water. Heat the boiling flask to distill the distilled water and collect the distillate.

(*ii*) Perchloric acid-70% (AR-grade).

(*iii*) Distilled conc. Nitric Acid. (AR-Grade). Set up a completely glass distillation assembly. Fill the boiling flask with 1 l of concentrated HNO_3(AR-grade). Heat the contents so as to boil and allow it to distill. Discard the initial 250 ml distillate and collect the subsequent 500 ml distillate.

(*iv*) Diacid Plant Sample Digestion Mixture. Mix distilled conc. Nitric acid and Perchloric acid -70% (AR-Grade) in the ratio of 2 : 1 (v/v).

(*v*) Distilled Hydrochloric acid-Approx. 6N. Set up a completely glass distillation assembly. Fill the boiling flask with 500 ml of deionized–distilled water and then with 500 ml of conc. HCl (AR-grade). Heat contents so as to boil and allow it to distill. Discard the initial 250 ml distillate and collect the subsequent 500 ml distillate.

(*vi*) Sodium carbonate, 0.1 N. Transfer 0.530 g of dry Na_2CO_3 (AR-grade) into 100 ml volumetric flask. Add distilled water. Shake to dissolve the contents and make the volume to 100 ml mark with distilled water.

(*vii*) Methyl orange, 0.1%. Transfer 0.100 g of methyl orange indicator into a 100 ml volumetric flask. Add distilled water. Shake to dissolve the contents and make the volume to 100 ml mark with distilled water.

(*viii*) Distilled hydrochloric acid, 1.0 N. Determine the normality of distilled hydrochloric acid approx. 6N against 0.1 N Na_2CO_3 solution using 0.1% methyl orange indicator. 1000/normality of distilled HCl will give ml of distilled HCl to be diluted to 1000 ml mark with deionized distilled water to obtain 1.0 N HCl.

(*ix*) Stock solutions of Zinc, Copper, Manganese and Iron

Zinc-1000 ppm solution. Place 1.000 g of zinc metal (AR-grade) into a 100 ml corning or pyrex glass beaker. Cover the beaker with a pyrex watch glass. Slowly add 50 ml of distilled 6N HCl slowly. Thereafter boil the contents gently until zinc becomes completely dissolved. Transfer the zinc solution to 1 l volumetric flask, and make the volume to 1000 ml mark with deionized distilled water.

Zinc-100 ppm solution. Pipette 10.0 ml of 1000 ppm Zn solution into a 100 ml volumetric flask. Then add 1N HCl to make the volume to 100 ml mark.

Copper 100 ppm solution. Place 0.100 g of copper metal (AR-grade) into a 100 ml capacity corning beaker. Cover the beaker with a pyrex watch glass. Then add 50 ml of 1 : 1 distilled nitric acid (v/v) carefully.

Boil the contents gently for few minutes until brown fumes of nitrous oxide cease to evolve. Then, allow the beaker to cool. Transfer the copper solution into 1 l volumetric flask. Rinse the beaker with deionized distilled water and transfer the rinsings to the same volumetric flask. Make the volume to 1000 ml mark with deionized-distilled water.

Manganese-1000 ppm solution. Place 3.0762 g of $MnSO_4 \cdot H_2O$ (AR-grade) into a 100 ml beaker. Add 50 ml of 1 : 1 distilled HCl (v/v). Stir the contents to dissolve the reagent. Transfer the Mn solution to 1 l volumetric flask. Rinse the beaker with deionized-distilled water, and transfer the rinsings to the same volumetric flask. Make up the volume to 1000 ml mark with deionized-distilled water.

Iron-100 ppm solution. Place 0.100 g of electrolytic iron into a 100 ml pyrex beaker. Cover with a watch glass. Then add 50 ml of 1 : 3 distilled concentrated HNO_3 (v/v). Boil the contents gently for few minutes until brown fumes of nitrous oxide cease to evolve. Allow the beaker to cool. Transfer the Fe-solution to 1 l volumetric flask. Rinse the beaker with deionized-distilled water and transfer the rinsings to the same volumetric flask. Make the volume to 1000 ml mark with deionized-distilled water.

(*x*) Working standards of Zinc, Copper, Manganese and Iron.

Zinc, 0 to 1.00 ppm solution. Transfer 0, 0.2, 0.4, 0.6, 0.8 and 1.0 ml of 100 ppm Zn solution into separate 100 ml volumetric flasks, and add 1N HCl to each flask to make the volume to 100 ml mark.

Copper 0 to 1.6 ppm solution. Transfer 0, 0.4, 0.8, 1.2 and 1.6 ml of 100 ppm Cu-solution into separate 100 ml volumetric flasks, and add 1N HCl to each to make the volume to 100 ml mark.

Manganese, 0-20 ppm solution. Transfer 0, 0.5, 1.0, 1.5 and 2.0 ml of 1000 ppm Mn-solution into separate 100 ml volumetric flasks. Then, add 1N HCl to each flask and make the volume to 100 ml mark with 1N HCl.

Iron, 0 to 20 ppm solution. Transfer 0, 5, 10, 15 and 20 ml of 100 ppm Fe solution into separate 100 ml volumetric flasks. Add 1N HCl to each to make the volume to 100 ml mark.

3.18.1.3 Procedure

1. Preparation of plant sample diacid digest. Transfer 1.000 g of dried and ground plant sample into a 250 ml pyrex or corning Erlenmeyer flask. Then add 20 ml of diacid digestion mixture-conc. HNO_3 + 70% $HClO_4$ (2 : 1 v/v) wetting the plant sample thoroughly. Place a funnel on the mouth of the flask. Mix the contents of the flask by swirling, and let it stand for an hour.

Thereafter, place the Erlenmeyer flask containing plant sample treated with diacid digestion mixture on a hot plate at low heat. At this stage a vigorous reaction would occur, and evolution of copious red fumes will occur. Continue heating gently until the digest clears up. Then increase the heat and continue digestion even after fumes of perchloric acid begin to condense on the neck of the flask. If sign of charring is observed, remove the flask from the hot plate and let it cool. Then, add 2-3 ml of distilled concentrated HNO_3 and heat again to complete the digestion. Repeat treatment with conc. HNO_3 on sign of charring.

When the digest is reduced to about 2-3 ml and appears to be a faint yellow green fluid which becomes colourless on cooling, remove the flask from hot plate and let it cool. After cooling the flask, add about 3 ml of distilled 6 N HCl. Then, heat the contents to boil in order to remove excess HCl and water. Then, allow the Erlenmeyer flask to cool. Add 25 ml of deionized-distilled water, swirl the contents, and transfer the solution into a 50 ml volumetric flask. Rinse the Erlenmeyer flask with portions of deionized distilled water, and transfer the rinsings to the same 50 ml volumetric flask. Make the volume to 50 ml mark with deionized-distilled water. Mix the contents and filter the test sample solution through a funnel lined with Whatman No. 42 filter paper. Collect the filtrate into a 100 ml polyethylene bottle. Similarly prepare the blank test solution.

2. Extraction of nutrient elements from plant sample by 1N HCl for elemental analysis by atomic absorption spectrophotometer. Transfer 1.000 g of dried and ground plant sample in a 100 ml polystyrene bottle which has been rinsed with 1N HCl. Add 25 ml of 1N HCl into the bottle through a burette wetting the plant sample thoroughly. Let the sample remain in the acid for 24 hours without shaking.

Thereafter, shake the bottle for a brief period, then filter the contents through Whatman No. 1 filter paper. Collect the filtrate in a 100 ml polystyrene bottle. Similarly prepare a blank test solution.

3. Preparation of the standard graph for each micro nutrient element by the atomic absorption spectrophotometer. Let the operator run the atomic absorption spectrophotometer as per specification for the desired element given in the instruction manual. Run the instrument for a period of 20 minutes to allow it to warm up.

Thereafter, first run a series of standards of the desired element beginning with the concentration of 0 ppm, and raising the concentration of the element by increment of 0.2 ppm in case of Zn and by increment of 0.4 ppm in case of Cu. However, in the case of Fe and Mn, the element concentration be raised by increment of 5 ppm in the range of 0-20 ppm. The Zn concentration ranges from 0 to 1.0 ppm whereas Cu concentration ranges from 0 to 1.6 ppm.

The absorbance recorded on atomization of solution of known concentration of an element is noted and tabulated corresponding to the concentration of element. Then on a linear scale graph paper with absorbance scale on the ordinate, and the concentration scale on the abscissa, the value of absorbance corresponding to each concentration be plotted, and then the points be connected by a straight line to obtain the calibration curve. With the attachment of a reading system the same is achieved by a computer.

4. Determination of concentration of an element in the plant sample and blank digest or extracts. When the atomic absorption spectrophotometer is in operation, after running the working standards, the plant sample and blank digests or extracts be atomized, and the values of absorbance recorded by the atomic absorption spectrophotometer be noted. The concentration of element in the digest or extract is determined by reference to the calibration curve for each element.

The value of concentration of desired element is substituted in the formula given under the heading "Calculation" for estimating the concentration of element in the plant sample. With the attachment of a computerized reading system, the data recording processing and print out of the results is simultaneously done without any loss of time.

3.18.1.4 Calculation

$$\text{Micronutrient Element in Plant Sample (ppm)} = \frac{(S - B) \times VE}{M}$$

Where

S is ppm concentration of the element in the plant sample digest or extract as estimated from the recorded absorbance by reference to the element's calibration curve.

B is ppm concentration of the element in the blank digest or extract as estimated from the recorded absorbance by reference to the element's calibration curve.

VE is the total volume of the plant digest or extract prepared with the plant sample.

M is the mass of dried and ground plant sample taken for the preparation of the plant digest or else plant extract.

3.18.1.5 Comments

If the concentration of an element in the plant digest or extract exceeds the maximum concentration of the working standard, the plant digest or extract be diluted so that the absorbance recorded by the atomic absorption spectrophotometer fall within the range of the working standard. In such a case the extent of dilution be noted, and the concentration of element estimated by reference to the calibration curve be multiplied by the extent of dilution to obtain the concentration of the element in the undiluted digest or extract.

Kushizaku (1968) prepared plant extract by equilibration with 1N HCl for the rapid determination of Zn, Ca, Mn and Mg by atomic absorption spectroscopy. He found the results reproducible. Besides, the HCl extraction procedure apart from being simple, it compared well with the standard acid digestion procedure.

The critical content of micro nutrient elements for deficiency and toxicity in rice plant are given below :

Critical content for deficiency

Element	Plant part	Growth stage	Critical content
Fe	Leaf blade	Tillering	70 ppm
Mn	Shoot	Tillering	20 ppm
Zn	Shoot	Tillering	10 ppm
Cu	Straw	Maturity	$<$6ppm

Critical content for toxicity

Element	Plant Part	Growth Stage	Critical Content
Fe	Leaf blade	Tillering	300 ppm
Mn	Shoot	tillering	$>$2500 ppm
Zn	Straw	Maturity	$>$ 1500 ppm
Cu	Straw	Maturity	30 ppm

The critical level of micronutrient element in tissues of other plants have been summarized under the heading "comments" in micronutrient determinations by the colorimetric procedures. A guide to micronutrient concentration in tissues of selected plant species is given in Table 3.18.1.

Table 3.18.1.

Plant	Plant to sample	Micronutrient Content-ppm			
		Fe	Mn	Zn	Cu
			Sufficient Ranges		
1. Pine trees	Current year needles near terminals	20-100	50-600	20-50	2-6
2. Apple	Leaf from base of nonfruiting shorts	50-250	35-100	25-50	8-20
3. Alfalfa	Upper third of plant at first flower	50-250	25-100	25-70	30-80
4. Soyabean	Recently matured trifoliali flowering stage	50-250	30-200	25-50	8-20
5. Maize	Ear leaf of maize at flowering stage	10-300	20-200	20-70	6-50
6. Wheat	Ist leaf blade from top of the plant	35-55	30-50	20-35	6-10
7. Rice	3rd leaf from top	68	20	10	2
8. Citrus	Leaf blade	36-120	20-90	20-50	5-16

Data from different sources

REFERENCES

Amma, M.K. 1989. Plant and Soil Analysis. Rubber Research Institute, Rubber Board, Kottayam– 686009, Keral India.

Kushizaki, M. 1968. An extraction procedure of plant for the rapid determination of Mn, Cu, Zn and Mg by the Atomic Absorption Analysis J. Sci. Soil Manure Japan 39 : 489-490.

Perkin-Elmer Corp. 1971. Analytical Methods for Atomic Absorption Spectrophotometry. Norwalk, Connecticut, U.S.A.

Robinson, I.W. 1966. Atomic Absorption Spectroscopy. Marcel Dekker Inc., New York, USA.

Slavin, W. 1968. Atomic Absorption Spectroscopy. Inter Science Publishers, John Wiley and Sons, New York, USA.

Yoshida, S., D.A. Forno, J.H. Cock and K.A. Gomez-1972. Laboratory Manual For Physiological Studies of Rice-International Rice Research Institute, P.O. Box. 583, Manila, Philippines.

3.18.2 Calcium, Magnesium Determination in Plant by AAS

3.18.2.1 Principle

See section 3.18.1.1.

3.18.2.2 Equipment/Reagents

Equipment

See section 3.18.1.2.

Reagents

(*i*) Deionized-distilled water or double distilled water.

(*ii*) Hydrochloric acid. 1.0N

(*iii*) Lanthanum, 5% solution. Weigh 58.64 g La_2O_3 and wet it with 100 ml of deionized distilled water. Then slowly add 250 ml of 1 : 1 (v/v) HCl. When the Lanthanum has dissolved, transfer the solution into a 1 l volumetric flask, and make up the volume to 1000 ml mark with deionized distilled water.

(*iv*) Stock solution for Calcium and Magnesium

Calcium-1000 ppm solution. Place 2.5022 g $CaCO_3$ (AR-grade) into a 100 ml tall beaker. Cover the beaker with a pyrex watch glass and add 50 ml of 1N HCl slowly and very carefully as the reaction is very vigorous, and so care be taken to prevent any loss by splattering. Gently boil the solution for 3 minutes. Then allow the beaker to cool. Transfer the Ca solution to 1 l volumetric flask. Make up the volume to 1000 ml mark with deionized distilled water.

Magnesium-1000 ppm solution. Place 1.000 g of unoxidized Mg metal (reagent grade) into 100 ml tall beaker. Cover the beaker with a pyrex watch glass and carefully add 50 ml of 1 : 1 (v/v) HCl. Boil the solution gently for few minutes to ensure complete dissolution of Mg. Then, transfer the solution to a 1 l volumetric flask. Make up the volume to 1000 ml mark with deionized distilled water.

(*v*) Working Standards for calcium and magnesium

Calcium 0 to 20 ppm in 1% La. Transfer 0, 0.5, 1.0, 1.5 and 2.0 ml of 1000 ppm Ca solution into separate 100 ml volumetric flasks. Add 20 ml of 5% La solution into each flask and make the volume to 100 ml mark with deionized distilled water.

Magnesium 0 to 10 ppm. Transfer 0, 0.2, 0.4, 0.6, 0.8 and 1.0 ml of 1000 ppm Mg solution into separate 100 ml volumetric flasks and make the volume to 100 ml with deionized distilled water.

3.18.2.3 Procedure

1. Preparation of plant sample extract with 1N HCl. Transfer 1.0 g of dried and ground plant sample into a 100 ml polystyrene bottle rinsed with 1N HCl. Then add 25 ml of 1N HCl through a burette wetting the sample. Let it stand for 24 hours. Then, after shaking briefly, filter the solution through Whatman No.1 filter paper collecting the filtrate in a 100 ml polystrene bottle. Similarly run a blank.

2. Preparation of sample test solution for Ca determination. Pipette 1.0 ml of the plant sample extract into a 10 ml graduated tube. Pipette 2.0 ml of the 5.0% La solution into the tube and dilute it to 10 ml mark with 1N HCl. Seal the tube with plastic. Similarly prepare a blank test solution for Ca determination.

3. Preparation of sample test solution for Mg determination. Pipette 2.0 ml of the plant sample extract into a vial graduated at 40 ml. Dilute the 2 ml aliquot to 40 ml with 1N HCl. Seal the vial with a plastic lid.

4. Determination of absorbance of working standard and sample and blank test solution. Following the instruction manual set the AAS for Ca determination. Also set zero knob position as per operation manual. Run the instrument for 20 minutes to allow it to warm up..

Then atomize 0, 5, 10, 15 and 20 ppm Ca solution in sequence noting down the corresponding absorbance recorded by the instrument. Thereafter, atomize the plant sample and blank test solution for calcium and note the corresponding values of absorbance recorded by the instrument.

Thereafter, set the AAS for Mg determination as per instruction manual. Also set zero knob position as per operation manual. Then atomize 0 to 10 ppm Mg working standard and note the absorbance recorded. Then atomize the plant sample and blank test solution and note the absorbance recorded by the instrument.

5. Preparation of standard graph, and estimation of concentration of the element in sample and blank test solution. On a linear scale graph paper with absorbance scale on the ordinate and the concentration scale on the abscissa, plot the absorbance of the working standard corresponding to the concentration of the element. Connect the points plotted by a straight line to obtain the calibration curve.

The concentration of the element in the plant sample and the blank test solution is estimated from the corresponding absorbance value by reference to the calibration curve. The values of the concentration of the element in the sample and blank test solution are substituted in the formula given under the heading "Calculation". The formula is then simplified to estimate the concentration of the element in the plant sample. With the attachment of a computerized reading system, the data recording, processing and print out of the results is simultaneously done without any loss of time.

3.18.2.4 Calculation

$$\text{Element in plant sample (g / 100g sample)} = \frac{(S - B) \times VST \times VE \times 100}{A \times M \times 10^6}$$

Where

S is ppm element in the plant sample test solution by reference to the calibration curve.

B is ppm element in the blank test solution by reference to the calibration curve

VST is volume of the sample test solution in ml.

VE is the volume of extractant in ml employed for preparation of the plant extract.

A is ml aliquot of plant extract taken for preparation of the plant sample test solution

M is the mass of dried and ground plant sample in g taken for the preparation of the plant extract.

3.18.2.5 Comments

The critical content of Ca and Mg in rice plant at deficient level is summarized below :

Element	–	Plant Part	–	Growth Stage	–	Critical content
Ca	–	Straw	–	Maturity	–	0.15%
Mg	–	Straw	–	Maturity	–	0.10%

REFERENCES

David, D.J. 1959. Determination of calcium in plant material by Atomic absorption spectrophotometry. Analyst. 84 : 536–54.

Kushizaki, M. 1968. An extraction procedure of plant materials for the rapid determination of Mn, Cu, Zn and Mg by the atomic absorption analysis. J. Sci. Soil Manure, Japan, 39 : 489-490.

Perkin-Elmer Corp. 1971. Analytical methods for atomic absorption spectrophotometry. Norwalk, Connecticut, U.S.A.

Yoshida, S., D.A. Forno, J.H. Cock, and K.A. Gomez, 1972. Laboratory Manual for Physiological Studies of Rice. The International Rice Research Institute. P.O. Box. 583, Manila Philippines.

Chemical Analysis of Water Sample

4.1 WATER SAMPLING-RICHARDS ed. (1954) & TATA (1987)

There are different sources of water such as rain, streams, canals, lakes, tanks, ponds, large reservoirs, shallow and deep wells, shallow and deep tube wells, as well as sewage effluents and sea water.

The composition of the different sources of water vary with the season, rate of flow, the extent of effluents from the industrial units, and sewage effluents from the municipal area during rainy season, the river, tank and pond water become silty and muddy due to soil erosion. The depth of ground water also vary with the seasons.

The sources of water available to the farmer for irrigation of crops are generally few such as canal, shallow and deep wells, pumping setts, tube wells, and run off water collected in excavations along roads, natural drains and adjoining areas under impeded drainage.

The source of water vary with the season. The run off water is usually available in rainy season and is used for irrigating paddy by lifting with swing basket. The composition of ground water also vary with the season.

Whatever may be sources of water, the water sample be taken along with the soil sample to soil testing laboratory for its analysis by way of assessing the suitability of water for irrigation after evaluation of its quality.

For drawing maximum benifit, the samples taken be representative of the water in bulk, and it should be collected in a clean vessel. One to three litres capacity plastic bottles duly washed with a detergent are satisfactory. However, before filling with water samples, the bottles must be rinsed with water samples three times. The information such as farmer's name, owner of pumping set, date and month of sampling, depth of ground water, crops and soils being irrigated, any specific agricultural or human problems encountered be briefly recorded along with the name of village block, tehsil and district. The water sample should be properly labelled and submitted to soil testing laboratory for its analysis.

In a laboratory the water samples may be analysed for pH, EC, concentration of cations and anions such as Ca, Mg, Na, K, CO_3, HCO_3, Cl, SO_4, B and in case of sewage effluents and industrial effluents the concentration of toxic heavy metals such as Cr, Cd, Ni, Hg etc. The salinity and alkalinity may be tested in terms of EC, SAR, RSC, etc depending upon the problem of the tract. The water sample tests may highlight the deficiencies and implications of usage of water in crop and soil management.

REFERENCES

Richards, L.A. *et. al* (ed). 1954. Diagnosis and Improvement of Saline and Alkali Soils. USDA Hand Book 60. Oxford & IBH Publishing Co., New Delhi.

Tata, S.N. (Chief ed). 1987. Hand Book of Agriculture I.C.A.R, New Delhi.

4.2 pH OF WATER SAMPLE-JACKSON (1958)

4.2.1 pH of water sample by a pH meter

4.2.1.1 Principle

pH is defined as the log of the reciprocal of hydrogen ion activity or else as the negative log of hydrogen ion activity expressed in g moles/l.

pH measurement by a pH meter require use of a glass and reference calomel electrode. pH meter is a potentiometer, and it measures the potential difference developed between the two electrodes dipped in the sample solution. As the pH decrease by one unit, there is increase in the voltage by 59.16 millivolts. The instrument dial is usually calibrated in pH unit, but in some instruments, the corresponding potential difference is also printed below the pH scale.

4.2.1.2 Reagents

Buffer solution of pH 4.0, pH 7.0 and pH 9.2

Transfer one buffer tablet of each pH mentioned above into separate 100 ml volumetric flask. Add about 50 ml of distilled water in each volumetric flask. Dissolve the tablets, and then dilute the solution in each flask to 100 ml with distilled water. Indicate pH of the buffer on each volumetric flask. Place the stopper, and mix the contents by turning the flask upside down holding the stopper firmly.

4.2.1.3 Procedure

Immerse the glass and reference calomel electrode in distilled water atleast 24 hours before taking measurements on pH. Connect the leads of the two electrodes into the appropriate socket provided in the instrument. Switch on the instrument and let it run for 10 minutes to allow it to warm.

For the calibration of the pH meter, first set the temperature correction knob to the temperature of the sample solution. Now, immerse the two electrodes in buffer solution of pH 7.0 and adjust the pH recorded by the instrument to 7.0 by the pH regulator knobs. Then, take out the two electrodes, rinse it with a jet of distilled water and then wipe it dry with a good quality blotting paper or else with ordinary filter paper. Then, immerse the two electrodes in buffer solution of either pH 4.0 or else pH 9.2 and check the pH recorded by the pH meter. If the instrument records pH of the buffer exactly, the pH meter is calibrated for the measurement of the sample solution.

Now, remove the glass and reference calomel electrodes from the buffer solution. Rinse the electrodes with a jet of distilled water sufficiently, and wipe it dry with a blotting or filter paper. Thereafter, rinse the electrodes with water sample and discard the rinsings. Then, place the electrodes in beaker filled with water sample. Let the lower portion of the electrodes remain immersed in water sample. Note down the pH meter reading at which the pH needle rest.

4.2.2 pH of water sample by universal pH indicator or else by pH paper

4.2.2.1 Principle

Composite organic dyes sensitive to pH range from 4 to 11 is available from BDH in a liquid form along with the pH colour chart. Paper clippings soaked in composite organic dye solution sensitive to different pH range along with the pH colour chart is also available.

4.2.2.2 Reagents

(*i*) Universal pH indicator bottle pasted with pH colour chart.

(*ii*) pH paper of different pH range.

4.2.2.3 Procedure

Fill a clean test tube of 10 ml capacity with water sample filtered through whatman No. 1 filter paper. Add 2 drops of the pH indicator solution. Mix the contents, and let it stand for 5 minutes to develop colour depending on pH of the solution. Then, match the colour developed with the pH colour chart supplied by the manufacturer, and note down the pH of the sample.

If pH paper of different pH range is available, remove a small strip of pH paper, and moisten or wet a portion of pH paper strip with the water sample. Note the change in the colour of the wetted portion of the pH paper, and match its colour with the pH colour chart and note down the pH of the water sample.

4.2.2.4 Calculation : Nil

4.2.2.5 Comments

It is very convenient to measure pH with a pH paper or with universal pH indicator solution, but accurate measurement of pH is possible with reliable pH meter supplied with reliable glass and reference calomel electrodes. Now portable and sturdy as well as field pH meter are available in which electrodes are well protected inside a metal covering. It is quite hardy and reliable if used carefully.

The pH of solution reflect on its ionic composition also. As pH increases above 8.2, there is progressive increase in CO_3 and HCO_3 primarily associated with Na. As pH decrease below pH 8.2, there is increase in Cl and SO_4 associated with basic cations such as Ca, Mg, K, Na. At pH below 7.0, basic cations & phosphates decrease, but the concentration of H & Al and micro nutrient elements such as Fe, Mn, Zn & Cu increase. The presence of soluble salts in higher concentration tend to depress the pH. If irrigation water is strongly alkaline. It's use may result in more severe problem of sodicity

REFERENCES

Jackson, M.L. 1973. Soil Chemical Analysis Prentice Hall of India Private Limited, New Delhi.

4.3 ELECTRICAL CONDUCTIVITY OF WATER SAMPLE-WILCOX (1950)

4.3.1 EC of water sample by use of non digital type of conductivity meter

4.3.1.1 Principle

In equation E = IR, E is the electric potential in volt, I is the current in amperes, R is the resistance offered to the flow of current in ohms. The reciprocal of resistance in ohms is termed conductance in mho. The conductance measured between a pair of platinized electrode $1 cm^2$ in cross sectional area placed 1 cm apart at 25°C is termed specific conductance which is numerically equal to ECt × fl × K. where ECt is EC in mho/cm at "t" temperature, ft is the temperature correction factor to express the conductance at 25°C, and K is the cell constant of conductivity cell which is estimated from the equation $K = 1411.8 \times 10^{-6} \times Rt/ft$ by measuring the resistance of .01N KCl at t temperature, ft is the temperature correction factor, and 1411.8×10^{-6}mho/cm is the specific conductance of .01N KCl solution at 25°C. The Rt can be also estimated by taking the reciprocal of ECt.

The electrical conductivity of water sample is measured by a conductivity meter in units of micromho/cm preferably. Some time EC of water is also expressed in millimho/cm. The electrical conductivity is also expressed in dSm^{-1} which is numerically equal to conductivity expressed in milli mho/cm.

The conductivity meter is either digital or non digital. In non digital, the instrument measures ECt which is equal to dial reading × the multiplication factor in micromho/cm. K value of the conductivity cell is indicated on the cell by the manufacturers, but it can be experimentally determined when the electrodes are replatinized on long usage. The digital type conductivity meter require calibration of the instrument with .01 N KCl solution before measurement of the conductivity of water samples.

4.3.1.2 Reagents

Normally no reagents are required unless need is felt for either replatinization of the conductivity cell or for checking K value of the conductivity cell.

4.3.1.3 Procedure

Follow instructions of the manufacturer. Following procedure may be followed in case of ELICO conductivity meter provided with cathode ray as null point indicator termed as magic eye.

Before use the conductivity cell should remain immersed in pyrex glass beaker filled with distilled water for atleast 24 hours.

Connect the leads of the conductivity cell into the socket provided in the conductivity meter and provide electrical connection to the conductivity meter through a voltage stabilizer. Then switch on the conductivity meter, and let it run for 10 minutes time to allow it to warm.

In the mean time, remove the conductivity cell immersed in distilled water, wipe it dry with a blotting paper or else with an ordinary filter paper. Thereafter, rinse the conductivity cell once with water sample and discard the rinsing. Then, immerse the conductivity cell in water sample.

Now, rotate the multiplier factor switch one point at a time towards the conductivity range in micromho/cm until widest gap is attained in the magic eye. Note down the multiplication factor.

Then rotate the meter dial knob towards the left or right which ever widens the gap in the magic eye. Stop rotating the knob when widest gap is attained in the magic eye. Note down the dial reading accurately.

Dip a thermometer in the water sample, and note down its temperature after 1 minute. Find out the ft value *i.e.* the temperature correction factor at the observed temperature of water sample from the table giving value of ft at different temperature.

4.3.1.4 Calculation

$$EC_{25} \times 10^6 = ECt_{\text{(micro mho/cm)}} \times ft \times k \text{ micromho/cm}$$

$$= \text{Dial Reading} \times \text{Multiplication factor} \times ft \times k \text{ micromho/cm}$$

$$EC_{25} \times 10^3 = \frac{DR \times MF \times ft \times K}{10^3} \text{ millimho/cm}$$

where

DR is dial reading

MF is multiplication factor in micromho/cm

ft is temperature correction factor

K is the cell constant

ECt is Electrical conductivity in micromho/cm at 't' temperature.

4.3.2 EC of water sample by use of digital type of conductivity meter

4.3.2.1 Principle as given in section 4.3.1.1

4.3.2.2 Reagents

Potassium chloride - .01N solution. Dissolve 0.7456 g of dry KCl (AR) in distilled water and dilute it to 1000 ml mark with distilled water in 1l volumetric flask. Stopper and mix the contents. This solution has a conductivity of 1411.8×10^{-6} mho/cm at 25°C.

4.3.2.3 Procedure

Connect the leads of the conductivity meter to a voltage stabilizer. Switch on the meter and let it run for 10 minutes to allow it to warm.

Put the conductivity range knob at its maximum position and put the CAL/READ knob at the CAL position and now move the CAL regulator knob until the instrument records a numerical value of 199.

Now immerse the conductivity cell in .01N KCl solution after rinsing it with the same solution. Connect the terminals of the conductivity cell into socket provided in the conductivity meter.

Measure the temperature of the .01N KCl solution and find out the estimated conductivity of .01N KCl solution at the observed temperature from the table provided in the instrument operation manual. Convert the value from mho/cm to millimho/cm by multiplying mho/cm by 1000. The electrical conductivity increases by 2% for each degree centigrade rise in temperature.

Now, change the CAL/READ knob from CAL position to READ position. Then, adjust the meter reading to the estimated value of electrical conductivity of .01N KCl solution in millimho/cm by regulating the conductivity knob in the millimho/cm range and also by regulating the cell constant knob. Once the desired meter reading is obtained, do not touch any knob.

Thereafter, remove conductivity cell from .01N KCl solution wash it with distilled water. Wipe it dry with a filter paper. Rinse the conductivity cell with the water sample and discard the water sample used for rinsing. Then fill a 100 ml beaker with water sample and immerse the conductivity cell in the water sample.

Now take the temperature of the water sample and note it down. Also note down the meter reading in milli mho/cm. Find out the ft value at the observed temperature of water sample from the table giving temperature correction factor at different temperature.

4.3.2.4 Calculation

$EC_{25} \times 10^3$ of water sample = Digital meter reading × ft millimho/cm
(millimho/cm)

$EC_{25} \times 10^6$ of water sample = Digital meter reading × ft × 10^3 micromho/cm
(millimho/cm)

Where EC_{25} is specific conductance in mho/cm at 25°C, ft is the temperature correction factor.

4.3.2.5 Comments

The water sample be collected in a thoroughly detergent washed plastic bottle after rinsing the bottle with the water sample. The month, date, time, location and full particulars of the source be noted. Container of ordinary glass should not be used as a small fraction of glass gets dissolved in water.

The platinizing solution is prepared by dissolving 0.5 g of platinum chloride and .006 g of lead acetate in 50 ml distilled water. Keep the platinizing solution in pyrex flask and store in a refrigerator.

When the platinum black coating of conductivity cell has flaked off or when the conductivity meter readings appear to be erratic the electrodes need to be platinized.

Transfer 40 ml of platinizing solution into a 100 ml beaker, and immerse the electrodes of the conductivity cell in the platinizing solution. Then pass a current from 1.5 volt dry battery through the cell for about 15 seconds. The current flow be such that a small quantity of gas is evolved. Then reverse the direction of current flow for 15 seconds. Repeat the procedure until a uniform black coating is found on either sides of the two electrodes. After platinization of the conductivity cell, the cell constant be experimentally determined.

In case of non digital type of conductivity meter, if it is desired to determine K value *i.e.*, cell constant experimentally, place the conductivity cell in .01N KCl solution, and rotate the multiplication factor knob towards the resistance in ohms/cm until widest gap is attained in the magic eye. Note down the multiplication factor. Then, rotate the conductivity meter dial knob until widest gap is again attained in the magic eye. Note down the dial reading. The product of dial reading and the multiplication factor in the resistance range give Rt. Take the temperature of .01N KCl solution, and find out the ft value at the observed temperature from table giving temperature correction factor at different temperature. Substitute the value of Rt and ft in the equation for K, and simplify to obtain K value of the conductivity cell.

The salinity rating of irrigation water as judged from its electrical conductivity in micromho/cm at 25°C is as follows :

C_1	C_2	C_3	C_4
< 250	250-750	750-2250	>2250

The suitability of water for irrigation is actually ascertained considering EC of water, sensitivity of the crop to salinity, soil texture and permeability of the soil to water. On coarse textured soil with higher permeability to water, it is possible to use water of higher salinity without adverse effect on crop yields but on soils clayey in texture with very low permeability to water, water of lower salinity would be suitable. Otherwise salts added by irrigation water will accumulate in the soil and crop yields would be adversely affected particularly in salt sensitive crops.

The soluble salts in meq/l may be estimated from the equation log S = log 10.37 + 1.065 log $EC_{25} \times 10^3$, when EC is in milli mho/cm and S is salt in meq/l. The relationship between specific conductance of solution in millimho/cm and other units of salt concentration or osmotic pressure are given below

$EC_{25} \times 10^3 \times 640$ = ppm salt in solution

$EC_{25} \times 10^3 \times 12.5$ = meq salt/l

$EC_{25} \times 10^3 \times 0.064$ = % salt in solution

$EC_{25} \times 10^3 \times 0.36$ = Osmotic Pressure in atmosphere

REFERENCES

Jackson, M.L. 1973. Soil Chemical Analysis, Prentice Hall of India Private Limited, New Delhi.

Richards, L.A. *et al.* (ed). 1954. Diagnosis And Improvement of Saline And Alkali Soils. USDA Hand Book 60, Oxford & IBH Publishing Co, New Delhi.

Wilcox, L.V. 1950. Electrical Conductivity. Amer. Water Works Assoc. J. 42 : 775-776.

4.4 Ca AND Mg IN WATER SAMPLE BY EDTA TITRATION-CHENG & BRAY (1951) AND DIEHL *et. al.* (1950)

4.4.1 Ca + Mg determination by EDTA titration using Mg indicator

4.4.1.1 Principle

EDTA form complexes with Ca, Mg and divalent heavy metals. The interference from heavy metals is prevented by introducing heavy metal complexing reagents *e.g.* disodium diethyl dithiocarbamate, KCN, K.Na-ferrocyanide, hydroxylamine hydrochloride and triethanolamine. EDTA titration in the presence of Ca-indicator (ammonium purpurate) determines Ca and EDTA titration in the presence of Mg indicator (Eriochrome Black T) determines Ca + Mg as Ca-EDTA complex has a higher stability than Mg-EDTA complex. Mg is conveniently determined by subtraction of Ca from Ca + Mg.

4.4.1.2 Reagents

(*i*) NH_4Cl–NH_4OH Buffer.

(*ii*) NaOH–4N.

(*iii*) $CaCl_2$ .01N.

(*iv*) $MgCl_2$ approx. .01N.

(*v*) Eriochrome Black T.

(*vi*) Ammonium purpurate.

(*vii*) EDTA-.01N.

Heavy Metal Complexing reagents.

(*viii*) KCN solution - 1.0%.

(*ix*) K.Na. ferrocyanide.4.0%.

(*x*) Hydroxylamine hydrochloride 5.0%.

(*xi*) Triethanolamine (1 : 1).

(*xii*) Disodium diethyl dithio carbamate.

The preparation of reagents has been described in Ca and Mg determination in soil extracts (section 2.14.1.2).

4.4.1.3 Procedure

Pipette 10.0 ml aliquot of filtered water sample into 150 ml Erlenmeyer flask. Use 5.0 ml aliquot of water sample if water sample is high in Ca + Mg. Then add 15 or 20 ml of distilled water and one drop of dilute $MgCl_2$.

Now, add 5 drops each of four heavy metal complexing reagents filled in indicator bottle

Rinse 10.0 ml micro burette with EDTA solution filled in a wash bottle provided with a jet, and then fill it with EDTA solution up to 0-ml mark, and set the burette on a burette stand. Place one white porcelain plate below the burette.

At this stage, add 2 ml or more of NH_4Cl—NH_4OH buffer to raise the solution pH to 10.0 with a 10 ml pipette. Add 2 drops of EBT indicator.

Immediately titrate with small portions of EDTA solution and swirl the Erlenmeyer flask. Continue to titrate slowly and carefully until the solution colour changes from pink to purple on addition of just one drop of EDTA.

Now, add one drop of EDTA at a time and then swirl the Erlenmeyer flask. When the solution colour changes from purple to clear blue or green. Note down the titre in two places after decimal.

Similarly repeat sample titration until the difference between the titre is not more than 0.05 ml. Run blank in duplicate following the entire procedure without taking the sample aliquot.

Also pipette 5.0 ml of standard $CaCl_2$ solution in 150 ml Erlenmeyer flask. Then, add 20.0 ml distilled water and then follow the above described procedure beginning with addition of 1 drop of dilute $MgCl_2$. Perform the titration in duplicate or triplicate. Run blank without taking 5 ml aliquot of standard $CaCl_2$ solution.

4.4.1.4 Calculation

(*i*) Normality of EDTA solution $= \dfrac{5.0 \times .01}{(S^1 - B^1)}$

(*ii*) meq of Ca + Mg/l in water sample $= \dfrac{(S - B) \times N \times 1000}{A}$

where

S^1 is EDTA titre for 5.0 ml of .01N $CaCl_2$

B^1 is EDTA titre without 5.0 ml of .01N $CaCl_2$

S is EDTA titre for sample aliquot in ml.

B is EDTA titre without sample aliquot in ml (*i.e.* blank).

N is the normality of EDTA against standard $CaCl_2$ using Mg indicator

A is the aliquot of water sample in ml taken for EDTA titration.

4.4.2 Ca determination by EDTA titration using Ca-indicator

4.4.2.1 Principle. As given in section 4.4.1.1

4.4.2.2 Apparatus/Reagents. As given in section 4.4.1.2 and 2.14.1.2

4.4.2.3 Procedure

Pipette 10.0 ml aliquot of filtered water sample into 150 ml Erlenmeyer flask. Use 5.0 aliquot of water sample if water sample is high in Ca. Then add 15 or 20 ml of distilled water.

Now add a pinch (about 50 mg) of disodium diethyl dithio carbamate crystals and swirl the flask to dissolve the reagents.

Then, rinse one 10.0 ml microburette with EDTA solution filled in wash bottle provided with a jet and then fill it with EDTA solution up to 0 ml mark. Set the burette on a burette stand. Place one white porcelain plate below the burette.

At this stage add 5 drops or more of 4N NaOH solution into the Erlenmeyer flask in order to raise the solution pH to 12.0.

Add a pinch (50 mg) of Ca indicator (ammonium purpurate in powdered K_2SO_4). Swirl the Erlenmeyer flask and observe the solution colour. The solution colour after adding Ca-indicator should neither be deep nor very faint pink. It should be just easy visible pink. One may miss the end point if Ca indicator added is less.

Now, titrate with EDTA solution each time adding small portion of EDTA. Swirl the flask after each addition. Continue to titrate slowly and carefully one drop

EDTA at a time near the end point until the solution colour changes from pink to purple on adding just one drop. There is no further change in colour. A white background is desirable for easy detection of colour from pink to purple. This titration be done on sufficient day light. Note down the titre in two places after decimal.

Similarly repeat the sample titration until the difference between two titre is not more than 0.05 ml. Run blank in duplicate following the above described entire procedure without taking the sample aliquot.

Also pipette 5.0 ml aliquot of standard $CaCl_2$ solution in 150 ml Erlenmeyer flask. Add 20 ml distilled water and follow the entire procedure described above beginning with the addition of a pinch (50 mg) of disodium diethyl dithiocarbamate crystals. Ferform the titration in duplicate or triplicate. Run blank without taking 5.0 ml aliquot of standard $CaCl_2$ solution.

4.4.2.4 Calculation

(*i*) Normality of EDTA solution $= \dfrac{5.0 \times .01}{(S^1 - B^1)}$

(*ii*) meq of Ca/l in water sample $= \dfrac{(S - B) \times N \times 1000}{A}$

where,

S^1 is EDTA titre for 5.0 ml of .01N $CaCl_2$ using Ca-indicator.

B^1 is EDTA titre without 5.0 ml of .01N $CaCl_2$ using Ca-indicator.

S is EDTA titre for sample aliquot in ml.

B is EDTA titre without sample aliquot in ml (i.e. blank).

N is the normality of EDTA against standard $CaCl_2$ using Ca-indicator.

A is the aliquot of water sample in ml taken for EDTA titration.

4.4.3 Mg determination in water sample

Mg concentration in water sample is conveniently determined by substraction of Ca concentration from Ca + Mg concentration obtained by EDTA titration of water sample using Ca indicator and Mg indicator respectively.

meq Mg/l in water = meq Ca + Mg/l in water sample-meq Ca/l in water sample.

4.4.4 Comments

In water sample Ca can be determined also by flame photometer or else by its precipitation as CaC_2O_4 and then separating CaC_2O_4 and titrating it in acid medium by standard $KMnO_4$ solution. Mg can be determined also by gravimetric method after its precipitation as Mg-phosphates and weighing after ignition in muffle furnace as Mg-pyrophosphate. The above methods are described in Ca and Mg determination in plant digest (Section 3.6.2, 3.6.3, 3.6.4 and 3.7.2).

The EDTA titration is convenient as well as quite sensitive. The glass wares must be washed with the detergent. After washing with tap water, it should be rinsed with conc. HCl and then washed three times with sufficient volume of tap water, and thereafter, it should be rinsed twice with distilled water.

The distilled water used in EDTA titration should be free from calcium.

The buffer solution as well as Ca and Mg indicators should be freshly prepared. These reagents should not be stored longer than three weeks. EDTA & $CaCl_2$ should be stored in polythene bottles.

Water samples in humid regions are normally low in basic cations *e.g.*, Ca + Mg. Water samples in arid regions are high in Ca + Mg, but water samples high in SAR are associated with low Ca + Mg relative to Na concentration. In soils having pH above 9.0, Ca + Mg in soil extracts seldom exceed 2 meq./l. The soluble Ca + Mg is usually low in solution having appreciable concentration of CO_3 + HCO_3.

REFERENCES

Jackson, M.L. 1973. Soil Chemical Analysis, Prentice Hall of India Private Limited, New Delhi.

Richards, L.A. *et. al.* (ed). 1954. Diagnosis And Improvement of Saline And Alkali Soils. USDA Hand Book 60, Oxford & IBH Publishing Co., New Delhi.

Cheng, K.L. and R.H. Bray. 1951. Determination of calcium and magnesium in soil and plant material. Soil. Sci. 72. 449 - 458.

Diehl H. *et. al.* 1950. The versenate titration for total hardness. Amer. Water Works Jour. 42 : 42 - 48.

4.5 Na IN WATER SAMPLE – TOTH *et. al.* (1948)

4.5.1 Principle

Na in solution is atomized for spraying into air-propane blue flame. Na on being heated begin to dissipate energy by emission of 589 millimicron wave length. The emission is proportional to the concentration of Na.

The emission from Na after passing through Na filter strikes a photo cell which is connected to a galvanometer. The photo cell converts light energy into electrical energy which is measured by galvanometer in micro amperes. The galvanometer needle or spot light, however deflect on the meter dial in the range of 0 to 100 or more. In digital type of flame photometer, digits increases with the increase in emission which is adjusted and kept within a certain range by regulating the elemental concentration and the instrument sensitivity.

The instrument is first calibrated with Na standards in the range of 0 to 20 ppm Na solution. Then, the sample solution is atomized after desired dilution so that the sample reading lies within the specified range. The concentration of Na in sample solution is determined by reference to the calibration curve.

4.5.2 Reagents

1. Standard Sodium solution, 1.0 g Na/l. Dissolve 2.5421 g of pure dry NaCl (AR) in distilled water and dilute it to 1000 ml mark with distilled water. Store the solution in polythene bottle. This solution has Na concentration of 1000 ppm.

2. Standard Sodium solution, 0 to 20 ppm. Transfer 0, 2.5, 5.0, 7.5 and 10.0 ml of 1000 ppm Na solution into separate 500 ml volumetric flasks. Then, add distilled water in each flask to make the volume to 500 ml. Add 2-3 drops of butyl alcohol. Stopper and mix the contents. The resulting solution will have Na concentration of 0, 5, 10, 15 and 20 ppm. Store the solution in polythene bottles.

If internal $LiCl_2$ standard is desirable, Na standards be prepared keeping uniform concentration of $LiCl_2$ as specified by the manufacturer.

4.5.3 Procedure

1. Operation of the flame photometer. Operate the flame photometer following the procedure given in the instruction manual. The essential feature of the operation of the instrument is given below :

The gas tubing from gas cylinder be connected to gas inlet tube of the instrument. Na filter be placed against the burner flame by rotating the filter disc. Air tubing from air compressor be connected to air inlet tube. Switch on the instrument and then switch on the air compressor. Set the desired pressure by regulating the air inlet knob and checking the air pressure recorded by the pressure gauze.

Now place a lighted match stick above the built-in-gas burner and immediately get the gas valve of the gas regulator opened. When gas burner get lighted, regulate the gas inlet knob to obtain blue flame of sufficient length. Now the instrument is ready for calibration.

2. Calibration of the instrument. First atomize to spray 0 ppm Na solution and adjust the flame photometer reading to zero. Then, atomize 20 ppm Na solution and adjust the flame photometer reading to 100. Atomize 0 and 20 ppm Na solution again and adjust the reading to 0 and 100 by regulating the sensitivity knob. Thereafter, atomize 5, 10 and 15 ppm Na solution and note the corresponding flame photometer reading.

3. Preparation of Calibration Curve. On a linear scale graph paper, set the flame photometer reading scale on the ordinate and Na concentration scale in ppm on the abscissa. Then, plot flame photometer reading on the ordinate corresponding to Na concentration in ppm on the abscissa. Connect the flame photometer reading for 0 and 20 ppm Na by a straight line to obtain a calibration curve for Na.

4. Na determination in diluted water sample. Dilute water sample 50 to 500 times with distilled water and atomize to spray diluted water sample solution after calibrating flame photometer reading with 0 and 20 ppm Na solution. As the diluted water sample give flame photometer reading within the range of 0 to 100, note down the flame photometer reading and the corresponding dilution of the water sample.

Estimate Na concentration in diluted water sample by reference to the calibration curve. Locate the flame photometer reading of water sample on the ordinate and draw an imaginary line parallel to the abscissa intersecting the calibration curve. From the point of intersection, draw a perpendicular to the abscissa, and read Na concentration in ppm given at the base *i.e.* abscissa. Alternatively divide flame photometer reading recorded by 5 to obtain Na concentration in ppm.

4.5.4 Calculation

(*i*) ppm Na in water sample. = ppm Na in diluted sample × dilution times

(*ii*) meq Na/l in water sample. $= \dfrac{\text{ppm Na in water sample}}{\text{equivalent weight of Na i.e. 22.9898}}$

4.5.5 Comments

In flame photometric determination, direct readings without the internal $LiCl_2$ standard give satisfactory results if air propane blue flame is used.

If interference from metallic cations is anticipated, the interfering cations be added to the standards in amounts equal to their concentration in the diluted sample test solution. K interferes if K/Na ratio is 5.0 or greater, Ca interfere if Ca/Na ratio is 10 or greater and Mg interfere when Mg/Na ratio is in excess of 100.

Na concentration in water sample is judged either in relation to concentration of CO_3^{--} and HCO_3^- or else in relation to concentration of $Ca^{++} + Mg^{++}$. For the assessment of suitability of water for irrigation, it would be desirable to estimate. Residual Sodium Carbonate, and Sodium Adsorption Ratio of water sample.

REFERENCES

Jackson, M.L., 1973. Soil Chemical Analysis. Prentice Hall of India, Private Limited, New Delhi.

Richards, L.A. (ed). 1954. Diagnosis and Improvement of Saline And Alkali Soils-USDA Hand Book 60, Oxford & IBH Publishing Co, New Delhi.

Toth, S.J., A.I. Prince, A. Wallace and D.S. Mikkelsen, 1948. Rapid quantitative determination of eight mineral elements in plant tissue by a systematic procedure involving use of a flame photometer, Soil Sci. 66 : 459-466.

4.6 CARBONATES AND BICARBONATES IN WATER SAMPLE

A.O. A.C. (1950)

4.6.1 Principle

The titration of an alkaline solution with a strong acid to pH 8.2 or phenolphthalein end point provide a measure of CO_3^{--}, and further titration to pH

4.5 or methyl orange end point provide a measure of HCO_3^-. The presence of borate, phosphate and silicates will contribute to CO_3^{--} & HCO_3^- titration, but their concentration is seldom in significant amounts except at high pH. The concentration of hydroxyl ions is also generally low except in mineral water or polluted stream water.

4.6.2 Reagents

I. Phenolphthalein indicator-0.25%. Dissolve 0.25 g of phenolphthalein in 50 ml ethyl alcohol and dilute it to 100 ml with distilled water. Stopper and mix the contents for 10 minutes and then filter through whatman No 1 filter paper.

II. Methyl orange indicator, 0.1%. Dissolve 0.1 g of methyl orange in distilled water and dilute it to 100 ml with distilled water.

III. Sodium carbonate, 0.1 N. Dissolve 0.530 g of dry, Na_2CO_3 in distilled water and make the volume to 100 ml mark with distilled water.

IV. Sulphuric acid, approx. 1.0 N. Transfer 2.8 ml of conc. H_2SO_4 (sp. gr. 1.84, Assay 96%) into 100 ml volumetric flask. Then dilute the solution to 100 ml mark with distilled water.

V. Sulphuric acid, 0.05 N. Transfer 50 ml of approx., 1.0 N H_2SO_4 into 1l volumetric flask and dilute the contents to 1000 ml mark with distilled water. Standardize the reagent against standard 0.1 N Na_2CO_3 solution using few drops of methyl orange indicator.

4.6.3 Procedure

Transfer 50 ml of filtered water sample into a 150 ml Erlenmeyer flask. Add 2 drops 0.25% phenolphthalein indicator.

If pink colour develops, titrate the solution with 0.05N H_2SO_4 filled in a 10 ml micro burette adding one drop of acid every 2-3 seconds. Continue titration until the pink colour disappears. Note down the acid titre and designate it as "P".

Now add 1 or 2 drop of methyl orange indicator and titrate the above solution with the standard acid without refilling the burette. Continue titration adding few drops at a time every 2-3 seconds until the colour of the solution changes from yellow to persistent orange. Note down the total acid titre, and designate it as "T".

The solution remaining in the Erlenmeyer flask after titrating to the methyl orange end point may be preserved for the determination of Cl^- in water sample.

Repeat the titration of water sample for CO_3^{--} plus HCO_3^- atleast in duplicate. Similarly run a blank taking 50 ml of CO_2 free distilled water.

The titre value of standard acid for each anion OH^-, CO_3^{--} and HCO_3^-, be estimated from the following relationship.

I. Acid titre for OH^- determination

It is equal to 2P-T when P is greater than 1/2 T. It is equal to T, when P is equal to T.

II. Acid titre for CO_3^{--} determination

It is equal to 2P, when P is less than or equal to 1/2 T. It is equal to 2 (T — P) when P is greater than 1/2 T.

III. Acid titre for HCO_3^- determination

It is equal to T when P is 0. It is equal to T—2P when P is < 1/2 P.

Where P denote mineral acid titre to the phenolphthalein end point.

T denote total mineral acid titre to the methyl orange end point.

4.6.4 Calculation

$$\text{meq } OH^- \text{ or } CO_3^{--} \text{ or } HCO_3^- /l = \frac{(TA - TB) \times N \times 1000}{A}$$

Where

T_A is the mineral acid titre estimated for the anion concerned on its relationship with P and T titre.

T_B is the corresponding acid titre for the blank.

N is the normality of acid determined against standard. Na_2CO_3 solution to the methyl orange end point.

A is aliquot of water sample taken in ml for titration with standard acid.

4.6.5 Comments

The results of the titrations are reproducible and accurate to about 0.1 meq/l.

The titratable alkalinity to the phenolphthalein end point give a measure of soluble carbonates. At high pH, soluble sodium silicates, phosphates and borates also occur in small amounts which contribute to titratable alkalinity and also result in unfavourable chemical and physical characteristics.

The absence of HCO_3^- and CO_3^{--} is not known to have any detrimental effect on plant growth. However, these ions have toxic effects when occur in soil solution in appreciable concentration, HCO_3^- associated with Na cause Fe chlorosis in plants HCO_3^- reduced elongation of radish seedlings apart from depressing protein synthesis. Apples showed marked decline in vigour on irrigation with water high in HCO_3^-. CO_3^{--} even in smallest amounts results in deficiency of Ca + Mg. Besides, it renders soil water caustic and corrosive and it also prevents development of good tilth.

Higher concentration of CO_3^{--} + HCO_3^- relative to Ca^{++} + Mg^{++} in irrigation water favour accumulation of exchangeable Na. Eaton (1950) found a correlation between Residual Sodium Carbonate (RSC) and occurence of sodic soils. He defined RSC as the sum of CO_3^{--} + HCO_3^- minus the sum of Ca^{++} + Mg^{++} each ion expressed in meq/l in the water sample. Wilcox *et. al.* (1954) observed adverse effect when RSC exceeded 1.25 meq/l. The RSC value in water sample is interpreted as follows:

Residual Sodium Carbonate	—	Interpretation
Less than 1.25 meq/l	—	water is safe to use
1.25 to 2.50 meq/l	—	water is marginal
greater than 2.50 meq/l	—	Water is unsafe to use.

The concentration of CO_3^{--} and HCO_3^- in irrigation water can be controlled by adding acid or acid forming fertilizer.

REFERENCES

Amer. Public Health Assoc. & Amer. Water Works Assoc. 1946. Standard Methods For The Examination of Water And Sewage. Ed. 9. 286 pp. illus. New York.

Assoc. Officicial Agric. Chemist, 1950. Official and Tentative Methods of Analysis of The Association of Offcial Agricultural Chemists. Ed. 7, 910 pp. Washington.

Eaton, F.M. (1950). Significance of carbonates in irrigation water. Soil Sci, 69 : 123 - 133.

Jackson, M.L., 1973. Soil Chemical Analysis, Prentice Hall of India Private Limited, New Delhi.

Pratt, P.F. 1975. Carbonates And Bicarbonates pp. 93-97. In H.D. Chapman (ed.) Diagnostic Criteria For Plants and Soils, Eurasia Publishing House (P) Ltd., Ram Nagar, New Delhi.

Richards, L.A. et. al (ed). 1954. Diagnosis And Improvement of Saline And Alkali Soils., USDA Hand Book 60, Oxford & IBH Publishing Company, New Delhi.

Wilcox, L.V., G.Y. Blair and C.A. Bower (1954). Effect of bicarbonate in irrigation water for irrigation, Soil Sci. 77 : 259 - 265.

4.7 SAR AND RSC DETERMINATION IN WATER SAMPLE – RICHARDS ed. (1954) AND EATON (1950)

4.7.1 Principle

The Sodium Adsorption Ratio is a ratio of Na concentration in meq/l to the square root of half the concentration of Ca + Mg in meq/l *i.e.* $SAR = Na^{+} / \sqrt{Ca^{++} + Mg^{++} / 2}$

The Residual Sodium Carbonate is the sum of concentration of carbonate plus bicarbonate in meq/l minus the concentration of calcium plus magnesium in meq/l

i.e. $RSC = (CO_3^{--} + HCO_3^{-}) - (Ca^{++} + Mg^{++})$

The water sample has to be analysed for the determination of the concentration of Na, Ca + Mg and $CO_3^{--} + HCO_3^{-}$ in meq/l to estimate SAR and RSC by way of assessing sodicity hazard.

4.7.2 Reagents

I. For Na determination

Sodium solution - 0, 5, 10, 15 and 20 ppm

II. For Ca + Mg determination

EDTA solution, .01N

Calcium chloride solution, .01N

Magnesium chloride solution, .01N

Ammonium Chloride-Ammonium Hydroxide Buffer

Erochrome Black T indicator

Heavy metal complexing reagents

1% KCN solution

4% Na. K ferrocyanide solution

5% Hydroxylamime hydrochloride

(1 + 1) Triethanoalamine

III. For $CO_3 + HCO_3$ determination

Sulphuric acid, .05 N

Sodium carbonate solution, .1N

Phenolphthalein indicator, 0.5%

Methyl orange, 0.1%

The preparation of reagents are described in appropriate section on element analysis in water sample *i.e.* in section *i.e.* 4.5.2, 4.6.2 and 2.14.1.2.

4.7.3 Procedure

1. Na determination in water sample. Adjust flame photometer reading to 0 by atomizing o ppm Na after placing Na filter opposite air-propane blue flame, and then atomize 20 ppm Na solution and adjust flame photometer reading to 100. Repeat atomization of 0 and 20 ppm Na solution, and check the flame photometer reading. If necessary again adjust the flame photometer reading. Then atomize the intermediate Na standards. Thereafter, atomize to spray water sample diluted 50 to 500 times. If flame photometer reading is found in the range of 0 to 100, note down the flame photometer reading and the corresponding dilution. Determine Na concentration (ppm), in diluted water sample by the following equation.

$$\text{meq Na / l in water sample} = \frac{\text{ppm Na in diluted sample} \times \text{times of dilution}}{23}$$

2. Ca + Mg determination in water sample. Transfer 5.0 ml aliquot of water sample and 20 ml of distilled water into 150 ml Erlemeyer flask. Then add 1 drop $MgCl_2$ (.01 N), 5 drops each of four heavy metal complexing reagent, 2 ml of NH_4Cl-

NH_4OH buffer and 2 drops of EBT indicator. Titrate with .01N EDTA until the reaction mixture colour changes from pink to purple and then from purple to clear blue or green with 1 drop of EDTA. Similarly run a blank taking distilled water instead of sample aliquot.

$$\text{meq Ca + Mg/l in water sample} = \frac{\text{EDTA titre (Sample - Blank)} \times 0.01 \times 1000}{5}$$

3. CO_3^{--} and HCO_3^- determination in water sample. Transfer 50.0 ml aliquot of filtered water sample into a 150 ml Erlenmeyer flask. Add 3 drops of 0.5% phenol lphthalein indicator and titrate with 0.05 N H_2SO_4 until the solution colour changes from purple to colourless. Note down the acid titre and designate it as "P".

Then add 2 drops of 0.1% methyl orange indicator, and titrate with 0.05 N H_2SO_4 without refilling the burette until the colour changes from yellow to orange. Heat the contents to boil gently for 1 minute. If the orange colour turn yellow on heating, and it does not turn orange on cooling add few drops of .05N H_2SO_4 till the solution turn orange. Note down the total acid titre, and designate it as "T"

4.7.4 Calculation

$$\text{meq } CO_3^{--}\text{/l in water sample} = \frac{2P \times .05 \times 1000}{50} \text{ when } P < \frac{1}{2}T \text{ or when } P = T$$

$$\text{-do-} = \frac{2(T - P) \times .05 \times 1000}{50} \text{ when } P > \frac{1}{2}T$$

$$\text{meq } HCO_3^- \text{/l} = \frac{T \times .05 \times 1000}{50} \text{ when } P = 0$$

in water sample

$$\text{-do-} = \frac{(T - 2P) \times .05 \times 1000}{50} \text{ when } P < \frac{1}{2}T$$

where P is acid titre for phenolphthalein end point and T is total acid titre *i.e.* 'P' plus acid titre to methyl orange end point.

$$\text{Sodium Adsorption Ratio} = \frac{Na^+}{\sqrt{\frac{Ca^{++} + Mg^{++}}{2}}}$$

$$\text{Residual Sodium Carbonate} = (CO_3^{--} + HCO_3^-) - (Ca^{++} + Mg^{++})$$

where the concentration of each ion is expressed in meq/l.

4.7.5 Comments

The sodium hazard classes of irrigation water on the basis of SAR is given in Table 4.7.1.

Table 4.7.1 Sodium hazard clauses

Class	S_1	S_2	S_3	S_4
SAR	<10	10 - 18	18 - 26	> 26
Sodium hazard	Low	Medium	High	Very High

The water of SAR class 1 can be safely used on almost all soils. The water of SAR class 2 can be safely used only on coarse textured soils with good soil permeability. There can be appreciable sodium hazard on fine textured soil with low soil permeability. The water of SAR class 3 will cause much greater Na hazard and its use will demand regular application of sufficient water in excess of water requirement to flush out sodium salt deposited by previous irrigation. This would be feasible under conditions of satisfactory drainage. The water of SAR class 4 is unsatisfactors for irrigation. The treatment of water with H_2SO_4 or $CaSO_4.2H_2O$ reduces the SAR value.

The quality of irrigation water as judged by RSC expressed in meq/l is given below

CLASS	1	2	3
RSC	< 1.25	1.25 to 2.50	> 2.50
Interpretation	safe	marginal	unsuitable

Water sample with RSC greater than 2.5 are considered unsuitable. One can add commercial H_2SO_4 in irrigation stream to reduce RSC.

REFERENCES

A.O.A.C. 1950. Official and Tentative Methods of Analysis of the Association of Official Agricultural Chemists. Ed. 7, 910 pp., Washington.

Cheng, K.L. and R.H. Bray. 1951. Determination of calcium and magnesium in soil and plant material. Soil Sci. 72 : 949-458.

Eaton F.M. 1950. Significance of carbonates in Irrigation water Soil Science. 69 : 123-133.

Jackson, M.L. 1973. Soil Chemical Analysis. Prentice Hall of India Private Limited. New Delhi.

Richards, L.A. (ed). 1954. Diagnosis And Improvement of Saline And Alkali Soils. USDA Hand Book 60., Oxford & IBH Publishing Co., New Delhi.

Subramoney, H.S. and Y.S. Verma. 1969. Analysis of irrigation water samples and its interpretation. Directorata of Extension, Krishi Bhavan, New Delhi.

Toth, S.J., A.I. Prince, A. Wallace and D.S. Mikkelsen. 1948. Rapid quantitative determination of eight mineral elements in plant tissues by a systematic procedure involving use of a flame photometer. Soil Sci. 66 : 459-466.

4.8 K DETERMINATION IN WATER SAMPLE BY FLAME PHOTOMETER-STANFORD AND ENGLISH (1949)

4.8.1 Principle

Water sample is diluted and is then atomized to spray in air-propane blue flame of a flame photometer. K ions on gaining energy from the flame begin to dissipate energy in the form of 767 millimicron wave length which is directly proportional to the concentration of K. The emission from K is made to pass through K. filter before striking a photocell connected to a galvanometer. The photocell converts light energy into electrical energy which is measured by the galvanometer. The instrument is first calibrated by atomizing K standards and then water sample is atomized and K in water sample is determined by reference to a calibration curve.

4.8.2 Reagents

I. Standard potassium solution, 1.0 g K/l. Dissolve 1.9066 g pure dry KCl (AR) in distilled water and dilute it to 1000 ml mark with distilled water. The concentration of K in this solution is 1000 ppm. Store the solution in polythene bottle.

II. Standard potassium solution, 0 to 20 ppm. To obtain 0, 5, 10, 15 and 20 ppm K solution, transfer 0, 2.5, 5.0, 7.5 and 10.0 ml of 1000 ppm K solution into separate 500 ml volumetric flask. Add distilled water to make the volume to 500 ml.

If internal lithium standard is specified by the manufacturer the estimated amount of lithium/500 ml K standard be added to each flask before making the volume to 500 ml mark with distilled water.

If interference from metallic cations is anticipated the interfering cations be also added to the K standards in amounts equal to their concentration in the diluted sample test solution prior to making the volume of K standard to 500 ml. Store each K standards in 500 ml polythene bottles.

4.8.3 Procedure

1. Operation of the flame photometer. Place K-filter against the burner by rotating the filter disc and then operate the flame photometer as described in Na determination in water sample.

2. Calibration of the instrument. Atomize 0 and 20 ppm K solution and adjust the flame photometer reading to 0 and 100 respectively. Then atomize the intermediate K standards, and note down the corresponding flame photometer reading.

3. Preparation of calibration curve for K. On a linear scale graph paper, set flame photometer reading scale on the ordinate and K concentration scale on the abscissa. Then, plot flame photometer reading on the ordinate corresponding to K concentration in ppm on the abscissa. Connect flame photometer reading for 0 and 20 ppm K solution by a straight line to obtain the calibration curve for K.

4. K determination in diluted water sample. Dilute water sample by transfering 4.0, 2.0 & 1.0 ml of water sample into separate 100 ml volumetric flask and dilute each to 100 ml to obtain water sample diluted 25, 50 and 100 times.

Atomize to spray 0 and 20 ppm K solution into the flame of the flame photometer, and set the flame photometer reading to 0 and 100. Then atomize diluted water sample. As diluted water sample give flame photometer reading within the range of 0 and 100, note down the flame photometer reading and the corresponding times of dilution.

Estimate K concentration in diluted water sample by reference to the calibration curve or else divide flame photometer reading by 5 to obtain K concentration in ppm.

4.8.4 Calculation

$$\text{ppm K in water sample} = \text{ppm K in diluted water by reference to calibration curve} \times \text{dilution times}$$

$$\text{meq K / l in water sample} = \frac{\text{ppm K in undiluted water sample}}{\text{Equivalent weight of K i.e. 39.102}}$$

K supply from irrigation = meq K /l in irrigation water × 3.91

water in kg K/ha. × Depth of irrigation in cms

4.8.5 Comments

Addition of few drops of butyl alcohol in sample test solution and standards improve the spraying properties of the solution.

In flame photometric determination of K, the interference by Na occurs if Na : K ratio is 5 or greater, the interference by Ca occur if Ca : K ratio is 10 or greater and the interference by Mg occur if Mg : K ratio is in excess of 100. The interference in K determination is very rarely encountered.

K is an essential element for plant nutrition. The potassium needs of crops in general are high, and yet in some areas crops do not respond to fertiizer K. The soils may not be deficient in K, but K present in irrigation water may be contributing in maintaining adequate K in soil. Irrigation to a depth of 5 cm by water containing 1.0 meq K/l would add 19.55 Kg K/ha.

Well water rich in KNO_3 are highly valued in Gujarat for growing high quality tobacco which also fetches high price.

REFERENCES

Jackson, M.L., 1973. Soil Chemical Analysis. Prentice Hall of India Private Limited, New Delhi.

Richards, L.A. (ed). 1954. Diagnosis and Improvement of Saline And Alkali Soils, USDA. Hand Book 60, Oxford and IBH Publishing Co., New Delhi.

Standford, S. and L. English. 1949. Use of the flame photometer in rapid soil tests for K and Ca. Agron. J. 4 : 446-447

Toth, S.J., A.I. Prince, A. Wallace, and D.S. Mikkelsen, 1948. Rapid quantitative determination of eight mineral elements in plant tissues by a systematic procedure involving use of a flame photometer. Soil Sci. 66 : 459-466.

4.9 CHLORIDE DETERMINATION IN WATER SAMPLE-A.O.A.C. (1950)

4.9.1 Principle

The solution containing Cl is titrated by standard $AgNO_3$ in the presence of CrO_4 in an alkaline medium (pH 8.2). As long as some Cl persist in the solution formation of red Ag_2CrO_4 is momentary. When Cl in solution is exhausted through precipitation as AgCl, the red precipitate of Ag_2CrO_4 sharply signals the end point.

4.9.2 Reagents

(*i*) Sulphuric acid, approx. 0.1N.

(*ii*) Sodium carbonate, 0.1 N.

(*iii*) Phenolphthalein indicator, 0.5%.

(*iv*) Sodium Chloride, .05 N.

(*v*) Silver Nitrate, .05 N.

(*vi*) Potassium chromate, 5%.

The preparation of above reagents is described in chloride determination in plant sample (section 3.16.2).

4.9.3 Procedure

Transfer 50 ml aliquot of water sample into a 150 ml Erlenmeyer flask. Adjust the solution pH to 8.2 just colourless to the phenolphthalein end point by adding first 0.1N Na_2CO_3. When the solution colour becomes purple, add few drops of 0.1 N H_2SO_4 to change the solution colour from purple to just colourless *i.e.* pH 8.2.

Now add 1 ml of K_2CrO_4 indicator, and titrate the solution with standard $AgNO_3$ solution till the appearance of first permanent red colouration due to precipitation of Ag_2CrO_4. Similarly run a blank with the same volume of Cl free distilled water.

4.9.4 Calculation

$$\text{meq Cl/l in water sample} = \frac{(S-B) \times N \times 1000}{A}$$

Where,

S is ml of $AgNO_3$ titre used for precipitating Cl in aliquot of water sample

B is ml of $AgNO_3$ titre used for precipitating Cl in same volume of distilled water.

N is the normality of $AgNO_3$ solution as determined by titration against standard NaCl solution

A is ml aliquot of water sample taken for titration.

4.9.5 Comments

The CrO_4^{--} react with Ag^+ to form red coloured Ag_2CrO_4 only in alkaline medium (pH 8.0) on the exhaustion of Cl due to precipitation as AgCl.

The results of $AgNO_3$ titration of Cl are reproducible and accurate to the order of .05 meq/l. In case of highly coloured solution, the Cl may be titrated potentiometrically using Ag. AgCl half cell.

The irrigation water high in Cl may cause injury in Cl sensitive crops such as citrus, peaches and grape vines. Water containing 6 meq Cl/l was found to be injurious to citrus orchards in california. In general Cl as $CaCl_2$is considered to be more toxic than Cl present as NaCl. No definite limits has been established for classifying irrigation water on the basis of Cl concentration. Generally water containing more then 10 meq Cl/l is not considered to be safe for irrigation.

Eaton (1942) reported a slight decrease in dry weight of wheat plant at Cl concentration of 10 meq/l in the substrate but at Cl concentration of 100 meq/l, the growth was reduced to 30%.

Pearson (1959) reported relative yield of rice at Cl concentration of 22 and 120 meq/l to be 100 and 20, whereas Ehrler and Bernstein (1958) reported relative growth of rice at Cl concentration of 6 and 56 meq/l in the substrate to be 100 and 23. Cl at higher concentration induces sterility in plants.

REFERENCES

A.O.A.C. 1950. Official And Tentative Methods of Analysis of The Association of Official Agricultural Chemists. Ed. 7, 910 pp., Washington.

Eaton, F.M. 1942. Toxicity and accumulation of chloride and sulfate salts in plants. J. Agric. Res. 64 : 357-399.

Eaton, F.M. 1975. Chlorine pp. 98-135. In H.D. Chapman (ed) Diagnostic criteria For Plants And Soils. Eurasia Publishing House (P) Ltd. New Delhi.

Ehrler, W. and L. Bernstein. 1958. Effects of root temperatures, mineral nutrition, and salinity on the growth and composition of rice. Bot. Gaz. 120 : 67-74.

Jackson, M.L. 1973. Soil Chemical Analysis. Prentice Hall of India, (Pvt) Ltd., New Delhi.

Pearson G.A. 1959. Factors Influencing salinity of submerged soils and growth of caloro rice. Soil Sci. 87 : 198-206.

4.10 SULPHUR IN WATER SAMPLE

4.10.1 Sulphate sulphur in water sample by titrimetric method-modified from Manager *et. al* (1950)

4.10.1.1 Principle

SO_4-S in irrigation is precipitated as $BaSO_4$ with $BaCl_2$ solution and the $BaCl_2$ added in excess is back titrated by standard EDTA solution using EBT indicator in presence of $MgCl_2$ and enough NH_4Cl-NH_4OH buffer.

A separate aliquot of the same volume of irrigation water is directly titrated by standard EDTA solution without the addition of $BaCl_2$ to determine meq of Ca + Mg present in the sample aliquot. A blank is run with standard $CaCl_2$ solution equivalent to Ca + Mg present in aliquot of sample solution and the same volume of $BaCl_2$ and $MgCl_2$ as taken in sample titration.

The blank sample titre gives a measure of SO_4-S present in the sample aliquot.

4.10.1.2 Apparatus/Reagents

Erlenmeyer flask 125 and 150 ml capacity, microburette, graduated pipettes 10, 5, 2 and 1 ml capacity. volumetric flasks and hot plate.

Reagents

(*i*) Ammonium chloride-ammonium hydroxide buffer. Dissolve 67.5 g of NH_4Cl in 570 ml of conc NH_4OH and make the volume to 1l.

(*ii*) 0.01 N calcium chloride solution. Dissolve 0.500 g of pure Ca CO_3 (calcite crystals) in 10 ml of approx (1 + 3) HCl and dilute it to 1000 ml with distilled water.

(*iii*) 0.01N Barium chloride solution. Dissolve 1.2212 g of $BaCl_2 \cdot 2H_2O$ in distilled water and dilute the solution to 1000 ml mark with distilled water.

(*iv*) 0.01N Magnesium chloride solution. Dissolve 1.0161 g of $MgCl_2 \cdot 6H_2O$ in distilled water and dilute the solution to 1000 ml mark with distilled water.

(*v*) 0.01 N Ethylene diamine tetra acetic acid (EDTA) solution. Dissolve 2.00 g of disodium dihydrogen ethylene diamine tetra acetic acid and 0.05 g of $MgCl_2.6H_2O$ in distilled water and dilute to a volume of 1000 ml mark with distilled water.

(*vi*) Eriochrome black T indicator. Dissolve 0.5g of Eriochrome black T and 4.5 g of hydroxylamine hydrochloride ($NH_4Cl \cdot NH_2OH$) in 100 ml of 95% ethanol. Prepare fresh reagent every month.

(*vii*) 1.0% Potassium cyanide solution. Dissolve 1.0 g of KCN in distilled water and dilute the solution to 100 ml with distilled water.

(*viii*) 5.0% Hydroxylamine hydrochloride solution. Dissolve 5.0 g of NH_4Cl. NH_2OH in distilled water and dilute the solution to 100 ml.

(*ix*) 50% Triethanolamine solution. Transfer 25ml of triethanol amine into 50 ml volumetric flask and add distilled water to dilute the reagent to 50 ml mark. Shake to mix the contents.

(*x*) 0.1% methyl orange indicator. Dissolve 0.1g of methyl orange indicator in 100 ml of distilled water.

(*xi*) 10.0 N Hydrochloric acid. Dilute 218.2 ml of conc. HCl (AR) to 250 ml with distilled water.

(*xii*) 2.0N Hydrochloric acid. Dilute 20 ml of 10 N HCl to 100 ml with distilled water.

(*xiii*) 0.02N Hydrochloric acid. Dilute 2.5 ml of 2N HCl to 250 ml with distilled water.

4.10.1.3 Procedure

1. Standardization of EDTA solution against standard .01N $CaCl_2$. Pipette 5.0 ml of 0.01N $CaCl_2$ into 125 ml Erlenmeyer flask. Add 10 ml of distilled water, and four drops each of 1.0% KCN, 5.0% NH_4Cl . NH_2OH, 50% triethanolamine and 0.01N $MgCl_2$. Thereafter add 2 ml of NH_4Cl . NH_4OH buffer, and 1 drop of EBT indicator. Then titrate the solution by EDTA until the colour changes from wine red to purple. Then add 1 or 2 drop more until the colour changes to clear blue or green. Similarly run a blank. Perform the titrations in triplicate and take the mean of two closest titre. Divide .05 by the (sample-blank) titre to estimate the normality of EDTA.

2. Standardization of EDTA against standard 0.01N $BaCl_2$ solution. Pipette 5.0 ml of standard 0.01N $BaCl_2$ solution into 125 ml Erlenmeyer flask and proceed as described in step I beginning with the addition of 10 ml of distilled water. Perform the titration in triplicate and take the mean of the closest two titre. Similarly run a blank. Divide 0.05 by (sample-blank) titre to estimate the normality of EDTA against .01N $BaCl_2$ solution.

3. Total alkalinity in sample of irrigation water. Pipette 10.0 ml aliquot of sample of irrigation water into 125 ml Erlenmeyer flask. Then add 10 ml of distilled water, a pinch of $CaCO_3$ and 2 drops of 0.1% methyl orange indicator, and titrate by 0.02N HCl until the colour changes from yellow to orange. Heat to boil the contents gently for 30 seconds and remove the flask from the hot plate and let it cool. On cooling if the solution remains orange the end point has reached. Perform the titration in triplicate. Take the mean of the closest titre. Similarly run a blank. Substract blank titre from sample titre and multiply by the normality of HCl to estimate total alkalinity of the sample aliquot.

4. Ca + Mg in aliquot of sample of irrigation water. Pipette 10.0 ml aliquot of sample of irrigation water into 125 ml Erlenmeyer flask. Then proceed as described in step I beginning with the addition of 10 ml of distilled water. Perform the titration in triplicate and take the mean of two closest titre. Similarly run a blank. Substract blank titre from sample titre, and multiply by the normality of EDTA estimated against standard 0.01 N $CaCl_2$ to obtain the meq of Ca + Mg in 10 ml aliquot of sample of irrigation water.

5. Precipitation of SO_4-S in aliquot of sample of irrigation water with $BaCl_2$ and back titration of $BaCl_2$ added in excess by EDTA titration. Pipette 10.0 ml aliquot of sample of irrigation water into a 150 ml Erlenmeyer flask. Then, add 10 ml of distilled water and neutralize the alkalinity of the sample solution by adding equivalent volume of 0.02 N HCl. Add 1 ml of 0.02 N HCl in excess and heat the contents on a hot plate to boil gently for 30 seconds. Remove the Erlenmeyer flask from the hot plate and add 10.0 ml of 0.01N $BaCl_2$. Again place the Erlenmeyer flask on a hot plate to boil the contents gently for 30 seconds.

Remove the Erlenmeyer flask from the hot plate and let it cool. Thereafter, add 10 ml of ethanol, 3.0 ml of 0.01N $MgCl_2$ and 10 ml of NH_4 Cl–NH_4OH buffer, and 6 drops of EBT indicator. Immediately titrate the contents by 0.01N EDTA until the colour of the solution changes from wine red to purple. Then add EDTA drop by drop and shake the contents after each addition until the colour changes from purple to clear blue or green. Then, add 2.0 ml of 0.01N $MgCl_2$ solution and 2 drops of EBT indicator and further titrate the contents by standard EDTA solution slowly until the colour of the contents changes from wine red to purple, and then add EDTA drop by drop and shake the contents with each addition of EDTA until the colour changes from purple to clear blue or green. Note down the EDTA titre. Perform the titration in triplicate and take the mean of the closest two titre.

6. Determination of the blank titre. Pipette 10.0 ml of distilled water into 150 ml of Erlenmeyer flask. Then, add 0.01N $CaCl_2$ solution equivalent to the meq of Ca + Mg estimated in 10.0 aliquot of sample of irrigation water. Thereafter, proceed as described in step 5 beginning with the addition of 1.0 ml of 0.02N HCl. Perform the titration in triplicate and take the mean of the closest two titre.

4.10.1.4 Calculation

$$SO_4 - S \text{ in meq /l} = \frac{(A - B) \times C \times 1000}{D}$$

$$\begin{array}{l}\text{ppm } SO_4 - S \\ \text{in water sample}\end{array} = \frac{\text{meq } SO_4\text{–S/litre} \times \text{milligram Eq. wt. of } SO_4\text{-S} \times 10^6}{1000}$$

$$= \text{meq } SO_4 - S \text{ /litre} \times 16.032$$

where A is blank EDTA titre for titrating 10 ml distilled water + $CaCl_2$equivalent to Ca + Mg in 10 ml aliquot of irrigation water + 10 ml of .01N $BaCl_2$ + 5 ml of .01N Mg Cl_2.

B is the sample EDTA titre for titrating 10.0 ml irrigation water + 10 ml dist. water + 10.0 ml of 0.01N $BaCl_2$ + 5.0 ml of .01 N $MgCl_2$.

C is the normality of EDTA determined against standard 0.01N $BaCl_2$ using EBT- as Mg indicator.

D is ml aliquot of sample of irrigation water taken for the determination of SO_4–S

4.10.1.5 Comments

The plants absorb sulphur as sulphates from soil, and so sulphates in irrigation water partly meets the sulphur requirement of the crop plants.

The addition of ethanol on precipitation of SO_4 as $BaSO_4$ decrease the solubility of $BaSO_4$.

There is deterioration of the EBT indicator prepared in the medium of ethanol on storage and it is therefore desirable to prepare the EBT indicator fresh every three weeks.

In titration of both sample and blank solution even with sufficient $MgCl_2$ by EDTA using EBT indicator in presence of excess of ethanol, the first end point is not sharp. Therefore, $MgCl_2$and EBT is added for the second time and the solution is titrated again by EDTA to the clear blue or green end point.

If water sample is passed through resin column to remove interfering cations prior to precipitation of SO_4 by $BaCl_2$, the determination of Ca + Mg in aliquot of sample of irrigation water as well as addition of 0.01N $CaCl_2$ to the blank equivalent to Ca + Mg in aliquot of sample of irrigation water will not be required.

The ion exchange resin column for removing interfering cations may be prepared as described below :

Fit up a resin column made from pyrex glass tubing 225 mm long and 7 mm diameter. Place half inch layer of glass wool at the bottom and pour down the slurry consisting of Amberlite 120 H^+ form (30-50 mesh) cation exchange resin suspended in water so that the resin column measures about 80 mm.

Wash the column with 2 M sodium hydroxide followed by water and then 2M nitric acid and water. Repeat this process three to four times. The resin column is now ready.

Regenerate the column each time with 2M nitric acid and wash it down completely with deionized water before passing the sample of irrigation water. The column, when not in use, should always be kept moist with deionized water which should stand about 5 to 6 cm on its top. It should also be free from any air bubbles.

REFERENCES

Biedermann and Schwarzenback (1948). Chimia 2 : 56.

Cheng, K.L. and R.H. Bray (1951). Determination of Calcium and Magnesium in soil and plant material. Soil. Sci. 72 : 449-458.

Chopra, S.L. and J.S. Kanwar (1976). Analytical Agricultural Chemistry. Kalyani Publishers Ludhiana, New Delhi.

Diehl, H., C.A. Goetz and C.C. Hauch (1950). The versene titration for total hardness Amer. Water Works Assoc. J. 42 : 40-48.

Jackson, M.L. (1958). Soil Chemical Analysis. Prentice Hall of India Private Limited. New Delhi 1973.

Munger et. al. 1950. Anal. Chem. 22 : 1455.

4.10.2 SO_4–S in water sample by gravimetric method after precipitation as $BaSO_4$-A.O.A.C. (1950)

4.10.2.1 Principle

A sample of irrigation water (100 ml) is transferred into 300 ml tall beaker. The solution is acidified with HCl and then, it is diluted to about 200 ml. Thereafter, the solution is heated to boil and $BaCl_2$solution is added in excess drop by drop with continuous stirring. The beaker is placed on a water bath and heating is continued until the solution is reduced to a volume of about 50 ml. The suspension after cooling is filtered through a ignited and weighed crucible. The precipitate of $BaSO_4$is quantitatively transferred into Gooch crucible and it is made free from chloride by washing with distilled water. Then the Gooch crucible is placed again in a muffle furnace at low red heat. The crucible is weighed after cooling to estimate SO_4-S as $BaSO_4$.

4.10.2.2 Apparatus/Reagents

Apparatus

Tall beaker 300 ml cap, watch glass, burette 50 ml cap. hot plate, water bath, electric muffle furnace, Gooch crucible and desiccator.

Reagents

(*i*) Conc. Hydrochloric acid (AR).

(*ii*) 10% Barium chloride solution. Dissolve 10 g of $BaCl_2 . 2H_2O$ in 100 ml distilled water.

(*iii*) 0.1% Methyl orange indicator. Dissolve 100 mg of methyl orange indicator in 100 ml of distilled water.

4.10.2.3 Procedure

1. Precipitation of sulphate as $BaSO_4$. Take an aliquot of sample of irrigation water (100 ml) containing between 0.2 and 5 meq of sulphate and dilute it to a volume of 200 ml. Add few drops of 0.1% methyl orange indicator and 1 ml of conc. HCl. Heat the solution to boil and add an excess of 10% $BaCl_2,2H_2O$ solution drop by drop with constant stirring. Place the beaker on a water bath and evaporate the contents until the volume is reduced to about 50 ml. Remove the beaker from the water bath and let it cool.

2. Washing, Ignition and weighment of the $BaSO_4$ precipitate. Transfer the $BaSO_4$ precipitate in beaker in small portions into an ignited and preweighed Gooch crucible. Transfer the precipitate quantitatively into the Gooch crucible using distilled water. Thereafter, wash the precipitate in Gooch crucible with distilled water until the precipitate becomes free from chloride. Test filtrate (leachate) with few drops of dilute $AgNO_3$ solution.

When the precipitate has become free from chloride, dry the contents of the Gooch crucible, and then ignite the Gooch crucible with $BaSO_4$ precipitate inside an electric muffle furnace. Transfer the Gooch crucible from the electric muffle furnace into a desiccator, and let it cool further. Thereafter, weigh the Gooch crucible plus $BaSO_4$ precipitate on a chemical balance up to fourth place after decimal, and substract the mass of empty Gooch crucible to find out the mass of $BaSO_4$ precipitate.

4.10.2.4 Calculation

$$\text{S in irrigation water (ppm)} \left[\frac{A \times B}{D}\right] \times 10^6 = \left[\frac{A \times 0.137376}{100}\right] \times 10^6$$

$$SO_4^{--} \text{ / in meq / 1000 ml irrigation} = \frac{A \times B^1 \times C}{D \times F} = \frac{A \times 0.4116 \times 1000}{100 \times 1.0480}$$

water

where

A is mass of ignited $BaSO_4$ obtained from aliquot of the sample of irrigation water.

B is mass of sulphur per gram of barium sulphate.

B^1 is mass of sulphate per gram of barium sulphate.

C is total volume of irrigation water *i.e.*, one litre

D is ml aliquot of sample of irrigation water taken for the precipitation of sulphate as barium sulphate.

F is milligram equivalent weight of SO_4.

4.10.2.5 Comments

The gravimetric method of total S determination in irrigation water may be used as a basis for assesing indirect method of S determination in irrigation water. The sulphur needs of the crop plants is partly met from sulphur supplied by irrigation water.

REFERENCES

A.O.A.C. (1950). Official and Tentative Methods of Analysis of the Association of Official Agricultural Chemists. Ed.. 7, 910 pp. Illus. Washington.

4.10.3 SO_4–S in water sample by turbidimetric method of Chesnin and Yien (1950)

4.10.3.1 Principle

$BaSO_4$ turbidity is developed in aliquot of sample of irrigation water in medium of sodium acetate-acetic acid buffer (pH 4.8). The turbidity developed is stabilized by adding gum acacia solution. The absorbance of tubidity is measured on a spectrophotometer at 440 mμ wavelength or else on a Klett photoelectric colorimeter with 440 mμ (blue) filter. The concentration of SO_4-S in the aliquot of sample solution is estimated by reference to the standard calibration curve.

4.10.3.2 Apparatus/Reagents

Apparatus

Volumetric flasks, graduated pipettes, spectrophotometer or Klett photoelectric colorimeter with 440 mμ (blue) filter.

Reagents

(*i*) Sodium acetate-acetic acid buffer (pH 4.8). Dissolve 25g of sodium acetate (AR) in distilled water. Add 7.8 ml of glacial acetic acid (AR) and make the volume to 250 ml

(*ii*) 0.25% gum acacia solution. Dissolve 0.25 g of powdered gum acacia in distilled water. Make the volume to 100 ml and allow it to stand over night. Thereafter filter the solution through whatman No 1 filter paper.

(*iii*) Barium chloride crystals (50 mesh sieve).

(*iv*) 50 ppm SO_4–S solution. Dissolve 0.2722 g of K_2SO_4 (AR) in distilled water and dilute the solution to 100 ml mark with distilled water.

4.10.3.3 Procedure

1. Development of $BaSO_4$ turbidity with aliquot of the sample of irrigation water. Transfer 5.0 ml of irrigation water into a 25 ml volumetric flask. Then add 10 ml of sodium acetate-acetic acid buffer (pH 4.8) and 1 ml of 0.25% gum, acacia solution into the same vol. flask. Mix the contents and add 1g of $BaCl_2$ crystals (< 50 mesh size). Allow it to stand for a minute. Then, swirl the contents frequently until the crystals are dissolved. Dilute the solution to 25 ml mark with distilled water. Similarly run a blank with 5 ml of distilled water.

2. Measurement of the absorbance of $BaSO_4$ turbidity developed with aliquot of irrigation water. Within 5 to 12 minutes of developing $BaSO_4$ turbidity, measure the absorbance on a spectrophotometer at 440 mμ wavelength or else on a Klett photoelectric colorimeter with 440 mμ (blue) filter. Set 0 absorbance with the turbidity developed with the blank, and measure the absorbance of $BaSO_4$ turbidity developed with aliquot of irrigation water. Prior to measurement of the absorbance, the instrument should be switched on to allow it to warm for 10 minutes. The cuvette or colorimeter tube be filled with distilled water and placed in the socket provided. Besides, the appropriate knob be adjusted to the desired wave length or else the desired filter be fitted before the instrument is switched on.

3. Development of $BaSO_4$ turbidity with the SO_4–S standards. Pipette 0, 2, 4, 6 and 8ml of 50 ppm SO_4-S solution into separate 25 ml volumetric flasks. Then, add to each 10 ml of sodium acetate-acetic acid buffer (pH 4.8) and 1 ml of 0.25% gum acacia solution. Mix the contents and add 1g of $BaCl_2$ crystals (50 mesh size). Allow it to stand for a minute. Then, swirl the contents until the crystals are dissolved. Dilute the solution to 25 ml with distilled water.

4. Measurement of $BaSO_4$ turbidity developed with the SO_4–S standards and preparation of the standard calibration curve. Measure the absorbance within 5 to 12 minutes of developing $BaSO_4$ turbidity with SO_4–S standards. Set 0 absorbance with $BaSO_4$ turbidity developed with 0 µg SO_4–S and then measure the absorbance of the subsequent SO_4–S standards. Prior to the measurement of the absorbance, switch on the spectrophotometer or Klett photoelectric colorimeter and allow it to warm for 10 minutes. The cuvette or colorimeter tube be filled with distilled water and placed in the socket provided, and the appropriate knob be adjusted to the desired wave length or else the desired filter be fitted before the instrument is switched on.

Record the absorbance of the turbidity corresponding to the µg SO_4–S taken. On a linear scale graph paper, let ordinate be the scale for absorbance, and let the abscissa be the scale for µg SO_4–S taken for developing turbidity. Plot the value of absorbance corresponding to the µg SO_4–S taken for developing turbidity. Connect 0 absorbance with the maximum absorbance by a straight line to obtain a standard calibration curve.

5. Estimation of SO_4–S in $BaSO_4$ turbidity developed with aliquot of sample of irrigation water. Divide the absorbance of turbidity developed with the aliquot of sample of irrigation water by the slope of the straight line standard calibration curve in units of absorbance/μg SO_4–S to obtain the content of μg SO_4–S in the aliquot of sample of irrigation water.

4.10.3.4 Calculation

(1) ppm $SO_4 - S$ in irrigation water = $\frac{A}{B} = \frac{A}{5}$

(2) meq $SO_4 - S$ /l in irrigation water = $\frac{A}{B \times C} = \frac{A}{5 \times 48.03}$

Where A is μg SO_4–S in aliquot of sample of irrigation water as estimated by reference to the standard calibration curve.

B is ml aliquot of sample of irrigation water taken for developing $BaSO_4$ turbidity.

C Equivalent weight of SO_4–S.

4.10.3.5 Comments

The sulphur needs of the crop plants is partly met from the SO_4–S present in irrigation water.

The irrigation water may be passed through resin column as it would decatinize the sample of irrigation water and remove the interfering ions.

REFERENCES

Chesnin, L. and C.H. Yien (1950). Tubidimetric determination of available sulphate. Soil. Sci. Soc. Amer. Proc. 15 : 149-151.

4.10.4 SO_4–S in water sample by turbidimetric method-Tabatabai (1974)

4.10.4.1 Principle

In aliquot of sample of irrigation water, $BaSO_4$ turbidity is developed in an acidic medium by adding 1 ml of 0.5% $BaCl_2 . 2H_2O$ reagent in 0.3% gelatin solution. The $BaSO_4$ turbidity developed is stabilized by the gelatin and its absorbance is measured 30 minutes after developing turbidity at 420 mμ wave length on a spectrophotometer or photo-electric colorimeter.

4.10.4.2 Apparatus/Reagents

Apparatus

Volumetric flasks, graduated pipette, 500 ml capacity Beaker, spectro photometer or else photoelectric colorimeter.

Reagents

(*i*) Distilled water-free from sulphur. Use deionized water or else use water distilled through completely glass distill after adding few grains of $KMnO_4$ crystals in the distillation flask.

(*ii*) 10.0 N Hydrochloric acid. Dilute 218.2 ml of conc. HCl (AR) to 250 ml with distilled water.

(*iii*) 0.5 N Hydrochloric acid. Dilute 5.0 ml of 10 N HCl to 100 ml with distilled water

(*iv*) 0.5% Barium chloride solution in 0.3% gelatin solution. Dissolve 0.6g of gelatin (Difco Bacto gelatin, Difco Lab, Detroit, Michigan or its substitute) in 200 ml of warm (60-70°C) distilled water. Thereafter, place the solution in a refrigerator at 4°C, and allow it to stand for 16 hours. Then, remove the semi-gelatin from the refrigerator and allow it to cool to room temperature and add 1.0 g $BaCl_2.2H_2O$ crystals and shake the contents until $BaCl_2$ crystals are dissolved. Store the solution in a refrigerator. Remove the reagent from the refrigerator and let it stand for 2 hours at room temprature before use.

(*v*) 1000 ppm SO_4–S solution. Dissolve 5.4438g of K_2SO_4 in distilled water and dilute the solution to 1000 ml mark with distilled water.

(*vi*) 0 to 5 ppm SO_4–S standards. Pipette 0, 1, 2, 3, 4 & 5 ml of 1000 ppm SO_4–S solution into separate 1000 ml volumetric flasks and dilute each SO_4–S standards to 1000 ml mark with distilled water to obtain 0 to 5 ppm SO_4–S solution.

4.10.4.3 Procedure

1. Development of $BaSO_4$ turbidity with an aliquot of sample of irrigation water. Pipette 20.0 ml aliquot of irrigation water into a 50 ml volumetric flask. Then add 2.0 ml of 0.5 N HCl. Swirl to mix the contents, and add 1.0 ml of $BaCl_2$ gelatin reagent. Stopper and mix the contents. Similarly run a blank taking 20.0 ml of distilled water and following the procedure as for sample aliquot beginning with the addition of 2 ml of 0.5 N HCl.

2. Measurement of absorbance of $BaSO_4$ turbidity developed with aliquot of irrigation water. Prior to measurement of absorbance, switch on the spectrophotometer adjusting wavelength knob to 420 mμ or photoelectric colorimeter fitted with 420 mμ (blue) filter and the cuvette or test tube filled with distilled water placed in the socket provided and allow the instrument to warm for 10 minutes.

Thereafter, measure the absorbance of $BaSO_4$ turbidity developed in aliquot of irrigation water adjusting 0 absorbance with the turbidity developed in blank with 20 ml of distilled water.

3. Development of $BaSO_4$ turbidity with SO_4-S standards containing 0 to 100 µg SO_4–S. Transfer 0, 20, 40, 60, 80 and 100 μg SO_4–S by pipetting 20 ml of 0–5 ppm SO_4—S standards into separate 50 ml volumetric flasks and then proceed as described in step 1 of procedure beginning with the addition of 2.0 ml of 0.5 N HCl.

4. Measurement of absorbance of $BaSO_4$ turbidity developed with 0–100 μg SO_4–S. Adjust 0 absorbance with $BaSO_4$ turbidity developed with 0 μg SO_4–S on a spectrophotometer or else on a photoelectric colorimeter at 420 μg wave length, and measure the absorbance of the subsequent standards *i.e.* $BaSO_4$ turbidity developed with 20 to 100 μg SO_4–S.

5. Preparation of the standard calibration curve. On a linear scale graph paper, let the ordinate be the scale for absorbance and abscissa be the scale for μg SO_4–S taken for developing turbidity. Then, plot the value of absorbance corresponding to the μg SO_4–S taken for developing $BaSO_4$ turbidity. Connect the point of 0 absorbance and the point of maximum absorbance by a straight line to obtain a straight line standard calibration curve.

6. Estimation of SO_4–S in the $BaSO_4$ turbidity developed with the aliquot of irrigation water. Divide the absorbance of $BaSO_4$ turbidity developed with aliquot of irrigation water by the slope of the standard calibration curve expressed in unit of absorbance/μg SO_4–S taken for developing $BaSO_4$turbidity to obtain ug SO_4–S present in aliquot of irrigation water.

4.10.4.4 Calculation

(1) SO_4–S in irrigation water (ppm) $= \dfrac{A}{D} = \dfrac{A}{20}$

(2) SO_4–S in irrigation water (meq/l) $= \dfrac{A}{D \times E} = \dfrac{A}{20 \times 48.03}$

Where A is μg SO_4–S in aliquot of irrigation water as estimated by reference to the standard calibration curve.

D is ml aliquot of irrigation water taken for developing $BaSO_4$ turbidity.

E is the equivalent weight of SO_4–S i.e. 48.03

4.10.4.5 Comments

When water sample contains more than 5 ppm SO_4–S reduce the size of the sample aliquot proportionately so that μg SO_4–S taken for developing $BaSO_4$turbidity does not exceed 100 μg. Add distilled water to make the volume to 20 ml prior to acidification with 2 ml of 0.5 N HCl.

The $BaCl_2$ gelatin reagent be prepared fresh and it be not stored for more than a week even when stored in a refrigerator.

REFERENCES

Tabatabai, M.A. (1974). A rapid method for the determination of SO_4 in water sample. Environmental Letters 7 : 237–243.

4.10.5 Sulphur in water sample by the reduction method of Johnson and Nishita (1952)

4.10.5.1 Principle

2 ml aliquot of sample of irrigation water is transferred to 50 ml reduction-distillation flask, and then it is treated with the reduction mixture and the flask is attached to a distillation apparatus. Purified N_2 gas is passed through the reduction mixture. The reduction-distillation flask is heated by a micro burner to distill H_2S.

H_2S distilled is made to pass through a gas washing column and then it is absorbed in zinc acetate-sodium acetate buffer solution. The H_2S absorbed is treated with acid solution of p-amine dimethyl aniline sulphate and ferric ammonium sulphate solution to form methylene blue colour, the absorbance of which is measured at 670 mμ wave length on a spectrophotometer or on a photoelectric colorimeter.

4.10.5.2 Apparatus/Reagents

Almost the same apparatus as described in soluble sulphate by the reduction method (section 2.21.5.2.2).

Reagents

3 to 12 as described in the determination of total S in plant by the reduction method (section 3.9.3.2).

4.10.5.3 Procedure

1. Treatment of the aliquot of irrigation water with the reduction mixture, and distillation of H_2S formed. Transfer 2 ml of irrigation water into the 50 ml of reduction-distillation flask. Add 4 ml of reduction mixture and immediately attach the flask to the condenser assembly, and follow steps 2 and 3 of procedure described in total S in plant sample by the reduction method (section 3.9.3.3).

2. Development of methylene blue colour, measurement of its absorbance preparation of standard calibration curve, and estimation of S in aliquot of the sample of irrigation water. Follow steps 4 to 9 of procedure described in total S determination in plant sample by the reduction method (section 3.9.3.3).

4.10.5.4 Calculation

$$\text{S in irrigation water (ppm)} = \frac{A}{D} = \frac{A}{2}$$

where A is μg S in aliquot of irrigation water as estimated by reference to the standard calibration curve by the methylene blue method.

D is ml aliquot of irrigation water taken for the reduction and subsequent distillation of H_2S for the development of methylene blue colour.

4.10.5.5 Comments

The procedure for the determination of sulphur in irrigation water will not distinguish between sulphate, sulphite, sulphide and possibly very labile organic sulphur compounds. Besides, a number of organic sulphur compounds do not liberate hydrogen sulphide in this procedure.

REFERENCES

Johnson, C.M., and Hideo Nishita (1952). Micro estimation of sulfur in plant materials, soils and irrigation water. Anal. Chem. 24 : 736-742.

4.11 B DETERMINATION IN WATER SAMPLE

4.11.1 B determination in water sample by forming boron-carmine coloured complex-Hatcher and Wilcox (1950)

4.11.1.1 Principle

B in 2 ml water sample is treated with few drops of conc. HCl to eliminate interference from NO_3 and then it is treated with 10 ml of conc. H_2SO_4 and 10 ml of .05% carmine solution in sigcol or low boron corning Erlenmeyer or volumetric flask. The reaction mixture is kept inside a desiccator for 45 minutes for development of the maximum colour. Thereafter, absorbance or % transmittance of the coloured complex is measured at 585 mμ wave length adjusting 0 absorbance with the coloured complex similarly prepared with 2 ml distilled water. B in water sample is estimated by reference to a calibration curve.

4.11.1.2 Reagents

(*i*) Sodium hydroxide 1.0 N NaOH. Dissolve 4.0 g of NaOH pellets (AR) in distilled water and dilute the solution to 100 ml mark in a sigcol or low boron corning volumetric flask. Do not use pyrex or borosilicate glasswares for B determination.

(*ii*) Conc. Hydrochloric acid. AR, Sp. gr. 1. 18, Assay 35.4%.

(*iii*) Hydrochloric acid, 1.0 N. Dilute 8.7 ml of conc. HCl to 100 ml with distilled water

(*iv*) Conc. sulphuric acid AR, Sp. gr. 1.84, Assay 98.0%.

(*v*) Carmine solution, .05% in conc. H_2SO_4 by weight. Dissolve 0.125 g carmine in 250 ml conc. H_2SO_4 (sp. gr. 1.84).

(*vi*) Concentrated B solution, 100 μg B/ml. Transfer 0.5716 g of oven dry H_3BO_3 (AR) into 1 l volumetric flask. Add distilled water and shake to dissolve the reagent and then dilute it to 1000 ml mark with distilled water.

(*vii*) Dilute B solution 0 to 10 μg B/ml. Transfer 0, 2.5, 5.0, 7.5 and 10.0 ml of B solution containing 100 μg B/ml into separate 100 ml volumetric flasks and dilute each to 100 ml mark with distilled water to obtain 0, 2.5, 5.0, 7.5 and 10.0 ppm B solution.

4.11.1.3 Procedure

1. Development of carmine B coloured complex with 0 to 10 ppm B solution. Transfer 2.0 ml of 0, 2.5, 5.0, 7.5 and 10.0 ppm B solution into separate Erlenmeyer or volumetric flask. Then add 2 drops of conc. HCl to eliminate interference from NO_3. Now, add 10 ml of conc. H_2SO_4 to each flask. Swirl to mix the contents and let it cool. Then add 10 ml of .05% carmine solution to each flask, swirl to mix the contents and allow it to stand at least for 45 minutes in a desiccator containing conc. H_2SO_4.

2. Measurement of absorbance or % transmittance of coloured complex by use of spectrophotometer or photoelectric colorimeter. Adjust the wavelength knob to

585 mμ in a spectrophotometer or else fit 585 mμ green filter into the filter holder and place it into the socket provided in the photoelectric colorimeter.

Now place cuvette or colorimeter tube filled with conc. H_2SO_4 and place it into the socket provided. Connect the instrument to a voltage stabilizer and switch on the instrument to allow it to warm for 10 minutes before measurement of the absorbance.

Remove the coloured complex from the dessicator after 45 minutes. Fill a clean cuvette with coloured complex obtained with 2 ml distilled water. The cuvette be rinsed first with the same coloured complex. Now, remove the cuvette filled with H_2SO_4 from the instrument and replace it by the cuvette filled with coloured complex developed with 2 ml distilled water *i.e.* with 0 μg B.

Adjust the absorbance to 0 with the adjustment knob. Take cuvette which was filled with H_2SO_4 and fill it with coloured complex developed with 2 ml of 2.5 ppm B solution. Place this cuvette into the cuvette socket after removing the cuvette filled with colour complex developed with 2 ml distilled water *i.e.* with 0 μg B.

Note down the absorbance value at which the meter needle rest. In photoelectric Klett colorimeter, the galvanometer needle will deflect from 0 and so the dial knob has to be slowly rotated until the needle moves back to zero. At this point read the dial value coinciding with the zero line and note it down as Klelt reading which is equal to absorbance × 500.

Now remove the cuvette from the instrument and immediately replace it by cuvette filled with coloured complex developed with 0 μg B. Now move back the Klett dial knob until the galvanometer needle rest at 0 line and the Klett dial reading is also zero. In spectrophotometer, if the dial needle is not at 0 absorbance, set it at zero with the adjustment knob.

Similarly measure the absorbance of coloured complex developed with 2 ml of 5.0, 7.5 and 10.0 ppm B solution after adjusting 0 absorbance with coloured complex developed with 2 ml distilled water i.e. 0 μg B/ml. Record the absorbance or Klett Reading corresponding to μg B taken for the development of coloured complex. In a spectrophotometer, below the absorbance scale % transmittance scale is printed. Absorbance is equal to log 100 - log % transmittance.

3. Calibration curve. When linear scale graph paper is used, set absorbance or klett reading scale on the ordinate and μg B taken for the development of coloured complex on the abscissa. Then plot the absorbance or Klett reading corresponding to the μg B taken for colour development. Connect the absorbance point for 0 and 20 μg B by a straight line to obtain the calibration curve.

When semilog graph paper is used % transmittance is plotted on the log scale and μg B taken for colour development on the linear scale.

4. B in water sample. Pipette 2 ml aliquot of distilled water and water sample into separate Erlenmeyer flasks. Add 2 drops of conc. HCl, 10 ml of conc H_2SO_4 and 10 ml of .05% carmine in conc. H_2SO_4 in each Erlenmeyer flask and transfer into a desiccator for atleast 45 minutes.

Switch on the spectrophotometer 10 minutes before taking measurement on absorbance. Adjust wave length knob at 585 mμ. Place cuvette filled with conc. H_2SO_4.

After allowing 45 minutes time for the development of coloured complex in aliquot of water sample and distilled water, adjust 0 absorbance with colour complex developed with 0 μg B, and then measure the absorbance of colour complex developed with 2 ml water sample as described in section on measurement of absorbance. Then estimate the content of boron in 2 ml aliquot of water sample.

If the absorbance reading of water sample exceeds the absorbance obtained with 2 ml of 10 ppm B, B determination be repeated after dilution of the water sample.

If B concentration in water sample is too low, take a much larger aliquot of water sample in an evaporating dish. Add 1.0N NaOH in excess of amount required to make the solution alkaline. Evaporate the contents to dryness on a water bath. Transfer the residue into conical centrifuge tube with 5 or 10 ml of 1N HCl. Centrifuge at 1500 rpm for 10 minutes. Pipette 2 ml aliquot of supernant clear solution into Erienmeyer flask and proceed beginning with the development of coloured complex.

If sample is coloured due to dissolved organic matter, 1N NaOH be added in excess of amount required to make the sample aliquot alkaline. Evaporate the contents to dryness on water bath. Allow it to dry and then transfer it to a muffle furnace at 525°C for one hour. Remove the evaporating dish from the muffle furnace and let it cool. Now transfer the residue into conical centrifuge tube with 5 or 10 ml 1N HCl. Pipette 2 ml aliquot of supernant clear solution into Erlenmeyer flask and proceed beginning with the development of coloured complex. Simultaneously and simillarly run a blank.

4.11.1.4 Calculation

$$\text{ppm B in water sample} = \frac{\mu\text{g B in aliquot of water sample by reference to calibration curve}}{\text{Volume of aliquot in ml}}$$

4.11.2 B in water sample by forming boron–curcumin coloured complex–Naftel (1939) and Dible *et al.* (1954)

4.11.2.1 Principle

An aliquot of water sample containing < 4.5 μg of B is made alkaline with Ca $(OH)_2$ suspension in an evaporating dish. The content is evaporated to dryness on a water bath. The residue is made distinctly acidic with 2.5 N HCl before treating with curcumin-oxalic acid solution.

The solution is evaporated to dryness at low heat, and residue is baked for 15 minutes and is then dissolved in 10 ml of 95% ethyl alcohol and transferred to a 15 ml conical centrifuge tube. The suspension is centrifuged for 10 minutes at 1500 rpm. Then most of the clear solution is transferred to a 25 ml volumetric flask and the volume is made to 25 ml with 95% ethyl alcohol.

The curcumin react with B to form red coloured complex called "rosecyanine" which is dissolved in 95% ethyl alcohol, the absorbance of which is measured at

540 mμ wave length. A calibration curve is prepared by taking absorbance of curcumin-B coloured complex with graded amount of boron (0 to 4.5 μg)

The boron content in water sample is determined by developing red coloured complex "rosecyanine" on reacting curcumin with boron in water, and by reference to the calibration curve on measuring its absorbance by use of spectrophotometer or photoelectric colorimeter.

4.11.2.2 Reagents

(*i*) Ethyl alcohol 95% (AR).

(*ii*) Hydrochloric acid, 2.5 N. Dilute 21.8 ml of conc. HCl (sp gr. 1.18, Assay 35.4%) to 100 ml with distilled water.

(*iii*) Calcium hydroxide solution, 0.4%. Transfer 0.4 g $Ca(OH)_2$ into 100 ml volumetric flask and add distilled water and shake to mix the contents. Make the volume to 100 ml.

(*iv*) Curcumin-Oxalic acid solution. Dissolve 0.04 g of curcumin (1-7 bis (4 hydroxy -3 methoxy phenyl) -1.6 heptadiene-3.5 dione), and 5.0 g of oxalic acid in 100 ml of 95% ethyl alcohol. Store the solution in low boron glass stoppered bottle (sigcol or low boron corning) or in polythene bottle inside a refrigerator.

(*v*) Phenolphthalein indicator, 0.05%. Dissolve 0.05 g of phenolphthalein in 50 ml of 95% ethyl alcohol, and dilute the solution to 100 ml with distilled water.

(*vi*) Concentrated B solution, 100 μg B/ml. Transfer 0.5716 g of oven dry H_3BO_3 (AR) into 1000 ml volumetric flask. Add distilled water and shake to dissolve the reagent. Then make the volume with distilled water.

(*vii*) Dilute B solution, 0 to 4.5 μg B/ml. Transfer 0, 1.5, 3.0 and 4.5 ml of 100 µg B/ml solution into separate 100 ml volumetric flask and dilute each to 100 ml mark with distilled water.

4.11.2.3 Procedure

1. Development of curcumin-boron coloured complex with B standards. Pipette 1.0 ml of dilute B solution (1.5, 3.0 & 4.5 ppm) and 1.0 ml of distilled water into separate porceclain, silica or platinum dish. Add 5 ml of 0.4% $Ca(OH)_2$ suspension to each evaporating dish, and evaporate the contents to dryness on a water bath. Remove the evaporating dish from water bath and let it cool.

Now add 1 drop of phenolphthalein indicator and add 2.5 N HCl drop by drop until the indicator colour become colourless. Then, add 0.5 ml of 2.5 N HCl in excess.

Thereafter, add 4 ml of curcumin-oxalic acid solution. Rotate the dish so that the curcumin-oxalic acid solution comes in contact with the entire residue. Now, evaporate the contents at 55°± 3°C in an oven to dryness and then allow the contents to bake for 15 minutes.

Moisten the red coloured product known as rosecyanine with 95% ethyl alcohol taken in a wash bottle provided with fine jet. Then, transfer the residue into 15 ml conical centrifuge tube with 10 ml of 95% ethyl alcohol in portions, each time triturate the evaporating dish with a rubber policeman.

Run the centrifuge at 1500 rpm for 10 minutes, and decant the clear solution into 25 ml volumetric flask leaving traces of solution remaining in the centrifuge tube. Make up the volume to 25 ml by adding 95% ethyl alcohol.

2. Measurement of the absorbance of coloured complex at 540 mμ wave length. Setting 0 absorbance with coloured complex developed with 1 ml distilled water *i.e,* .0 μg B, determine the absorbance of coloured complex developed with 1.5, 3.0 and 4.5 μg B as described in section on measurement of absorbance in method 1 (section 4.11.1.3).

3. Calibration curve. Set absorbance scale on the ordinate of a linear scale graph paper and the amount of B taken in μg on the abscissa. Then plot the absorbance value corresponding to the amount of boron taken. Connect the point of absorbance of 0 & 4.5 μg B by a straight line to obtain the calibration curve.

4. B determination in water sample. Transfer 1 to 10 ml aliquot of water sample containing less than 4.5 μg of B and proceed with the development of coloured complex as for dilute B solution. Simultaneously run a blank taking distilled water in place of water sample.

Set 0 absorbance with the blank and determine the absorbance of coloured complex developed with aliquot of water sample at 540 mμ wave length. Determine the content of B in aliquot of water sample by reference to the calibration curve.

If sample solution is coloured with dissolved organic fraction, pipette a suitable aliquot into an evaporating dish and treat it with 5 ml of 0.4% $Ca(OH)_2$ suspension to make it alkaline. Simultaneously run a blank taking distilled water in place of sample aliquot. Evaporate the contents to dryness on a water bath, and then ignite for an hour at 525°C in a muffle furnace. Remove the evaporating dish from the muffle furnace and let it cool. Then add 1 drop of phenolphthalein indicator and proceed further as described in the development of coloured complex with 0 to 4.5 μg of B (step 1, section 4.11.2.3).

4.11.2.4 Calculation

$$\text{ppm B in water sample} = \frac{\mu\text{g B in aliquot of water sample by reference to calibration curve}}{\text{Volume of aliquot in ml}}$$

4.11.3 B in water sample by curcumin spot test – Yoshida and Yoshida (1964)

4.11.3.1 Principle

The dry residue remaining on evaporation of 1 ml water sample and 0.1 ml of 0 to 5 ppm B solution in circular cavities of spot plate pretreated with 2% solution of silicon oil, is treated with 0.25 ml of curcumin-oxalic acid after adding 0.1 ml of 0.5 N HCl.

The reaction mixture is evaporated to first dryness and then it is further baked for 15 minutes. The baked residue is dissolved in 0.5 ml of 20% glycerin in 95%

ethyl alcohol by stirring with a polythene rod. The curcumin–B coloured complex developed with water sample and dilute B standards are matched to ascertain content of B in water sample.

4.11.3.2 Reagents

(*i*) Silicone oil, 2%.

(*ii*) Hydrochloric acid, 0.5 N. Dilute 4.4 ml of conc. HCl (Sp. gr. 1.18, Assay, 35.4, AR) to 100 ml with distilled water.

(*iii*) Glycerin solution, 20%. Transfer 10 ml glycerin into 50 ml volumetric flask and make the volume to 50 ml with 95% alcohol.

(*iv*) Curcumin-oxalic acid solution. Dissolve 0.04 g of curcumin and 5 g of oxalic acid in 100 ml of 95% ethyl alcohol.

(*v*) Concentrated B solution, 100 μg B/ml. Transfer 0.5716 g of oven dry H_3BO_3 into 1l volumetric flask. Add distilled water and shake the contents to dissolve the reagent. Then make the volume to 1000 ml mark with distilled water.

(*vi*) B solution, 10 μg B/ml. Dilute 10 ml of 100 ppm B solutions to 100 ml with distilled water.

(*vii*) B solution, 0-4.5 μg B/ml. Transfer 0, 2.5, 5.0, 7.5, 15.0 and 22.5 ml of 10 ppm B solution into separate 50 ml volumetric flasks and dilute each to 50 ml mark with distilled water to obtain 0, 0.5, 1.0, 1.5, 3.0 and 4.5 ppm B solution.

4.11.3.3 Procedure

Wash a white porcelain spot plate having 1 ml capacity circular cavities with detergent and rinse it thoroughly with distilled water.

Dry the spot plate, and then give a thin coating of 2% silicon oil to the surface of the spot plate and then dry it in an oven at 120°C for 30 minutes.

Remove the spot plate and let it cool. Then transfer 1 ml of water sample and 0.1 ml of B standards 0 to 4.5 μg B/ml and evaporate the contents to dryness in an oven or else under infra red lamp.

Remove the spot plate and let it cool. Then add 0.1 ml of 0.5 N HCl and 0.25 ml of curcumin-oxalic acid solution to each cavity in which water sample or B standards (0-4.5 μgB/ml) was added in the beginning.

Transfer the spot plate to an oven at 55°C to evaporate the contents to dryness and then to bake it for 15 minutes.

Remove the spot plate from oven and let it cool. Then add 0.5 ml of 20% glycerin in 95% ethanol to each cavity in which curcumin-oxalic acid solution was added. Stir with a polythene rod to dissolve the red coloured product of curcumin-boron known as rosecyanine.

Match the coloured complex developed with the residue from 1 ml water sample and 0.1 ml of boron standards and ascertain the content of boron in 1 ml water sample.

4.11.3.4 Calculation

$$\text{ppm B in water sample} = \frac{S \times dt}{10 \times A}$$

where

S is Boron standards in ppm with which coloured complex from water sample is matching

dt is times of dilution of water sample if any.

A is aliquot of water sample taken for colour development in ml.

4.11.3.5 Comments

Use sigcol or corning glassware labelled low in boron. Use distilled water from a metal distill instead of distilled water from a borosilicate or pyrex glass distillation unit. It is safe to store reagents in a polythene bottle.

As heating an acid solution above 55°C cause loss of B by voltalization, the sample be made alkaline before heating.

In carmine method, the carminic acid in conc. H_2SO_4 changes its colour from a bright red to a bluish red or blue depending upon the concentration of B present. The colour remains stable from 45 minutes to 4 hours. The carmine method is more suitable for samples high in B concentration.

After adding carmine, the solution containing B be preferably kept in a desiccator or stoppered properly for colour development. Otherwise, the solution is likely to get diluted as a result of adsorption of water from the atmosphere.

Few drops of conc. HCl is added to the sample aliquot prior to the addition of carmine just to eliminate the possible interference from oxidizing agents such as NO_3. Nitrates above 20 μg per aliquot interfere in B determination.

The curcumin reacts with B to form red coloured product called rosecyanine. The red colour is intensified by oxalic acid. The rosecyanine is solubilized in ethyl alcohol for the measurement of absorbance. It has the advantage of avoiding corrosive solvent such as conc. H_2SO_4. Besides, the absorbance of the coloured complex is less sensitive to temperature.

The concentration of B in irrigation water even at the level of 1 ppm may cause injury to B sensitive crops such as Lemon. However, alfalfa makes maximum growth with irrigation water containing 1-2 ppm B. The level of B concentration as proposed by Scofield (1936) for assessing the quality of irrigation water is given in Table 4.11.1.

Table 4.11.1 Permissible limits of B for several classes of irrigation water

	ppm B in irrigation waterm		
Boron Classes	**Sensitive Crops**	**Semi tolerant crops**	**tolerant crops**
1	< 0.33	< 0.67	< 1.00
2	0.33 - 0.67	0.67 - 1.33	1.00 - 2.00
3	0.67 - 1.00	1.33 - 2.00	2.00 - 3.00
4	1.00 -1.25	2.00 - 2.50	3.00 - 3.75
5	> 1.25	> 2.50	> 3.75

REFERENCES

Diblc, *et. al.* 1954. Anal. chem. 26 : 418.

Hatcher, J.J., and L.V. Wilcox (1950). Colorimetric determination of boron using carmine. Anal. Chem. 22 : 250–252.

Jackson, M.L. 1973. Soil Chemical Analysis. Prentice Hall of India, Private Limited, New Delhi.

Naftel, 1939. Ind. Eng. Chem. A.E. 11 : 407.

Richards, L.A. (ed). 1954. Diagnosis and Improvement of Saline and Alkali Soils. USDA Hand Book 60. Oxford and IBH Publishing Co. New Delhi.

Scofield, C.S. 1936. The Salinity of Irrigation Water. Smithan Inst. Ann. Rept. 1935 : 275-287 illus.

4.12 NO_3–N IN WATER SAMPLE

4.12.1 NO_3–N determination in water by phenol disulphonic acid method-Harper (1924) and Prince (1945)

4.12.1.1 Principle

Nitration of position 6 of 2,4.–phenol disulphonic acid in fuming H_2SO_4 yields a product which is colourless in acidic medium and yellow in neutral and alkaline medium. The intensity of colour developed is proportional to the concentration of NO_3.

Nitrate phenoldisulphonic colour is developed with 0 to 100 μg NO_3–N. The coloured complex is diluted to 100 ml and its absorbance is read at 420 mμ wave length in a spectrophotometer. The absorbance of coloured complex corresponding to μg NO_3–N taken is plotted on a linear scale graph paper to draw a calibration curve. The absorbance of nitrate-phenol disulphonic colour developed with aliquot of water sample is read, and the content of μg NO_3–N in aliquot of water sample is estimated by reference to the calibration curve.

4.12.1.2 Reagents

(*i*) Phenol 2, 4-Disulphonic acid. Dissolve 25 g of crystal clear pure phenol in 150 ml of conc. H_2SO_4(AR) in 500 ml capacity flat bottom round flask. Add 75 ml of fuming H_2SO_4(minimum 20% SO_3) under a fume hood. Immerse the flask in a hot water bath, and keep it immersed for 2 hours. Store in a amber coloured bottle.

(*ii*) NO_3–N, 100 ppm solution. Transfer 0.7221 g of KNO_3 in 1 litre volumetric flask. Add distilled water. Shake to dissolve the reagent. Make the volume to 1000 ml with distilled water.

(*iii*) NO_3–N, 0 to 10 ppm solution. Transfer 0, 2, 4, 6, 8 and 10 ml of 10 ppm NO_3–N solution into separate 100 ml volumetric flasks and make the volume to 100 ml with distilled water to obtain 0 to 10 ppm NO_3–N solution.

(*iv*) Silver sulphate solution, .02 N. Dissolve 3.12 g Ag_2SO_4 in 1 litre distilled water.

(*v*) Silver nitrate solution, .05N. Dissolve 8.494 g of $AgNO_3$ in distilled water, and dilute the solution to 1000 ml mark with distilled water.

(*vi*) Sodium chloride solution .05 N. Dissolve 1.923 g of oven dry NaCl (AR) in distilled water and dilute the solution to 1000 ml mark with distilled water.

(*vii*) Potassium chromate solution 5.0%. Dissolve 5.0 g of K_2CrO_4 in 80 ml distilled water. Then add saturated $AgNO_3$ solution dropwise with stirring until a permanent red precipitate is produced. Filter the solution and dilute the filtrate to 100 ml with distilled water.

(*viii*) Ammonium hydroxide, approx. 7N. Transfer 50 ml NH_4OH (Analer sp. gr. 0. 90) in a 100 ml volumetric flask containing 40 ml distilled water. Dilute the solution to 100 ml with distilled water.

(*ix*) Calcium Carbonate powder (AR).

(*x*) Darco G-60

4.12.1.3 Procedure

1. Clarification of water sample. Filter water sample through whatman No. 1 filter paper to remove suspended constituents. Discard the initial filtrate after rinsing the receiving flask and collect the subsequent filtrate.

Transfer 100 ml of filtered water sample and 100 ml of distilled water into separate 250 ml Erlenmeyer flask. Add to each 1g of Darco G-60. Stopper the flasks and shake the contents on a mechanical shaker for 15 minutes. Then filter the contents of two flasks separately through dry whatman No. 44 filter paper. Discard the initial 10 ml filtrate after rinsing the 150 ml cap. receiving Erlenmeyer flask and collect the clear subsequent filtrate for determination of NO_3–N.

2. Precipitation of chloride in sample and blank test solution. Titrate 20 ml of sample and blank test solution with .05 N $AgNO_3$ solution on adding 4 drops of 5% K_2CrO_4 solution until the colour of the reaction mixture changes to permanent red. Note down the $AgNO_3$ titre for the aliquot of sample and blank test solution.

Now, transfer 20 ml of sample and blank test solution into separate 25 ml. volumetric flasks. Add .02N Ag_2SO_4 equal to 2.5 times of the $AgNO_3$ titre into the respective volumetric flasks and make the volume to 25 ml with distilled water. Allow it to stand for 10 minutes.

Then, transfer the contents of each 25 ml volumetric flasks into 50 ml capacity centrifuge tubes and run the centrifuge at 1500 rpm for 10 minutes. Switch off the centrifuge and allow the centrifuge to come to rest. Decant off the supernant clear sample and blank test solution free from chloride into separate clean and dry 50 ml beaker.

3. Development of Nitrate phenol disulphonic yellow colour with NO_3–N standards. Transfer 10.0 ml aliquot of 0, 2, 4, 6, 8 and 10 ppm NO_3–N solution into separate 8 cm evaporating dish and add to each 50 mg of $CaCO_3$ powder. Then,

evaporate the contents to dryness on a water bath. Remove the evaporating dish from water bath, and allow it to cool.

Add 3 ml of phenol disulphonic acid into each evaporating dish rapidly. Rotate the evaporating dish to dissolve the entire residue in phenol disulphonic acid. Allow it to stand for 10 minutes.

Then, add 10 ml distilled water to each evaporating dish and transfer the contents to separate 100 ml volumetric flasks. Add another 10 ml portions of distilled water into each evaporating dish and then transfer the solution to the respective 100 ml volumetric flasks.

Now, add approx. 7N NH_4OH to make the solution alkaline as indicated by the development of yellow colour and then add 3 ml of 7N NH_4OH in excess. Dilute the solution to 100 ml mark with distilled water. Stopper and mix the contents by turning the flask upside down holding the stopper firmly. Allow 10 minutes time for maximum development of the yellow colour.

4. Measurement of the optical density/absorbance of the nitrate phenol disulphonic colour with the NO_3–N standards. Fit 420 mμ blue filter in the filter holder of Klett colorimeter or else set the wave length knob of a spectrophotometer at 420 mμ. Then place cuvette/test tube filled with distilled water in the socket provided. Switch on the instrument and allow it to warm for 10 minutes.

Thereafter adjust 0 absorbance with colour complex developed with 0 μg NO_3–N at 420 mμ wave length and determine the absorbance of coloured complex developed with 20, 40, 60, 80 and 100 μg NO_3–N.

5. Calibration curve for NO_3–N. On a linear scale graph paper, set absorbance scale on the ordinate and μg NO_3–N taken for developing nitrate phenoldisulphonic colour on the abscissa. Then plot absorbance corresponding to NO_3–N taken in μg and connect the points plotted to obtain the calibration curve.

6. Development of nitrate-phenol disulphonic colour with the sample and blank test solution.

Transfer 10.0 ml aliquot of chloride free sample and blank test solution into separate 8 cm evaporating dish and then add to each 50 mg of $CaCO_3$ powder. Evaporate the contents to dryness on a water bath.

Thereafter, proceed as for NO_3–N standards beginning with the addition of 3 ml of phenol disulphonic acid solution (step 3 of section 4.12.1.3)..

7. Measurement of absorbance of nitrate-phenoldisulphonic colour and determination of NO_3-N concentration in aliquot of sample and blank test solution.

Adjust 0 absorbance with the coloured complex developed with the aliquot of blank test solution at 420 mμ wave length and read the absorbance of coloured complex developed with the aliquot of sample test solution.

Determine the concentration of NO_3–N in aliquot of the sample test solution by reference to the calibration curve for NO_3–N.

4.12.1.4 Calculation

$$\text{ppm } NO_3\text{–N in water sample} = \frac{S}{A}$$

$$\text{meq } NO_3\text{–N/l of water sample} = \frac{\text{ppm } NO_3\text{–N in water sample}}{EW}$$

where,

S is NO_3–N in µg in aliquot of sample test solution by reference to the calibration curve for NO_3–N.

A is ml aliquot of sample test solution taken for the development of NO_3.

$$\text{phenol disulphonic colour} = \frac{20}{25} \times 10 = 8 \text{ ml}$$

EW is the equivalent weight of NO_3–N *i.e.* 14.0g.

4.12.2 NO_3–N in water sample by Kjeldahl distillation in the presence of Devarda alloy-A.O.A.C. (1950)

4.12.2.1 Principle

Filtered water sample in large volume (600 ml) is transferred to 800 ml Kjeldahl flask. The water sample is boiled to evaporate the contents till the volume decreases from 600 ml to 500 ml. Then, MgO is added, and boiling is continued till the volume of water sample decreases from 500 ml to 200 ml. NH_4–N present in water sample is also lost by volatalization. Boiling is discontinued and the contents are allowed to cool.

At this stage paraffin wax and glass beads are added into the Kjeldahl flask, and the Kjeldahl flask is attached to the condenser unit through the glass adapter provided with ground joints. The feeding funnel is also fitted into the glass adapter. The condenser outlet tube is immersed in boric acid solution containing mixed indicator.

Now Devarda alloy and 40% NaOH is added through the feeding funnel. The reaction mixture made strongly alkaline is boiled so that NO_3 reduced to NH_3 in presence of Devarda alloy is distilled. The boiling is continued until 125 ml distillate is collected. NH_3distilled into boric acid form ammonium borate which is back titrated with a standard dilute mineral acid in the presence of brom cresol green-methyl red indicator.

4.12.2.2 Reagents

(*i*) Chromic acid.

(*ii*) Paraffin wax.

(*iii*) MgO.

(*iv*) Devarda alloy.

(*v*) Sodium hydroxide solution, 40%.

(*vi*) Boric Acid solution, 2%.

(*vii*) Brom cresol green-methyl red indicator.

(*viii*) Standard Hydrochloric acid, .02 N.

The preparation of reagents (*v*) to (*viii*) are described in section 2.8.2.

4.12.2.3 Procedure

1. Concentration of water sample. Transfer 600 ml of filtered water sample into 800 ml Kjeldahl flask and heat to boil for evaporating the contents. When the volume decreases from 600 ml to 500 ml add 1 g MgO and resume heating the Kjeldahl flask and continue boiling the contents till the volume decreases from 500 ml to 200 ml. Stop heating and allow the content and the Kjeldahl flask to cool. NH_4–N present in water sample will be lost by volatalization and the NO_3–N will get concentrated in the remaining water sample.

2. Reduction of NO_3–N in remaining water sample and its distillation. Add few glass beads and 1g paraffin wax into the Kjeldahl flask and attach the Kjeldahl flask to a condenser unit through a glass adapter provided with ground joints. Also fit the feeding funnel into the second glass joint. Lower the end of condenser outlet tube inside a 300 ml Erlenmeyer flask filled with 40 ml of 2% H_3BO_3containing brom cresol green-methyl red indicator. Let the end of condenser outlet tube remain dipped in boric acid solution.

Check that the distillation assembly is air tight. Fill feeding funnel with 10 ml distilled water and then lift the glass rod stopper of feeding funnel so that a portion of distilled water flow into the Kjeldahl flask. If it results in bubbling of air through boric acid solution due to displacement of air, the joints are air tight and it will not permit escape of NH_3 through the joints during distillation.

Now add 2g Devarda alloy and 50 ml of 40% NaOH through the feeding funnel into the Kjeldahl flask. Then fill feeding funnel with distilled water and allow 10 ml distilled water to flow into the Kjeldahl flask to wash away conc. NaOH adhering to the feeding funnel drain. Stopper the drain after letting distilled water to pass.

Thereafter heat the Kjeldahl flask to boil the contents, and continue boiling to distill NH_3 formed as a result of reduction of NO_3 in the presence of Devarda alloy. Collect 150 ml distillate in the receiving Erlenmeyer flask. Then lift the condenser outlet tube out of the distillate plus boric acid and rinse it with distilled water collecting the rinsing also in the same receiving Erlenmeyer flask. Remove the Erlenmeyer flask containing the distillate. Place a rubber stopper and label it.

Stop heating the Kjeldahl flask and allow it to cool. Remove the Kjeldahl flask when it is no longer hot. Throw away the contents and wash it thoroughly with tap water, rinse once with chromic acid. Again wash it with tap water and then rinse it with distilled water.

Now run a blank following the entire procedure described above taking 600 ml of distilled water in place of filtered water sample.

3. Titration of sample and blank distillate with a standard mineral acid. Titrate the sample and blank distillate with a standard mineral acid one at a time slowly

adding 2-3 drops of acid and then swirling the contents after each addition. Continue titration until the reaction mixture changes its colour from green to faint pink. Add few drops more to attain a distinct pink colour. Titrate sample and blank distillate to the same end point and note down the sample and blank titre.

4.12.2.4 Calculation

$$\text{ppm } NO_3\text{–N in water sample} = \frac{(S-B)\times N\times ME\times 10^6}{A}$$

$$\text{meq } NO_3\text{–N/l water sample} = \frac{\text{ppm } NO_3 - N \text{ in water sample}}{EW}$$

where, S is acid titre in ml used in titration of sample distillate

B is acid titre in ml used in titration of blank distillate

N is the normalility of mineral used taken for titration

ME is the milligram equivalent weight of NO_3–N *i.e.* .014 g

A is ml aliquot of water sample taken for NO_3–N determination *i.e.* 600ml.

EW, is the equivalent weight of NO_3-N *i.e.* 14.0 g.

4.12.2.5 Comments

In the determination of NO_3–N by phenol disulphoric acid method, Cl concentration greater than .03% cause loss of NO_3 during the reaction between phenol disulphonic acid and NO_3. It is desirable that for the colorimetric determination of NO_3 –N, the water sample is low or free from Cl.

The Cl in water sample can by determined by titration with $AgNO_3$ in the presence of K_2CrO_4 as indicator and the equivalent amount of Ag_2SO_4 may be added to an aliquot of water sample to precipitate Cl. The precipitate settle on centrifugation and supernant clear solution free from chloride can be decanted for the analytical work.

Water sample contains dispersed particles which can be removed by filtration. The clarification of water sample by use of Darco G-60 is also desirable. The Darco G-60 adsorb constituent which impart colour to water sample.

The use of phenol disulphonic acid for NO_3–N determination by colorimetric procedure is adapted from Harper (1924) and Prince (1945).

Devarda alloy is composed of 50% Cu, 45% Al and 5% Zn. In the presence of Devarda alloy, NO_3 is reduced to NH_3 which can be determined by Kjeldahl distillation method. The method is particularly convenient and preferable when water sample is high in chloride.

Care be taken that the lab is free from nitric acid fumes when NO_3–N determination is in progress.

Nitrates are readily absorbed by plants as well as it is readily leached from soil as NO_3 is not as much adsorbed by soil as other ions. NO_3 is so mobile that it may reach the sub soil and it may be carried away further with drainage eventually reaching ground water. Well water in Gujarat containing nitrates particularly associated with potassium are very much preferred by farmers growing quality tobacco. Chloride impart earthly flavour to tobacco whereas KNO_3 improves burning quality of tobacco. Water containing nitrates can supplement nitrogen needs of the crop. But NO_3 in plant is reduced by molybdenase activity before it is assimilated. If Mo is deficient, NO_3 remain unutilized and plant show characteristic systems of nitrogen deficiency due to Mo deficiency inspite of accumulation of nitrates.

REFERENCES

A.O.A.C. 1950. Official And Tentative Methods of Analysis of The Association of Official Agricultural Chemists. Ed. 7, 910 pp, Washington.

Harper 1924. Ind. Engr. Chem. 16 : 180

Jackson, M.L. 1973. Soil Chemical Analysis. Prentice Hall of India Private Limited New Delhi.

Prince, 1945. Soil Sci. 59 : 47.

APPENDIX 1

Temperature factors (f_1) for correcting resistance and conductivity data on soil extracts to the standard temperature of 25°C.

$$EC_{25} = EC_t \times f_t \; ; \; EC_{25} = (k/R_t) \times f_t \; ; \; R_{25} = R_t / f_t$$

°C	°F	t_1	°C	°F	f_t	°C	°F	f_t
3.0	37.4	1.709	22.0	71.6	1.064	29.0	84.2	0.925
4.0	39.2	1.660	22.2	72.0	1.060	29.2	84.6	.921
5.0	41.0	1.613	22.4	72.3	1.055	29.4	84.9	.918
6.0	42.8	1.569	22.6	72.7	1.051	29.6	85.3	.914
7.0	44.6	1.528	22.8	73.0	1.047	29.8	85.6	.911
8.0	46.4	1.488	23.0	73.4	1.043	30.0	86.0	.907
9.0	48.2	1.448	23.2	73.8	1.038	30.2	86.4	.904
10.0	50.0	1.411	23.4	74.1	1.034	30.4	86.7	.901
11.0	51.8	1.375	23.6	74.5	1.029	30.6	87.1	.897
12.0	53.6	1.341	23.8	74.8	1.025	30.8	87.4	.894
13.0	55.4	1.309	24.0	75.2	1.020	31.0	87.8	.890
14.0	57.2	1.277	24.2	75.6	1.016	31.2	88.2	.887
15.0	59.0	1.247	24.4	75.9	1.012	31.4	88.5	.884
16.0	60.8	1.218	24.6	76.3	1.008	31.6	88.9	.880
17.0	62.6	1.189	24.8	76.6	1.004	31.8	89.2	.877
18.0	64.4	1.163	25.0	77.0	1.000	32.0	89.6	.873
18.2	64.8	1.157	25.2	77.4	.996	32.2	90.0	.870
18.4	65.1	1.152	25.4	77.7	.992	32.4	90.3	.867
18.6	65.5	1.147	25.6	78.1	.988	32.6	90.7	.864
18.8	65.8	1.142	25.8	78.5	.983	32.8	91.0	.861
19.0	66.2	1.136	26.0	78.8	.979	33.0	91.4	.858
19.2	66.6	1.131	26.2	79.2	.975	34.0	93.2	.843
19.4	66.9	1.127	26.4	79.5	.971	35.0	95.0	.829
19.6	67.3	1.122	26.6	79.9	.967	36.0	96.8	.815
19.8	67.6	1.117	26.8	80.2	.964	37.0	98.6	.801
20.0	68.0	1.112	27.0	80.6	.960	38.0	100.2	.788
20.2	68.4	1.107	27.2	81.0	.956	39.0	102.2	.775
20.4	68.7	1.102	27.4	81.3	.953	40.0	104.0	.763
20.6	69.1	1.097	27.6	81.7	.950	41.0	105.8	750
20.8	69.4	1.092	27.8	82.0	.947	42.0	107.6	.739
21.0	69.8	1.087	28.0	82.4	.943	43.0	109.4	.727
21.2	70.2	1.082	28.2	82.8	.940	44.0	111.2	.716
21.4	70.5	1.078	28.4	83.1	.936	45.0	113.0	.705
21.6	70.9	1.073	28.6	83.5	.932	46.0	114.8	.694
21.8	71.2	1.068	28.8	83.8	.929	47.0	116.6	.683

Source : Richard, L.A. (ed.) 1954. Diagnosis and Improvement of Saline and Alkali Soils. Agriculture Handbook 60, U.S. Dept. of Agriculture. Oxford & 1 BH Publishing Co., New Delhi.

APPENDIX 2

Density of water in grams per cubic centimeter

Degrees	0	0.1	0.2	0.3	0.4	0.5	0.6	0.7	0.8	0.9
0	0.999841	847	854	860	866	872	878	884	889	895
1	0.999900	905	909	914	918	923	927	930	934	938
2	0.999941	944	947	950	953	955	958	960	962	964
3	0.999965	967	968	969	970	971	972	972	973	973
4	0.999973	973	973	972	972	972	970	969	968	966
5	0.999965	963	961	959	957	955	952	950	947	944
6	0.999941	938	935	931	927	924	920	916	911	907
7	0.999902	898	893	888	883	877	872	866	861	855
8	0.999849	843	837	830	824	817	810	803	796	789
9	0.999781	774	766	758	751	742	734	726	717	709
10	0.999700	691	682	673	664	654	645	635	625	615
11	0.999605	595	585	574	564	553	542	531	520	509
12	0.999489	486	475	463	451	439	427	415	402	390
13	0.999377	364	352	339	326	312	299	285	272	258
14	0.999244	230	216	202	188	173	159	144	129	114
15	0.999099	084	069	054	038	023	007	*991	*975	*959
16	0.998943	926	910	893	877	860	843	826	809	792
17	0.998744	757	739	722	704	686	668	650	632	613
18	0.998595	576	558	539	520	501	482	463	444	424
19	0.998405	385	365	345	325	305	285	265	244	224
20	0.998203	183	162	141	120	099	078	056	035	013
21	0.997992	970	948	926	904	882	860	837	815	792
22	0.997770	747	724	701	678	655	632	608	585	561
23	0.997538	514	490	466	442	418	394	369	345	320
24	0.997296	271	246	221	196	171	146	120	095	069
25	0.997044	018	*992	*967	*941	*914	*888	*862	*836	*809
26	0.996783	756	729	703	676	649	621	594	567	540
27	0.996512	485	457	429	401	373	345	317	289	261
28	0.996232	204	175	147	118	089	060	031	002	*973
29	0.954994	914	885	855	826	796	766	736	706	676
30	0.995646	616	586	555	525	494	464	433	402	371

*Where values are designated with an asterisk, the first three figures are to be supplied from the zero column in the next before lower row.

Source : Daniels, F. *et al.* 1970. Experimental Physical Chemistry. McGraw-Hill Book Company, NY.

APPENDIX 3

Dynamic Viscosity (CP) of water at different temperatures.

Viscosity	Temperature–°C							
	0	5	10	15	20	30	40	50
Poise × 10^2	1.792	1.519	1.308	1.140	1.005	0.801	0.656	0.549

Divide the viscosity value given in the table by 100 for obtaining viscosity in poise.

APPENDIX 4

Gypsum requirement (tonnes/ha) in relation to soil pH

pH (1 : 2)	Soil texture		
	Sandy	**Sandy-loam**	**Laom**
<9.0	-----------------not needed-----------------		
9.2	1.7	2.5	3.4
9.4	3.4	5.0	6.8
9.6	5.0	7.5	10.00
9.8	6.8	10.0	14.6
10.0	8.5	12.5	15.0
10.2 and more	10.0	15.0	15.0

APPENDIX 5

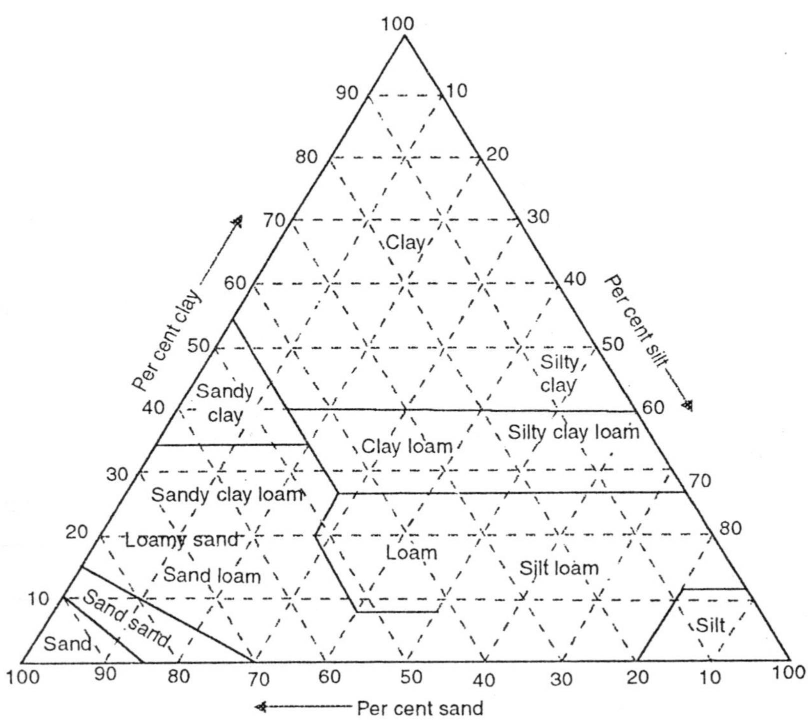

Guide for textural classification of soils. (Soil Survey Staff, 1960)

Procedure :

On the clay line, locate % clay corresponding to clay in soil sample, and draw a line from the point parallel to the base of the textural triangle.

Then, on the silt line, locate % silt corresponding to silt in soil sample, and draw a line from the point parallel to the clay line.

Mark the compartment in which the above two lines intersect, and note down the textural name of the soil. If the intersect is found on the periphery of the compartment, note the textural class name of the adjoining compartment.

A line drawn from the point of % sand corresponding to sand in soil sample on the sand line, parallel to the silt line will also meet the above intersect.

Example :

Textural name of soil containing 30% clay and 40% silt will be clay loam.

Textural name of soil containing 40% clay and 30% silt will be clay.

APPENDIX 6

Table of Relative Atomic Weights 1961[a]

Based on the atomic mass $^{12}C = 12$

[2]The value for atomic weighs given in the table apply to elements as they exist in nature, without artificial alternations to their isotopic composition, and further, to natural mixtures that do not include isotopes of radiogenic origin.

Name	Symbol	Atomic number	Atomic weight	Name	Symbol	Atomic number	Atomic weight
Aluminium	Al	13	26.9815	Molybednum	Mo	42	95.94
Antinimy	Sb	51	121.75	Neodymium	Nd	60	144.24
Argon	Ar	18	39.948	Neon	Ne	93	20.183
Arsenic	As	33	74.9216	Nickel	Ni	28	58.71
Barium	Ba	56	137.34	Niobium	Nb	41	92.906
Beryllium	Be	4	9.0122	Nitrogen	N	7	14.0067
Bsimuth	Bi	83	208.980	Osmium	Os	76	190.2
Boron	B	5	10.811	Oxygen	O	8	15.9994
Bromine	Br	35	79.904	Palladium	Pd	46	106.4
Cadmium	Cd	48	112.40	Phosphorus	P	15	30.9738
Calcium	Ca	20	40.08	Platinum	Pt	778	195.09
Carbon	C	6	12.01115	Potassium	K	19	39.102
Cerium	Ce	58	140.12	Praseodymium	Pr	59	140.907
Cesium	Cs	55	132.905	Rhenium	Re	45	186.2
Chlorine	Cl	17	35.453	Rhodium	Rh	37	102.905
Chromium	Cr	24	51.996	Rubidium	Rb	37	85.47
Cobalt	Co	27	58.9332	Ruthenium	Ru	44	101.07
Copper	Cu	29	63.546	Samarium	Sm	62	150.35
Dyspromium	Dy	66	162.50	Scandium	Sc	21	44.956
Erbium	Er	68	167.26	Selenium	Se	34	78.96
Europium	Eu	63	151.96	Silicon	Si	14	28.086
Flourine	F	9	18.9984	Siliver	Ag	47	107.868
Gadolinium	Gd	64	157.25	Sodium	Na	11	22.9898
Gallium	Ga	64	69.72	Strontium	Sr	38	87.62
Germanium	Ge	32	72.59	Sulphur	S	16	32.064
Gold	Au	79	196.967	Tantalum	Ta	73	180.948
Halfnium	Hf	72	178.49	Tellurium	Te	52	127.60
Helium	He	2	4.0026	Terbium	Tb	65	158.924
Hydrogen	H	1	1.00797	Thallium	Tl	81	204.37
Indium	In	53	114.82	Thorium	Th	90	232.038
Iodine	I	49	126.9044	Thulium	Tm	69	168.934
Iridium	Ir	77	192.2	Tin	Sn	50	118.69
Iron	Fe	26	55.847	Titanium	Ti	22	47.90
Krypton	Kr	36	83.80	Tungesten	W	74	183.85
Lanthanum	La	57	138.91	Uranium	U	92	238.03
Lead	Pb	82	207.19	Vanadium	V	23	50.942
Lithium	Li	3	6.939	Xenon	Xe	54	131.30
Lutetium	Lu	71	174.97	Ytterbium	Y	39	88.905
Magnesium	Mg	12	24.312	Yttrium	Y	39	88.905
Manganese	Mn	25	54.9380	Zinc	Zn	30	65.37
Mercury	Hg	80	200.59	Zirconium	Zr	40	91.22

A.E. Cameron and Edward Wichers, *J. Am. Chem. Soc.*, 84: 4192 (1962). The table includes the revision of the atomic weights adopted by the International Commision on Atomic Weights in July, 1965.

APPENDIX 7

Numericals on Physical Analysis of Soil

1. Soil separates in soil sample by Bouyoucos soil hydrometer method.

An organic C free soil sample weighing 39.5 g completely dispersed to prepare one litre of soil-water suspension was transferred to a tall soil testing cylinder for studies on sedimentation of mineral particles with time. The Bouyoucos hydrometer reading & temperature of the soil-water suspension and the salt solution was taken after sedimentation period of 4 and 120 minutes. The observation taken are presented in the under given Table 1.1 for estimation of % sand, % silt and % clay in the sample.

Table 1.1

Bouyoucos soil hydrometer reading & temperature of soil-water suspension and salt solution after sedimentation period of 4 & 120 minutes.

Sedimentation time in minutes	Bouyoucos hydrometer reading		Suspension temperature °F
	R_{susp}	R_{salt} at suspension temp	
4	12.0	2.8	87
120	07.0	2.6	88

Ans. The soil sample contains 67.1 % sand, 11.6 % silt and 21.3 % clay.

(See section 14.1)

2. Soil separates in soil sample by pipette method.

2.1 Completely dispersed and homogenized soil water suspension filled inside a tall soil testing cylinder upto 1000 ml mark is allowed to sediment. Estimate the time which will be taken for the particles of diameter 0.02 and .002 mm to settle to a depth of 5 cm from the surface of the soil-water suspension at 20^0 C *i.e.* 68^0 F.

Answer

Particle size	Time required for the particle to settle to a depth of 5 cm.		
	Hours	Minutes	Seconds
0.02 mm	—	2	19.2
0.002 mm	3	51	57.2

The above given answer was obtained by solving under given equation derived from Stokes' Law

$$T = \frac{0.03059\ \eta L}{(pS - pL)\ g \times d^2}$$

Where,

T is time in minutes.

η is viscosity in poise (.01002 at 20^0 C).

L is settling depth in cm *i.e.* 5 cm in the present case.

pS is the particle density, 2.65 g/ml.

pL is the density of water, 0.998203 g/l at 20^0 C.

g is acceleration due to gravity 980.7 cm.

d is the diameter of particles in mm.

Check the above given answer, and estimate the time required for the mineral particles of the diameter 0.2 mm to settle to a depth of 5 cm.

2.2 Soil separates in soil sample by pipette method.

Organic C free H-saturated soil weighing 9.9 g completely dispersed to prepare 1L of soil-water suspension was transferred into a tall soil testing cylinder for the sedimentation of the mineral particles with time.

10 ml aliquot by pipette from a depth of 5 cm was drawn at pre determined time to collect mineral particles of the diameter 0.02 mm & less i.e. < .02 mm, and then 0.002 mm and less i.e. < .002 mm. The aliquots taken were evaporated to dryness for determining the mass of mineral fraction plus salt. 10 ml aliquot of the salt solution was also evaporated to dryness to determine the content of salt.

The observations collected are summarized below :

PARTICULARS	MASS (g)
Organic C free H-saturated soil taken for preparation of soil-water suspension	9.9
Mineral fraction (.02 mm & less) + salt in 10 ml aliquot	0.0800
Mineral fraction (.002 mm & less) + salt in 10 ml aliquot	0.0600
Salt (Sodium hexa meta phosphate) is 10 ml aliquot	0.0050

Estimate the percentage of sand, silt and clay in the soil sample.

Answers : % sand 24.2, % silt 20.2 and % clay 55.6

Mineral fraction in the soil sample was estimated by the under given general equation.

% Mineral fraction in soil sample

$$= \frac{\left\{\left(\begin{matrix}\text{Mass of mineral fraction} \\ \text{+ salt in aliquot taken in g.}\end{matrix} - \begin{matrix}\text{Mass of salt in the aliquot} \\ \text{taken in g.}\end{matrix}\right) \times 1000\right\} \times 100}{\begin{matrix}\text{Volume of aliquot} \\ \text{in ml}\end{matrix} \times \begin{matrix}\text{Mass of organic C free,} \\ \text{H-saturated soil in g.}\end{matrix}}$$

(also see section 1.4.2)

3. Particle size distribution Analysis.

The hydrometer reading R of soil-water suspension, suspension temperature, R-salt solution corresponding to each sedimentation time in minutes are presented in the 1st to 4th column of the Table 3.1 on particle size distribution analysis. Process the data so as to obtain percentage of minerals of different mean diameter in the soil sample. (see section 1.4.3 also).

The further processing of the data require.

1. Estimation of R^*_{susp} :

   In column 5, the value of R^*_{susp} at 68° F is presented after solving equation 1.

   $R^*_{susp} = (R_{susp} - R^*_{salt}) - (68 - \text{susp. temp.}^{0}\text{F}) \times 0.2$ ...equation(1)

   Where R^*_{salt} is the hydrometer reading of the salt solution at suspension temperature which is equal to,

   $R^*_{salt} = R_{salt} - (\text{Susp. temp. }^{0}\text{F} - \text{Salt solution temperature }^{0}\text{F}) \times 0.2$

   Note : The hydrometer stem reading is actually calibrated at 68° F

2. The depth of immersion of hydrometer (h) given in column 6 has been estimated by solving equation 2.

   $$h = H + \left(R^*_{susp} \times \frac{L}{60}\right) \qquad \text{...equation(2)}$$

   Where

   H is the distance in cm from the centre of hydrometer bulb to zero reading of the hydrometer stem.

   R^*_{susp} is the adjusted hydrometer reading of the soil-water suspension at 68° F.

   L is the distance from 0 to 60 reading of the hydrometer stem in cm.

   In Bouyoucos hydrometer (ASTM) of Taylor Industrial, Rochester, NY, the value of H and L measure 17.9 cm and 10.2 cm respectively.

3. θ value given in column 7 was estimated by the equation 3 for each hydrometer reading as the depth of immersion of hydrometer h vary with the hydrometer reading.

   $$\theta = 1000\left[\frac{30\,\eta\,h}{(pS - pL)g}\right]^{1/2} \qquad \text{...equation(3)}$$

   Where θ is sedimentation parameter, η is viscosity of liquid in poise,

   h is depth of immersion of the hydrometer, pS is the particular density of the soil, pL is the density of water in g/ml, and g is acceleration due to gravity.

   The above equation may be presented as follows :

Table 3.1 Particle Size Distribution Analysis.

Sedimentation time in minutes	R_{susp}	Suspension temperature °F	R_{salt} at susp. temp.	R^*_{susp} $(R_{susp}-R_{salt})$ at 68^0 F	h depth of immersion of hydrometer at 68^0 F in cm	θ $= 1000\left[\frac{30\,\eta\,h}{(pS - pL)g}\right]^{1/2}$ at 68^0 F	Mean diameter $\overline{X} = \frac{\theta}{t^{1/2}} \bullet 10^{-3}$ mm	Mineral particles %
0.67	23.0	87	2.8	24.0	13.82	50.6406	0.0619	48.4
1.0	20.0	87	2.8	21.0	14.33	51.5669	0.0516	42.4
2.0	19.0	87	2.8	20.0	14.50	51.9606	0.0367	40.3
3.0	17.0	87	2.8	18.0	14.84	52.4769	0.0303	36.3
4.0	15.0	87	2.8	16.0	15.18	53.0122	0.0265	32.3
30.0	12.0	87	2.8	13.0	15.69	53.9590	0.0098	26.2
60.0	10.0	88	2.6	11.4	15.96	54.4208	0.0070	23.0
120.0	9.0	88	2.6	10.4	16.13	54.7096	0.0050	21.0
270.0	8.5	85	3.2	8.7	16.42	55.2000	0.0034	17.6
720.0	8.0	82	3.8	7.0	16.71	55.6849	0.0021	14.1

$$= \left[\frac{30\,\eta}{(pS - pL)g}\right]^{1/2} .h^{1/2} \times 1000 \qquad ...(i)$$

By substituting the value of η, pS, pL at 68^0 F and g, we have

$$= \left[\frac{30 \times .01002}{(2.65 - .998203) \times 980.7}\right]^{1/2} .h^{1/2} \times 1000 \qquad ...(ii)$$

$$= [0.013622225] \times .h^{1/2} \times 1000. \qquad ..(iii)$$

The equation (*iii*) enables rapid estimation of θ

4. The mean diameter of particles $\overline{x}$ in mm given in column 9 was estimated by equation 4

$$\overline{x} = \frac{\theta}{t^{1/2}} \times 10^{-3} \text{ in mm} \qquad ...\text{equation}(4)$$

Where t is the sedimentation time in minutes.

5. The percentage of particles of $\overline{x}$ diameter given in column 9 was estimated by solving equation 5

$$\% \text{ Minerals in soil sample} = \frac{R^{*}_{susp} \times 100}{Ms} \qquad ...\text{equation}(5)$$

Where R^{*}_{susp} is the adjusted hydrometer reading of the soil-water suspension at 68^0 F as estimated by equation 1.

Ms is the mass of organic carbon free soil sample taken for the preparation of 1L of soil-water suspension *i.e.* 49.57 g in the present study.

Graphical Presentation

On a 4 cycle semilog graph paper, the diameter of mineral particles (.002, .02 .2 and 2.0 mm) was set on the log scale (Figure 3.1) and the percentage of mineral particles (0, 20, 40, 60, 80, & 100) was set on the linear scale. Then, the percentage of mineral particles corresponding to each mean diameter in mm was plotted. Thereafter, a smooth line was drawn along the plotted points so that most of the points were either located on/or near the particle distribution smooth line curve.

The reader is advised to check the entire steps to become well versed in the processing of the data and its graphical presentation.

4.1 Estimation of specific gravity & particle density of soil.

Equations :

Specific gravity of soil

$$= \frac{\text{Mass of loose oven dry soil}}{\text{Mass of water displaced by the soil}} = \frac{W_2 - W_1}{(W_4 - W_1) - (W_3 - W_2)}$$

...equation(1)

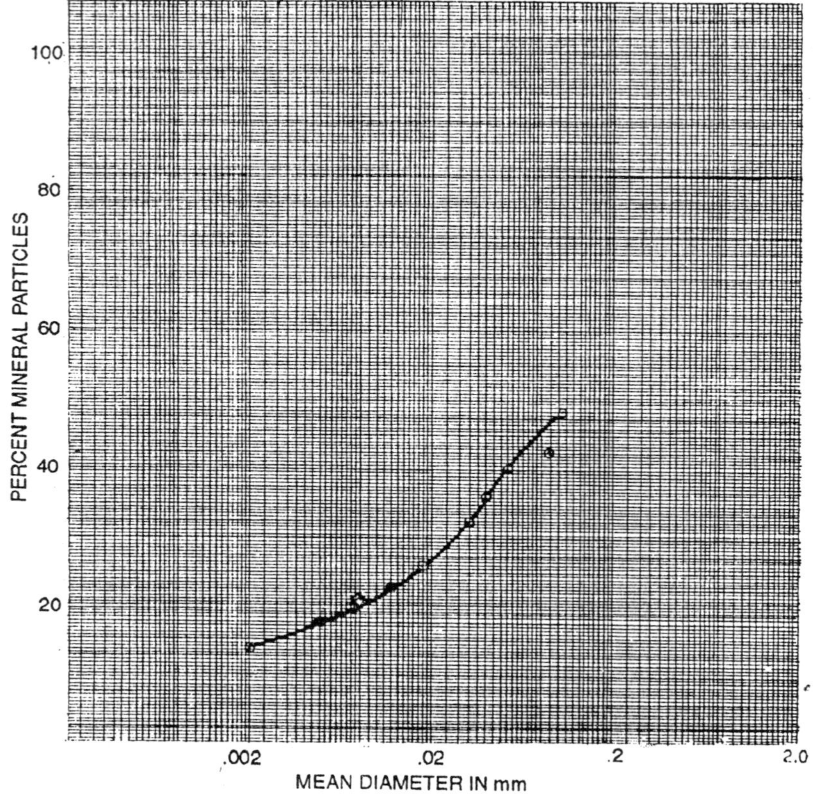

Fig. 3.1 Particle Size Distribution

Particle density of soil

$$= \frac{\text{Mass of loose oven dry soil}}{\text{Volume of water displaced by the soil}} - \frac{(W_2 - W_1) \times dw}{(W_4 - W_1) - (W_3 - W_2)}$$ ..equation(2)

(see section 1.5 also)

Observations

1. W_1 – mass of empty 100 ml volumetric flask – 60.0 g
   W_2 – mass of 100 ml vol. flask + half filled with dry soil – 116.1 g
   W_3 – mass of 100 ml vol. flask + half filled with soil
   + remaining space filled with water up to 100 ml mark – 194.0 g
   W_4 – mass of 100 ml volumetric flask
   + distilled water up to 100 ml mark – 159.0 g
   dw – density of water at the temperature of measurement – 0.998203
2. m – mass of loose oven dry soil – 40.0 g
   v – volume of water displaced by 40 g soil in a 100 ml graduated cylinder filled up to 50 ml mark – 15.1 ml
   Answers of first observation

Specific gravity of soil = 2.659

Particle density of soil = 2.654 g/ml

Answers of second observation

Particle density of soil = 2.649 g/ml

Specific gravity of soil = 2.654

4.2 Estimation of particles density (dp) of soil deriving from pertinent parameter by numerical relationship.

Observations

(*i*) Mass of moist soil *i.e.* $Mt_{(S+W)} = 800$ g

(*ii*) Ratio of mass of water (M_W) to mass of soil (M_S)

i.e. $M_W/M_S = 0.25$ g w/g soil

(*iii*) Volume of soil (V_S) = 235 cm^3

Calculations

$$\therefore \quad dp = \frac{M_S}{V_S}$$

As V_S is given, M_S has to be estimated

$$M_W/M_S = 0.25 \text{ g water / g soil}$$

$$= \frac{M_t - M_S}{M_S}$$

$$0.25 = \frac{M_t - M_S}{M_S}$$

$$0.25\, M_S = M_t - M_S$$

$$0.25\, M_S + M_S = M_t$$

$$M_S\,(0.25 + 1.00) = M_t$$

$$\therefore \quad M_S = \frac{M_t}{1.25} = \frac{800}{1.25} = 640 \text{ g}$$

$$\because \quad dp = \frac{M_S}{V_S} \text{ when } M_S = 640 \text{ g and } V_S = 235 \text{ cm}^3$$

$$\therefore \quad = \frac{640}{235} = 2.72 \text{ g/cm}^3$$

5. Estimation of apparent specific gravity and bulk density of soil.

Equations :

Apparent Specific gravity

$$= \frac{\text{mass of a given volume of oven dry soil}}{\text{mass of water equal to the volume of soil}} \quad \text{...equation 1}$$

$$\text{Bulk density of soil} = \frac{\text{mass of a given volume of oven dry soil}}{\text{volume of the soil in c.c}}$$

...equation 2

(see section 1.5 also)

Observations

1. Mass of a hollow metal cylinder = 283.8 g
   (dia. 7.48 cm, height 7.64 cm)
   Mass of metal cylinder + oven dry soil = 865.5 g
   Volume of cylinder = $\pi r^2 h$
   Where π is 3.14, r is the radius in cm and h is the height in cm.
   Density of water at temperature of measurement = 0.998203 g/ml
2. Mass of soil excavated (after drying in an oven to a constant weight) = 21.8g
   Volume of soil excavated as determined by the volume of fine sand used to fill the excavated portion uniformly = 15.0 c.c.
   Density of water at temperature of measurement = 0.998203 g/ml
   Answer

	Apparent specific gravity	Bulk density g/c.c
1st observation	1.7366	1.7335
2nd observation	1.4566	1.4533

6. Estimation of air capacity of soil.

1. Definition

Kopecky (1917) defined air capacity of soil as the quantity of air remaining in soil on saturation to the point of absolute water holding capacity *i.e.* water retained by the soil 24 hours after saturation.

Air capacity of soil is synomynous to non-capillary pore space or else air filled pore space at field capacity.

(see section 1.8 also)

2. Equations

% Total pore space = 100 – % solid space ...(*i*)

$$= 100 - \frac{\text{bulk density of soil}}{\text{particle denosity of soil}} \times 100$$

$$= 100 - \frac{\text{Apparent specific gravity of soil}}{\text{Specific gravity of soil}} \times 100$$

(see section 1.7 also)

$$\text{Volume of total pore space} = \text{Volume of soil} \times \frac{\text{percent total pore space}}{100} \quad ...(ii)$$

% capillary pore space

$$= \frac{\text{Volume of water filled pore space at field capacity}}{\text{Volume of the soil}} \times 100 \quad ...(iii)$$

% air filled pore space

$$= \frac{\text{Volume of air filled pore space at field capacity}}{\text{Volume of the soil}} \times 100 \quad ...(iv)$$

% capillary water on over dry basis (P_w) = % moisture in soil at field capacity – % moisture in soil at hygroscopic coefficient ...(*v*)

% capillary water on volume basis (P_v)

$$= P_W \times \text{apparent specific gravity of soil} \quad ...(vi)$$

$$\text{Apparent specific gravity} = \frac{\text{mass of a given volume of oven dry soil}}{\text{mass of water equal to the volume of soil}} \quad ...(vii)$$

Volume of capillary water *i.e.* water filled pore space

$$= \text{volume of soil} \times \frac{\text{\% capillary water on volume basis}}{100} \quad ...(viii)$$

Volume of air filled pore space = volume of total pore space – volume of water filled pore space at field capacity. ...(*ix*)

Air capacity of soil *i.e.* % air filled pore space

$$= \frac{\text{volume of air filled pore space at field capacity}}{\text{volume of the soil}} \times 100$$

Numerical relationship between field capacity and other soil moisture constants ...(*x*)

Field capacity = 2.62 + moisture equivalent × 0.865

Moisture equivalent = (Field capacity – 2.62) / 0.865

Wilting coefficient = Moisture equivalent ÷ 1.84

Hygroscopic coefficient = Wilting coefficient × 0.68

3. Observations

Mass of hollow metal cylinder (dia. 7.50 cm, height 7.56 cm) = 286.0 g

Mass of metal cylinder + moist soil at field capacity = 901 .6 g

Mass of metal cylinder + oven dry soil = 798.2 g

Density of water at temperature of measurement = 0.998203 g/ml

Particle density of soil = 2.65 g/cc

Specific gravity of soil = 2.65

4. Calculations

Mass of moist soil at field capacity = 615.6 g

Mass of oven dry soil =512.2 g

Mass of moisture in soil core at field capacity = 103.4 g

Volume of soil core ($\pi r^2 h$) = 333.69 cm^3

Bulk density of soil = 1.535 g/cc

% total pore space =42.1 %

% moisture in soil at field capacity (P_w)

$$= \frac{\text{Mass of moisture in soil at field capacity}}{\text{oven dry mass of soil}} \times 100$$

= 20.19 %

Apparent specific gravity of soil = 1.5377

% moisture on volume basis (P_v) at field capacity = 31.046

Hygroscopic coefficient of the soil on volume basis = 11.54 %

% capillary water on volume basis = 19.51 %

Volume of capillary water = 65.13 cm^3

Volume of total pore space = 140.48cm^3

Volume of air filled pore space = 75.38 cm^3

% air filled pore space *i.e.* air capacity of soil = 22.59

7. Soil compaction measurement with Penetrometer

Penetrometer measures resistance offered by the soil to penetration of a metal cone or a thin metal circular rod designated as needle.

(see section 1.9 also)

1. Cone index

Definition : The kilogram of force that must be applied to the handle of cone penetrometer per square centimeter of the end area of the cone tip in order to force it to the ground.

Particulars of cone

30^0 circular cone with 3 cm diameter base to eliminate frictional force.

$$\text{Cone index} = \frac{\text{load in kg}}{\text{cone base area}}$$

Where,

Load in kg = 14.8368 × penetrometer reading

Cone base area = πr^2 where r is the radius of cone base *i.e.* 1.5 cm

Observations :

Penetrometer readings for estimating cone index

Particulars	Soil conditions	
	Moist	Dry
Ploughed land	0.12	0.33
Unploughed land	1.44	3.50

Answer – Cone index as estimated from above observations

Particulars	Soil conditions	
	Moist	Dry
Ploughed land	0.252	3.024
Unploughed land	0.693	7.350

2. Needle index : Expressed in load in kg/cm^2 needle tip but needle penetrometer measures load in lbs/inch^2.

$$\text{Load in lbs/inch}^2 \times 0.07045 = \text{load in kg/cm}^2$$

Observations

Needle penetrometer readings in lbs/inch^2

Particulars	Soil conditions	
	Moist	Dry
Unploughed land	1.00	2.25
Ploughed land	0.75	1.31

Answer : Needle index in kg/cm^2

Particulars	Soil conditions	
	Moist	Dry
Unploughed land	0.0704	0.1585
Ploughed land	0.0528	0.0923

8.1 Estimate Mean Weight Diameter (MWD) of aggregates size distribution after wet sieving summarized in the 1–3rd column of Table 8.1.1.

Table 8.1.1

Crop	Aggregates class size mm	Mass of Aggregates g	W_i Fraction of total sample weight	$\overline{X}_i$ * Aggregates Mean Diameter mm	$W_i \times \overline{X}_i$
under continous blue grass	8.0–2.0	31.1	0.311	5.0	1.555
	2.0–1.0	15.7	0.157	1.5	0.236
	1.0–0.5	14.0	0.140	0.75	0.105
	0.5–0.25	15.9	0.159	0.375	0.060
	0.25–0.10	12.8	0.128	0.175	0.022
	0.10–0.00	12.5	0.125	0.05	0.066
	Total	100.0	NR	NR	1.984
under maize for 3 years	8.0–2.0	6.7	0.067	5.0	0.335
	2.0–1.0	5.7	0.057	1.5	0.086
	1.0–0.5	14.1	0.141	0.75	0.106
	0.5–0.25	24.8	0.248	0.375	0.093
	0.25–0.10	27.0	0.270	0.175	0.047
	0.10–0.00	21.7	0.217	0.05	0.011
	Total	100.0	NR	NR	0.678

The 1st three column as reported by Gardener,W.R.,1956.Soil Sci. Soc. Proe.,U.S.A.,151–153 $\overline{X}_i^*$, adapted from Dakshinamurti, C. and R.P. Gupta. 1968. Practicals in soil Physics Indian Agricultral Research Institute,. New Delhi–12.

NR,denote not required.

$$MWD = \sum_{i=1}^{n} W_i \times \overline{X}_i$$

Where, n is the number of fractions *i.e.* mass of aggregates within a certain range of aggregates size (class size).

$\overline{X}$ is the mean diameter of the fraction in mm–estimated by taking the sum of the upper and lower limit of the class size and dividing the sum obtained by 2 to obtain the mean diameter.

W_i is the proportion of the total sample weight occuring in the fraction i.

Σ is for the summation of the product of $\overline{X}_i$ W_i to $\overline{X}_n$ W_n. Thus, three values viz W_i, $\overline{X}_i$ and $W_i \times \overline{X}_i$ need to be worked out. The values estimated has been tabulated in the 4th, 5th and 6th column.The MWD of soil aggregates under blue grass and maize work out to 1.984 and 0.678 respectively. Each step of data processing be checked and confirmed. (see also section 1.10.1 and 1.10.2 also).

8.2 Estimate Geometric Mean Diameter (GMD, and log GMD of aggregate size distribution of two soil samples presented in the table given below :

Table 8.2.1. Aggregate size distribution of silt loam soil under continous blue grass and under maize for 3 yrs.

Crop	Aggregates class size mm	W_i Mass of Aggregates g	$\overline{X}_i$ Mean* diameter mm	Percent oversize
Under continuous blue grass	8.0–2.0	31.1	2.0	31.1
	2.0–1.0	15.7	1.0	46.8
	1.0–0.5	14.0	0.5	60.8
	0.5–0.25	15.9	0.25	76.7
	0.25–0.10	12.8	0.10	89.5
	0.10–0.00	12.5	0.00	100.0
	Total	100.0	NR	NR
Under Maize for 3 years	8.0–2.0	6.7	2.0	6.7
	2.0–1.0	5.7	1.0	12.4
	1.0–0.5	14.1	0.5	26.5
	0.5–0.25	24.8	0.25	51.3
	0.25–0.10	27.0	0.10	78.3
	0.10–0.00	21.7	0.00	100.0
	Total	100.0	NR	NR

*Source, Gardner,W.R. 1956, Soil. Sei. Soc. Proc, USA, pp 151–153 NR denote not required.

a. Estimation of GMD

$$GMD = \exp\left[\frac{\sum_{i=1}^{n} W_i \log \overline{x}_i}{\sum_{i=1}^{n} W_i}\right]$$

Where, W_i is the mass of aggregates in a aggregate class size with a mean diameter $\overline{X}_i$, and $\sum_{i=1}^{n} W_i$ is the total mass of the aggregates in 1 to nth fraction.

(see section 1.10.1 and 1.10.2 also)

Working out product of $W_i \log \overline{X}_i$ for estimation of GMD

Table 8.2.2 Product of $W_i \log \overline{X}_i$ for each class size of aggregates from silt loam soil under continous blue grass and under maize for 3 yrs.

$\overline{X}_i$ Aggregates mean diameter mm	Log $\overline{X}_i$	Under continuous blue grass.		Under three years Maize	
		W_i mass of aggregates g	$W_i \log \overline{X}_i$	W_i mass of aggregates g	$W_i \log \overline{X}_i$
2.00	0.3010	31.1	9.3611	6.7	2.0167
1.00	0.0000	15.7	—	5.7	– 1
0.50	–0.3010	14.0	–4.2140	14.1	–4.2441
0.25	–0.6021	15.9	–9.5734	24.8	–14.9321
0.10	–1.0000	12.8	–12.8000	27.0	–27.0000
		$\sum_{i}^{\eta} W_i$ 89.5		$\sum_{i}^{\eta} W_i$ 78.3	
0.00	—	10.5	—	21.7	—
		$\sum_{i=1}^{n}$	–17.2263	$\sum_{i=1}^{n}$	–44.1595

By substitution of values $\sum_{i}^{n} W_i \log \overline{x}_i$ and $\sum_{i}^{n} W_i$ in equation for GMD

We have,

(*i*) $\text{GMD} = \exp\left[\frac{-17.2263}{89.5}\right]$

$= \exp[-0.192473]$

$= 0.8075$

(*ii*) $\text{GMD} = \exp\left[\frac{-44.1595}{78.3}\right]$

$= \exp[-0.563978]$

$= 0.4360$

Result : (i) GMD of soil under continuous blue grass = 0.8075

(ii) GMD of soil under Maize for 3yrs =0.4360

b. Determination of Log GMD of soil aggregates after wet Sieving.

On a 3 cycle semi log probability graph paper, mark mean diameter of soil aggregates (0.01, 0.1, 1.0 & 10.0 mm) on the log scale, and 0 to 100 percent over size of soil aggregates on the probability scale.

There after, plot percent over size of soil aggregates corresponding to mean diameter of each class size. Then, draw a straight line along the points of each set of data on aggregate analysis plotted on the graph so that most points either fall on the line or else close to it. (Fig.1).

Now ascertain the value of mean diameter of aggregates at 50 percent oversize. Mark the vertical line at 50 percent oversize on the probability scale, where it intersect the aggregate size distribution line. From the point of intersection draw an imaginary line parallel to the base and mark as it intersect the line for mean diameter on log scale. Read the value of mean diameter, and designate it as log GMD. The log GMD value for the soil under continous blue grass is observed to be 0.96 as against log GMD of 0.28 for 3 yrs maize rotation (Figure 8.2.1).

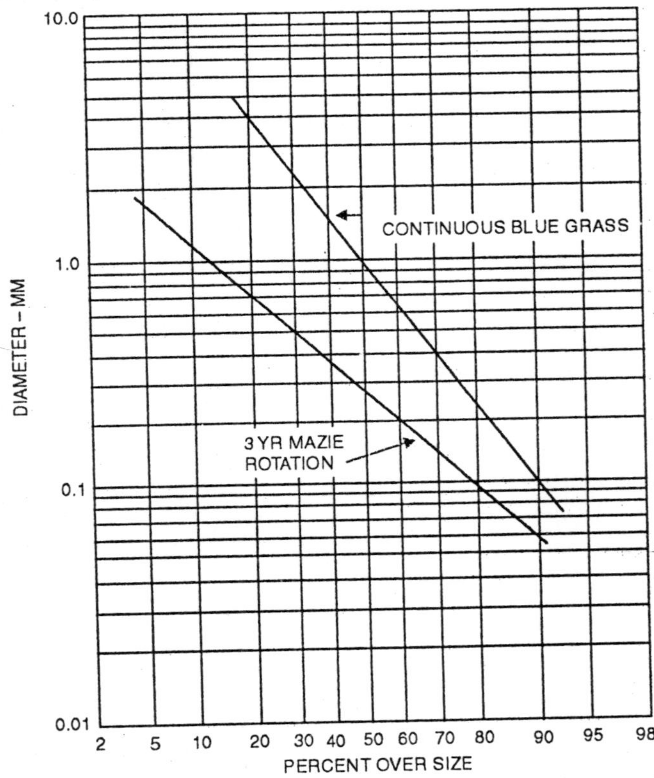

Figure 8.2.1 Log probability plot of aggregate size distribution of silt loam soil under continous blue grass and under 3 yrs. maize rotation.

9. Hydraulic conductivity of soil by constant head method.

1. Equation :

Water flow through a soil column takes place in accordance with the equation

$$\frac{Q}{t} = K\,A\frac{\Delta H}{\Delta L} \qquad ...(i)$$

Where,

Q is the discharge *i.e.* volume of water in cc passing through the soil column in time t in hours.

A is the cross sectional area of the soil column.

K is the average hydraulic conductivity in the soil in cm/hour.

ΔL is the depth of soil column in cm.

ΔH is the hydraulic head measured from surface of water standing on the soil surface upto the bottom of the soil column.

Solving the above equation for K, we have

$$K = \frac{Q \times \Delta L}{t \times A \Delta H} \quad ...(ii)$$

(see section 1.11 also)

2. Observations

The discharge through a 3 cm deep wet soil column having a radius of 3.5 cm with 5.8 cm high free water above soil surface was found to be 32.4 cm per hour at a temprature of 20^0 C.

3. Calculation

Estimate K, the hydraulic conductivity by substituting the values in the above equation and setting it for the K.

Answer : Hydraulic conductivity of the soil is 0.2869 cm/hour.

10. Moisture content of soil by gravimetric method.

Method

Moist soil sample after weighment is dried in an oven at 105^0 C for 24 hours or longer to a constant soil mass and then loss of moisture on drying in oven is expressed as percentage of the oven dry mass of the soil.

(see section 1.13.1 also)

1. Equations for estimation of moisture content in disturbed sample of moist soil

% moisture in soil

$$= \frac{\text{mass of moisture in a given sample of moist soil in g} \times 100}{\text{mass of oven dry soil in the sample in g.}} \quad ...(i)$$

$$= \left[\frac{\text{mass of moist soil in g}}{\text{mass of oven dry soil in g}} - 1.000\right] \times 100 \quad ...(ii)$$

Note : Sample be weighed with the accuracy of one place after decimal.

Observations :

m_1 – mass of empty metal container = 23.4 g

m_2 – mass of metal container + moist soil = 34.2 g

m_3 – mass of metal container + oven dry soil = 32.8 g

Calculation : By substituting values in equations and solving for % moisture in soil.

Answer : % moisture in soil = 14.894 %

2. Equations for estimation of moisture content in undisturbed sample of moist soil.

moisture content in soil on even dry basis Δ (P_w)

$$P_w = \frac{\text{mass of moisture in sample of moist soil in g}}{\text{oven dry mass of the soil sample}} \times 100 \qquad ...(iii)$$

moisture content in soil on volume basis, P_V

$$P_v = \frac{\text{volume of moisture in soil in cc}}{\text{volume of oven dry soil in cc}} \times 100 \qquad ...(iv)$$

$$= P_w \times \text{apparent specific gravity of soil} \qquad ...(v)$$

Apparent Specific Gravity (ASG)

$$\text{Where, ASG} = \frac{\text{mass of a given volume of oven dry soil}}{\text{mass of water equal to the volume of soil}} \qquad ...(vi)$$

Moisture content of soil expressed in cms of water column per unit depth of soil column, P_d

$$P_d = \frac{P_v}{100} \times \text{soil depth in cms} \qquad ...(vii)$$

Observations

mass of empty hollow metal cylinder = 283.8 g

(dia. 7.48 cm, height 7.64 cm)

mass of metal cylinder + moist soil = 952.9 g

mass of metal cylinder + oven dry soil = 865.5 g

density of water at temperature of measurement = 0.998203 g/ml

Calculations

By solving equations (*iii*) to (*vii*) after substitution of pertinent observations in the equation.

Answer

P_w = 15.08 %

P_v = 26.09 %.

P_d = 7.83 cm H_2O/30 cm soil depth.

3. Estimation of mass and volume of water in soil by deriving pertinent values from numerical relationships.

3.1 Estimation of mass of water in soil *i.e.* M_w from the following observations.

mass of moist soil *i.e.* $Mt_{(s+w)}$ = 835.0 g

ratio of mass of water (M_w) to mass of soil (M_s) *i.e.* M_w/M_s = 0.22 g water per g soil

Calculation

$$M_w = M_t - M_s$$

as M_t is known, M_s be estimated.

We know that $M_w/M_s = 0.22$

$$= \frac{M_t - M_s}{M_s}$$

$$0.22 = \frac{835 - M_s}{M_s}$$

$$0.22\ M_s = 835 - M_s$$

$$0.22\ M_s + M_s = 835$$

$$M_s\ (0.22 + 1.00) = 835$$

$$M_s = \frac{835}{1.22} = 684.42$$

Again $M_w = Mt - M_s$

$$= 835 - 684.42 = 150.58$$

2. Estimation of volume of water from the following observations.

Volume of moist soil *i.e.* $Vt_{(w+s)} = 455\ cm^3$

ratio of volume of water (V_w) to volume of soil (V_s) *i.e.* $V_w/V_s = 0.27\ cm^3$ water/cm^3 soil

Calculation

$$V_w = Vt - V_s$$

As $Vt_{(w+s)}$ is known, V_s be estimated

Since $V_w/V_s = .27$

$$= \frac{Vt - V_s}{V_s}$$

$$0.27 = \frac{455 - V_s}{V_s}$$

$$0.27\ V_s = 455 - V_s$$

$$0.27\ V_s + V_s = 455$$

$$V_s\ (0.27 + 1.00) = 455$$

$$V_s = \frac{455}{1.27} = 358.27\ cm^3$$

$\because V_w = Vt - V_s$

$$= 455 - 358.27 = 96.73\ cm^3$$

11. Water content of soil by indirect methods.

Method 1. proposed by Papadakis (1914) and tested by Singh & Rajput (1970).

Equation

$$y = 82.8 - 5.10x \qquad r = -0.892^{**} \qquad ...(i)$$

Where, y is percent moisture in soil on oven dry basis, P_w

$$x = W_1 - W_2$$

Where,

W_1 = mass of 100 ml vol. flask + 25 g moist soil + water filled upto 100 ml mark.

W_2 = mass of 100 ml vol. flask + water filled upto 100 ml mark.

Note : The $W_1 - W_2$ is found less than 25 g. It is numerically equal to $W_3 - \frac{W_3}{D_s}$ where W_3 is mass of soil present in 25 g moist soil and D_s is the particle density of soil in gcm^{-3}.

The loss in weight is thus equal to the volume of water displaced by the soil. (see section 1.13.3 also)

Observations

W_1 = 169.0 g

W_2 = 159.0 g

Calculation

by solving equation (*i*), after substituting the value of x.

Answer

P_w = 31.8 %

Method 2. proposed by Gardner (1965) and tested by Singh & Rajput (1970). (see section 1.13.2 also)

Equation

$$y = 0.989\,x - 0.80 \qquad r = -0.999^{**} \qquad ...(ii)$$

Where,

x is percent moisture in soil by burning 20 g moist soil in a container twice on wetting moist soil with 10 ml alcohol or spirit.

y is the percent moisture in soil on oven dry basis *i.e.* P_w

Observations – mass of container – 23.4 g

mass of container + 20 g moist soil = 43.4 g

soil container + dry soil after burning with alcohol = 38.4 g

Answer – % moisture in soil by spirit burning = 33.3

% moisture in soil by drying in oven at 105^0 C = 32.85

Method 3. Water content of soil by immersion method *i.e.* by increase in volume on displacement by 20 g moist soil. Prihar & Sandhu (1967).

(see section 1.13.4 also)

Equations

$$W_w = D_w (D_s V_i - W_{sm})/(D_s - D_w) \quad \text{... (i)}$$

$$P_w = \frac{W_w}{W_{sm} - W_w} \times 100 = \frac{W_w}{W_s} \times 100 \quad \text{...(ii)}$$

Where,

W_w is the water content of soil in g.

P_w is % moisture in soil on oven dry basis..

W_{sm} is the mass of moist soil.

V_i is the increase in volume of water on displacement of water by 20 g moist soil.

D_s is the density of soil *i.e.* 2.65 g/cc

D_w is the density of water at temperature of measurement = 0.998203 g/ml

Observations

W_{sm} – mass of moist soil = 20.0 g

V_i – increase in volume of water on displacement of water by 20 g of moist soil = 10.7 ml

D_s – density of soil = 2.65 g/cc

D_w – density of water at temperature of measurement = 0.998203 g/ml

Calculation

by substitution of values in equation (*i*) and (*ii*) and solving it for W_w and P_w

Answer : W_w = 5.0490 g

P_w = 33.77 %

Method 4. Water content of soil by working out soil moisture–tension relationship on installation of tensiometer.

The soil moiture–tension relationship of a sandy loam soil of Allahabad worked out on installation of Tensiometer at a depth of 20 cms are given in Table 1

Table 1. Soil moisture–tension relationship of a sandy loam soil of Allahabad.

	Soil moisture tension in bar			
	0.20	**0.28**	**0.32**	**0.48**
% moisture in soil	21.65	17.40	15.45	8.90

(see section 1.17.1 on soil moisture tension relationship by use of tensiometer).

The three ranges of soil moisture–tension and the corresponding % decrease in moisture per centibar increase in soil moisture–tension as well as increase in soil moisture tension per unit decrease in % moisture in soil are presented in Table 2.

Table 2. Relationship between decrease in soil moisture and increase in soil moisture–tension.

Range of soil moisture–tension bar	% decrease in moisture per centibar increase in soil moisture tension	increase in soil moisture–tension per unit decrease in % moisture
0.20–.28	0.5312	0.0178
0.28–.32	0.4875	0.0205
0.32–.48	0.4094	0.0244

Tensiometer may be used for scheduling irrigation in berseem and potatoes at 0.25 bar, and in wheat & sugarcane at 0.45 to 0.48 bar in sandy loam soil.

It is apparent from Table 2. that percent decrease in moisture per centibar increase in soil moisture–tension decrease progressively at higher soil moisture–tension range, and the increase in soil moisture–tension per unit decrease in % moisture increase at higher soil moisture–tension range.

If some one desires to estimate the moisture content of the soil from the soil moisture tension relationship, it would be desirable to first consider the range in which soil moisture tension falls, and then determine the moisture content within the range by interpolation.

Example

Estimation of moisture content of soil at moisture tension of 0.30 bar. It may be noted that in the range of 0.28 to 0.30 bar soil moisture tension. % decrease in soil moisture per centibar increase in soil moisture tension is 0.4875.

$\because$ 1 centibar increase in soil moisture tension $\cong$ decrease in % moisture by 0.4875

$\because$ 2 centibar increase in soil moisture tension

$\cong$ 0.4875 × 2 % decrease in moisture

$\cong$ 0.975 %

% moisture in soil at 0.30 bar = % moisture in soil at 0.28 bar – 0.975

= 17.40 – 0.975

= 16.42 %

Answer – Moisture content at 0.30 bar will be 16,42 %

Comments

As soil moisture tension relationship vary with the soil texture, mineralogical composition and organic matter content of the soil, the method would be applicable for soils of similar texture, mineralogical composition and organic matter content.

12. Soil moisture–suction relationship and available water.

The soil moisture characteristics of a sandy loam and a silty clay loam soil are presented in the table given below : –

Table Water content of soil on volume basis–cm^3 water/cm^3 oven dry soil.

Suction in bar	Soil Texture	
	Sandy loam	Silty clay loam
0.00	0.35	0.54
0.10	0.27	0.41
0.30	0.17	0.36
0.50	0.12	0.34
1.00	0.10	0.33
5.00	0.09	0.28
15.00	0.08	0.20

(see section 1.17.2 & 1.17.3 for working out soil moisture–suction relationship).

The soil moisture–suction relationship enable estimation of available water in the soil. The water held by the soil in the range of field capacity to permanent wilting point is termed as available water. The soil moisture suction at field capacity range from 0.1 bar to 0.33 bar depending on soil texture. The soil moisture suction at permanent wilting point is taken to 15 bar irrespective of the plant species. In the present example. the soil moisture suction at field capacity may be taken to be 0.2 bar in sandy loam soil and 0.3 bar in silty clay loam soil.

Equations :–

(*i*) % available water on volume basis(P_v) = % water in soil at field capacity on volume basis – % water in soil at permanent wilting point on volume basis.

(*ii*) Available water in soil *i.e.*

cm H_2O/30 cm soil depth $P_d = \frac{P_V}{100} \times$ soil depth i.e. 30 cm

Observations :– Water content of soil on volume basis cm^3 water/cm^3 oven dry soil.

	Soil Texture	
	Sandy loam	Silty clay loam
At field capacity	0.22	0.36
At permanent wilting point	0.08	0.20

Multiply by 100 to obtain % moisture on volume basis.

Calculations

By substituting values in eqnaton (*i*) and (*ii*) for solving the equation for % available water on volume basis, P_v and depth of available water, Pd in cms H_2O/30 cms soil depth.

Answer

Soil Texture	Available Water in Soil	
	% available water on volume basis, P_v	depth of available water Pd–cms H_2O/30 cm soil depth
Sandy Loam	14.00	4.20
Silty clay loam	16.00	4.80

APPENDIX 8

Optimum Nutrient Dose for Crops

1. Introduction

When graded amount of nutrient is applied to crops grown on nutrient deficient soils, the yield of crops tend to increase progressively with the increase in nutrient dose but at diminishing increment in crop yield.

The increase in crop yield by successive increment in nutrient doses may be depicted on a graph paper and the response curve may be quantitatively defined by numerical equation of the form $y = a + b\text{N} - \text{CN}^2$ where y is the crop yield, a is the yield without added nutrient, b and c are the regression coefficients of N and N is the amount of nutrient applied in kg/ha, when a is deleted, ($b\text{N} - \text{CN}^2$) give an estimate of crop yield above control. By multiplying crop yield by the price per kg of produce, the gross return could be estimated. The nutrient dose multiplied by the price per kg of nutrient plus the cost of fertilizer application (a_1) would provide an estimate of expenditure or cost. The benefit/cost ratio or the Return/cost ratio may be estimated by dividing return by cost. The expenditure may be substracted from the gross return to obtain Net Return (Table 1).

The net return increases progressively with the increase in nutrient dose, but the Gross Return/Cost ratio after initial increase at lower level of nutrient tend to decrease progressively at higher doses of nutrient. The economist advise that one should continue to increase the nutrient dose until the marginal return is equal to the marginal cost, *i.e.* when one rupee increase in return is obtained by increase in nutrient dose worth one rupee.

When yield equation is available, the return and cost function may be worked out. The profit function equal return function – cost function. The nutrient dose for maximizing profit can be estimated by setting the derivative of profit function with respect to N the nutrient dose equal to zero and solving the equation for N as shown in the illustration.

Algebraic Illustration

If P is a function of N, then for P to be maximum, the derivatives of P with respect to N *i.e.* $\frac{d\text{P}}{d\text{N}} = 0$.

Thus, when $\underset{\text{profit}}{\text{P}} = \underset{\text{Return function}}{\text{F}(a + b\text{N} - \text{CN}^2)} - \underset{\text{cost function}}{[a_1 + b_1\text{N}]}$

Where,

F = is price per unit produce

a = is the intercept, *i.e.* yield without added nutrients

b & c = are regression coefficients

N = is the amount of nutrient dose in kg/ha

a_1 = is the fixed cost *i.e.* cost of fertilizer application.

b_1 = is the price per unit of nutrient dose.

For P to be maximum

$$\frac{dP}{dN} = [F(a + bN - CN^2)] - [a_1 + b_1N] = 0$$

$$= F[b - 2CN) - [b_1] = 0$$

Again

If

$$P = [(0.7)(1929 + 47.1\,N - 0.145\,N^2) - [(20.0 + 2.7\,N)]$$

$$\frac{dP}{dN} = [(0.7)(47.1 - 2 \times 0.145\,N)] - [2.7] = 0$$

$$= 32.97 - 0.203\,N - 2.7 = 0$$

$$\therefore \quad 0.203\,N = 30.27$$

$$N = 149.11$$

Example

Given : F = Rs. 12.00 per kg wheat grain

Return function = $F(1929 + 47.1\,N - 0.145\,N^2)$

Cost function = $(a_1 + b_1N) = (600 + 12.17\,N)$

Where, Rs. 600 is the cost of fertilizer application ha.

Rs. 12.17 is the price in Rs. per kg of nitrogen as urea.

Table 3. Gross Return from wheat crop, cost of fertilizer N, Return/Cost Ratio, and Net Return (Rs/ha) at different levels of Nitrogen.

Nitrogen levels kg/ha	Increase in grain yield kg/ha ($47.1N - 145\,N^2$)	Value of grain yield (kg/ha) (grain yield × 12.00)	Cost of fertilizer Nitrogen (Rs/ha) (600 + 12.17N)	Return Cost	Net Return Rs/ha
25.0	1086.9	13042.8	904.25	14.42	12138.6
50.0	1992.5	23910.0	1208.50	19.78	22701.5
100.0	3260.0	39120.0	1817.00	21.53	37303.0
125.0	3621.0	43452.0	2121.25	20.48	41330.8
150.0	3802.5	45630.0	2425.50	18.81	43204.5
158.92	3823.0	45876.0	2534.06	18.10	43341.9
160.0	3824.0	45888.0	2547.20	18.02	43340.8

Where profit $P = [(12.00)(1929 + 47.1\,N - 0.145\,N^2)] - [(600 + 12.17\,N)]$

Profit function = Return function − Cost function

$$\frac{dP}{dN} = [(12.00)(47.1 - 2.x.145\,N)] - [12.17] = 0$$

$$= 565.2 - 3.48\ N - 12.17 = 0$$

$$\therefore \quad 3.48\ N = 553.03$$

$$N = \frac{553.03}{3.48} = 158.92$$

The monetary implications involved are presented in Table 1. Although Return/Cost ratio at nitrogen level of 100 kg/ha is found to be maximum *i.e.* 21.53 but the net return in comparison with nitrogen level of 158.92 kg/ha is found to be Rs. 6038.9/ha lower. One would not restrict to nitrogen level of 100 kg/ha as by investing Rs. 717.06/ha more on N fertilizer, one can obtain an additional retuin of Rs. 6038.9/ha which would maximize the profit/ha.

2. Exercise

In the above example, if price of grain/kg is raised to Rs. 16.00, and the price per kg nitrogen remains at Rs. 12.17, estimate.

(*a*) the nitrogen dose for maximizing profit/ha

(*b*) Work out the increase in net return over net return from 100 kg N/ha

Answer : (*a*) = 158.9 kg N/ha (*b*) Rs. 8293.00

3. References

Heady, E.C. and Ray, H.E., (1971). Intern. Symp. Soil Fert. Evalu. Proc. India, pp. 1073-1082.

Raychauduri, S.P., 1971. Intern. Symp. Soil Fert. Evalu. Proc. India. pp. 1093-1102.

APPENDIX 9

Activity of Ions in the Solution

1. Introduction

The activity of an ion is a product of its concentration in moles/litre and its activity coefficient. The activity is estimated by the Debye-Huckel equation given below :

$$-\log f_i = \frac{AZ_i^2 \sqrt{\mu}}{1 + B\, a_i \sqrt{\mu}}$$

Where,

f = is the activity coefficient of ion

Z = is the valence of the ion i

A = is the Debye-Huckel constant 0.509

B = is the Debye-Huckel constant 0.329

a_i = is an ion size parameter for each ion

Table 1. Individual ion-size parameters, a_i, in Ångstrom units for the Debye-Huckel equation for single ion activity coefficient.

Ion	a_i value
H^+, Al^{3+}, Fe^{3+}	5
Mg^{2+}	8
Li^{2+}, Ca^{2+}, Cu^{2+}, Zn^{2+}, Mn^{2+}, Fe^{2+}	6
Sr^{2+}, Ba^{2+}	4.5
SO_4^{2-}, HPO_4^{2-}, PO_4^{3-}	4
OH^-, F^-	3.5
K^+, Cl^-, NO_3^-	3
Rb^{2+}, Cr^{3+}, NH_4^+	2.5

other constants for the Debye-Huckel equation on the molar scale at 25°C are :

A = 0.509, B = 0.329 with a_i in Ångstrom unit.

μ is the ionic strength of the equilibrium solution containing the ion species as estimated by the equation.

$$\mu = \frac{1}{2} \Sigma C_i Z_i^2$$

Where,

C_i is the actual molar concentration of each ion in the solution and Z_i is its valence. It is necessary to know the concentration of all ions that will affect μ of the solution.

The ionic strength μ may be estimated from the EC expressed in millimho/cm as given by Graffin and Zarink (1973).

$$\mu = \text{EC in millimho/cm} \times .013$$

Where

Ionic strength μ is in moles/litre and EC is the specific conductance expressed in millimho/cm at 25°C. Utilizing soil extracts of flooded soils and electrolyte solution of ionic strength less than .06 moles/litre. Ponnamperuma *et. al.* (1966) derived the relationship μ = 16 E.C., where E.C. is the specific conductance in mho/cm at 25°C and μ is the ionic strength. The ionic strength is a measure of the intensity of the electrical field in an electrolyte solution.

The activity of ions may be calculated from measured concentrations on the basis of the second approximation to the Debye-Huckel equation in the form.

$$-\log f_i = \frac{0.5\sqrt{\mu} \times Z_i^2}{1 + 1.5\sqrt{\mu}}$$

Where, f_i is the activity coefficient of the ions whose valency is Z and μ is the ionic strength of the solution.

2. Illustration

Example 1. 10 *m*M $CaCl_2$ solution containing 0.01 moles Ca^{++}/litre and .02 moles Cl^{+}/litre.

2.1.1 $$\mu = \frac{1}{2}\Sigma C_i Z_i^2$$

(Ionic strength)

$$= \frac{1}{2}\Sigma(.01 \times 4) + (.02 \times 1)$$

$$= \frac{1}{2}\Sigma 0.06 = 0.03$$

2.1.2 Activity coefficient of Ca^{2+}

$$-\log f_{Ca^{2+}} = \frac{AZ_i^2\sqrt{\mu}}{1 + B\,a_i\sqrt{\mu}}$$

By substituting the value of μ, Z_i^2 and Debye-Huckel constants in the above equation.

$$-\log Ca^{2+} = \frac{(0.509)(4)(0.03)^{1/2}}{1 + (0.329)(6)(0.03)^{1/2}}$$

$$= \frac{0.35264538}{1 + 0.34190667}$$

$$= \frac{0.35264538}{1.34190667}$$

$$= 0.2628$$

Since $\quad -\log f_{Ca^{2+}} = 0.2628$

$$\log f_{Ca^{2+}} = -0.2628$$

$$f_{Ca^{2+}} = \text{antilog}\,(-0.2628)$$

$$= \text{antilog}\,(\bar{1}.7372)$$

$$= 0.5460$$

2.1.3 Activity of Ca^{2+}

= Concentrates of Ca^{2+} in moles/litre × Activity Coefficient of Ca^{2+}

= 0.01 × 0.5460

= 0.00546 moles/litre

2.1.3. Activity coefficient of Cl^-

$$\log f_{Cl^-} = \frac{(0.509)\,(1)\,(0.03)^{1/2}}{1 + (0.329)\,(3)\,(0.03)^{1/2}}$$

$$= \frac{0.088161385}{1 + 0.170953413}$$

$$= \frac{0.088161385}{1.170953413}$$

$$= 0.0753$$

Since $\quad -\log f_{Cl^-} = 0.0753$

$$\log f_{Cl^-} = 0.0753$$

$$f_{Cl^-} = \text{antilog}\,(-0.0753)$$

$$= \text{antilog}\,(\bar{1}.9247)$$

$$= 0.8408$$

2.1.3 Activity of Cl^-

= Concentrates of Cl^- in moles/litre × Activity Coefficient of Cl^-

= 0.02 × 0.8408

= 0.016816 moles/litre

Example 2. Concentration of K and Ca + Mg in the equilibrium soil solution.

K = 0.00022 moles/litre

Ca + Mg = 0.00310 moles/litre

2.2.1 Ionic strength of the equilibrium soil solution

$$\mu = \frac{1}{2}\,\Sigma C_i Z_i^2$$

$$= \frac{1}{2}\,\Sigma\,(0.00022 \times 1^2) + (0.00310 \times 2^2)$$

$$= \frac{1}{2} \Sigma (0.00022 + 0.01240)$$

$$= \frac{1}{2} \times 0.01262 = 0.00631$$

2.2.2. Activity coefficient of K^+ ions as per Debye-Huckel formula

$$- \log f_{K^+} = \frac{AZ_i^2 \sqrt{\mu}}{1 + B\, a_i \sqrt{\mu}}$$

By substituting the value of μ, Z_i^2 and Debye-Huckel constants in the above equation.

$$- \log f_{K^+} = \frac{0.509 \times 1^2 . \sqrt{\mu}}{1 + 0.329 \times 3 . \sqrt{\mu}}$$

$$= \frac{0.509 \times \sqrt{0.00631}}{1 + 0.987 \times \sqrt{0.00631}}$$

$$= \frac{0.509 \times 0.079436}{1 + 0.987 \times 0.079436}$$

$$= \frac{0.04043292}{1.07840332}$$

$$- \log f_{K^+} = 0.037493319$$

$$\log f_{K^+} = - 0.037493319$$

$$f_{K^+} = \text{antilog } (- .0374993319)$$

$$= \text{antilog } (\bar{1}0.962506681)$$

$$= 0.917290 \text{ moles/litre}$$

2.2.3 Activity of K^+

= Concentration of K in moles/litre × Activity coefficient of K^+

= 0.00022 × 0.917290

= 0.0002018 moles/litre

2.2.4 Activity coefficient of Ca + Mg ions as per Debye-Huckel formula

$$- \log f_{Ca + Mg} = \frac{0.509 \times 2^2 \sqrt{\mu}}{1 + 0.329 \times 7 \sqrt{\mu}}$$

$$= \frac{2.036 \times \sqrt{0.00631}}{1 + 2.303 \times \sqrt{0.00631}}$$

$$= \frac{0.16173696}{1 + 0.182939975}$$

$$\log f_{Ca + Mg} = 0.13672457 = .136719274$$

$$\log f_{Ca + Mg} = - 0.1367$$

$$f_{Ca+Mg} = \text{antilog}\ (-0.1367)$$

$$= \text{antilog}\ (\bar{1}.8633)$$

$$= 0.7299$$

2.2.5 Activity of Ca + Mg

= Concentration of Ca + Mg moles/litre × Activity coefficient

= 0.00310 × 0.7299

= 0.002262 moles/litre

2.2.6 Activity coefficient of K and Ca + Mg in the equilibrium soil solution on the basis of the second approximation to the Debye-Huckel equation.

2.2.6.1 Ionic strength.

$$\mu = \frac{1}{2} \Sigma C_i Z_i^2$$

= 0.00631 as estimated in 2.2.1.

2.2.6.2 Activity coefficient of K in the equilibrium soil solution,

$$-\log f_K = \frac{0.5 \times \sqrt{\mu} \times Z_i^2}{1 + 1.5\sqrt{\mu}}$$

$$= \frac{0.5 \times \sqrt{0.00631} \times 1^2}{1 + 1.5 \times \sqrt{0.00631}}$$

$$= \frac{0.039718}{1 + 0.119154}$$

$$-\log f_K = 0.035489307$$

$$\log f_K = -0.035489307$$

$$f_K = \text{antilog}\ (-0.035489307)$$

$$= \text{antilog}\ (\bar{1}.964510693)$$

$$= 0.921532578$$

2.2.6.3 Activity coefficient of Ca + Mg in the equilibrium soil solution.

$$-\log f_{Ca+Mg} = \frac{0.5 \times \sqrt{\mu} \times Z_i^2}{1 + 1.5\sqrt{\mu}}$$

$$= \frac{0.5 \times \sqrt{0.00631} \times 2^2}{1 + 1.5 \times \sqrt{0.00631}}$$

$$= \frac{0.158871}{1 + 0.119153}$$

$$-\log f_{Ca+Mg} = 0.141956461$$

$$\log f_{Ca+Mg} = -0.141956461$$

$$f_{Ca+Mg} = \text{antilog} (-0.141956461)$$
$$= \text{antilog} (\bar{1}.8580435)$$
$$= 0.7211797755$$
$$= 0.7212$$

3. Exercise

Estimate activity coefficient of Zn^{++} and activity of Zn^{++} in the equilibrium soil solution from the following observations.

Concentration of Zn in equilibrium soil solution = 3.8245×10^{-6} moles/litre

EC of the equilibrium soil solution at 25°C = 0.420×10^{-4} mho/cm

Answer : Activity coefficient of Zn = 0.900582 moles/litre

Activity of Zn = 3.4438×10^{-6} or 0.000003443 moles/litre

4. Comments

The activity of ions in the equilibrium soil solution is of significance when it is desired to estimate potash potential (pK – 1/2 *p* (Ca + Mg) or Ratio of activity of K to activity of (Ca + Mg) in working out PBCK (Potential Buffering Capacity of Potassium) in soils, phosphate potential (pH_2PO_4) + 1/2 *p* (Ca + Mg) and Lime Potential (pH - 1/2 *p* (Ca + Mg)).

5. References

Adams, Fred (1971). Ionic concentration and activities in soil solution, *Soil Sci. Soc. Amer. Proc.* Vol. 35, pp. 421-426.

Bear, F.E. (1964). Chemistry of soil, Oxford & IBH Publishing Co. Pvt. Ltd., 56, Janpath, New Delhi-110001, India.

Graffin, R.A. and Jurinak, J.J. (1973). Estimation of Activity Coefficient from the Electrical conductivity of Natural Aquatic Systems and Soil Extracts, *Soil Science*, Volume 116(1), pp. 26-30.

Ponamperuma, R.N. Tianco, E.M., and Loy, T.A. (1966). Ionic strengths of the solutions of flooded soils and other natural aqueous solutions from specific conductance, *Soil Science*, 102 : 408-413.

Taylor, A.W. (1958). Some Equilibrium Solution Studies on Rothamsted Soils. Soil *Science Society Proceedings,* pp. 511-513.

INDEX

C

D

N

O

P

Q

R